ENERGY SCIENCE, ENGINEERING AND

NUCLEAR POWER

CURRENT AND FUTURE ROLE IN THE WORLD ELECTRICITY GENERATION

ENERGY SCIENCE, ENGINEERING AND TECHNOLOGY

Additional books in this series can be found on Nova's website under the Series tab.

Additional E-books in this series can be found on Nova's website under the E-book tab.

ENERGY SCIENCE, ENGINEERING AND TECHNOLOGY

NUCLEAR POWER

CURRENT AND FUTURE ROLE IN THE WORLD ELECTRICITY GENERATION

JORGE MORALES PEDRAZA

Nova Science Publishers, Inc.

New York

NOTICE TO THE READER

The Publisher has taken reasonable care in the preparation of this book, but makes no expressed or implied warranty of any kind and assumes no responsibility for any errors or omissions. No liability is assumed for incidental or consequential damages in connection with or arising out of information contained in this book. The Publisher shall not be liable for any special, consequential, or exemplary damages resulting, in whole or in part, from the readers' use of, or reliance upon, this material. Any parts of this book based on government reports are so indicated and copyright is claimed for those parts to the extent applicable to compilations of such works.

Independent verification should be sought for any data, advice or recommendations contained in this book. In addition, no responsibility is assumed by the publisher for any injury and/or damage to persons or property arising from any methods, products, instructions, ideas or otherwise contained in this publication.

This publication is designed to provide accurate and authoritative information with regard to the subject matter covered herein. It is sold with the clear understanding that the Publisher is not engaged in rendering legal or any other professional services. If legal or any other expert assistance is required, the services of a competent person should be sought. FROM A DECLARATION OF PARTICIPANTS JOINTLY ADOPTED BY A COMMITTEE OF THE AMERICAN BAR ASSOCIATION AND A COMMITTEE OF PUBLISHERS.

LIBRARY OF CONGRESS CATALOGING-IN-PUBLICATION DATA

Pedraza, Jorge Morales.
  Nuclear power current and future role in the world electricity generation
/ Jorge Morales Pedraza.
     p. cm.
  Includes bibliographical references and index.
  ISBN 978-1-61728-504-2 (hardcover)
  1. Nuclear power plants. 2. Electric power production. I. Title.
  TK1078.P3872 2010
  333.792'4--dc22
                          2010025927

Published by Nova Science Publishers, Inc. ✝ *New York*

CONTENTS

PREFACE

It is an undisputed reality that the energy production and, particularly, the production of electricity and their sustained growth, constitute indispensable elements for the economic and social progress of any country. Without any doubt energy constitute the motive force of the civilization and it determines, in a high degree, the level of economic and social development of a country. To ensure an adequate economic and social growth of a country it is indispensable the use of all available types of energy sources for electricity production, including nuclear energy.

The generation of electricity using fossil fuels is a major and growing contributor to the emission of carbon dioxide – a greenhouse gas that contributes significantly to global warming and produces a significant change in the climate all over the world. These changes are affecting, in one form or another, almost all countries in all regions. By the contrary, nuclear energy is a clean technology that does not produce carbon dioxide and, for this reason, does not affect the climate.

One of the main problems that the world is now facing is how to satisfy the increase in electricity demand using all available energy sources in the most efficient manner and without increasing the emission of CO_2 to the atmosphere.

The preparation of a national energy policy in which priorities and preferences are identified should be one of the main responsibilities of national authorities in charge of the energy sector. It is important that every country's energy mix involves a range of national preferences and priorities reflected in national policies and strategies to be applied in order to satisfy the foresee increase in the electricity demand. These policies and strategies should represent a compromise between expected energy shortages, environmental quality, energy security, energy cost, public attitudes, safety and security, available skills, as well as production and service capabilities. Relevant national authorities must take all of these elements into account when formulating an energy policy and strategy.

One of the available energy sources that have probed in the past that can be effectively used for the generation of electricity is nuclear energy. The nuclear industry has proclaimed that nuclear energy for electricity generation played an important role in the 1970s and the 1980s and is ready to play again in the future a relevant role in the energy balance in several countries in North America, Latin America, Europe, Asia and the Pacific and Africa.

Keywords: Nuclear energy; nuclear power; nuclear power reactors; nuclear power plants; nuclear power sector; nuclear installations; nuclear facility; nuclear safety; nuclear waste; energy policy; safety authority; waste management

ACKNOWLEDGMENT

During the preparation of the present book different nuclear experts and other professionals in different fields reviewed some of its chapters. Eng. Hector Espejo (Argentina) and Dr. Alejandro Bilbao (Cuba) submitted their opinions, suggestions, comments and recommendations on different sections of the book. Their suggestions and comments allowed me to improve considerably the first draft prepared. Eng Hector Espejo reviewed the section of Chapter VI dedicated to the nuclear power programme in Argentina, as well as the section of Chapter I dedicated to the types of fission nuclear power reactors currently in operation in several countries. Dr. Alejandro Bilbao reviewed the section of Chapter IV dedicated to explain the safety of the nuclear power reactors currently in operation in several countries as well as the section of Chapter VI dedicated to the nuclear power programme in Cuba. Master Juan Carlos Marcos Portal (Cuba), from Ceiba Investment Ltd, UK, was very helpful in the selection of the tables in Chapter VI, specially the one related with the generation capacity in Cuba.

Lastly, I would like to say some words about the role of my family in the preparation of this book. Without any doubt the present book is a reality thanks to the valuable support of my lovely wife, Aurora Tamara Meoqui Puig, which had assumed other family responsibilities in order to give me the indispensable time and the necessary environment to write the book. My lovely daughter, MSc. Lisette Morales Meoqui, has been an extremely helpful assistant in collecting the necessary information and reference documents used by me in the preparation of the book, despite of her own Doctorate studies on Economics and her current job in a competent firm. My dear son, MSc. Jorge Morales Meoqui, has been also an extremely helpful assistant in the revision of the initial materials used in the preparation of the book.

ABOUT THE AUTHOR[*]

Jorge Morales Pedraza is now a Consultant on International Affairs and has a University Diploma in Mathematic and another in Economy Sciences. He was diplomatic of his country with the rank of Ambassador for more than 26 years and he was appointed as Ambassador and Permanent Representative to the IAEA in the 1980s and to the OPCW in the 1990s. He was also University professor in Mathematic Science and Invited Professor for International Relations in the Diplomatic Academy of his country. He worked as Senior Manager in the International Atomic Energy Agency (IAEA) in the Director's Office in the 1990s and 2000s. In the last three years he was involved in the preparation, as author and co-author, of more than 30 articles already published by international publisher houses as well as in the preparation of several chapters of different books related with the peaceful uses of nuclear energy, crisis management, biosaline agriculture, the use of radiation for the sterilization of tissues, tissue banking, among others. During this period he wrote a book on non-proliferation and disarmament already published by an international publisher house. He was also invited editor of international journals.

---

[*] E-mail: jmorales47@hotmail.com or jmorales547@yahoo.com

INTRODUCTION

The present book is the result of more than 26 years of work in different sectors associated with the peaceful uses of nuclear energy, including the nuclear power sector, first as Ambassador and Permanent Representative of my country to the IAEA during the 1980s and, later on, in the 1990s and 2000s, as IAEA Senior Manager in charge of promoting nuclear cooperation, first in the Latin American region and then in all regions of the world.

The book was prepared with the intention to participate actively in the current international debate on the future role that nuclear energy should play in the production of electricity. The book provides the latest available information regarding the current situation with respect to the use of nuclear energy for electricity generation in 63 countries in all regions of the world and singles out which are the perspectives of the use of this type of energy for the production of electricity in these countries in the future.

The book has nine Chapters, including the conclusions in Chapter IX. Chapter I provide a general overview on the current situation of the nuclear energy sector in the world. It stress the undisputed reality that the production of energy, particularly the generation of electricity and their sustained growth, constitutes an indispensable element for the economic and social progress of any country. One of the conclusions that could be reached after reading this chapter is the following: all available energy sources should be carefully study with the purpose to identify which is the best energy mix option for a given country. No energy sources available in the country should be excluded from these studies. Without any doubt the use of nuclear energy for the production of electricity is a realistic option for several countries all over the world with the aim to satisfy the foresee increase in the demand of electricity. For this reason, the use of nuclear energy for the generation of electricity is again under seriously consideration in a group of countries in all regions. However, it is important to understand the following: Many countries are not in a position to use nuclear energy for electricity generation due to its technological complexity and because it is a very expense technology that requires a certain basic infrastructure to be in place before this type of energy coudld be use for the generation of electricity. The country also need to have certain technological development, financial resources and an important number of professionals well prepared and trained that most of the countries do not have. In other countries this type of energy can be used only if there is an active international cooperation and assistance provided by some advanced countries as well as by some international organizations such as the International Atomic Energy Agency (IAEA).

From the information contained in Chapter I the following conclusions could be reached: the current contribution of nuclear energy to the total electricity generation varies according with the region. In 2006, the European region nuclear generated electricity accounts for 26.03% of the total electricity produced in that year; in North America it was approximately 19%, whereas in Africa and Latin America it was 1.84% and 2.61% respectively. In Asia and the Pacific nuclear energy accounts for 9.46% of total electricity generation but the use of this type of energy for the production of electricity is expanding rapidity in several countries of the region.

Chapter II provides information on the main elements associated with the introduction of a nuclear power programme in a country, bearing in mind that the use of nuclear energy for electricity production should be viewed as a medium to long-term electricity supply strategy within the develop plan of the country's energy sector. When deciding on how to expand the electricity generating system in a given country, government and energy industry would have to carry out comprehensive assessments of all energy options available in the country, in order to use them in the best possible way during the preparation of its energy balance. However, it is important to note that choosing a specific energy option will differ from country to country even in a specific region, because this will depend on local and regional energy resources available, technological capabilities, availability of appropriate financial resources and sufficient qualified personnel, environmental considerations and the country's overall energy policy.

Chapter II single out the following: the introduction of a nuclear power programme in a specific country requires the establishment of a basic infrastructure to deal with all aspects of this programme, particularly in the case of the construction for the first time of a nuclear power plant. The adoption of a nuclear law and others specific regulations and the creation of an appropriate infrastructure, particularly the establishment of the nuclear regulatory authority, are actions that need to be taken by the national competent authorities of a given country after the adoption, by the government, of the political decision to introduce a nuclear power programme. The nuclear law and other specific regulations, as well as the infrastructure that is needed to support the introduction of a nuclear power programme in a given country, should be in place well in advance of the initiation of the construction of the first nuclear power reactor. The selection of the type of reactor to be built and the site in which the nuclear power plant will be located are also important elements to be considered during the preparation of the country for the successful introduction of a nuclear power programme. Another relevant element included in this chapter associated with the construction of a nuclear power plant is the licensing procedures to be in place to ensure the operational safety of the plant, with the purpose to keep the construction cost within the budget approved, to reduce to the minimum possible the possibility of a nuclear accident and the correct management of nuclear waste.

Finally, Chapter II stresses the following: for the successful introduction of a nuclear power programme in a country, it is important to have a well-prepared and trained force for the production of electricity using nuclear energy and in all aspects related with the safety operation of the nuclear power plant and all pro and cons of the use of nuclear energy for the production of electricity should be evaluated with all available information

Chapter III provides detail information on the current situation and the perspectives of the use of nuclear energy for the production of electricity in the North American region (Canada and the USA). All countries of this specific region are adopting concrete measures to expand their nuclear power programmes in the near future, with the purpose to satisfy the foresee increase in the electricity demand and to reduce, as much as possible, the CO_2 emission to the atmosphere. In 2009, the number of nuclear power reactors in operation in Canada and the USA is 122 with a net capacity of 113 771 MWe. One nuclear power reactor is under construction in the USA with a net capacity of 1 180 MWe, and 40 planned or propose to be built in this region with a net capacity of 49 400 MWe (Canada (9) and the USA.

The construction of new nuclear power reactors in North America, particularly in the USA, was stop after the Three Mile Island and Chernobyl nuclear accidents occurred in Pennsylvania, USA in 1979 and in Ukraine in 1986 respectively. However, it is important to single out that the energy crisis the world in now facing, the real possibility of the extinction of the world oil reserves in the coming decades, the significantly increase in oil price in 2007 and 2008 and the negative impact in the environment due to the release to the atmosphere of a significant amount of CO2 as a result of an increase in the consumption of fossil fuel in the USA and Canada, among other relevant factors, are changing the negative perception of the public opinion in these two countries regarding the future use of nuclear energy for electricity generation.

Chapter IV provides detail information regarding the current situation and the perspectives of the use of nuclear energy for the generation of electricity in the European region (including Ukraine, Belarus and Russia). Due to different reasons, today's the debate about the use of nuclear energy for electricity generation started again in the European region. The first of these reasons is the high price of oil reached during 2007 and 2008 and the tendency to an increment of the gas price. The second reason is the need to reduce the CO2 emissions to the atmosphere due to the commitments of the European states to the Kyoto Protocol. The third reason is the dependency of the majority of the European countries to the import of fossil fuel from politically instable regions such as the Middle East.

During the consideration of the role that nuclear energy should have in the energy balance in the European region in the coming years, three main realities should be taken into account. These realities are the following: a) expectations of the use of nuclear energy for electricity generation are rising again in several European states; b) the answer to questions like 'is nuclear power economic" cannot be made using a single universal answer. The real answer is 'it depends' — sometimes yes, sometimes no; and c) economics comparison. Whether the use of nuclear energy for electricity generation is more economic or not than the use of other energy sources used for the same purpose, will depend on how cheap it is compared to other energy sources available in the country.

The European region generates around 31% of its electricity from 196 nuclear power reactors in operation in 18 countries. Until December 2009, there were 17 nuclear power reactors under construction, the majority of them in Russia [9]. A total of 102 units are planned or proposed in the region.

Based on the information included in Chapter IV the following conclusion can be reached: To satisfy the foresee growth in the demand of electricity in the European region in the coming years, an urgent investment should be made in the energy sector. In Europe alone, to meet expected energy demand and to replace ageing infrastructure, an investment of around one trillion Euros will be necessary over the next 20 years. At the same time, there should be no doubt that the European energy balance should included all available energy technology, including nuclear energy, if the foresee energy demand, particularly the foresee increase in the demand of electricity, is going to be fully satisfied in the future. For this reason, in many countries nuclear energy will be one of the energy alternatives to be used with the purpose to contribute to alleviate energy supply dependency, local air pollution and global climate change in the coming years. In 2004, the nuclear share of electricity production in the European region represented a 'non-emission' to the atmosphere of nearly 900 million tons of CO_2 per year. This is almost the same quantity of CO_2 produced annually in the European region by the transport sector.

There are other reasons why nuclear energy should be included in the European energy balance in the future. One of the reasons is the following: Unless the European States make domestic energy more competitive in the next 20 to 30 years, around 70% of the European energy requirements, compared with 50% today, will be met by imports products some from regions threatened by insecurity. If the European States does not increase the nuclear share in the coming years, then it is expected that the European energy dependence, particularly the EU dependency, would increase up to 60%-70% or even more by 2030; the reliance on imports of gas is expected to increase from 57% to 84% and reliance on imports of oil from 82% to 93%. The EU is consuming more and more energy and importing more and more energy products in order to satisfy it increase energy needs. External dependence for energy is constantly increasing and this situation is considered, by many politicians and experts, very danger from the political and economical point of view. If no measures are taken in the next 20 to 30 years, 70% or more of the European energy requirements will be covered by energy imported from outside the European region.

Chapter V provides detail information regarding the current situation and the perspectives of the use of nuclear energy in the Asian and the Pacific region. In this region, the situation regarding the use of nuclear energy for electricity generation is completely different from the situation in the European and the North American regions. Nuclear was, is and will be an important component of the energy balance of several countries in the region. In the near future, the major increase in nuclear energy capacities for electricity generation is foresee in China, Japan, India and the Republic of Korea. In 2030, it is expected that these four countries, plus the USA and Russia, will have around two-third of the world nuclear energy capacity, an increase of 16%-17% from the current level.

In Asia and the Pacific are now 103 nuclear power reactors in operation in six countries, including Taiwan, with a total net capacity of 170 432 MWe and there are 34 nuclear power reactors under construction in five countries, with a total net capacity of 21 921 MWe. The nuclear power reactors planned or proposed is 174. China and India alone have 25 nuclear power reactors under construction and have plans to construct 128 new nuclear power reactors in the coming decades with the purpose to elevate the proportion of nuclear energy in the generation of electricity in both countries. In the case of China, the new nuclear power reactors planned and proposed is 103, for the time being the biggest nuclear power programme in the world.

Without any doubt Asia and the Pacific is the only region in the world where electricity generating capacity, specifically nuclear power capacity, is growing significantly. The countries that have announced plans for significant expansion in the use of nuclear energy for electricity generation are China, India, Japan, Pakistan and the Republic of Korea. Other countries such as Indonesia, Vietnam, Bangladesh and the Democratic People Republic of Korea have plans for the introduction of nuclear energy for electricity generation in the coming years.

In 2007, Asia and the Pacific produced a total of 523 GWh from nuclear energy representing 9.4% of the total electricity produced in the region by all energy sources. The Southeast Asian economies, themselves beneficiaries of an oil and gas export bonanza through the 1970s-1990s, also find themselves in an energy crunch as once ample reserves run down and the search is on for new and cleaner energy supplies.

Until 2010, the projected new generating capacity in Asia and the Pacific is 38 GWe per year. For the period 2010 to 2020, the projected new generating capacity is 56 GWe. Much of the growth foresees will be in China, Japan, India and the Republic of Korea. In 2020, the nuclear share in the region is expected to be at least 39 GWe and maybe more, if environmental constraints limit fossil fuel expansion. The largest increase in installed nuclear generating capacity is expected in non-OECD Asia, where annual increases in nuclear capacity average 6.3% and account for 68% of the total projected increase in nuclear power capacity for the non-OECD region as a whole. Of the 58 GWe of additional installed nuclear generating capacity projected for non-OECD Asia between 2004 and 2030, 36 GWe is projected for China and 17 GWe for India.

Chapter VI provides detail information regarding the current situation and the perspectives in the use of nuclear energy for the generation of electricity in Latin America. In the case of this region, higher world market prices for fossil fuels, the reduction of the current world fossil fuels reserves and the climate changes affecting all Latin American countries, have put the use of nuclear energy for electricity generation again on the energy agenda of Chile, Uruguay and Venezuela countries without nuclear power programmes, and had revived interest in Argentina, Brazil and Mexico in the expansion of their nuclear power programmes.

It is important to note that, in general, the use of nuclear energy for electricity generation in Latin American region is very low. In 2009, only 2.3% of the electricity generated in the region comes from nuclear energy sources. However, if expansion plans approved by Argentina, Brazil and Mexico are implemented, and the intention of Chile, Venezuela and Uruguay of constructing for the first time nuclear power reactors are materialized in the near future, then that proportion could be more than double in a decade.

Until December 2009, there were six nuclear power reactors operating in three countries of the region, two in Argentina, two in Brazil and two in Mexico. However, the conclusion of the construction of one nuclear power reactor in Argentina (Atucha 2) is now underway and there are plans for the construction of two new units in the future. In the case of Brazil, there are plans for the conclusion of the construction of one nuclear power reactor (Angra 2) and for the construction of five new units in the coming years. Mexico has plans for the construction of two nuclear power reactors in the future, but there are no specific dates for the beginning of the construction works.

Chapter VII provides detailed information on the current situation and the perspectives in the use of nuclear energy for electricity generation in the Middle East and the North of Africa, where the use of nuclear energy for the production of electricity is widely use as it is in other regions. Until April 2009, there were no nuclear power reactors operating in the Middle East and natural gas and oil are the dominant sources of energy for the production of electricity in the region. According with the information included in this chapter, the projection of the participation of nuclear energy in the generation of electricity in the Middle East and in the North of Africa will continue to be very marginal in 2030, despite of the intention of some countries to use nuclear energy for the production of electricity. However, it is important to stress that the great instability that exist in the Middle East is an important obstacles for the introduction of a nuclear power programme in any country in the region in the near future. The world energy projection for the Middle East place natural gas as the dominant source of energy in the energy balance in the region for 2010, 2020 and 2030.

Chapter VIII provides detailed information on the current situation and the perspectives in the use of nuclear energy for electricity generation in Africa (Sub-Saharan region), where the electricity production is unacceptably low. Only a limited number of African countries located below the Sahara desert have large energy potential. Hydropower potential is the most evenly spread, but the highest concentration is on the Congo River. Oil and gas are mostly concentrated in Angola and Nigeria; coal is mostly found in Southern Africa and geothermal potential exists in Eastern Africa. Together with South Africa, the Maghreb countries (Morocco, Algeria, Tunisia, Libya and Mauritania) account for more than 80% of Africa's electricity generating capacity.

Until December 2009, there are only two nuclear power reactors in operation in South Africa. However, nuclear energy contribution could grow in the future, if plans for the use of this type of energy for the generation of electricity are approved and implemented by South Africa, Namibia and Nigeria, among others. In the case of South Africa, the government has plans for the construction of three new nuclear power reactors and is considering the possibility to built four new units in the future. The governments of the other countries are seriously considering the use of nuclear energy for the production of electricity for the first time in the future, but the possibilities to materialize such plans are very limited.

The possibilities to use nuclear energy for the generation of electricity in Africa are associated with the solution, among others, of the following important matters: a) the use of small and medium-size nuclear power reactor designs that should be available in the market in the short term; b) the integration of electric grids among neighboring countries; c) the lack of technological development of many African countries: d) the safety operation of the nuclear power plants; e) the management of the spent nuclear fuel; e) the political instability in some African countries, the interstates conflicts and ethnic strife; f) the lack of resources to finance the construction of a nuclear power plant; and g) the lack of well-trained professionals and technicians in the use of nuclear energy for electricity generation.

The attitudes of African decision makers, experts and public about nuclear power range from negative/cautious to positive/enthusiastic. Supporters perceive nuclear power as an opportunity to demonstrate both technical progress and technological competence and the commitment of Africa to reduce its vulnerability to climate change produced by the use of fossil fuels for the generation of electricity. Some of the serious consequences of climate change in Africa include desertification, food shortages, epidemics, insufficient water supply, coastal erosion and increased refugees.

There should be no doubt that the use of nuclear energy for the production of electricity in Africa will not be a reality, at least in some of the most advanced African countries, in the middle-long term, but this situation could change if small and medium size nuclear power reactors are available in the market in the near future, the regionalization of the electric grid can be materialized and sufficient resources can be allocated by a group of African countries to support a joint nuclear power programme.

Chapter 1

GENERAL OVERVIEW

It is an undisputed reality that the energy production, particularly the generation of electricity and their sustained growth, constitute indispensable elements to ensure the economic and social progress of any country. In other words, energy constitutes the motive force of the civilization and it determines, in a high degree, the level of economic and social development of any country. For this reason, the use of all type of energy sources for electricity production available in a country, including nuclear energy, should be considered during the preparation of the energy mix by any given country in order to ensure its future economic and social development. However, it is important to be aware that the generation of electricity using fossil fuels is a major and growing contributor to the emission of carbon dioxide – a greenhouse gas that contributes significantly to global warming–, produce a significant change in the climate all over the world affecting almost all countries in all regions in one way or another.

Considering the different available energy sources that the world can use now to satisfy the foresee increase in energy demand, particularly the electricity demand in the coming years, there should be no doubt that, at least for the next decades, there are only a few realistic options available to reduce further the CO_2 emissions to the atmosphere as a consequence of the electricity generation. These options are, among others, the following:

a) increase efficiency in electricity generation and use;
b) expansion in the use of all available renewable energy sources such as wind, solar, biomass and geothermal, among others;
c) massive introduction of new advanced technology like the capture carbon dioxide emissions technology at fossil-fueled (especially coal) electric generating plants, with the purpose to permanently sequester the carbon produced by these plants;
d) increase use of new types of nuclear power reactors inherent safe and proliferation risk-free;
e) increase energy saving.

The amount of total electricity produced and used per capita is increasing in several countries, particularly in emerging economies countries such as China, India, Republic of Korea, among others. The total world electricity requirements increased from 6 181 GW per year in 1970 to 15 311 GW per year in 2006, this means by a factor of 2.5. However, according with estimates made by the World Energy Council, the International Institute for Applied Systems, among other international organizations, the demand of electricity probably is triple from now until 2050. Why this significantly increase in the demand of electricity in the next 40 years? The following are, among others, the main reasons for this increase:

a) increase in the world population;
b) increase in the percentage of the world population living in big cities, raising the demand of electricity;
c) improve the quality of life of the world population bringing as consequence an increase in the demand of electricity and other forms of energy;
d) increase in the demand of electricity in the most advanced developing countries such as India, China, Republic of Korea, among others, due to fast economic and social development.

As can be seen from Figure 1, the regions with a relevant use of nuclear energy for the generation of electricity in 2008 were the following: North America, Europe and the Far East. However, in these regions the use of gas and fossil liquids and solids are the dominant type of energy.

The problem that the world is now facing is how to satisfy this increase demand of electricity using the available energy sources in the most efficient manner and without increasing the emission of CO_2 to the atmosphere and changes in the climate. How to achieve this? One of the most effective solutions is the elaboration of a national energy policy in which priorities and preferences are identified as well as the main responsibilities of the national authorities in charge of the energy sector. This energy policy should represent a compromise between expected energy shortages, environmental quality, energy security, cost, public attitudes, energy safety and security, available skills, and production and service capabilities. Relevant national energy authorities and the representatives of the energy industry must take all of these elements into account when formulating an energy policy and strategy for the development of the energy sector.

One of the available energy sources that has probed in the past that can be effectively used for the generation of electricity is nuclear energy. The nuclear industry has proclaimed that nuclear energy for electricity generation played an important role in the development of the energy sector in more than 30 countries in the 1970s and 1980s in North America, Latin America, Europe, Asia and the Pacific, including Africa, and they believe that nuclear energy will continuing to play a relevant role in the production of electricity in the future not only in these countries but in other countries as well.

Over the last three years, several international assessments of the possible future of nuclear power in the world have been adjusted to more optimistic prospects for the horizon of 2030. The World Energy Outlook 2007 presents a reference scenario, an alternative policy scenario and a 450 stabilization case that include respectively 415 GW, 525 GW and 833 GW of nuclear power. Electricity generation from nuclear power plants under the high scenario would more than double from current levels to reach 6 560 TWh in 2030. Under the reference scenario the share of nuclear power in the world commercial primary energy supply would drop from 6% to 5% in 2030. However, it is important to stress that nuclear power will only become more important if the governments of countries where nuclear power is part of the energy mix play a stronger role in facilitating private investment, especially in liberalized markets and if concerns about plant safety, nuclear waste disposal and the risk of proliferation can be solved to the satisfaction of the public. (Schneider, M. and Froggatt, A., 2007)

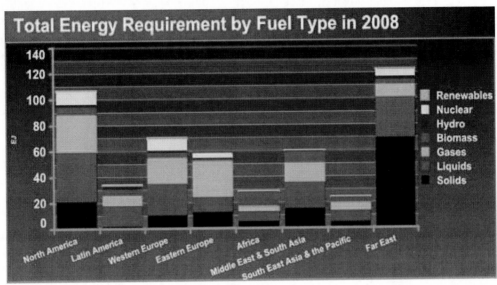

Source: IAEA.

Figure 1. Total energy requirement by fuel type in 2008.

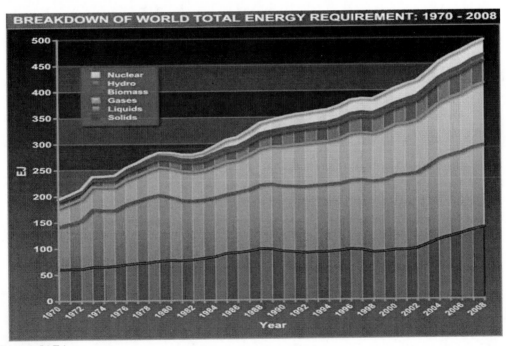

Source: IAEA.

Figure 2. Breakdown of world total energy requirement: 1970-2008.

According with the World Nuclear Association of Operators (WANO) and the IAEA, in 2009 there were 436 nuclear power reactors in operation in 31 countries, including Taiwan, with a net capacity of 366 008 MWe and 54 nuclear power reactors under construction in 14 countries with a net capacity of 46 288 MWe. During the period 1950-2007, 119 nuclear power reactors were shut down in 18 countries totalizing 35 150 MWe (net). The countries with the highest number of nuclear power reactors shut down in this period are USA with 28 units with a net capacity of 9 764 MWe, following by the UK with 26 units with a net capacity of 3 324 MWe, Germany with 19 units with a net capacity of 5 944 MWe and France with 11 units with a net capacity of 3 951 MWe.

From the 31 countries that in 2009 operates nuclear power plants, six of them, USA, France, Japan, Germany, Russia and South Korea, produce three quarters of the world nuclear electricity. However, the role of nuclear power in the overall energy sector remains very limited even in these six countries. In France, for example the country with the highest participation of nuclear energy in its energy balance in the world generated, in 2007, a total of 77% of its electricity with nuclear power plants[1], but the use of this type of energy only provides 17.5% of its final energy demand. Like most of the other countries, France remains highly dependent on fossil fuels reaching over 70% of its final energy consumption. None of the other five countries cover more than 7% of their final energy demand by nuclear power. In the specific case of the USA and Russia, the nuclear energy cover less than 4% of their final energy demand.

During 2007, three nuclear power reactors were connected to the grid in India, China and Romania adding 1 852 MWe to the electric grid in these countries; started the construction of seven nuclear power reactors totalizing 5 195 MWe in China, the Republic of Korea, Russia and France; and nine new orders for the construction of nuclear power reactors were approved in China, the Republic of Korea, Japan and Russia totalizing 11 660 MWe.

During 2008, no new nuclear power reactor was connected to the grid, while three old nuclear power reactors were closed. These reactors were Hamaoka 1 and 2 in Japan and Bohunice 3 in Slovakia. During that year, ten new nuclear power reactors begun to be constructed in China (six units), in the Republic of Korea (two units) and in Russia (two units), increasing the number of nuclear power reactors under construction from 33 in 2007 to 54 until April 2009. In Slovakia the works for the conclusion of Mochovce 2 and 3 began after years of paralization. India two small PHWR[2] units were expected to entry in service in 2008, but they could not make it due to lack of nuclear fuels. In 2009, two nuclear power reactors were connected to the electric grid: a) Tomari 3 (866 MWe, Japan) on 20 March; and b) Rajasthan 5 (202 MWe, India) on 22 December.

In the following tables and figures the most relevant information regarding the number of nuclear power reactors in operation, under construction and planned or proposed are included, as well as the types of nuclear power reactors per country, among others.

---

[1] According with the IAEA, in 2008, the nuclear share was 76.18%.
[2] Pressured Heavy Water Reactor.

Table 1. World nuclear power reactors in 2009

Country (Click name for Country Profile)	Reactors Operable 1 April 2009		Reactors under Construction 1 April 2009		Reactors Planned April 2009		Reactors Proposed April 2009	
	No.	MWe	No.	MWe	No.	MWe	No.	MWe
Argentina	2	935	1	692	1	740	1	740
Armenia	1	376	0	0	0	0	1	1 000
Bangladesh	0	0	0	0	0	0	2	2 000
Belarus	0	0	0	0	2	2 000	2	2 000
Belgium	7	5 728	0	0	0	0	0	0
Brazil	2	1 901	0	0	1	1 245	4	4 000
Bulgaria	2	1 906	0	0	2	1 900	0	0
Canada	18	12 652	2	1 900	3	3 300	6	6 600
China1	11	8 587	19	18 920	26	27 660	77	63 400
Czech Republic	6	3 472	0	0	0	0	2	3 400
Egypt	0	0	0	0	1	1 000	1	1 000
Finland	4	2 696	1	1 600	0	0	1	1 000
France	59	63 473	1	1 630	1	1 630	1	1 630
Germany	17	20 339	0	0	0	0	0	0
Hungary	4	1 826	0	0	0	0	2	2 000
India	17	3 779	6	2 976	10	9 760	15	11 200
Indonesia	0	0	0	0	2	2 000	4	4 000
Iran	0	0	1	915	2	1 900	1	300
Israel	0	0	0	0	0	0	1	1 200
Italy	0	0	0	0	0	0	10	17 000
Japan	53	45 199	2	2 285	13	17 915	1	1 300
Kazakhstan	0	0	0	0	2	600	2	600
Korea DPR (North)	0	0	0	0	1	950	0	0
Korea RO (South)	20	17 451	6	5 170	3	4 050	2	2 700
Lithuania	1	1 185	0	0	0	0	2	3 400
Mexico	2	1 310	0	0	0	0	2	2 000
Netherlands	1	485	0	0	0	0	0	0
Pakistan	2	425	1	300	2	600	2	2 000
Poland	0	0	0	0	0	0	5	10 000
Romania	2	1 310	0	0	2	1 310	1	655
Russia	31	21 743	9	5 980	11	12 870	25	22 280
Slovakia	4	1 688	2	840	0	0	1	1 200
Slovenia	1	696	0	0	0	0	1	1 000
South Africa	2	1 842	0	0	3	3 565	4	4 000
Spain	8	7 448	0	0	0	0	0	0
Sweden	10	9 014	0	0	0	0	0	0
Switzerland	5	3 220	0	0	0	0	3	4 000
Thailand	0	0	0	0	2	2 000	4	4 000
Turkey	0	0	0	0	2	2 400	1	1 200
Ukraine	15	13 168	2	1 900	2	1 900	20	27 000
UAE	0	0	0	0	3	4 500	11	15 500
United Kingdom	19	11 035	0	0	0	0	6	9 600

Table 1. Continued

USA	104	101 119	1	1 180	11	13 800	20	26 000
Vietnam	0	0	0	0	2	2 000	8	8 000
WORLD	436	366 008	54	46 288	110	121 595	253	266 905

Sources: WANO, IAEA and CEA, 8th Edition, 2008.

(1): The figures mentioned include six nuclear power reactors in operation in Taiwan with a net capacity of 4 921 MWe, and two nuclear power reactors under construction with a net capacity of 2 600 MWe.

Note: This table includes only those future nuclear power reactors envisaged in specific plans and proposals and expected to be operating by 2030.

Table 2. Type of nuclear power reactors in operation per country in 2007

Country	PWR No. MWe	BWR No. MWe	GCR No. MWe	PHWR No. MWe	LWGR No. MWe	FBR No. MWe	Total No. MWe
Argentina				2 935			2 935
Armenia	1 376						1 376
Belgium	7 5 824						7 5 824
Brazil	2 1 795						2 1 795
Bulgaria	2 1 906						2 1 906
Canada				18 12 589			18 12 589
hina	11 9 052	4 3 141		2 1 300			17 13 493
Czech Rep.	6 3 619						6 3 619
Finland	2 976	2 1 700					4 2 676
France	58 63 130					1 130	59 63 260
Germany	11 14 013	6 6 457					17 20 470
Hungary	4 1 829						4 1 829
India		2 300		15 3 482			17 3 782
Japan[1]	23 18 420	32 29 167					53 47 587
Korea Rep.	16 14 824			4 2 627			20 17 451
Lithuania					1 1 185		1 1 185
Mexico		2 1 360					2 1 360
Netherlands	1 482						1 482
Pakistan	1 300			1 125			2 425
Romania				2 1 300			2 1 300
Russia	15 10 964				15 10 219	1 560	31 21 743
Slovakia	5 2 034						5 2 034
Slovenia	1 666						1 666
South Africa	2 1 800						2 1 800
Spain	6 5 940	2 1 510					8 7 450
Sweden	3 2 787	7 6 227					10 9 014
Switzerland	3 1 700	2 1 520					5 3 220
UK	1 1 188		18 9 034				19 10 222
Ukraine	15 13 107						15 13 107
USA	69 66 697	35 33 885					104 100 582
Total	265243 429	94 85 267	18 9 034	44 22 358	16 11 404	2 690	437 372 182

Note: Two BWRs were shutdown after 2007.

Source: IAEA and CEA 8th Edition, 2008.

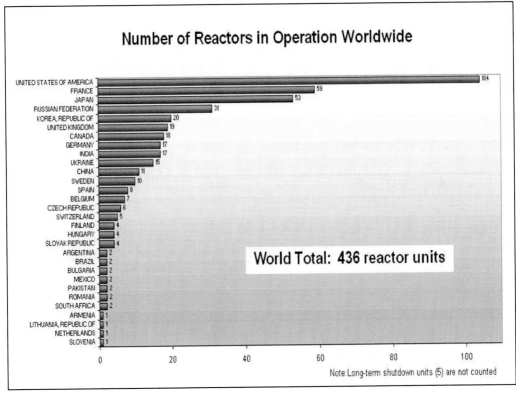

Source: IAEA, 2009.

Figure 3. Number of reactors in operation worldwide.

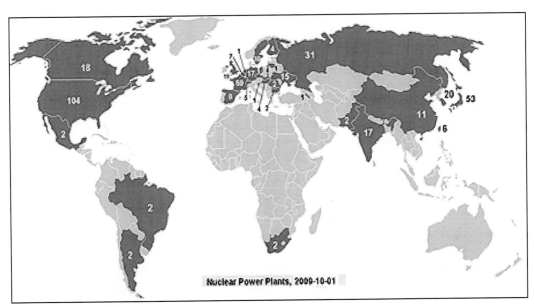

Source: IAEA and NEA, 2009.

Figure 4. Nuclear power reactors in operation worldwide.

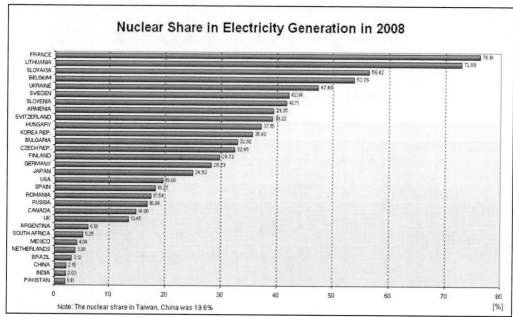

Source: IAEA, 2009.

Figure 5. Nuclear share of total electricity generation in 2008.

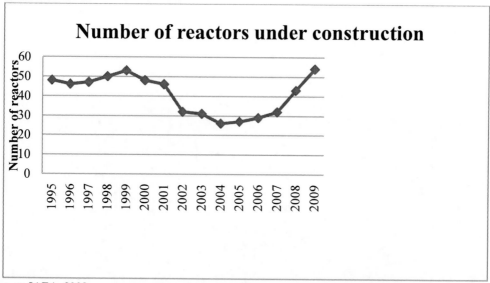

Source: IAEA, 2009.

Figure 6. Number of reactors under construction during the period 1995-2009.

From Table 2 can be stated that the pressurized water reactors (PWR) with 265 units and the boiling water reactors (BWR) with 94 units are the two types of reactors more used in the world. The most common reactor in the USA, France, Republic of Korea, the Russian Federation, Ukraine and Germany is the PWR. In the case of the BWR, the USA and Japan are the two countries with highest number of reactors of this type in operation.

Table 3. Shutdown reactors by country

Country	No of Units	Total MW
Armenia	1	376
Belgium	1	10
Bulgaria	4	1 632
Canada1	3	478
France	11	3 798
Germany	19	5 879
Italy	4	1 423
Japan2	5	1 618
Kazakhstan	1	52
Lithuania	1	1 185
Netherlands	1	55
Russia	5	786
Slovakia	3	909
Spain	2	621
Sweden	3	1 225
Ukraine	4	3 515
UK	26	3 324
USA	28	9 764
Total:	122	36 650

Note: (1) Canada has four other nuclear power reactors under long-term shut down. (2) Japan has another nuclear power reactor under long-term shutdown.

Source: IAEA, 2009.

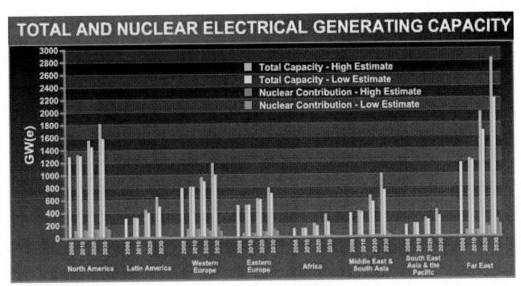

Source: IAEA.

Figure 7. Total and nuclear electrical generating capacity in 2008.

Table 4. Lifetime unit capability factor.
(Includes all operational and shutdown reactors from beginning of commercial operation up to 2008)

Country	No. of Reactors	Unit capability factor (%)
Argentina	2	81
Armenia	1	66.5
Belgium	8	86.6
Brazil	2	70.3
Bulgaria	6	73.3
Canada	25	76.1
China	11	84.9
Czech Republic	6	80.8
Finland	4	91.4
France	68	78.7
Germany	30	83.1
Hungary	4	84.7
India	17	66.7
Italy	4	73
Japan	57	72.5
Korea	20	87.2
LIthuania	2	71.5
Mexico	2	83.4
Netherlands	2	85
Pakistan	2	44.3
Romania	2	89.2
Russia	31	72.6
Slovakia	7	79.8
Slovenia	1	84.9
South Africa	2	76.5
Spain	10	84.7
Sweden	13	81.4
Switzerland	5	87.4
Ukraine	17	71.4
UK	29	73
USA	121	79
World Wide	517	78.2

Note: The following data from Taiwan, China is included in the totals: 6 units (83.8%).
Source: IAEA, 2008.

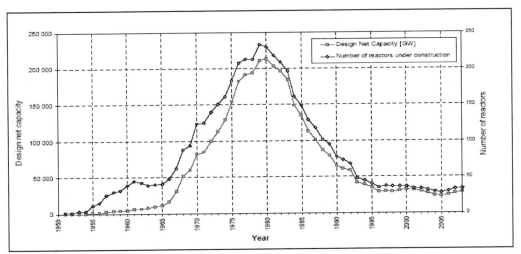

Source: IAEA document GC (52)/INF/6.

Figure 8. Number of nuclear power reactors and the total nuclear power reactor capacity under construction from 1951 to 2008.

With respect to the construction of nuclear power reactors it is important to single out the following. During the period of peak construction of nuclear power reactors in the 1970s and 1980s there were several major nuclear supply companies in Canada, France, Germany, Japan, the former Soviet Union (now the Russian Federation), Sweden, Switzerland, UK and USA. Today, the main nuclear system suppliers companies are located in Canada, China, France, India, Japan, Republic of Korea, Russian Federation and USA. There are other potential nuclear suppliers systems that have designs in development such as Argentina and South Africa, but the designers of currently available nuclear steam supply systems have reduced to a small group who increasingly are working very closely together. This group includes Areva and Mitsubishi, GE and Hitachi and Toshiba and Westinghouse.

As can be easily see from Figure 8 the number of nuclear power reactors under construction dropped significantly during the period 1999-2004. After 2005, there is a trend to increase slowly the total number of nuclear power reactors under construction each year. It is expected that this trend will continue in the coming years.

One of the major problems that several countries have to deal with in the coming years, particularly in the case of North America and Europe, is the increase ageing power-generation capacity, even in the field of electricity generation using nuclear energy. To overcome this problem both regions have an urgent need for major investment in the energy sector in order to meet the expected increase in the electricity demand and to replace ageing infrastructures in the energy sector. According to the World Energy Outlook for 2006, "around 800-900 GWe capacity will be required by 2030 to replace the existing capacity and to address increasing needs. It is reasonable to assume that out of these potential new 800-900 GWe, at least 100 GWe will be produced by Generation-III nuclear power reactors. This corresponds to the construction of 60 to 70 big nuclear power reactors, which represents an investment of €150 billion over 20 years (for an average overnight construction cost of €1,500 per kWe). These new nuclear power reactors to be constructed in the coming years should be designed to operate 60 years. In the following figure the ageing of the nuclear power reactors currently in operation is shown.

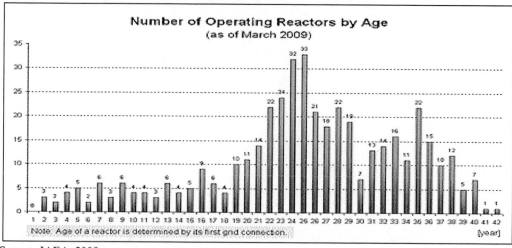

Source: IAEA, 2009.

Figure 9. Number of nuclear power reactors in operation in the world by age as March 2009.

From Figure 9 can be stated that 136 nuclear power reactors had between 20 and 25 years of operation representing 31.2% of the 436 nuclear power reactors that were in operation in 2009; 134 units had more than 30 years of operation, representing 30.7% of the total. Considering only these two groups of nuclear power reactors can be stated that 61.9% of the total nuclear power reactors in operation in 2009 has more than 20 years of service. Only 35 nuclear power reactors had 10 or less years of operation, representing around 8% of the total. The number of nuclear power reactors in operation in the world in 2009 with more than 20 years connected to the grid is 350, representing 80.3% of the total. For this reason, the age of the nuclear power reactors is one of the urgent problems that need to be faced by those countries in which these reactors are located, either by extending their lifetime or by initiating the process of shut down and decommissioning. Taking into account the current energy situation in the world, it is expected that the majority of the states currently operating nuclear power reactors will decide to extend the lifetime of the majority of its nuclear power reactor in order to satisfy the foresee increase in the demand of electricity in the coming years in the most cost effective manner.

The extension of the lifetime of the majority of the current nuclear power reactors in operation do not represent a risk that could affect the safety operation of these reactors? The answer of this question in much of the cases is no. The reason is the following: "The reactor technology available for use today is fundamentally based upon previous designs and takes into account the following design characteristics:

a)   60 year life;
b)   simplified maintenance - online or during outage;
c)   easier and shorter construction.
d)   inclusion of safety and reliability considerations at earliest stages of design;
e)   modern technologies in digital control and man-machine interface;
f)   safety system design guided by risk assessment;
g)   simplicity by reducing the number of rotating components;

h) increased reliance on passive systems (gravity, natural circulation, accumulated pressure etc.);

i) addition of severe accident mitigating equipment;

j) complete and standardized designs with pre-licensing". (IAEA GC(52)/INF/6, 2008)

It is foresee that in the longer term a new generation of nuclear power reactors, the so-called "Generation-IV system", will be in the market not before 2040 once they have reached technical maturity and met sustainable development criteria, particularly those pertaining to waste management and preservation of energy resources[3]. (EC report, 2007)

One of the main features of the Generation IV systems is that they would be much smaller in size (100 MW to 200 MW) and capital investment, representing a more flexible solution due to much shorter building times and a lower potential risk due to smaller radioactive inventories and passive safety features. (Schneider and Froggatt, 2007) Based on what has been said before, it is important to single out that the international community should be aware that each of the three previous generations of nuclear power reactors will coexist during the 21st century. All of these types of nuclear power reactors face specific technological challenges to be overcome on the path to sustainability, but all share the common goal of guaranteeing the highest level of safety.

Summing up can be stated the following: According with the IAEA information, during 2009 seven nuclear power reactor started to be constructed. These units were the following:

a) Hongyanhe 3 (1000 MWe, PWR, China). Construction officially started on 7 March;

b) Sanmen 1, (1000 MWe, PWR AP-1000, China). Construction officially started on 19 April;

c) Yangjiang 2 (1000 MWe, PWR, China). Construction officially started on 4 June;

d) Fuqing 2, (1000 MWe, PWR, China). Construction officially started on 17 June;

e) Novovoronezh 2-2 (1085 MWe, PWR-VVER, Russia). Construction officially started on 12 July;

f) Fangjiashan 2 (1000 MWe, PWR, China). Construction officially started on 17 July;

Shin-Kori 4 (1340 MWe, PWR-APR 1400, Republic of Korea). Construction officially started on 15 September.

In four units construction were reactivated. These units were the following: a) Akademik Lomonosov 1 and 2, (30 MWe, PWR-KLT40, Russia). Floating nuclear power plant will be finally located close to Vilyuchinsk instead of Severodvinsk; and b) Mochovce 3 and 4, (405 MWe, PWR-VVER, Slovakia). Construction officially reactivated on 11 June. Two units

---

[3] The first generation of nuclear power reactors was commercially available in the 1950s and 1960s. The second generation began in the 1970s in large commercial nuclear power plants and several of them still are operating today. Generation III was developed more recently in the 1990s with a number of evolutionary designs that offer significant advances in safety and economics. Several units have been built, primarily in East Asia. Advances to Generation III (the so-called "Generation III+") are underway resulting in several near-term deployable plants that are actively under development and are being considered for deployment in several countries. It is expected that the new nuclear power plants to be built between now and 2030 will likely be chosen from this type of reactor. Beyond 2030, the prospect for innovative advances through renewed research and development has stimulated interest worldwide in the Generation IV of nuclear energy systems.

were definitive shutdowns: Hamaoka 1 and 2, (515/806 MWe, BWR, Japan). These units were officially closed on 31 January.

PROJECTION OF THE CONTRIBUTION OF NUCLEAR ENERGY IN THE GLOBAL ENERGY BALANCE

The IAEA revised projection of the contribution of nuclear energy in the global energy balance for 2030 is shown in Table 5. According with the IAEA projections, until 2030 the total production of electricity using nuclear energy will be between 418 GWe and 640 GWe representing 8% and 8.9% respectively of the total electricity generation for that year. The reduction in the participation of nuclear energy in the total production of electricity in 2030 will be between 8.6% and 9.5% in comparison with 2009. Eastern Europe will be the region with major participation of nuclear energy in the production of electricity, between 12% and 13%, followed by the Far East with 11%-14%. In the case of Latin America, Africa, the Middle East, South Asia and South East Asia and the Pacific, the participation of nuclear energy in the production of electricity will be very small with a maximum of 5.3% in the case of the Middle East and South Asia.

Table 5. Low and high estimates of total electricity generation and contribution by nuclear power (Low estimate: first row in each region; High estimate: second row)

		2004 Total Nuclear% Elect. GW(e) % GW(e)			2010* Total Nuclear% Elect. GW(e) % GW(e)			2020* Total Nuclear% Elect. GW(e) % GW(e)			2030* Total Nuclear % Elect. GW(e) GW(e)		
North America		1055	111.3	10,6	1099	16	11	1194	118	10	1318	115	8,7
					1155	117	10	1279	128	10	1422	145	10
Latin America		264	4,1	1,6	303	4,1	1,4	383	6,1	1,6	483	5,8	1,2
					350	4,1	1,2	543	6,1	1,1	828	15,0	1,8
Western Europe		724	125,1	17,3	762	119	6 15	842	97	11 14	940	79	8,5
					816	125		951	130		1118	145	13
Eastern Europe		466	49,4	10,6	469	48	10	505	64	13	543	66	12
					496	51	10	605	78	13	736	97	13
Africa		105	1,8	1,7	115	1,8	1,6	143	2,1	1,5	181	2,1	1,2
					135	1,8	1,3	207	4,1	2	316	9,3	3,0
Middle East and South Asia		284	3	1	331	9	2,8	430	15	3,6	556	18	3,2
					370	10	2,8	555	27	4,9	811	43	5,3
South East Asia and the Pacific		143			169 184			213	0,9	0,4	264	0,9	0,3
								270	0,9	0,3	391	3	0,8
Far East		651	72,8	11,2	685	82	12	804	13 142	14 12	937	131	14
					840	85	10	1167			1589	183	11
World total	Low estimate	3693	367,5	10	3934	380	10	4515	416	9,2	5223	418	8
	High estimate				4347	395	9,1	5576	516	9,3	7210	640	8,9

Note: *Estimate figures.
Source: IAEA.

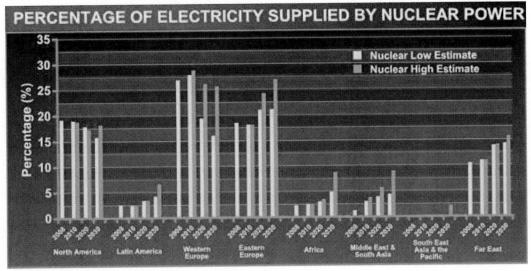

Source: IAEA.

Figure 10. Percentage of electricity supplied by nuclear power.

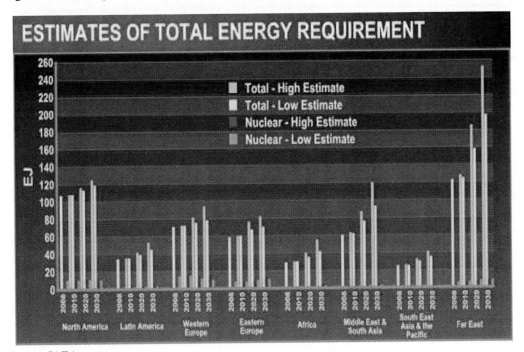

Source: IAEA.

Figure 11. Estimates of total energy requirement.

In Table 6 the latest IAEA estimates of nuclear electricity generating capacity per region up to 2030 is included.

Table 6. The IAEA latest estimates of nuclear electricity generating capacity (GWe)

Region	2007	2010		2020		2030	
		Low	High	Low	High	Low	High
North America	113.2	113.5	114.5	121.4	127.8	131.3	174.6
Latin America	4.1	4.1	4.1	6.9	7.9	9.6	20.4
Western Europe	122.6	119.7	121.3	92,1	129.5	73.9	150.1
Eastern Europe	47.8	48.2	48.3	72.1	94.7	81.2	119.4
Africa	1.8	1.8	1.8	3.1	4.5	4.5	14.3
Middle East and South Asia	4.2	7.6	10.1	12.5	24.3	15.9	41.5
South East Asian and the Pacific	0	0	0	0	0	1.2	7.4
Far East	78.5	81.3	83.1	129.2	151.8	155.7	219.9
World Total	372.2	376.3	383.1	437.4	541.6	473.2	747.5

Source: IAEA GC (52)/INF/6.

Table 6 shows that the greatest expansion of nuclear electricity generating capacity for the following 20 years is projected for the Far East. The increase will go from 78.5 GWe in 2007 to a minimum of 155.7 GWe in 2030 or to a maximum of 219.9 GWe; this means a minimum increase of 198.3% or a maximum increase of 280%. The region with the greatest uncertainty, i.e. the greatest difference between the low and high projections is Western Europe, region that is now debating the future role to be played by nuclear energy in the energy balance mix in several countries. Although approximately 20 new countries are included in the 2030 projections, the global increase in the high projection comes mainly from increases in the 30 countries already using nuclear energy for electricity production[4]. The low projection also includes approximately five new countries that might have their first nuclear power reactors in operation by 2030.

It is important to single out that the IAEA projections have changed over the past few years. In particular the high projection for the rate of increase in installed nuclear power plant capacity between 2020 and 2030 is higher than the projections presented in 2004, reflecting an increase in optimism about the use of nuclear energy for electricity production in some regions, particularly in North America and Europe. The low projection in 2004 showed a declining installed capacity as nuclear power plants in operation were taken out of service without replacement in North America and Europe.

The IAEA has identified several issues that could affect the future implementation of nuclear power programmes in a group of countries and hence the accuracy of the predictions of nuclear power used. According with this information, these issues are the following:

1) nuclear power has generated stronger political passions than any other energy alternatives source. Alternatives to nuclear power, natural gas, coal, hydropower, oil and renewables, have nothing comparable to the prohibitions and phase-out policies that several countries have adopted for nuclear power;

---

[4] In this number, Taiwan was included in China.

2) because of the front-loaded cost structure of a nuclear power plant, high interest rates, or uncertainty about interest rates, will weaken the business case for nuclear power more than for alternatives;

3) nuclear power's front-loaded cost structure also means that the cost of regulatory delays during construction is higher for nuclear power plants than for alternatives. In countries where licensing processes were relatively untested in recent years, investors face potentially more costly regulatory risks with the construction of nuclear power plants than with alternatives;

4) the strength, breadth and durability of commitments to reducing CO_2 emissions will also influence nuclear power's growth;

5) the nuclear industry is a global industry with good international cooperation and hence the implications of an accident anywhere will be felt in the industry worldwide;

6) similarly, nuclear terrorism may have a more far reaching impact than comparable terrorism directed at other fuels power plants;

7) while a nuclear power plant in itself is not a principal contributor to proliferation risk, proliferation worries can affect public and political acceptance of nuclear power;

8) among energy sources, high level radioactive waste is unique to nuclear power. The nuclear power industry might feel a disproportionately broad impact if major problems are encountered in any of the repository programmes that are most advanced in Finland, France, Sweden and USA. (IAEA GC(52)/INF/6, 2008)

If the world is to meet even a fraction of the economics and social aspirations of the developing world, energy supplies must expand significantly. If the increase needs of the industrialized countries for its economic and social development are considered as well, then the world energy supplies must expand even further.

Table 7. Level of participation of the different types of energies in the generation of electricity by region in 2006

Region	Thermal		Hydro		Nuclear		Renewables		Total	
	Use (EJ)	%	Use (EJ)	%	Use (EJ)	%	Use (EJ)	%	Use (EJ)	%
North America	22.21	65.71	2.43	14.53	9.61	18.99	0.73	0.77	34.98	100
Europe	32.92	67.02	2.84	5.82	12.71	26.03	0.55	1.13	49.02	100
Asia	52.84	84.70	2.94	4.72	5.90	9.46	0.70	1.12	62.38	100
Latin America	4.42	38.28	2.46	58.31	0.33	2.61	0.32	0.81	7.54	100
Africa	4.89	80.01	0.35	17.74	0.11	1.84	0.04	0.41	5.4	100
World total	117.28	73.66	11.02	6.92	28.66	18	2.24	1.42	159.22	100

Note: One EJ = 2.78 × 105 GWh or 31.7 GWa.

Source: IAEA.

CONTRIBUTION OF THE DIFFERENT TYPES OF ENERGY IN THE WORLD ELECTRICITY PRODUCTION

Table 7 includes information on the level of participation of the different types of energy in the generation of electricity by region.

From Table 7 can be concluded that, in 2006, the contribution of nuclear energy to the total electricity generation varies considerably among regions. In Europe, for instead, nuclear generated electricity accounts for 26.03% of the total electricity produced in that year[5]. In North America it is approximately 19%, whereas in Africa and Latin America it is 1.84% and 2.61% respectively[6]. In Asia, nuclear energy accounts, in 2006, for 9.46% of electricity generation. Based on the mentioned figure it is not difficult to concluded that the use of nuclear energy for electricity generation is concentrated in technologically advanced regions. For this reason:

1) Europe remained the largest nuclear energy user for electricity generation since 1980;
2) North America was the second largest user of nuclear energy for electricity production since 1980;
3) with the largest regional population and fast growing economies, Asia became the third largest nuclear energy user for electricity production since 1980 and is the region with the world most ambitious nuclear power plans to be implemented in the future;
4) Latin America registered almost the same growth in its nuclear energy use for electricity generation since 1980. Its share remained between 2.61% and 3.1%;
5) Africa is the region with the smallest nuclear share since 1980 and will continue in this position in the future.

Figure 12 includes relevant information regarding the energy availability factor in 2008.

The situation regarding the number of nuclear power plants in operation and under construction by region in 2009 is the following: In North America (USA and Canada), there are 122 nuclear power reactors in operation with a total net capacity of 113 771 MWe, one nuclear power reactor under construction with a total net capacity of 1 180 MWe and 40 nuclear power reactors planned or proposed with a total net capacity of 49 700 MWe.

In Asia and the Pacific, the number of nuclear power plants in operation is 103 in five countries (if Taiwan is included within China) with a total net capacity of 72 539 MWe and there are 34 nuclear power reactors under construction in 5 countries, with a total net capacity of 21 921 MWe. The nuclear power reactors planned or proposed is 174.

The situation regarding the number of nuclear power plants in operation and under construction by region in 2009 is the following: In North America (USA and Canada), there are 122 nuclear power reactors in operation with a total net capacity of 113 771 MWe, one nuclear power reactor under construction with a total net capacity of 1 180 MWe and 40 nuclear power reactors planned or proposed with a total net capacity of 49 700 MWe.

---

[5] In 2007, the contribution of nuclear energy in the European region was 31%.
[6] In 2009, the contribution of nuclear energy in Latin America was 2.3%.

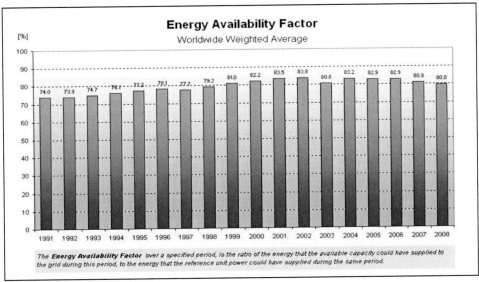

Source: IAEA.

Figure 12. Energy availability factor during the period 1991-2008.

In Asia and the Pacific, the number of nuclear power plants in operation is 103 in five countries (if Taiwan is included within China) with a total net capacity of 72 539 MWe and there are 34 nuclear power reactors under construction in 5 countries, with a total net capacity of 21 921 MWe. The nuclear power reactors planned or proposed is 174.

In the European region, there are 196 nuclear power reactors in operation in 18 countries with a total net capacity of 170 007 MWe and 17 nuclear power reactors under construction in 5 countries with a total net capacity of 13 850 MWe. The number of nuclear power reactors planed or proposed is 102 (including Belarus, Ukraine and Russia) with a total net capacity of 127 775 MWe.

In the Latin American region, there are six nuclear power reactors in operation with a total net capacity of 4 146 MWe, one nuclear power reactors under construction with a total net capacity of 692 MWe and nine nuclear power reactors planned or proposed with a total net capacity of 8 725 MWe.

Table 8. Evolution of nuclear power plants capacities connected to the electric grid during the period 1970 to 2009

Year	Number of countries	Units	MWe
1970	14	81	16 202
1980	24	243	136 164
1990	30	419	325 854
2000	31	436	351 550
2007	31	439	372 182
2009	31	436	366 008

Source: CEA, 8th edition, 2008, WANO and IAEA sources.

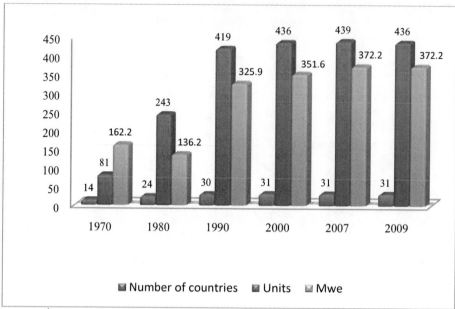

Source: CEA 8th edition, 2008 and IAEA source.

Figure 13. Number of nuclear power reactors in operation and the total capacities installed from 1970 to 2009.

Finally, Africa has two nuclear power reactors in operation in one country with a total net capacity of 1 842 MWe and nine nuclear power reactors planned or proposed with a total net capacity of 9 565 MWe.

In Table 8 the evolution of nuclear power plants capacities connected to the electric grid during the period 1970 to 2007 is shown.

From Figure 13 can be concluded that the number of countries using nuclear energy for the generation of electricity are not increasing since 1990 and the number of nuclear power reactors in operation is almost the same since 2000. However, there are a systematic increase in the total net capacity of the nuclear power plants built since 1970 due to uprates and modernization of several nuclear power reactors.

As can be clearly seen in Table 8 the major increase in the construction of nuclear power reactors occurred in the 1970s, 1980s and 1990s. The increase in the number of nuclear power reactors constructed from 1970s to 1980s jump from 81 to 243, this means an increase of 162 new units. From 1980s to 1990s the number of nuclear power reactors built increased from 243 to 419, this means an increase of 176 new units, ten more units than in the previous decade. However, in the last 17 years the number of nuclear power reactors built was only 20 a very small figure in comparison with the two previous decades. It is important to stress that despite of the significant reduction in the construction of new nuclear power reactors after the Chernobyl nuclear accident, nuclear power generation has been increasing continually as a result of improved performance of the nuclear power reactors under operation.

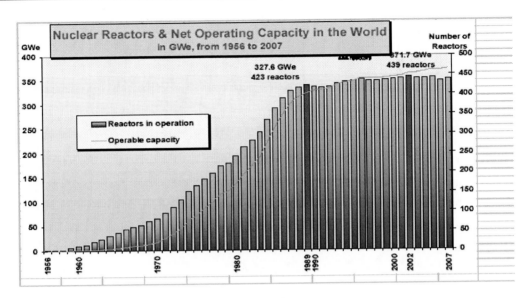

Source: IAEA SIPRIS, 2007.

Figure 14. Nuclear power reactors in operation and net operating capacity in the world from 1956 to 2007.

According with the IAEA information, in 1990, the world average annual capacity factor for nuclear power plants was 67.7%. In 2005, this figure increases up to 81.4%, which is the equivalent to the construction of some 74 new nuclear power reactors with a total net capacity of one GWe. Nevertheless, it is important to note that the number of nuclear power reactors under construction are not enough to satisfy the foresee increase in the demand of electricity in the coming years in almost all regions. According to conservative calculations, more than 60% of the increase in the demand of world primary energy up to 2030 will come from Third World countries, mainly from the Asia and the Pacific region, particularly China and India. Because of this situation, energy demand in the non-OECD Asia region is projected to grow at an average rate of 3.2% per year, more than doubling over the 2004 to 2030 period and accounting for more than 65% of the increase in energy use for the non-OECD region as a whole. In 2004, energy consumption in the countries of non-OECD Asia made up just over 48% of the non-OECD total; in 2030, its share is projected to be above 56%. (IEO, 2007)

The International Energy Outlook for 2009 indicated that "electricity generation from nuclear power is projected to increase from about 2.7 trillion kWh in 2006 to 3.8 trillion kWh in 2030, as concerns about rising fossil fuel prices, energy security and greenhouse gas emissions support the development of new nuclear generation capacity. Higher fossil fuel prices allow nuclear power to become economically competitive with generation from coal, natural gas and liquids despite the relatively high capital and maintenance costs associated with nuclear power plants. Moreover, higher capacity utilization rates have been reported for many existing nuclear facilities, and it is anticipated that most of the older nuclear power plants in the OECD countries and non-OECD Eurasia will be granted extensions to their operating lives".

The following figure illustrates changes in the regional distribution of global energy use during the 25 year period in terms of shares. Highlights of these changes are as follows:

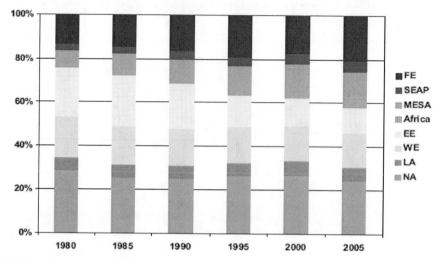

Source: IAEA.

Figure 15. Regional distribution of global energy use during the 25 year period in terms of shares.

From Figure 15 can be concluded that during the period 1980-2005:

1) the biggest decline (10%) was in the share of Eastern Europe;
2) the share of North America declined by 5% and the share of Western Europe dropped by 4%;
3) the share of the Far East increased significantly (7%);
4) there was no change in the share of Latin America;
5) the Middle East, South Asia, Southeast Asia and the Pacific and Africa had small increases in their shares.

In absolute terms:

1) North America remained the largest energy user among all regions throughout this period;
2) Eastern Europe was the second largest user of energy in 1980, but a rapid decline in energy use in the 1990s made it the fourth largest energy user in 2005, despite an increase in energy use from 2000 to 2005;
3) with the largest regional population and fast growing economies, the Far East became the second largest energy user in 2005, while it had been the fourth largest energy user in 1980;
4) Western Europe was the third largest energy user in 1980 and it remained so in 2005;
5) Latin America registered almost the same growth in its energy use as the world. As such, its share remained more or less unchanged at around 6% over the 25 year period.

Comparison of the growth rates of the regions shows that: a) the greatest growth was experienced in the Middle East and South Asia. Energy use in this region increased over the 25 year period by about 3.8 times, doubling its share of world energy use over the 25 year period. However, the region accounted for only 10% of world energy in 2005; and b) both Southeast Asia and the Pacific and Africa also had high growth in their energy use (between 4% and 5% per annum), but this growth was on a small base in 1980, i.e. only 8 EJ and 9 EJ, respectively. Their shares in global energy use increased from a mere 2.7% and 3.3% in 1980, respectively, to 4.8% and 6.3% in 2005.

In brief, energy use in all regions except Eastern Europe increased from 1980 to 2005. Two highlights of the regional distribution of world energy use in the 25 year period are: 1) declining shares of the regions that were major energy users; and 2) a significant increase in the share of the Far East, which grew from the fourth largest energy user of 1980 to the second largest energy user in 2005.

If a comparison of energy use on a per capita basis is made, then the following can be stated:

a) there was an increase in per capita energy use in all the regions except Western Europe and Eastern Europe, which registered declines in their per capita energy use over the 25 year period;

b) per capita energy use in North America, Western Europe and Eastern Europe was significantly higher than in other regions in both 1980 and 2005;

c) the ranking of the three largest energy users on a per capita basis did not change over the period 1980 to 2005;

d) per capita energy use in the Middle East and South Asia was the lowest among all regions in 1980, and it remained so in 2005.

However, in the specific case of nuclear energy for the production of electricity the situation is the following:

1) Europe remained the largest user of nuclear energy for electricity generation among all regions since 1980;

2) North America was the second largest user of nuclear energy for electricity production since 1980;

3) with the largest regional population and fast growing economies, Asia as a whole became the third largest nuclear energy user for electricity production since 1980;

4) Latin America registered almost the same growth in its nuclear energy use for electricity generation since 1980;

5) Africa is the region with the smallest nuclear share since 1980.

LIMITING FACTORS IN THE USE OF NUCLEAR ENERGY FOR ELECTRICITY GENERATION

If nuclear energy will be an important component of the energy balance in the Asia and the Pacific region in the coming years, then why the use of this type of energy cannot play the same role in other regions as well. The reason is the following: There are a number of limiting factors that impede that the nuclear energy plays an important role in the energy balance in other regions now and in the future. These limiting factors are, among others, the following:

a) management of the radioactive waste, particularly high–level nuclear waste;
b) proliferation security;
c) environment impact;
d) operational safety of the nuclear power plants;
e) economic competitiveness;
f) financial investment;
g) public acceptance;
h) human resources available, particularly high experience professionals and technicians;
i) transport of uranium;
j) infrastructure building;
k) size of the electricity grids;
l) reactor technology.

Some of these limiting factors are briefly explained in the following paragraphs.

Management of the Radioactive Waste

Management of the radioactive waste, particularly high–level nuclear waste, is an unresolved problem for the public opinion of many countries. The main irradiated fuel constituents discharged from LWRs are: uranium, plutonium, actinides and fission products. Uranium constitutes about 96% of the fuel unloaded from commercial power reactors. In the case of LWRs, the spent fuel on discharge still contains 0.90% enriched in the fissile isotope U-235, whereas natural uranium contains only 0.7% of this isotope. Plutonium constitutes of about 1% of the weight of discharged fuel; it is a fissile material, which can be used as fuel in present and future commercial nuclear power reactors. Minor actinides constitute about 0.1% of the weight of discharged fuel. They consist of about 50% neptunium, 47% americium and 3% curium, which are very radiotoxic. Fission products (iodine, technetium, neodymium, zirconium, molybdenum, cerium, cesium, ruthenium, palladium, etc.) constitute about 2.9% of the weight of discharged fuel. At the present stage of knowledge and technological capacity, they are considered as the final waste form of nuclear power production, unless a specific use is found for the non-radioactive platinum metals.

A typical 1000-MWe PWR unit operating at 75% load factor generates about 21 tons of spent fuel at a burn-up of 43 GWd/t; this contains about 20 tons of enriched uranium; 230 kg Pu; 23 kg minor actinides and 750 kg fission products.

It is important to note that the management of spent fuel should ensure that the biosphere is protected and the public must be convinced of the effectiveness of the methods used. Since the spent fuel contains very long-lived radionuclides, some protection is required for at least 100,000 years. There are two means to reach this goal. One of them is the following: Society can wait for the natural decay of the radioactive elements by isolating them physically from the biosphere through installation of successive barriers at a suitable depth in the ground. This strategy leads to deep geological disposal. The second one is the following: Society can make use of nuclear reactions that will transmute the very long-lived wastes into less radioactive or shorter-lived products. In the opinion of several experts, deep geological repository disposal is the most appropriate solution available today.

It is important to stress that the technology for the safe management of nuclear waste is now available and can be used by any countries with an important nuclear power programme. The USA; Finland and Sweden have achieved some progress regarding the final disposal of high radioactive nuclear waste and the technology used by these countries could represents a real and objective solution to this problem for other countries as well.

Proliferation Security

Nuclear power entails potential proliferation security, particularly for the possible misuse of nuclear technology, facilities or materials as a precursor to the production of a nuclear weapon. Due to this undisputed fact, it is extremely important to be aware that the future use of nuclear energy for electricity generation in several countries, will depends of the production of new type of nuclear power reactors that are proliferation risk-free.

Nuclear power should be expanded in the world only if the risk of proliferation from operation of the commercial nuclear fuel cycle is made acceptably small. How to achieve this goal? First, the international community should strengthen the application of the IAEA safeguards system to all states by putting into force, for all of them, the IAEA Additional Protocol. Second, the international community should adopt a multilateral approach to the nuclear fuel cycle. The international community must adopt all necessary measures to prevent the acquisition of weapons-usable material, either by diversion (in the case of plutonium) or by misuse of nuclear fuel cycle facilities (including related facilities, such as research reactors or hot cells) now operating in different countries. However, it is important to stress that the adoption of a multilateral approach to the nuclear fuel cycle should be done in a way that respect the right of any states to develop their own nuclear fuel cycle but under full IAEA safeguards, including the Additional Protocol.

There are three issues of particular concern for the international community when the nuclear energy option is considered by any government as a real alternative to satisfy the foresee increase in the demand of electricity in the country in the coming decades. These issues are the following:

a) the existing stocks of fissionable materials in the hands of several countries that are directly usable for the production of nuclear weapons;
b) the number of nuclear facilities with inadequate physical protection and controls. The lack of adequate physical protection of nuclear installations in several countries could be used by terrorist groups to have access to certain amount of fissionable materials in order to use them for the production of a nuclear weapon;
c) the transfer of sensitive nuclear technology, especially enrichment and reprocessing technology, to countries implementing a nuclear power programme that brings them closer to a nuclear weapons capability.

The proliferation risk due to the global growth in the use of nuclear energy for electricity production should be reduced to the minimum possible, if the nuclear option is going to be considered by the international community as a realistic alternative for the production of electricity in several countries in the coming years.

Environment Impact

Nuclear power has been perceived by the public opinion as very danger for the environment and for the health of people, heightened by the Three Mile Island and the Chernobyl nuclear accidents. Other accidents at fuel cycle facilities in the USA, Russia and Japan in the past has increase the fear of the public to the use of nuclear energy for the generation of electricity. Some countries decided to phase out their nuclear power programme or to prohibit the use of nuclear energy for electricity generation based on the consequences for the health and for the environment of some of these accidents, particularly the Chernobyl nuclear accident[7]. For this reason, the design of new nuclear power reactors with stringent safety requirements is an indispensable condition to spread to several countries the use of nuclear energy for electricity generation in the coming years[8].

Nuclear energy produces very few emissions of CO_2 to the atmosphere. If the whole production cycle is considered, this means from the construction of the nuclear power plant to their exploitation, the production of one kWh of nuclear origin electricity supposes less than six grams of CO2 emission to the atmosphere, mainly associated to the construction of the nuclear power plant and the transport of fuel. On the other hand, a combined cycle gas power plant generates 430 g of CO2 and a coal power plant between 800 g to 1 050 g of CO2, according with the type of technology used. Based on this facts can be stated that the use of nuclear energy for electricity generation is one of the cleanest type of energy available in the world, in comparison with any of the fossil fuel power plants currently in operation all over the world.

Economic Competitiveness and Financial Investment

Nuclear power is cost competitive with other forms of electricity generation, except where there is direct access to low-cost fossil fuels. Fuel costs for nuclear power plants are a minor proportion of total generating costs, though capital costs are greater than those for coal and oil fired power plants. The evolution of the price of uranium in the period 1968-2008 is shown in the following figure. From this figure can be concluded that the price of uranium increased significantly since 2004 from US$50 (both spot price and long term price) to US$250 spot price and to around US$190 long term price, an increase between 4 and 5 times in four years. However, this increase in the uranium price does not make more expensive the use of nuclear energy for electricity production in comparison with fossil fuels power plants. The reason is the following: Nuclear power is hardly sensitive to fluctuations in the price of

---

[7] Austria (1978), Sweden (1980) and Italy (1987) were the first countries in Europe to oppose or phase out their nuclear power programmes. The political and public opposition in Ireland prevented the government to develop a nuclear power programme in the country.

[8] However, it is important to single out that there is any nuclear power reactor design that is totally risk-free for the following two reasons: a) technical possibilities; and b) work-force problems. Safe operation of a nuclear power reactor requires effective regulation, a management committed to safety and a skilled work-force.

uranium, so that price shocks and market volatilities, as experienced recently, influence the generation price marginally.

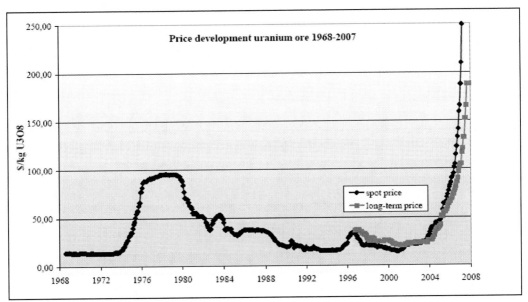

Source: Reference 12.

Figure 16. The evolution of the price of uranium in the period 1968-2008.

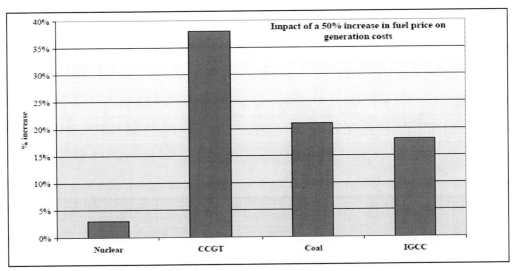

Source: IAEA.

Figure 17. Impact of a 50% increase in fuel price on generation costs.

In assessing the cost competitiveness of the use of nuclear energy for the production of electricity, decommissioning and waste disposal costs should be taken into account. According with some expert's calculations, decommissioning costs are about 10-20% of the initial capital cost of a nuclear power plant. This make the nuclear energy option more expensive than other sources of energy, because this type of costs are not included in the construction costs of any other electricity generating power plants. However, if the social, health and environmental costs of using fossil fuels for the generation of electricity are also taken into consideration, then the use of nuclear energy for electricity generation is outstanding.

The increase cost due to delay in the construction of a nuclear power plant is another important element that need to be taken into account, when the overall cost associated with the use of nuclear energy for electricity generation is analyzed. For all of these reasons, it can be stated that the future competitiveness of nuclear power will depend substantially on the additional costs, which may accrue to coal generating power plants and the cost of gas for gas-fired power plants. However, it is uncertain how the real costs of meeting targets for reducing sulphur dioxide and greenhouse gas emissions will be attributed to fossil fuel plants is going to be included in the generation cost of conventional power plants.

Public Acceptance

The public opinion is an important element that needs to be taken into consideration when planning the use of nuclear energy for electricity generation by any government. The future use of nuclear energy for the production of electricity in a particular country will depends on the public's perception of the importance of their use and the risks involved. The concern of the society arises because, although the probability of a nuclear accident is very low, the risks to have it can have important consequences on the environment and on the health of the population that lives in areas not only close to the plant, but in other areas far from the site as well. The Chernobyl nuclear accident is an example of an accident that affected not only the environment surrounded the plant and the health of the population located in the site, but the environment and the health of the population located far from the nuclear power plant site, including living in other countries as well. This concern is not based on an objective perception but in an intuitive judgement, due to the initial history on how nuclear energy was created in the past for its use for military purpose, the problems associated with the disposal of nuclear waste, the serious nuclear accidents occurred in some nuclear facilities and to the optimistic initial expectations associated with the use of nuclear energy for electricity generation, among others.

Different studies carried out so far on the use of nuclear energy for electricity generation in several countries indicate that there is a lack of information in the public on this important subject, as well as about the price of the different energy options available for the production of electricity and or their repercussion on the environment. The different nuclear actors, particularly the scientific community, technologists, operators, regulators and legislators, have not been able to convince the public opinion on which of the energy options available for electricity generation are the most convenient for a particular country. For this reason, it is extremely important to make additional efforts to provide reliable and impartial information about the use of the different types of energy sources available for electricity generation, including the nuclear energy option to the public opinion. The purpose of these efforts is to assist public opinion in a particular country to define their position regarding the use of the different energy options available and to identify which of them are the most appropriates for the country.

The increase in the costs of the construction of a nuclear power plant and the unforeseen delays in the construction of these type of plants due to new safety and security requirements demanded to the operators after the nuclear accidents of the Three Miles Island and Chernobyl, has stopped the construction of new nuclear power plants in the USA and in the European region. The repercussion of these accidents was a public opinion contrary to the use of nuclear energy for electricity generation in these two places and, as a consequence of this position, the rejection of the politicians to support any further development of the nuclear power sector in the USA and in several countries within the EU. In some of these countries the government adopted a nuclear phase out policy to shut down all nuclear power reactors in operation and to stop any future construction of new units. However, and despite of several other nuclear accidents of minor consequences occurred in several other countries, the opposition of the public opinion to the use of nuclear energy for the electricity production in Asia and the Pacific such as Japan, the Republic of Korea, China as well as in Russia, where the energy demand has grown considerably, are not so strong to stop the ongoing programmes for the construction of new nuclear power reactors.

It is important to single out that in many countries people are ready to support the nuclear energy option, if and only if problems related with the final disposal of high-level nuclear waste and the safety of the nuclear power reactors are properly solved. In the particularly case of the USA, the people are unlikely to support nuclear power expansion without substantial improvements in costs and technology.

The carbon-free character of nuclear power is one of the main elements used by those that support the use of nuclear energy for electricity generation, particularly in Europe. However, in the case of the USA, the cost-free character of nuclear power does not appear to be one of the main elements that motivate the public opinion to prefer the nuclear option for electricity generation, compare with other energy sources available in the country.

In some developed countries there is an impression that nuclear power is especially unpopular among available energy sources. In these countries, it is further assumed that unless and until this unpopularity can be overcome, nuclear power will not flourish, even in the event that the use of this type of energy for electricity generation has solid grounds to support it. Then the question to be answered is the following: Why public opinion in several countries is against the use of nuclear energy for electricity generation? Some of the factors behind the loss of public confidence in some developed countries were caused directly by the industry itself. The construction times and costs of many nuclear power plants were far higher than projected. The performance of many nuclear power plants was disappointing. The nuclear accidents at Three Mile Island and Chernobyl nuclear power plants also served to exacerbate growing mistrust on the nuclear industry and its often vocal supporters within governments. This mistrust had its origin, at least in part, in the secretiveness of nuclear spokesmen within the nuclear industry in many countries. The suspicion that the nuclear industry and its supporters were able, for example, to put undue pressure on regulators further damaged their public credibility. Critics of the nuclear industry often had no apparent vested interest to do so, while the industry's responses increasingly came to be discounted – 'they would say that, would not they?' The passion which has surrounded the nuclear debate in recent years is to a considerable degree a legacy of these factors. (Grimstom, 2002)

At the same time, perceptions of the availability of alternatives energy sources for electricity generation were changing. When global fossil fuel supplies were under apparent threat (notably in the 1950s and again in the 1970s until the beginning of the 1980s), nuclear power programmes were introduced in many countries with relatively little objections, at least by today's standards. The discovery of vast reserves of gas, as well as oil, coupled with low prices and the development of the highly efficient combined cycle gas turbine by the mid-1980s, reduced the apparent need for nuclear power in many developed countries. However, this perception was not shared by some developing countries, notably India and China and, for this reason, both countries continuing with the development of important nuclear power programmes.

Many develop and developing countries have anti-nuclear movements, some of them relative small and without great influence in policy decision-makers, while others have great influence at policy level. Environmental pressure groups are already consolidated within the industrialized world and are increasingly establishing themselves in the developing world.

A number of specific explanations have been suggested for the apparent special unease felt about nuclear power in many countries. They include:

1) links to the military, both real (the development of shared facilities) and perceptual;
2) secrecy, coupled sometimes with an apparent unwillingness to give 'straight answers' (in part, perhaps, because of links to military nuclear operations in some countries, and in part because of commercial issues);
3) the historical arrogance of many in the nuclear industry, dismissing opposition, however well-founded or sincerely held, as 'irrational';
4) the apparent vested interest of many nuclear advocates, to be contrasted with the apparent altruism of opponents who, for example, are often not funded to take part in public inquiries;
5) the perceived potential for large and uncontainable accidents and other environmental and health effects, notably those associated with the management of radioactive waste;
6) the overselling of nuclear technology, especially in its early days, in particular with regard to its economics, leading to a degree of disillusionment and distrust;
7) a general disillusionment with science and technology and with the 'experts know best' attitude of mind that was more prevalent in the years immediately after the Second World War;
8) the wider decline of 'deference' towards 'authority', including, for example, politicians and regulatory bodies. (Grimstom, 2002)

Other explanations to the public opposition to the introduction of a nuclear power programme or to an extended use of nuclear energy for other peaceful purposes are the following: a) radioactivity cannot be smelled, fell or seen; this means has no color and is not cold or warm; b) radioactivity can only be detected using special equipment, available only in special nuclear facilities and institutions.

Despite concerns against the use of nuclear energy for the production of electricity coming from the anti-nuclear establishment, as well as civil society organizations in several countries, a consensus is emerging that existing and future energy needs in several countries can be satisfied without increasing the emission of CO_2 to the atmosphere if the use of nuclear power is included in the energy balance of these countries. The use of nuclear energy for electricity generation should be one of the type of energy to be included in any energy plan and energy balance in a group of countries, including developed and developing countries alike. The reasons are very simple: Nuclear energy is a proved technology, provide electricity in a sustainable manner and is a carbon-free energy, among others. However, to consider the nuclear option as real alternative it is important to overcome the following four challenges: a) costs; b) operation safety; c) proliferation risk; and d) nuclear wastes. These challenges will escalate if a significant number of new nuclear power plants are built in a growing number of countries. The effort to overcome these challenges, however, is justified only if nuclear power can potentially contribute significantly to reduce global warming in a safe manner and to supply electricity in a systematic and economic way.

LOOKING FORWARD

In its 2008 edition, the World Energy Outlook projects an 85% increase in nuclear power capacity reaching 680 GWe by 2030, supplying double today's nuclear output, in order to address climate change concerns with determination. It also projects an enormous growth in non-hydro renewables by 1 469 GWe. If this increase is not feasible or cost-effective, then the nuclear outlook may be much larger than the nuclear capacity additions foresee. Most growth in electricity demand to 2030 will occurs outside the OECD countries. Investment in power plants will rises from US$6.1 trillion to US$9.7 trillion by 2030, and investment in demand-side efficiency is expected to be US$8.7 trillion. Together, this amount is 0.55% of world GDP for the period 2010-2030. In 2050, it is expected that the use of nuclear energy for electricity generation provides 23% of power, non-hydro renewables an ambitious 33% and fossil fuels 30%, mostly with carbon capture and storage (CCS).

An important part of this foresee energy demand will be satisfied, in the non- OECD Asia region, by an increase in the electricity production generating by nuclear energy.

Electricity generation from nuclear power worldwide it is expected to increases from 2.7 trillion kWh in 2006 to 3.0 trillion kWh in 2015 and to 3.8 trillion kWh in 2030, according with the IEO 2009 reference case, as concerns about rising fossil fuel prices, energy security and greenhouse gas emissions support the development of new nuclear generating capacity. Higher capacity utilization rates have been reported for many existing nuclear facilities and it is expected that most of the older plants now operating in OECD countries and in non-OECD Eurasia will be granted extensions to their operating lives.

It is important to note that nuclear power generation in the non-OECD countries is projected to increase by 4.0% per year from 2004 to 2030. The largest increase in installed nuclear generating capacity is projected for non-OECD Asia, particularly China and India, which accounts for 68% of the total projected increase in nuclear power capacity for the non-OECD region as a whole. Of the 58 GW of additional installed nuclear generating capacity projected for non-OECD Asia between 2004 and 2030, Russia is expected to add substantial nuclear generating capacity over the mid-term projection, increasing capacity by 20 GW. (IEO, 2007) In the other hand, several OECD nations are expected to increase their nuclear capacity in the future with the Republic of Korea adding a net of 16 GW to the existing nuclear capacity, Japan 14 GW, the USA 13 GW and Canada 6 GW.

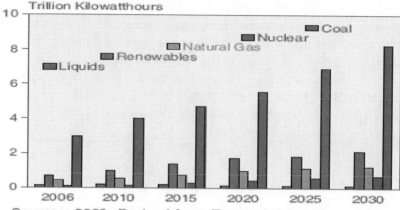

Sources: 2006: Derived from Energy Information Adminis-
tration (EIA), *International Energy Annual 2006* (June-
December 2008), web site www.eia.doe.gov/iea. Projections:
EIA, World Energy Projections Plus (2009).

Figure 18. Net electricity generation in non-OECD Asia by fuel during the period 2006-2030.

It is important to single out the following element to the positive impact that it will has in the development of the nuclear power sector all over the world: OECD members agreed in 2009 to support the export of nuclear and renewable equipment by offering better credits conditions endorsed by the respective governments. The new conditions will improve the viability of the commercial projects of the supplier's countries with the purpose to solve the growing demand of low carbon energy such as nuclear and renewable energies. The OECD describes it as a gentlemen agreement that establishes very favorable conditions for the participants in the programme. It is necessary to keep in mind that since 1984 the conditions for the export of goods and services for nuclear projects are almost the same. The new conditions are: 1) to prolong the payment of the loans from 15 years to 18 years; 2) to allow other options for a more flexible payment and mortgages; 3) to reduce the interest rates; 4) to increase from 15% to 30% the financing of national industry; and 5) to allow the financing of the interests to be paid during the construction.

It is expected that the new conditions will reduce between 15% and 25%, according to the technologies used, the cost of the electricity to be generated in the importers countries. For a nuclear power plant of about 1 500 MW, these changes will reduce the cost of the plant in almost 25%, from US$118.2 per MWh to about US$89.5.

Nuclear power is under serious consideration in several countries without a nuclear power programme. Thirty-one countries are already using nuclear energy for electricity generation and forty-one countries not currently using this type of energy are considering the introduction of a nuclear energy programme in the coming years. These countries are the following:

a) *in Europe*: Italy, Albania, Portugal, Norway, Poland, Belarus, Estonia, Latvia, Ireland and Turkey;

b) *in the Middle East and North Africa*: Iran, Gulf States, Yemen, Israel, Syria, Jordan, Egypt, Tunisia, Libya, Algeria and Morocco;

c) *in Central and Southern Africa*: Nigeria, Ghana and Namibia;

d) *in South America*: Chile, Ecuador, Peru, Uruguay and Venezuela[9];

e) *in Central and Southern Asia*: Azerbaijan, Georgia, Kazakhstan, Mongolia and Bangladesh;

f) *in South Eastern Asia*: Indonesia, Philippines, Vietnam, Thailand, Malaysia, Australia and New Zealand.

In all of the above countries, governments need to create the necessary environment to facilitate investment in nuclear power, including a regulatory regime, available sufficient experience and well-trained professional and technicians, adequate infrastructure and the adoption of policies on nuclear waste management, decommissioning and non-proliferation. The public opinion is an important factor to be taken into account when considering the introduction of a nuclear power programme in all of these countries.

It is important to single out the following: The institutional arrangements to be created vary from country to country. Usually governments in the industrialized world are heavily involved in planning the introduction of different types of energy in the balance mix, including nuclear energy and, in some cases, in the financing of the construction of nuclear power plants. In developing countries, governments are normally involved in all phases associated with the introduction of nuclear power programmes from planning till the operation of nuclear power plants.

In countries, in which nuclear energy for electricity generation is introduced for the first time there is always a lack of nuclear engineers and other scientists, professional and technicians duly prepared and, for this reason, the way in which the construction of the nuclear power plants are carried out is often on a turnkey basis. In this case, the supplier of the nuclear power reactors assume all technical and commercial risks in delivering a functioning plant on time and within the budget approved, or as an alternative, set up a consortium to build, own and operate the plant.

TYPES OF FISSION NUCLEAR POWER REACTORS[10]

Nuclear reactors are devices designed to produce and maintain a controlled chain nuclear reaction. There are two different types of nuclear reactors differentiated by their purpose and

---

[9] The situation of Peru and Ecuador regarding the introduction of a nuclear power programme in the future is less clear than the situation involving the other three countries.

[10] Fission occurs when a nucleus absorbs a neutron and splits it into two approximately equal parts, known as fission fragments, and ejects several high-velocity fast neutrons in the process. The reactors that use fission to produce heating are called fission nuclear power reactors. The fission process concerns only heavy nuclides. It could be spontaneous or a result of nuclear reaction (neutron-induced, or other light or heavier particles-induced). The most important for reactor applications is of course primarily neutron-induced fission reactions and, to less extend spontaneous fission. The international community is also developing the so-called "fusion nuclear power reactors". Fusion is the process by which two light atomic nuclei combine to form a heavier one. The long-term objective of fusion research is to harness this process to help meet mankind's future energy needs. It has the potential to deliver large-scale, environmentally benign, safe energy, with abundant and widely available fuel resources. No commercial fusion nuclear power reactors has been produced until today and it is expected, according with the results of the ongoing research in fusion activities, that there will be any of this type of reactors available in the market before 2050.

by their design features. Considering its purpose they can be classified in two groups: a) Nuclear research reactors, and b) Nuclear power reactors.

Nuclear power reactors are those devices found in nuclear power plants and are used for generating heat mainly for electricity production. However, this type of reactors can be used also for desalination of water and heating. In the form of smaller units, they also power ships.

There are many different types of nuclear power reactors but what is common to all of them is that they produce thermal energy that can be used for its own sake or converted into mechanical energy and ultimately, in the vast majority of cases, into electrical energy. In this type of reactors, the fission of heavy atomic nuclei, the most common of which is uranium-235, produces heat that is transferred to a fluid which acts as a coolant. The heated fluid can be gas, water or a liquid metal. The heat stored by the fluid is then used either directly (in the case of gas) or indirectly (in the case of water and liquid metals) to generate steam. The heated gas or the steam is then fed into a turbine driving an alternator, which produce the electricity.

Nuclear research reactors are devices that operate at universities and research institutions in many countries, including in countries where no nuclear power reactors are currently in operation. This type of reactors are used for multiple purposes, including the production of radiopharmaceuticals, medical diagnosis and therapy, testing materials and conducting basic research.

Nuclear power reactors can be classified according to the type of fuel they use to generate heat. These are: a) uranium-fuelled nuclear power reactors; and b) plutonium-fuelled nuclear power reactors.

Uranium–fuelled Nuclear Power Reactors

Uranium–fuelled nuclear power reactors are classified in: a) Pressurized water reactors (PWR), including the pressurized heavy water reactor (PHWR); b) Boiling water reactors (BWR); and c) Graphite-moderate, gas-cooled nuclear power reactors.

The only natural element currently used for nuclear fission as fuel is uranium. In the case of the PWRs, the fuel used is dioxide of uranium. In the case of the PHWRs the fuel used is the so-called "enriched uranium". Natural uranium is a highly energetic substance: one kilogram of uranium can generate as much energy as 10 tons of oil.

It is common practice to classify nuclear power reactors according to the nature of the coolant and the moderator plus, as the need may arise, other design characteristics. The light water reactors category comprises PWR[11] and BWR. Both types of nuclear power reactors use light water as moderator and coolant and enriched uranium as fuel.

The light water reactors operate in the following manner: The light water flows through the nuclear reactor core, a zone containing a large array of nuclear fuel rods where it picks up the heat generated by the fission of the uranium-235 present in the fuel rods. After the coolant has transferred the heat it has collected to a steam turbine, it is sent back to the reactor core, thus flowing in a loop called the primary circuit. In order to transfer high-quality thermal energy to the turbine, it is necessary to reach temperatures of about 300 °C. It is the pressure

---

[11] WWER in the case of the PWR produced in the former Soviet Union, now Russia.

at which the coolant flows through the reactor core that makes the distinction between PWRs and BWRs.

In PWRs, the pressure imparted to the coolant is sufficiently high to prevent it from boiling. The heat drawn from the fuel is transferred to the water of a secondary circuit through heat exchangers. The water of the secondary circuit is transformed into steam, which is fed into a turbine. The fission zone (fuel elements) is contained in a reactor pressure vessel under a pressure of 150 to 160 bar (15 to 16 MPa). The primary circuit connects the reactor pressure vessel to heat exchangers. The secondary circuit side of these heat exchangers is at a pressure of about 60 bar (6 MPa) - low enough to allow the secondary water to boil. The heat exchangers are, therefore, actually steam generators. Via the secondary circuit, the steam is routed to a turbine driving an alternator, which produces the electricity. The steam coming out of the turbine is converted back into water by a condenser after having delivered a large amount of its energy to the turbine. It then returns to the steam generator. As the water driving the turbine (secondary circuit) is physically separated from the water used as reactor coolant (primary circuit), the turbine-alternator set can be housed in a turbine hall outside the reactor building.

In BWRs, the pressure imparted to the coolant is sufficiently lower than in a PWR to allow it to boil. It is the steam resulting from this process that is fed into the turbine. This basic difference between pressurized and boiling water reactors dictates many of the design characteristics of the two types of light water reactors. Despite their differing designs, it must be noted that the two types of reactors provide an equivalent level of safety.

The heavy water reactors technology used in the PHWR was initially developed by the Atomic Energy of Canada Limited's (AECL's) from Canada and by Siemens and Kraftwerk Union (KWU) from Germany. In the first case the type of reactor produced is the so-called "CANDU" reactor. In the second case the reactor produced is the MZFR reactor[12], the first one built in the Karlsruhe Nuclear Research Center in Germany with a capacity of 65 MW. The MZFR was the type of reactor used for the construction of the first nuclear power reactor in Argentina (Atucha 1) in 1968. It has a pressure vessel, unlike any other existing heavy water reactor, and it now uses slightly enriched (0.85%) uranium fuel, which has doubled the burn-up and consequently reduced operating costs by 40%.

Now AECL is producing the Advanced CANDU reactor (ACR) design using slightly enriched uranium fuel to reduce the reactor core size, which at the same time reduces the amount of heavy water required to moderate the reactor and allows light water to be used as a coolant.

In 2005 and 2006, India connected the first two units using its new 540 MWe PHWR design at Tarapur. India is also designing an evolutionary PHWR with a capacity of 700 MWe and developing the Advanced Heavy Water Reactor (AHWR), a heavy water moderated boiling light water cooled, vertical pressure tube type reactor which has passive safety systems and optimized to use thorium as fuel. However, no AHWR units have been yet built in India.

The PWR is the most common nuclear power reactors operating in the world. There were 265 PWRs in operation all over the world in 2007 with a total net capacity of 243 429 MWe. The load factor of the PWRs in that year was 83.4% (first place). The USA (69 units) and

---

[12] Multipurpose research reactor (Mehrzweckforschungreaktor) built by the Karlsruhe Research Center in Germany.

France (58 units) are the countries with the highest number of PWRs in operation in the world. The main components of the PWRs are shown in Figure 19.

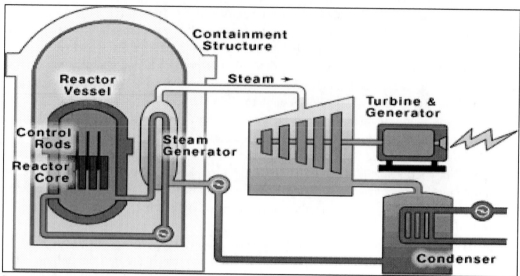

Source: International Nuclear Safety Center at Argonne National Laboratory, USA.

Figure 19. Pressurized water reactors (PWRs).

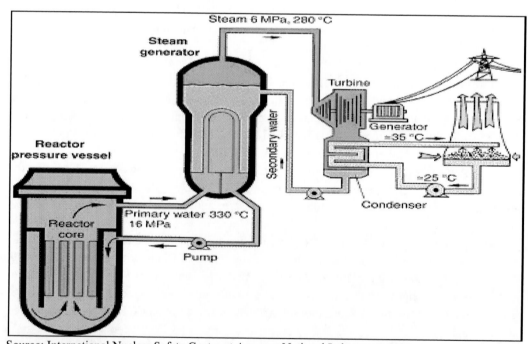

Source: International Nuclear Safety Center at Argonne National Laboratory, USA.

Figure 20. PWRs connected to an electrical grid.

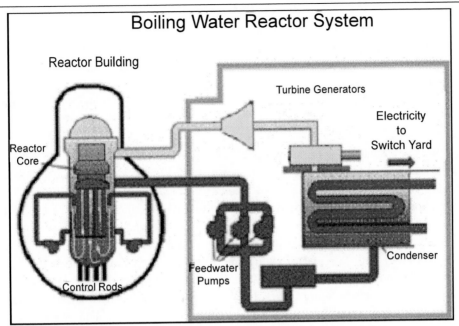

Source: International Nuclear Safety Center at Argonne National Laboratory, USA.

Figure 21. Boiling water reactor components.

In France and Germany, Areva NP has developed a new large pressurized water reactor called the "European Pressurized Water Reactor (EPR)" to meet European utility requirements and benefit from economies of scale through a higher power level relative to the latest series of PWRs in France (the N4 series) and Germany (the Konvoi series). (IAEA GC (51)/INF/3, 2007) The USA is also working in a design for a large advanced PWR type, the so-called "Combustion Engineering System 80+" with the purpose to build several units in the country and abroad.

In the Russian Federation, evolutionary versions of the current WWER-1000 (V-320) reactor, the Russian version of the Western PWR type, including the 1 200 MWe AES-2000 and WWER-1000 (V-392) designs. The first WWER-1000 was connected to the grid at Tianwan, China in 2006. Additional units are under construction in China, India and the Islamic Republic of Iran. Two units are planned at Russia's Novovoronezh site. Russia has also begun development of a larger WWER-1500 design. In July 2009, Russia and Kazakhstan created a joint venture to complete the design of a 200 - 400 MWe VBER-300 reactor for use in either floating or land-based co-generation plants. (IAEA GC (51)/INF/3, 2007)

In the case of the PHWR type, there were 44 units in operation around the world in 2007. The load factor of the PHWRs in that year was 77.5% (second place). In 2007, the net capacity installed of PHWRs was 22 358 MWe. Canada (18 units) and India (15 units) are the countries with the highest number of PHWRs in operation in the world.

The fission zone of the BWRs is contained in a reactor pressure vessel, at a pressure of about 70 bar (7 MPa). At the temperature reached 290 °C approximately, the water starts boiling and the resulting steam is produced directly in the reactor pressure vessel. After the separation of steam and water in the upper part of the reactor pressure vessel, the steam is routed directly to a turbine driving an alternator which produces the electricity. Since the steam produced in the fission zone is slightly radioactive, mainly due to short-lived activation products, the turbine is housed in the same reinforced building as the reactor.

In 2007, there were 94 BWRs in operation in 9 countries all over the world. The net capacity installed at that time was 85 267 MWe. The load factor was 75.3% (fourth place). The USA (35 units) and Japan (32 units) were the two countries with the highest number of BWRs in operation in the world in 2007. The main components of the BWR are shown in Figure 21.

In Germany Areva NP, with international partners from Finland, France, the Netherlands and Switzerland, is developing the basic design of the SWR-1000, an advanced BWR with passive safety features. In Japan, advanced boiling water reactor (ABWR) design benefit from standardization and construction in series. The first two ABWRs began commercial operation in 1996 and 1997, and two more began commercial operation in 2005 and 2006. Additional two ABWRs are under construction in Taiwan. A development programme was started in 1991 for ABWR-II with the goal of significantly reducing generation costs, partly through increased power and economies of scale. Commissioning of the first ABWR-II is foreseen for the late 2010s. In the USA, a large BWR (General Electric's ABWR) was certified in 1997. Westinghouse's AP-600 and AP-1000 designs with passive safety systems were certified in 1999 and 2006 respectively. An international team led by Westinghouse is developing the modular integral 360 MWe "International Reactor Innovative and Secure (IRIS)" with a core design capable of operating on a four-year fuel cycle. Design certification is targeted for 2008-2010. General Electric is designing a large European simplified boiling water reactor (ESBWR) combining economies of scale with modular passive safety systems. Both the IRIS and the ESBWR are currently subject to regulatory review. (IAEA GC (51)/INF/3, 2007)

Other light water reactors in the market are the "Korean Standard Nuclear Plant (KSNP)" series, the Chinese AC-600 design and the CNP-1 000 for electricity production. China is also developing the QS-600 for electricity production and seawater desalination. Until 2008, eight KSNPs are in commercial operation. Based on the accumulated experience in the operation of the KSNPs the Republic of Korea is now developing an improved KSNP, the so-called "Optimized Power Reactor" (OPR), with the first units planned for commercial operation in 2010 and 2011. The Korean next generation of nuclear power reactor, for which development began in 1992, is now named the "Advanced Power Reactor 1400 (APR-1 400)" and will be bigger to benefit from economies of scale. The first APR-1400 is scheduled to begin operation in 2012.

According with IAEA and CEA information, there were 18 operating gas cooled reactors (GCR) cooled by carbon dioxide plus two test reactors cooled by helium worldwide in 2007. All of the 18 GCR units are located in the UK with a net capacity of 9 034 MWe. In China, work continues on safety tests and design improvements for the 10 MWth "High Temperature Gas Cooled Reactor (HTR-10)", and plans are in place for the design and construction of a power reactor prototype (HTR-PM).

Other type of nuclear power reactors under development are the following. The South African Pebble Bed Modular Reactor Company (Pty) Ltd is developing a 165 MWe "Pebble Bed Modular Reactor (PBMR)", which is expected to be commissioned around 2010. The South African government has allocated initial funding for the project and orders for some lead components have already been made. In Japan, a 30 MWth "High Temperature Engineering Test Reactor (HTTR)" began operation in 1998, and work continues on safety testing and coupling to a hydrogen production unit. A 300 MWe power reactor prototype is also under consideration. The Russian Federation and the USA continue research and development on a 284 MWe "Gas Turbine Modular Helium Reactor (GT-MHR)" for plutonium burning. France has an active research and development programme on both thermal as well as fast gas reactor concepts and, in the USA, efforts by the Department of Energy (DOE) continue on the qualification of advanced gas reactor fuel. To demonstrate key technological aspects of Gas Cooled Fast Reactors, an experimental reactor in the 50 MWth range is planned for operation around 2017 in France. (IAEA GC (51)/INF/3, 2007)

Graphite-Moderated and Gas-Cooled nuclear power reactors, formerly operated in France and still operated in the UK, are not built any more in spite of some advantages that this type of reactors have. RBMK-reactors (Pressure-tube Boiling-Water Reactors, LWGR), which are cooled with light water and moderated with graphite, are now less commonly operating in some former Soviet Union bloc countries. In Russia, there were 15 RBMK in operation in 2007 and one under construction. In Lithuania, one RMBK was still in operation in 2007. Following the Chernobyl nuclear accident the construction of this type of reactors outside the former Soviet Union and now Russia ceased.

Plutonium-fuelled Nuclear Power Reactors

Plutonium (Pu) is an artificial element produced in uranium-fuelled nuclear power reactors as a by-product of the chain reaction. It is one hundred times more energetic than natural uranium: one gram of Pu can generate as much energy as one tone of oil. As it needs fast neutrons in order to fission, moderating materials must be avoided to sustain the chain reaction in the best conditions. The current plutonium-fuelled nuclear power reactors, also called "Fast Breeder Reactors", use liquid sodium which displays excellent thermal properties without adversely affecting the chain reaction. This type of reactors is in operation in France and Russia.

According with IAEA information, in China, the 25 MWe sodium cooled pool type Chinese Experimental Fast Reactor is under construction, with a grid connection expected by the middle of 2010. The next two stages of development will be a 600 MWe prototype of fast breeder reactor, for which design work started in 2005, and a 1 000 - 1 500 MWe demonstration fast breeder reactor. In India, the 500 MWe "Prototype Fast Breeder Reactor (PFBR)" is now under construction at Kalpakkam. It is scheduled for commissioning by September 2010.

In the Republic of Korea, the Korea Atomic Energy Research Institute has conducted research, technology development and design work on the 600 MWe KALIMER-600 advanced fast breeder reactor concept. The conceptual design was completed in 2006. BN-600 in Russia is the world's largest operating fast breeder reactor and has now been in operation for 26 years. The 800 MWe BN-800 is under construction with commissioning planned for 2012. Russia is also developing various concepts for advanced sodium cooled fast breeder reactors and for heavy liquid metal cooled reactors, specifically the "Lead Cooled BREST-OD-300" and the "Lead Bismuth Eutectic Cooled SVBR-75/100 systems".

In the USA, within the framework of the Global Nuclear Energy Partnership (GNEP), initial research and developing planning is underway for an "Advanced Burner Test Reactor (ABTR)" to demonstrate actinide transmutation in a fast spectrum, as well as innovative technologies and design features important for subsequent commercial demonstration plants.

Within the GIF framework, USA activities are focused on GFRs), LFRs), and "Small Modular Sodium Cooled Fast Reactors (SMFRs)". (IAEA GC (51)/INF/3, 2007)

The future fission nuclear power reactors are expected to have the following advantages over the present generation: a) lower investment costs and construction times; b) simpler designs; c) modular units; d) more passive safety features; and e) low proliferation risks.

For the time being, fission technology will be the main nuclear technology used for the production of the new nuclear power reactors that are going to be built in the world at least until 2040.

THE NEXT GENERATION OF NUCLEAR POWER REACTORS

The majority of the nuclear power reactors today in operation in the world are from the second generation of nuclear power reactors (the so-called "Generation II"). This generation of nuclear power reactors began to be built in the 1970s and is still operating in large commercial power plants in several countries. Most of the countries expanding their nuclear power programmes are constructing nuclear power reactors of the third generation (the so called "Generation III"), which are more reliable and with a number of built-in safety features. Generation III of nuclear power reactors was developed in the 1990s with a number of evolutionary designs that offer significant advances in safety and economics. Advances to Generation III are underway, the so-called "Generation III+", resulting in several near-term deployable plants that are actively under development and are being considered for deployment in several countries. New nuclear power reactors to be built between now and 2030 will likely be chosen using this type of reactors.

It is important to note that there is no clear definition of what constitutes a Generation III design, apart from it being designed in the last 15 years. However, the main common features quoted by the nuclear industry are: a) a standardized design to expedite licensing, reduce capital cost and construction time; b) a simpler and more rugged design, making them easier to operate and less vulnerable to operational upsets; c) higher availability and longer operating life, typically 60 years; d) reduced possibility of core melt accidents; e) minimal effect on the environment; f) higher burn-up to reduce fuel use and the amount of waste; and g) burnable absorbers ('poisons') to extend fuel life.

These characteristics are clearly very imprecise and do not define very well what a Generation III reactor is. Generation III reactors are evolved from existing designs of PWR, BWR and CANDU. Until there is much more experience with the use of these technologies, any figures on power generation cost from these designs should be treated with the utmost caution. (Thomas, 2005)

However, the future belongs to the fourth generation of nuclear power reactors (the so-called "Generation IV"). This new generation of nuclear power reactors is a revolutionary type of reactors with innovative fuel cycle technologies. Why a new generation of nuclear power reactors is needed? The answer is the following: Generation IV initiative is the recognition that maintaining global nuclear capacity at its current level of roughly 400 GWe will be insufficient to reduce and stabilize CO_2 emissions to the atmosphere in the longer term, taking into account a foresee increase in the energy demand all over the world and the impossibility to a significantly increase in the use of renewables energy for electricity production. For this reason, the international community needs new of nuclear power reactors which could deliver the highest power capacity in a manner which would be regarded as long term sustainable.

According with reference 6, it is expected that Generation IV reactors may be available for commercial application around 2030-2040. Generation IV systems will respond to the following main sustainability criteria and future market conditions:

1) *Sustainability:* Generation IV systems will provide sustainable energy generation that meets clean air objectives and promotes long-term availability of energy. This system will has effective fuel utilization for worldwide energy production and will minimize and manage their nuclear waste with the purpose to reduce the long-term stewardship burden, thereby improving protection for the public health and the environment.

2) *Economic competiveness:* Generation IV systems will have a clear life-cycle cost advantage over other energy sources and will have a level of financial risk comparable to other energy projects.

3) *Safety and reliability:* Generation IV systems operations will excel in safety and reliability, will have a very low likelihood and degree of reactor core damage and will eliminate the need for off-site emergency response.

4) *Proliferation resistance and physical protection:* Generation IV systems will increase the assurance that they are a very unattractive and the least desirable route for diversion or theft of weapons-usable materials and will provide increased physical protection against acts of terrorism.

The following are the designs of Generation IV systems already selected on the basis of the set of criteria that have been established:

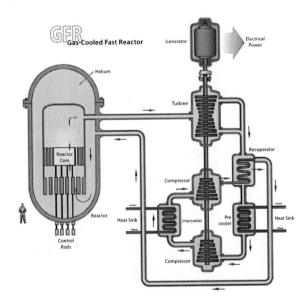

Source: Reference 6.

Figure 22. Gas-Cooled Fast Reactor System (GFR).

a) Sodium Cooled Fast Reactor (SFR).
b) Very High Temperature Gas Reactor (VHTR).
c) Super Critical Water Cooled Reactor (SCWR).
d) Lead Cooled Fast Reactor (LFR).
e) Gas Cooled Fast Reactor (GFR).
f) Molten Salt Reactor (MSR).

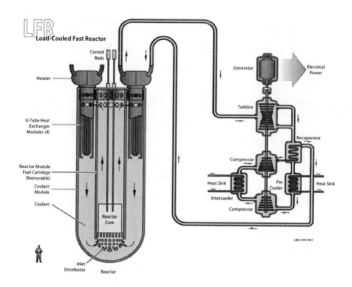

Source: Reference 6.

Figure 23. Lead-Cooled Fast Reactor System (LFR).

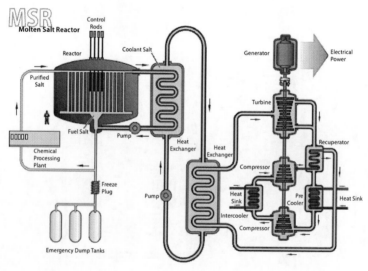

Source: Reference 6.

Figure 24. Molten Salt Reactor System (MSR).

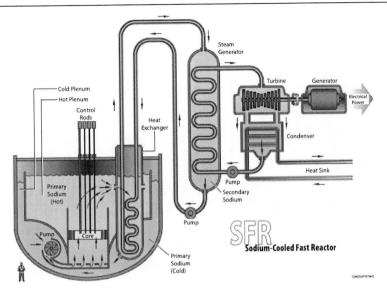

Source: Reference 6.

Figure 25. Sodium-Cooled Fast Reactor System (SFR).

The above six Generation-IV system designs are very different and also present different challenges that need to be solved in the ongoing research and development programmes in order to have all, or at least some of them, available in the market as soon as possible. It is expected that some of the Generation-IV designs could be available in the market not before 2030. Others designs may still need significant additional research and development work before they can be considered ready for the production electricity. The design of the different Generation IV systems is shown in Figures 22-27.

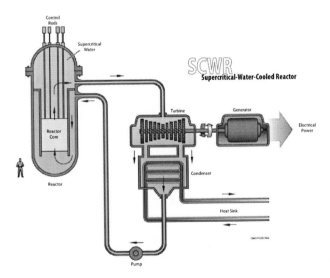

Source: Reference 6.

Figure 26. Supercritical-Water-Cooled Reactor System (SCWR).

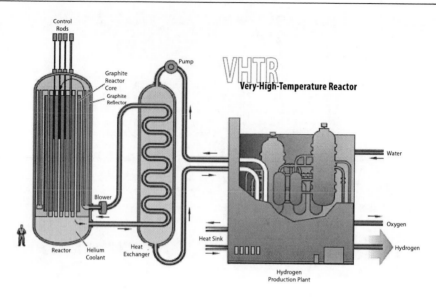

Source: Reference 6.

Figure 27. Very-High-Temperature Reactor System (VTHR).

Chapter 2

SOME RELEVANT ELEMENTS ASSOCIATED WITH THE INTRODUCTION OR THE EXPANSION OF A NUCLEAR POWER PROGRAMME

A reliable and adequate supply of energy, especially in the form of electricity, is an indispensable element for the economic and social development of any country. Based on this undisputed fact, providing a safe reliable energy in economically acceptable way is an essential political, economic and social requirement. A country cannot growth, from an economic and social point of view, if the energy system, particularly the electricity generating system, is not capable to provide the energy that the country need.

However, it is important to note that the use of nuclear energy is limited to a small number of countries in the world. In 2009, only 30 countries, or 16% of the 192 United Nations Member States, operate nuclear power plants. Six of them, USA, France, Japan, Germany, Russia and the Republic of Korea, produce almost three quarters of the nuclear electricity in the world. At the same time, half of the world's nuclear countries are located in Western and Central Europe and count for over one third of the world's nuclear production.

When deciding on how to expand the electricity generating system, the government and the private and public energy industry of any country would have to carry out comprehensive assessments of all the energy options available in the country in order to have the best possible energy balance. The reasons for choosing a specific option will differ from country to country, depending on local and regional energy resources, technological capabilities, availability of finance and qualified personnel, environmental considerations and the country's overall energy policy. (IAEA STI/PUB/1050, 1998)

Governments may have different reasons for considering the introduction of a nuclear power programme to achieve their national energy needs. These reasons could be: a) a lack of available indigenous energy resources; b) the desire to reduce dependence upon imported energy, particularly if the imported energy is done from unstable political regions; c) the need to increase the diversity of energy resources and/or the mitigation of carbon emission increases with the purpose to reduce the CO_2 emission to the atmosphere, among others.

It is important to note that if a country is considering the use of nuclear energy for electricity generation, then it should have a well defined energy policy and a plan for the whole energy system and, in particular, the electricity supply system. The plan should be prepared with the objective to achieve an overall energy optimization that includes not only a secure supply of different energy sources, including the import of energy needs, but also environmental considerations, cost associated, the management of nuclear waste, energy conservation and efficiency improvements, among others.

According with the IAEA, when a country is studying the introduction of a nuclear power programme, the following basic criteria are fundamental to be considered:

1) nuclear power should be considered only when it is technically feasible and when it would be part of an economically viable long-term energy and electricity supply expansion strategy, considering all alternatives and relevant factors;

2) a nuclear power programme should be launched only when it - and in particular the first project - has a definite likelihood of being successful, i.e. it can be executed within the planned schedule and predicted financial limits and can be operated safely and reliably once in service;

3) a nuclear power project should be finally committed only on the basis of comprehensive planning and after steps have been taken to meet all necessary supporting infrastructure requirements, including assurance of financing. (IAEA STI/PUB/1050, 1998)

Other issues that must be considered in the framework of the introduction of a nuclear power programme are the following:

1) the economic competiveness of the use of nuclear energy for the production of electricity in comparison with other energy sources available in the country;

2) the adoption of all necessary laws and regulations to ensure the successful introduction of a nuclear power programme;

3) the safety aspects related with the licensing of the nuclear power plant and the development of a safety culture;

4) the size of the electricity grids[13];

5) proliferation considerations;

6) environmental impact;

7) the cost involved in the construction of a nuclear power plant;

8) the need for trained personnel and how this training should be provided and by whom;

9) the technological capability to assimilate an advanced and demanding technology;

10) the safe management of the nuclear waste, particularly high-level nuclear waste;

---

[13] It is important to take into account that normally no single nuclear power reactor should account for more than 10% of the installed capacity of the entire electricity network of the country. According with IAEA recommendations, countries expanding or thinking to introduce a nuclear power programme are advised to consider their electric grids as part of their planning process, particularly as the grid impacts the size and type of nuclear power reactor that can be deployed. Specific issues that should be considered in the early phases of a nuclear power programme include grid capacity and future growth, historical stability and reliability, and the potential for local and regional interconnections. Assessment of the current electric grid and plans for improving it should, therefore, be developed to be consistent with plans regarding the use of nuclear energy for electricity generation.

11) the need to gain public acceptance;

12) the international and regional cooperation in the field of nuclear technology.

The introduction of a nuclear power programme requires the establishment of a basic infrastructure to deal with all aspects of a nuclear power project, particularly for the construction of the first nuclear power reactor. The creation of an appropriate infrastructure, particularly the establishment of the nuclear regulatory authority, should start immediately after the adoption by the government of the decision to introduce a nuclear power programme and should be in place well in advance of the initiation of the construction work of the first nuclear power reactor. The different stages of the development of the basic infrastructure include:

1) development of nuclear power policy and its formal adoption by the government;
2) confirmation of the feasibility of implementing a nuclear power project;
3) establishment of institutional components of the infrastructure, including a nuclear regulatory authority;
4) establishment of physical component of the infrastructure;
5) development, contracting and financing of the first nuclear power project;
6) construction of the first nuclear power reactor according with the established safety, quality and economic requirements;
7) safe, secure and efficient operation of the first nuclear power plant.

Government and utility funding is required to establish major components of the basic nuclear infrastructure. These may include:

a) national legal framework and international agreements;
b) nuclear safety and environmental regulatory authority;
c) physical facilities;
d) finance/ economics;
e) human resources, education and training;
f) operational practices and processes to assure safety and performance throughout the life of the nuclear power plant;
g) public information and acceptance. (IAEA-TECDOC-1513, 2006)

The introduction of a nuclear power programme in any country should be viewed within a medium to long-term electricity supply strategy adopted by the government to develop the country energy sector. Different tasks should be implemented with the purpose to ensure the successful introduction of a nuclear power programme. According with the IAEA, theses tasks should:

1) develop a comprehensive nuclear legal framework covering all aspects of the peaceful uses of nuclear energy, i.e. safety, security, safeguards and liability, in addition to the commercial aspects related to the use of nuclear material;
2) establish and maintain an effective regulatory system;

3) develop the human resources for the state organizations and also for the operating organizations required to effectively supervise and implement the introduction of the nuclear power programme adopted;

4) ensure adequate financial resources for the construction, sustained safe operation and decommissioning of the nuclear power reactor, as well as for the long term manage of nuclear materials;

5) communicate in an open and transparent manner with the public and the neighboring states about the considerations behind the introduction of a nuclear power programme. (GOV/INF/2007/2, 2007)

A nuclear power programme should be implemented with the purpose to produce stability in power generation and electricity price and to achieve an important impact in the development of the domestic industry. The potential economic benefits of the introduction of a nuclear power programme in a given country are reflected in maintaining a long-term stability of electricity prices. An important consideration in many developing countries has been the influence of a national nuclear power programme in increasing the technological level of the country and enhancing the global competitiveness of the domestic industry. (IAEA STI/PUB/1050, 1998)

One of the main elements that need to be considered to ensure the successful introduction of a nuclear energy programme in a given country, particularly in a developing country, is the level of participation of the domestic or national industry within the programme. The participation of national industry in the development of a nuclear power programme could be materialized in one or more of the following manners:

1) local labor and some construction materials could be used for non-specialized purposes on-site, especially civil engineering works associated with the construction of the nuclear power plant;

2) local contractors could take full or partial responsibility for the civil engineering work related with the construction of the nuclear power plant, including some design work assigned by the main constructor;

3) locally manufactured components from existing national factories could be used for non-critical parts of the nuclear power plant;

4) local manufacturers could extend their normal product line to incorporate nuclear designs and standards, possibly under licensing arrangements with foreign suppliers;

5) factories could manufacture heavy and specialized specific nuclear components, possibly under licensing arrangements with foreign suppliers.

However, the economic viability of such undertakings would have to be assessed carefully in view of the future domestic market and the availability of such equipment internationally with a better price and quality. (IAEA STI/PUB/1050, 1998)

Well-designed, constructed and operated nuclear power plants have proved to be a reliable, economic under certain conditions, environmentally acceptable and safe source of electrical energy all over the world, including in Africa. The economic competiveness of nuclear power arises from cost reductions in construction and plant operations. Still further reduction can be achieved in costs associated with waste management and decommissioning. Construction costs per kW for nuclear power plants have fallen considerably due to standardized design, shorter construction times and more efficient generating technologies. Further gains are expected as nuclear technology becomes even more standardized around a few globally-accepted designs. Meanwhile, recent new-build experience has demonstrated that new nuclear power plants can be built on time and on budget. Financing costs for new nuclear power plants, a critical component of nuclear economics, are expected to fall as new approaches are developed and tested to increase certainty and to lower investor risk. [132]

It is important to single out that in many countries, licensing procedures are being streamlined with the purpose to increase the operational safety of the new nuclear power plants and to reduce the possibility of a nuclear accident. Streamlined licensing will retain rigorous standards but reduce regulatory cost and uncertainty by establishing predictable technical parameters and timescales, from design certification through to construction and operating licenses. Operating costs of nuclear power plants have fallen steadily over the past twenty years as capacity factors have increased, squeezing far more output from the same generating capacity[14]. As marginal costs of electric generation from nuclear plants have fallen below prices of most other generating modes, owners have found it worthwhile to invest in nuclear power plant refurbishment and capacity up-rates. Nuclear power's low marginal cost and its high degree of price stability and predictability have also encouraged nuclear power plant owners to seek operating license extensions for nearly all operating reactors. Waste and decommissioning costs, which are included in the operational costs of nuclear power plants, represent a tiny fraction of the lifetime costs of a reactor's operation. Nuclear power plant economics are thus largely insensitive to these costs and will become even less so as fuel efficiency continues to increase and as waste and decommissioning costs are spread over reactor lifetimes that are becoming even longer. [132]

With regard the increase of fuel efficiency it is important to note that up-rating the power output of current nuclear power reactors in operation is a highly economic source of additional generating capacity. According with WANO opinion, the refurbishment of the plant turbo generator combined with utilizing the benefits of initial margins in reactor designs and digital instrumentation and control technologies can increase plant output significantly by up to 15-20%.

One of the most important decisions that need to be adopted by the national competent authority in a given country after the approval of the introduction of a nuclear power programme is the type of nuclear power reactor that should be selected. There are two main groups of nuclear power reactors available in the market: a) the PWRs, including the PHWRs, and b) the BWRs. Most of the nuclear power reactors available in the market for purchase and construction are of the PWRs type. There were 265 PWRs operating in 24 countries all over the world in 2008 with a net capacity of 243 429 MWe. In the case of BWRs, there were 85 BWRs operating in 10 countries in 2008 with a net capacity of 85 267 MWe. Finally, there were 44 PHWRs operating in seven countries in 2008 with a net capacity of 22 358 MWe.

The above mentioned type of reactors are generally available in sizes of about 1 000 MW or greater electrical output. However, smaller reactors of 600–700 MW output are also available. If a smaller unit is required due to the capacity of the national electrical grid network, then the available technology is limited, although reactors of 200–400 MW output are being operated and developed in some countries. Several designs are being developed for

---

[14] In the USA, operating costs per kWh shrank by 44% between 1990 and 2003.

future applications although a major challenge is to achieve an economic design at a smaller size.

It is important to stress that the High Temperature Gas Cooled Reactor with a net capacity between 160 and 270 MW and several small water cooled reactors are being developed which may reach design approval over the next ten years. In addition, a barge mounted moveable 70 MW output plant is currently under construction. (GOV/INF/2007/2, 2007) It is important to stress that the selection of the type of nuclear power reactor to be built will define the nuclear fuel cycle to be used.

The decommissioning of nuclear power plants means the removal of some or all of the regulatory controls that apply to a nuclear site whilst securing the long-term safety of the public and the environment. Underlying this there are other practical objectives to be achieved, including release of valuable assets such as site and buildings for unrestricted alternative use, recycling and reuse of materials and the restoration of environmental amenity. In all cases, the basic objective is to achieve an endpoint that is sensible in technical, social and financial terms, that properly protects workers, the public and the environment and, in summary, complies with the basic principles of sustainable development. [131] In other words, the term decommissioning covers all of the administrative and technical actions associated with cessation of operation and withdrawal from service. It starts when a facility is shut down and extends to eventual removal of the facility from its site (termed dismantling). These actions may involve some or all of the activities associated with dismantling of plant and equipment, decontamination of structures and components, remediation of contaminated ground and disposal of the resulting wastes.

There are three main decommissioning strategies that can be applied. These are the following: a) immediate dismantling; b) deferred dismantling, also called safe enclosure; and c) entombment. In the first case, a facility is dismantled right after the removal of materials and waste from the facility. In the second case, after the removal of materials and waste, the facility is kept in a state of safe enclosure for 30-100 years followed by dismantling. In the third case, a facility is encapsulated on site and kept isolated until the radionuclides decayed to levels that allow a release from nuclear regulatory control. The present trend is in favor of immediate dismantling. [120]

The selection of the correct decommissioning strategy depends of several factors that can be grouped into the following three categories: a) policy and socio-economic factors; b) technological and operational factors; and c) long-term uncertainties. Policy and socioeconomic factors are dominated by the national and/or the local situation, which varies from country to country. Countries with important nuclear power programmes tend to dismantle obsolete nuclear power plants immediately in order to use the sites for the construction of new facilities. Decommissioning costs associated to a nuclear power reactor include, among other elements, the following components: a) dismantling the nuclear power plant; b) waste treatment; c) disposal of all types of radioactive waste; d) security; e) site cleanup; and f) project management.

Dismantling and disposal represents a major share, each accounting for approximately 30% of the total decommissioning cost. The average cost estimates are in the range of US$320 to US$420/kWe for most nuclear power reactor types. In general, GCR are more expensive to decommission than water-cooled reactors, because they must dispose of large quantities of graphite. The cost for dismantling the older, smaller 160 MWe nuclear power reactor at Zorita nuclear power plant in Spain has recently been estimated by Union Fenosa at

€850 Euro/kW and the dismantling of the German plant Obrigheim was estimated at €1,400/kWe.

National policy may influence decommissioning strategy to be selected either directly or indirectly. If a national decommissioning policy is reflected in legislation, direct influence is exerted by way of the legal framework, and the extent of this influence depends on the degree to which laws are either prescriptive or enabling. Policies and regulations vary from country to country and affect some or all of the issues associated with public and occupational health and safety, environmental protection, the definition of end-state, waste management, reuse and recycling of materials, arrangements for release of materials from regulatory control and matters concerning regional development. However, national policy may influence decommissioning strategy indirectly. In this case, influence may be by way of national policies that are not concerned specifically with the process of decommissioning but may be linked to it by way of wider issues. These may include matters such as the future use of nuclear power, economic and societal issues associated with the effects of shutting down major industrial facilities, safety issues and broad financial issues concerned with costs, the use of available funds and the timing of their deployment. Although perhaps not associated with national policy, as such, the prospects for continued availability of qualified and trained staff may also have such an influence. [120]

Despite of the fact that decommissioning technology is available, technological and operational factors will influence the choice of strategy to be followed. Long-term uncertainties are of particular importance when a decommissioning strategy is selected. Although the radiological hazards decrease, the uncertainties increase with time. Policies and legal/regulatory frameworks are subject to change. The direction of change is uncertain although regulatory standards have tended to become more stringent with time. It is important to single out that the assessment of the above factors is a challenge, particularly when long time periods are involved and because most of these factors are not of a quantitative nature and need subjective assessment not always very objectives.

Techniques for decontaminating and dismantling nuclear facilities are already available and many nuclear facilities, including several nuclear power plants in different countries, have already been successfully decommissioned and dismantled using these techniques. Some sites have already been returned to a condition suitable for unrestricted reuse. It is recognized that provisions for funding dismantling and decommissioning of nuclear facilities need to be made during the operating lifetime of a facility. The challenges are to ensure that dismantling and decommissioning costs are calculated correctly and that sufficient funds will be available, when required. Waste management costs are a significant element of the overall costs of dismantling and decommissioning and may dominate in some cases depending on how the costs of residual spent fuel management, for example, are assigned. Hence, it is important not only that waste quantities are minimized but also that the costs of waste treatment, storage and disposal are separately identified and assigned.

Early planning is an important element in any decommissioning project, due to its complexity. The shift from operations to decommissioning requires a well-defined programme of work, similar to the methodologies used in the engineering industry. For a successful outcome, decommissioning must be treated as an engineering project with modern project management. A dedicated decommissioning organization is also required. This new 'mind-set' often poses difficulties, as the nature of the forward is radically different, requiring both new technical skills and the need to control and manage budgets proactively, to achieve cost and time targets. Such changes create tensions as the order of priorities change. The decommissioning phase can lead to the loss of experienced and younger staff as they may face redundancy or significant changes in their jobs.

Decommissioning and dismantling of nuclear facilities are the responsibility of the operator and must be conducted under license. A separate license is often required for decommissioning. The key points in decommissioning and dismantling of nuclear facilities are:

1) the purpose of decommissioning and dismantling is to allow the removal of some or all of the regulatory controls that apply to a nuclear site;
2) there is no unique or preferable approach to decommissioning and dismantling of nuclear facilities;
3) techniques for decommissioning and dismantling are available and experience is being fed back to plant design and decommissioning plans;
4) many nuclear facilities have been successfully decommissioned and dismantled in Germany, Belgium, France and the UK;
5) current institutional arrangements for decommissioning and dismantling (policy, legislation and standards) are sufficient for today's needs;
6) current systems for the protection of the safety of workers, the public and the environment are satisfactory for implementation and regulation of decommissioning and dismantling;
7) arrangements are in place for the funding of decommissioning and dismantling, but evaluation of costs requires further attention;
8) local communities are increasingly demanding involvement in the planning for decommissioning and dismantling.

Another important element for the successful introduction of a nuclear power programme is the existence of a well-prepared and trained force in several areas of the peaceful uses of nuclear energy, particularly for the production of electricity and nuclear safety. In the last two decades, particularly after the Chernobyl nuclear accident, there was an increase lack of interest in several countries, particularly in Europe and North America, in the development and use of nuclear energy for electricity generation. This lack of interest, in addition to the retirement of the specialized work-force working in the nuclear sector and its lack of perspectives, produced a significantly reduction in the number of well-trained professional and specialized force available in the nuclear sector in several countries. The situation is even more seriously regarding the new force that eventually will substituted the current specialized force working in the nuclear energy sector in the future. To overcome the lack of specialized trained work-force in different countries, several States had taken specific initiatives to increase the number of professional and specialized work-force that will be available in the nuclear sector in the coming years to ensure the development of this sector. In several of these countries there is an increase in the recruitment and formation of new work-force in the nuclear sector in the last years.

Several initiatives have been promoted by companies that are building, or they are about to build, new nuclear power plants and supported by different government institutions and agencies, including the IAEA. The two biggest exponents of the French nuclear industry, EdF and Areva that are building new nuclear power plants inside and outside France, hired a total of 15,500 new workers in 2008. However, it is important to single out that not only nuclear companies in charge of the construction of new nuclear power plants are requiring specialized personnel in the nuclear energy sector. In the USA, for example, the NRC should hire 600 new specialists in the next three years in order to assimilate the work increase foresee for this period, particularly related with new orders to built nuclear power reactors or for the extension of the operational life of several of the current nuclear power reactors in operation. The UK and the Russian Federation recently decided to provide support for the establishment of university and institutes specifically dedicated to prepare professionals and technicians in nuclear sciences and technologies in order to satisfy the foresee demand of specialized work-force in their respective nuclear sectors in the future.

At the same time, the IAEA has been also promoting the preparation of new professionals and technicians in sciences and nuclear technologies in its member states, in order to satisfy the foresee demand of specialized work-force in their respective nuclear sectors in the future. If the situation of the lack of specialized force is not reverse in the near future, then the nuclear sector will face the following problem: There will be an insufficient replacement of specialized force duly prepared to assume the responsibility to significantly increase the number of nuclear power reactors to be constructed and in operation in the future[15].

The introduction of a nuclear power programme in any given country could be facilitated by the consideration of three important elements. These elements are the following: 1) the elaboration and adoption of a national energy policy; 2) the elaboration and adoption of a regional energy policy[16]; and 3) the correct planning for the introduction of a nuclear power programme.

The main elements of such policies and planning process are described in some details in the following paragraphs.

A NATIONAL ENERGY POLICY

It is important to note that energy policy is strongly shaped and influenced by particular national and even regional situations. Thus, lesson from a country's energy policies and practices should prove less useful, by themselves, in formulating policy for others countries and indeed the entire globe. But not attempting to learn from the practice of others and develop more effective energy policy, bearing in mind the scale and the speed of the needed energy transitions, is irresponsible. (IAEA-TECDOC-1513, 2006)

A national energy policy is comprised by set of actions involving country's laws, regulations, treaties and agency directives. The national energy policy to be adopted by any given country may include one or more of the following elements:

---

[15] In the USA around 40% of the specialized forces now working in the nuclear sector are expected to retire in the next five years. In France almost 40% of the specialized force working in EdF will retire before 2015.

[16] This policy should take into account the energy situation of the different countries, the possibilities of cooperation among them in generating electricity in the most efficient manner and with the less negative impact in the environment.

a) statement of national policy regarding energy generation, transmission and usage;
b) legislation on commercial energy activities (trading, transport, storage, etc.);
c) legislation affecting energy use, such as efficiency and emission standard;
d) instructions for state owned energy sector assets and organizations;
e) fiscal policies related to energy products and services (taxes, exemptions, subsidies, etc.);
f) environmental issues.

There are a number of elements that are naturally contained in a national energy policy. The main elements intrinsic to a national energy policy are, among others, the following:

a) what is the extent of energy self-sufficiency for the country;
b) where future energy sources will be obtained;
c) how future energy will be consumed (e.g. among sectors);
d) what are the goals for future energy intensity and the ratio of energy consumed in relation to GDP;
e) what environmental externalities are acceptable and are forecast;
f) what form of portable energy is forecast (e.g. sources of fuel for motor vehicles);
g) how will energy efficient hardware (e.g. hybrid vehicles, household appliances) be encouraged;
h) how can the national energy policy drive province, state and municipal functions;
i) what specific mechanisms (e.g. taxes, incentives, manufacturing standards) are in place to implement the national energy policy.

Any country considering the introduction of a nuclear power programme should have a national energy plan specifying which the objectives are associated with the national energy policy to be implemented. This programme should identify which are the available energy sources that the country have, the possible role to be played by each of them in the energy balance of the country in the future and the cost associated with the use of the different energy sources for electricity generation. According with the IAEA, some of these objectives are, among others, the following:

1) improved energy independence;
2) development of indigenous energy resources;
3) economic optimization of energy and electricity supply;
4) stability of the national electric grid;
5) security of electricity supply;
6) environmental protection;
7) privatization of electric utilities[17].

---

[17] This specific objective will depend on the energy policy adopted by a given country. Some countries will prefer to keep the energy sector in the public sector while others will prefer that the private sector take it over.

Improved Energy Independence

One of the main goals of any energy policy are the following: a) to ensure the energy independency of the country by developing all national available energy sources; and b) to ensure the energy supply under any circumstances. The oil price shocks of 1973 and 1979 led many governments to draw the conclusion that too strong dependence on one or two imported primary energy forms was undesirable and, in some cases, very dangerous from the economic and political point of view. (IAEA STI/PUB/1050,1998) Therefore, one of the primary objectives for many governments is the diversification of energy supply sources.

Despite of the position of several governments regarding the use of nuclear energy for electricity production, the potential use of this type of energy for this purpose should not be excluded from any future energy balance studies to be carried out by the national competent authorities of these countries. The purpose of these studies is to find out the most adequate energy mix balance considering all available energy sources in the country, in order to satisfy any future increase in electricity demand with the lowest possible investment, in the most effective and efficiency manner and with the lest negative impact in the environment and the population. Is a great mistake not to consider the possible use nuclear energy for electricity generation without having a deep study on the possible role to be played by this type of energy.

Without any doubt the vulnerability of a country may be linked to a strong dependency on energy imports. The rate of dependency is measured by the ratio of net energy imports to total (primary) energy consumption. The energy dependency rate (ED) is naturally the complement of the rate of energy independence. The latter, often mentioned in various statistics, is the ratio of domestic production to total primary energy consumption. For example, the energy independence rate in 2005 for certain countries is as follows: a) 50% for France, relatively stable since 1990, due to nuclear power, against 24% in 1973 and 62% in 1960; b) 92% for the UK against 129% in 1999; c) 14% for Italy against 20% in 1995; and d) 35% for Germany against 40% in 1995.

Development of Indigenous Energy Resources

The development of the energy sector by any country is a costly process that requires significant investment. The construction of a nuclear power plant is very costly and requires not only a significantly initial capital investment but a considerable period to build the plant. The decision to introduce a nuclear power programme and the entry into commercial operation of a nuclear power reactor could be between 10 to 15 years and, in some specific cases, this period could be even longer. The construction time of a fossil fuel-fired power plant is much less that the construction time of a nuclear power plant. For this reason, it is natural that, when available, governments give priority to development of non-nuclear national energy resources such as coal, oil, gas, hydropower for electricity generation, because this will give an assurance of supplies at least to some extent. However, if the domestic production is more costly than imported fossil fuel, this will influence prices throughout the energy sector of the country. (IAEA STI/PUB/1050, 1998)

Without any doubt this situation will have a negative effect in other economic sectors of the country. At the same time, if fossil fuel reserves are very limited and the use of renewable energy sources available in the country is not sufficient to satisfy the future demand of electricity, or cannot provide electricity in a sustainable manner, then the use of nuclear energy for the generation of electricity cannot be excluded from any future energy balance without having a comprehensive study on the possible role to be played by this type of energy carried out by the national competent authorities.

Economic Optimization of Energy and Electricity Supply

Before the decision to built new power plants is adopted by the national competent authorities of a country, it is important that all necessary measures are adopted to achieve the economic optimization of the use of all energy sources available in the country for the generation of electricity. This is particularly important, if the decision adopted is to build a nuclear power plant due the level of capital investment involved to carry such undertaking. The cost of a nuclear power plant depends of its output, experience of the construction company, resources available, the type of contract signed, among others. Normally, the construction cost of a nuclear power plant of 1 000 MW could be around US$1.5- US$2.5 billion. If there is a delay in the construction process, then these figure could go up significantly. The current cost of the construction of the first EPR in Finland is around US$4.1 billion due to the delay in the construction of the reactor by different reasons.

Regarding the economic optimization of the energy electricity supply, it is assumed that energy and electricity supply systems should be economically optimized using different types of models, including computer models[18]. The planning period should be between 10 and 15 years to make the optimization meaningful[19]. The objectives of security and diversity of energy supplies could be factored in through appropriate assumptions on future prices of fossil fuel produced domestically or imported. In addition to the consideration of future prices of fossil fuel, energy policy decisions have to be made taking into account other costs, such as environmental and health costs and infrastructure development costs. (IAEA STI/PUB/1050, 1998) In the specific case of the use of nuclear energy for electricity generation, the decommissioning and management of spent nuclear fuel costs should be also included.

Stability of the National Electric Grid

The stability of the national electric grid is an important element that needs to be examined very carefully by the national competent authorities in charge of the energy sector, because it is an important prerequisite for the successful introduction of a nuclear power programme. Necessary measures to minimize any fluctuations of frequency and voltage during normal operation of the energy power plants have to be identified, adopted and implemented. (IAEA STI/PUB/1050, 1998) The adoption of these measures should create the necessary conditions to introduce a nuclear power programme successfully. At the same time, the government should ensure that the capacity of the electrical grid is the adequate for the

---

[18] The IAEA Secretariat has a computer model to carry this type of studies called "Wien Automatic Planning System" (WASP) available to all of its member states.
[19] In the specific case of the USA, this period could be of 15 years.

size of the nuclear power reactor to be built. According with the IAEA, the capacity of the electrical grid should be 7-10 times the capacity of the nuclear power reactor to be built. The adoption of specific measures to ensure the successful introduction of a nuclear power programme may improve, in some specific cases, the quality of the electricity supply system of the country.

Finally, it is important to single out the following. In addition to assuring that the electric grid will provide reliable off-site power to nuclear power plants, there are other important factors to consider when a nuclear power plant will be the first plant on the electric grid and, most likely, the largest one. If a nuclear power plant is too large for the size of an electric grid, then the operators of the plant may face several problems. These problems are the following: a) off-peak electricity demand might be too low for a large nuclear power plant to be operated in base load mode, i.e. at constant full power; b) there must be enough reserve generating capacity in the electric grid to ensure grid stability during the nuclear power plant planned outages for refueling and maintenance; and c) any unexpected sudden disconnect of the nuclear power plant from an otherwise stable electric grid could trigger a severe imbalance between power generation and consumption causing a sudden reduction in grid frequency and voltage. This could even cascade into the collapse of the electric grid, if additional power sources are not connected to the grid in time.

According with the IAEA, "grid interconnectivity and redundancies in transmission paths and generating sources are key elements in maintaining reliability and stability in high performance electric grids. However, operational disturbances can still occur even in well maintained electric grids. Similarly, even a nuclear power plant running in base load steady-state conditions can encounter unexpected operating conditions that may cause transients or a complete shutdown in the plant's electrical generation. When relatively large nuclear power plants are connected to the electric grid, abnormalities occurring in either can lead to the shutdown or collapse of the other. The technical issues associated with the interface between nuclear power plants and the electric grid includes: a) the magnitude and frequency of load rejections and the loss of load to nuclear power plants; b) grid transients causing degraded voltage and frequency in the power supply of key safety and operational systems of nuclear power plants; c) a complete loss of off-site power to a nuclear power plant due to electric grid disturbances; and d) a nuclear power plant unit trip causing an electric grid disturbance resulting in severe degradation of the grid voltage and frequency, or even to the collapse of the power electric grid.

Security of Electricity Supply

A secure supply of energy is a crucial element for the economic and social development of any country. For this reason, if a country is planning to introduce an important plan for economic and social development, then it has to guarantee the supply of electricity to ensure its implementation using all available energy sources. It is important to be aware that security of electricity supply and the implementation of an effective industrial development plan are two indispensable development components of any economic development and social policy.

Environmental Protection

The international community is well aware that the use of fossil fuel for different purpose, particularly for the generation of electricity, is one of the main causes for climate change. For this reason, any energy policy adopted for the development of the energy sector by any government should include an assessment of the impact of this policy in the environment. There is no doubt that fossil-fuelled power plants are sources of pollution and climate changes and, for this reason, they have become an essential factor in considerations of environmental protection, not only at the local level but also regionally and internationally as well. It is becoming increasingly unlikely that a country can avoid considering environmental issues when setting a national energy policy for electricity production in the future. (IAEA STI/PUB/1050, 1998)

Privatization of Electric Utilities

Many countries had privatized government owned energy utilities, following a privatization policy of public companies promoting by the International Monetary Fund (IMF) and the World Bank (WB) within the current economic development policy promoted by these two organizations. Others have decided to keep the energy sector within the public sector of the country. One of the objectives to be achieved with the privatization of public utilities, according with the opinion of those that support a privatization policy, is to make them operate under conditions that are more commercial. However, the adoption of this privatization policy could limit these utilities to seek funds only in the commercial money markets, with their stronger emphasis on short-term returns on investment. This may means that capital-intensive plants such as nuclear power plants will not be favored by utilities planning to construct these types of plants without government support. This is an important element that needs to be carefully study when the nuclear option is being under consideration for its possible inclusion in the energy balance of any country.

Summing up can be stated that some of the above objectives are, of course, overlapping and may lead to the adoption of the same energy policy. One element that cannot be ignored is the fact that in some countries the energy policy options under consideration exclude the use of nuclear energy for electricity generation due to specific political situation, particularly because of a strong and active public opinion against the use of this type of energy for the production of electricity. For example, in a country with high level of indigenous energy reserves this situation would not favor the introduction of a nuclear power programme. In the other hand, if a country does not have an adequate develop in the field of science and technology, or their electric grid is relative small, or there is a lack of resources to finance the construction of a nuclear power plant, or a lack of specialized and well-trained force in areas associated with nuclear power, then the use of nuclear energy for electricity generation could not be considered as a realistic option.

For all of the above reasons, it would be extremely important to have a well-defined role for all available types of energy sources within the overall energy policy adopted by the government, before a decision is adopted regarding the introduction of a nuclear power programme in the country.

A REGIONAL ENERGY POLICY

Besides to have a clear national energy policy for the introduction of a nuclear power programme it is important to take into account, during the consideration of the energy mix balance of the country, the status of the relations with neighboring countries and its energy policy and level of development. In today's world this is an important factor to be considered, particularly in the case when the country decides to introduce a nuclear energy programme.

The growing importance of a regional policy is reflected by the number of regional associations and alliances being formed for various purposes, some of them to ensure energy supply in the future and at affordable price. Regional cooperation policy applies also in the case of nuclear power programmes, within which many topics related with regional cooperation could yield direct benefits, not only for those countries that are using now nuclear energy for electricity generation, but also for those that are thinking to introduce this type of energy for the same purpose in their energy mix in the future.

According with reference 1, the following are important components of any regional energy policy related with the introduction of a nuclear power programme: 1) electric grid integration policy; 2) nuclear safety; 3) environmental issues; 4) sharing of power plant services; and 5) social economics and cultural issues.

Electric Grid Integration Policy

The adoption of a regional electric grid integration policy could have direct benefit for those countries supporting such policy by increasing the security of electricity supply and by improving the reliability of electricity supply and economics of operation. It is important to single out that the adoption of a regional electric grid integration policy will demand a cooperative planning effort for the expansion of electricity generation. Without any doubt, the planning process of such integration policy could be facilitated as result of the wider experience in the elaboration of such policy available in the region.

A regional electric grid integration policy may also permit not only the use of large units for the generation of electricity, but also an increase in the electric capacity of the units current in operation in different countries to a size greater than the ones that any national electric grid could accept. For these and others reasons, the adoption of a regional electric grid integration policy could be very important for those countries with a relative small national electric grid that are planning the introduction of a nuclear power programme. However, the adoption of a regional electric grid integration policy is not an easy task and requires, among others, the adoption of national energy policies that creates the necessary political conditions to promote and supports such regional integration.

Nuclear Safety

Nuclear safety is one of the most important components of any national and regional policy associated with the introduction of a nuclear power programme. Close cooperation at regional level in nuclear safety matters is an indispensable component of any nuclear power programme introduced or to be introduced by a group of countries of the same region. This cooperation can help to give additional assurances about the safety of the nuclear power plant in operation, or are going to entry into operation in the future, by providing immediate access to information about nuclear incidents and accidents that could occur in the plant and for the coordination of emergency plans. It can also provide the necessary assistance in case of a nuclear accident or incident, particularly the use of available facilities for medical treatment to those people affected, among others indispensable assistance.

Environmental Issues

Transboundary effects of pollution, such as acid rain and the emission of CO_2 to the atmosphere, are increasing the concern of many countries about the use of fossil fuels for the generation of electricity and the need to use nuclear energy for the same purpose. For this reason, the consideration of environmental issues is an indispensable component of any regional energy policy that supports the adoption of a nuclear power programme by a group of countries inside a region.

Sharing of Plant Services

If more than one country in a specific region has nuclear power plants in operation, there are obvious advantages in trying to share plant services, such as plant maintenance and repair and spare parts, where this is feasible option. However, in some regions there are still many difficulties and problems that impede sharing plant services due to different political and technological considerations. In the same region countries could use different type of nuclear power reactors supply by different supplier under specific conditions that could complicate the adoption of a regional cooperation policy in this area.

Social-Economics and Cultural Issues

Associated with the introduction of a nuclear power programme in a given country, there are a number of socio-economic and cultural issues that have to be considered. These issues are, among others, the following:

1. population (including relevant demographic characteristics);
2. economic base;
3. community infrastructure and services;
4. existing and planned land use;
5. heritage, cultural or archaeological sites;
6. recreation areas;
7. use of lands and resources for traditional purposes by aboriginal persons.

For the successful introduction of a nuclear power programme it is crucial to deal with these issues in an adequate manner and to find acceptable solutions to the problems that could be encountered in dealing with them.

PROS AND CONS OF THE USE OF NUCLEAR ENERGY FOR THE PRODUCTION OF ELECTRICITY

The growing dependency of many countries on energy imports due to a lack of available domestic energy sources, the systematic reduction of fossil fuel reserves and the increases in certain energy prices has reinforced the concerns of many governments about their possibilities to meet the energy demand in the future. It is well recognized that ensuring secure and reliable energy supplies at affordable and stable prices is vital to ensure the economic and social development of any country and should constitute an integral part of a sound and consistent national energy policy. This is particularly valid in the current process of market liberalization and competition in which several countries are involved.

In all nations, most people reject the view that shifting to alternative energy sources would hurt the economy, believing instead that it would save money in the long run. It is quite remarkable that there is such unanimity around the world that governments should address the problem of energy by emphasizing the use of alternative energy sources and by increasing efficiency. In several cases, governments are seriously considering the expansion of their ongoing nuclear power programmes or are considering the introduction of such programmes in the future to satisfy the foresee increase in energy demand and to reduce the negative impact in the environment produced by the use of fossil fuel for electricity generation. Why nuclear energy for electricity generation is again under consideration by several governments and the private energy sector in several countries all over the world? The answer is the following: Nuclear power is the only technology used for electricity generation that, from the very beginning of its development, took the environmental impacts into consideration. It is one of the human activities in which research on safety was developed together with its technology. Nuclear power plants are licensed from a safety point of view by independent governmental organizations and are also subject to regional and local site approval procedures. Participation of public and nongovernmental organizations in both licensing and environmental procedures is allowed and encouraged in many countries. Its impacts on the environment are almost non-existent if well managed. It occupies only small surfaces of land and consumes small amounts of fuel; its waste is small, confined and isolated from the environment. [21]

Since the beginning of the use of nuclear energy for electricity production two different groups of positions were clearly identified. One group was in favor in using nuclear energy for electricity generation, while the other was strongly against. The position of those who promoted the use of nuclear energy for the production of electricity in the past is quite different from the position of those who promote it today. At the beginning of the 1950s, the use of nuclear energy for the production of electricity was considered to be competitive, safe and clean with regards to the use of fossil fuel for the same purpose. Today this position has been forgotten and the defense of the nuclear option orbits around the urgent need to stop greenhouse emissions and climate change. (Fernandez-Vazquez and Pardo-Guerra, 2005)

The Pros

There are a number of strong arguments in favor of the use of nuclear energy for electricity generation. These arguments are the following:

1) it brings technological development in advanced areas from the technological point of view in comparison with any other form of energy;
2) it is a proven technology that can meet large-scale energy demands in the coming years;
3) it provides a continuous supply of energy. Other available technologies such as hydroelectric, solar and wind power depend on environmental factors difficult to predict;
4) there are no supply problems, at least in the medium and long terms, with regards the nuclear fuel. Global stocks of uranium are more than enough to satisfy any future increase in the world energy demand.
5) the proven reserves of uranium are not located in politically sensitive regions of the world;
6) the international cost of the nuclear fuel at this moment is acceptable and can be afforded by those countries with nuclear power programme and also by those thinking to introduce this type of programme for the first time in the future.

However, it is important to be aware of the following: The supporters of the nuclear option forget, nevertheless, that a series of characteristics of nuclear energy require being highly cautious when dealing with atoms. First, it generates dangerous waste that is difficult to isolate, cannot be reprocessed by nature's cycles, and lasts for several thousand years — therefore posing a tremendous threat to the environment and human health, if not well handle. Even though the conventional sources and methods used in electricity production generate residues that have to be managed, none of them pose as many risks as nuclear energy nor do they require such a long-term management programme. Second, nuclear energy is not entirely secure, as demonstrated by the accidents in the Three Mile Island and Chernobyl nuclear power plants. Though no energy source is inherently secure, an oil spill is not the same as a radioactive spill. Third, the construction, dismantling and decommissioning of nuclear facilities is extremely expensive. Finally, and above all, nuclear energy is intrinsically linked to the shadow of nuclear proliferation, which humanity has sought to exorcise without success since the 1950s. (Fernandez-Vazquez and Pardo-Guerra, 2005)

The international community should be aware that by the first decades of the 21th century, all forms of primary energy for electricity production will be needed, if sustainable development is to be achieved. In this context, we have the moral obligation to utilize those energy resources that lead to the lowest possible environmental impacts. [21]

One of the available energy source that does not emit any greenhouse gas (carbon dioxide, methane, nitrous oxide and others) or any gas causing acid rain or photochemical air pollution (sulfur dioxide and nitrogen oxides) when it is for the production of electricity is nuclear energy. This type of energy does not emit also any carcinogenic, teratogenic, or mutagenic metals (As, Hg, Pb, Cd, etc.). The utilization of nuclear energy does not release gases or particles that cause urban smog or depletion of the ozone layer. At the same time, nuclear power is the only energy technology that treats, manages, and contains its wastes in a way that is complete and segregated from the public and the environment and does not require large areas for resettling large populations because it is a highly concentrated form of energy. Hence, its environmental impact on land, forests and waters is minimal. In 2007, nuclear energy generated 15% of world electricity production; in 2008, according with the IAEA data, this percentage was 14%. The use of nuclear energy for electricity production avoids some 10% of additional CO_2 emissions to the atmosphere, considering all economic sectors, and about one-third in the power sector. However, it is important to stress that nuclear power alone cannot solve the environmental load created by the emissions of greenhouse gases, but without the use of nuclear power, no other solution for this crucial problem exists within a reasonable time span and the state of the art of energy generation technologies. [21]

The Cons

Let consider now the cons in the use of nuclear energy for electricity generation. The following are, among others the main cons of the use of nuclear energy for electricity generation:

a) the negative impact and the long-term consequences of a nuclear accident;
b) high initial investment;
c) dismantling and decommissioning costs;
d) management of the spent nuclear fuel and its costs;
e) proliferation risk.

The use of this type of energy for electricity generation requires being highly cautious when dealing with this energy option, because is not an accident-free industry and, in case of a serious accident, the consequences for the environment and the population could be not only very high but could have a long-term negative effect. The Chernobyl nuclear accident, despite of the fact that could be a very rare event, is an example of the negative consequences for the environment and the population with a long-term negative effect of a nuclear accident.

Chapter 3

THE CURRENT SITUATION AND PERSPECTIVES IN THE USE OF NUCLEAR ENERGY FOR ELECTRICITY GENERATION IN NORTH AMERICA

North America is one of the regions of the world with more nuclear power reactors currently in operation. In 2009, there were 122 nuclear power reactors in operation in the USA and in Canada with a net capacity of 113 771 MWe; one unit under construction in the USA with a net capacity of 1 180 MWe. In the specific case of the USA, there were 104 nuclear power plants in operation[20] in 2009. In Canada the current number of nuclear power plants in operation in 2009 was 18.

The construction of new nuclear power reactors in North America, particularly in the USA, was halted after the Three Mile Island and Chernobyl nuclear accidents occurred in Pennsylvania, USA in 1979 and in Ukraine in the former Soviet Union in 1986 respectively. It is important to single out that in the specific case of the USA, the construction of new nuclear power reactors almost stop after the Three Miles Island nuclear accident, despite of the fact that this accident had no serious consequences for the population and for the environment.

The energy crisis the world is now facing, the real reduction of the world oil reserves and the possibilities to its extinction in the coming decades, the significantly increase in oil price occurred in 2007 and 2008 and the negative impact in the environment due to the release to the atmosphere of a significant amount of CO_2 as a result of an increase in the consumption of fossil fuel in the USA and Canada, among other relevant factors, are somehow changing the negative perception of the public in these two countries regarding the future use of nuclear energy for electricity generation. According with public information disclose during the last US presidential campaign, the USA should plan the construction of around 45 new nuclear power reactors for 2030 and around 55 more units after that year in order to satisfy the foresee increase in the energy demand in the coming decades. This is, without any doubt, an

---

[20] The first nuclear power reactor in the USA was built in 1960. However, a prototype f a BWR system started its operation in 1957 and was decommissioned in 1963.

important nuclear power programme to be development by the USA in the future, in case that the new US Administration approved it.

Why the construction of new nuclear power reactors is so important for North America and particularly for the USA? The reason is the following: North America had an aging base of nuclear and fossil fuel power plants for electricity generation. In the specific case of nuclear power plants, the last unit to enter into commercial operation in the USA was TVA's Watts Bar Unit 1 in June 1996, this mean more than 13 years ago. The last successful order for a US commercial nuclear power plant was presented in 1973, this means 36 years ago.

The IAEA has foresee that the participation of nuclear energy in the worlds energy balance will drop from 16% to 8-10% in 2030, if no decision is adopted by the EU, Canada and the USA to build more nuclear power reactors for electricity generation in the coming years.

The USA is a pioneer of nuclear power development. Annual nuclear electricity generation in the USA has more than tripled since 1980 reaching 4 260 billion kWh of electricity in 2006, half of it from coal-fired power plant, 19% from nuclear (809.8 billion kWh), 19% from gas and 7% from hydro. In 2007, nuclear energy generated 808.9 billion kWh and achieved an average 91.8% in the load capacity factor, which is a very high capacity factor in the nuclear energy sector. The net load factor per type of nuclear power reactor in operation in the USA in 2007 was 90.4% for BWR and 92.9% for PWR. With all nuclear power reactors in operation in 2008, nuclear power accounts for 19.7% of total electricity generation in the USA. US annual electricity demand is projected to increase from 4 000 billion kWh in 2008 to 5 000 billion kWh in 2030; this means an increase of 25% in 22 years. From 2006 to 2030, the USA is expected to add 12.7 GW of capacity at newly built nuclear power plants and 3.7 GW from up rates of existing plants—offset in part by the retirement of 4.4 GW of capacity at older nuclear power plants.

Table 9. The production of electricity using nuclear energy in North America

Country	Number of nuclear units connected to the grid. April 2009	Nuclear electricity generation (net million kWh) 2007	Nuclear percentage of total electricity supply (%) 2007
United States of America	104	808.9	19.7
Canada	18	88.6	14.7
Total	122	897.5	-

Source: WANO, 2009.

In Canada, nuclear energy produced 88.8 MWe of electricity. It is expected to increase by 1.5% per year during the period 2006-2030. Until April 2009, three nuclear power reactors planned with a total net capacity of 3 300 MWe and six nuclear power reactors proposed with a total net capacity of 6 600 MWe. For most of this coming decade, the construction of new nuclear power reactors in Canada are uncertain based on the most recent electricity market outlooks. According with the Canadian authorities, the return to service of the remaining laid-up nuclear units and the completion of gas-fired power units already under construction should satisfy the foresee energy demand in the coming years. While market prospects for the construction of new nuclear power reactor sales in the near- to medium-term are not too promising, the refurbishment of existing units holds more promise and would avoid, at least in the medium-term, the replacement of nuclear generating capacity with fossil-fuelled power plants increasing the CO_2 emission to the atmosphere.

THE THREE MILE ISLAND NUCLEAR ACCIDENT

The accident at the Three Mile Island Unit 2 (TMI-2) nuclear power plant near Middletown, Pennsylvania, USA, on March 28, 1979, was the most serious nuclear accident in US commercial nuclear power plant operating history, even though it led to no deaths or injuries to plant workers or members of the nearby community. What caused this nuclear accident? The nuclear accident was caused by a sequence of events such as equipment malfunctions, design-related problems and worker errors, which led to a partial meltdown of the TMI-2 reactor core but only very small off-site releases of radioactivity.

The accident began about 4:00 a.m. on March 28, 1979, which a failure in the secondary, non-nuclear section of the nuclear power plant. The main feed water pumps stopped running, caused by either a mechanical or electrical failure, which prevented the steam generators from removing heat. First the turbine, then the reactor automatically shut down. Immediately, the pressure in the primary system, which is the nuclear portion of the nuclear power plant, began to increase. In order to prevent that pressure from becoming excessive, the pilot-operated open a valve located at the top of the pressurizer. The valve should have closed when the pressure decreased by a certain amount, but it did not. As a result, cooling water poured out of the stuck-open valve and caused the core of the reactor to overheat.

As coolant flowed from the core through the pressurizer, the instruments available to reactor operators provided confusing information. There was no instrument that showed the level of coolant in the core. Instead, the operators judged the level of water in the core by the level in the pressurizer, and since it was high, they assumed that the core was properly covered with coolant. In addition, there was no clear signal that the pilot-operated relief valve was open. As a result, as alarms rang and warning lights flashed, the operators did not realize that the plant was experiencing a loss-of-coolant accident. They took a series of actions that made conditions worse by simply reducing the flow of coolant through the core.

Because adequate cooling was not available, the nuclear fuel overheated to the point at which the long metal tubes which hold the nuclear fuel pellets ruptured and the fuel pellets began to melt. Although the TMI-2 plant suffered a severe core meltdown, the most dangerous kind of nuclear power accident, it did not produce the worst-case consequences that reactor experts had long feared. In a worst-case accident, the melting of nuclear fuel would lead to a breach of the walls of the containment building and release massive quantities of radiation to the environment. Hopefully, this not happen as a result of the Three Mile Island nuclear accident.

Summing up can be stated that the Three Mile Island nuclear accident was caused by a combination of personnel error, design deficiencies and component failures. There is no doubt that the accident at Three Mile Island permanently changed both the nuclear industry and the NRC. Public fear to the use of nuclear energy for the generation of electricity and distrust increased, NRC's regulations and oversight became broader and more robust, and management of the nuclear power plants was scrutinized more carefully. The problems identified from careful analysis of the events during those days have led to permanent and sweeping changes in how NRC regulates its licensees – which, in turn, has reduced the risk to public health and safety.

SHUT DOWN OF NUCLEAR POWER REACTORS

Until 2008, in North America the number of nuclear power reactors that has been shut-down reached 31. Canada shut-down three nuclear power reactors between 1977 and 1987, reducing the net nuclear power capacity of the country in 478 MWe. Four additional nuclear power reactors are in long-term shutdown until December 2007, reducing further the net power capacity of the country in 2 588 MWe. In the case of the USA, the number of nuclear power reactors that were shut-down between 1957 and 1998 was 28, reducing the net nuclear power capacity of the country in 9 764 MWe.

The need of nuclear energy for the production of electricity

Which is the current situation respect to the use of nuclear energy for electricity generation and the energy infrastructure in North America? The USA today consumes about 100 quads—100 quadrillion BTUs—of raw thermal energy per year. Three basic things are done with it: a) generate electricity (about 40% of the raw energy consumed); b) move vehicles (30%); and c) produce heat (30%). Oil is the fuel of transportation. Natural gas is used to supply raw heat, though it's now making steady inroads into electric power generation. Fueling electric power plants are mainly (in descending order) coal, uranium, natural gas, and rainfall, by way of hydroelectricity. (Huber, and Mills)

In North America, specifically in the USA, existing electricity generating plants continued to pollute the atmosphere with greenhouse gases in a significant manner, due to the use of fossil fuel particularly coal. However, there is a great difference between the USA and Canada on this important subject[21]. What is the reason of such difference? The main reason is the following: The reaction of the US and Canada governments with respect the contamination of the atmosphere due to the consumption of fossil fuel is totally different. The Canadian government became increasingly concerned about the emission of large quantity of CO_2 to the atmosphere and, for this reason, signed the Kyoto Protocol committing the energy sector to reduce the level of emission of this gas to the atmosphere. By the contrary, and despite the fact that the USA is the most pollute country in the world, the US government refused to signed the Kyoto Protocol and had said many times that has no intention to sign it in the future.

LIMITING FACTORS

The main limiting factor in raising the use of nuclear energy for electricity production in North America is the increase number of nuclear power reactors that are reaching the point where choices must be made between extensions of their useful life and decommissioning,

---

[21] The contamination is greater in the USA than in Canada.

particularly in the USA. However, there are other limiting factors that need to be considered as well. These limiting factors are the following:

1) in the case of the USA, one the main limiting factors to be considered is the limited incentives and measures until now adopted by the US government in order to encourage the participation of the private energy industry in the construction of new nuclear power plants;

2) the opposition of the majority of the banks in the region to provide to the private energy industry the necessary funds for the construction of new nuclear power reactors, particularly now due to the present economic and financial crisis that the world is facing and the lack of credit for fresh money to support large investment in risk project such as the construction of a nuclear power plant;

3) the difficulties facing designers to significantly improve the current nuclear power reactors design in order to meet the increase stringent safety requirements adopted by different governments;

4) the problem causing by the final disposal of high-level nuclear waste. In the particular case of the USA, the delay in entry into operation of the national nuclear waste repository at Yucca Mountain, Nevada and the limited capacity of this site is one of the limiting factors to be taken into account for increasing the participation of nuclear energy in the energy balance of the country;

5) the high construction cost and the time needed for the construction of the current type of nuclear power reactors, despite of all measures adopted to reduce the construction time and the time needed to receive the approval of the operation license requested.

It is easy to verify that there are several nuclear power plants in North America that are operating successfully and in an efficiency manner and this is one of the reasons why nuclear energy should continue to be a generally competitive profitable source of electricity. However, for the construction of new nuclear power plants, the following factors should be considered: a) the USA and Canada are rich in other energy sources making very competitive the use of any of them in comparison with nuclear energy; b) the overall current and future electricity demand, how fast it is growing and the level of the energy reserves of the different energy sources available in the region; and c) the market structure and investment environment, particularly the position of the government to support the construction of new nuclear power plants.

Banks were somehow reluctant to fund new projects for the construction of new nuclear power plants in the region, particularly in the USA, and only companies with their own capital resources are ready to go ahead with the construction of new nuclear power plants, while the existing ones becomes much older. Taking into account that the energy infrastructure in the North America region is relative old, particularly in the case of the nuclear infrastructure, it was only a matter of time before a massive blackout occurred in the USA and Canada. That happens in the spring of 2000 in California and in August 2003 in the USA and Canada when the electrical grid collapses during a high demand period. According with the IAEA, the collapse of the electric grid was caused in this case by a combination of human errors and technical challenges: power plant outages, overextended controllers, transmission line failures, the overheating of alternate transmission lines causing lines to sag into trees, an insufficient ability to repair or replace sensors and relays quickly, poor maintenance of control room alarms, poor communications between load dispatchers and power plant operators, insufficient understanding of transmission system interdependencies, and the electric grid operating very near its transmission capacity. As a consequence, nine nuclear power plants in the USA and eleven nuclear power plants in Canada were disconnected from the electric grid because of electrical instabilities[22].

Soon after the electricity crisis happened in the USA in California in 2000, the problems related with the supply of energy were one of the most important topics included in the new National Energy Plan, which begun its implementation in May of 2001. The foresee growth of the electricity demand and the increase of the external dependence that this implies have modified the government and the public opinion regarding the use of nuclear energy for electricity generation. The plan intends to add 12 000 MWe from nuclear origin to the existent nuclear power plants in the 2010 and to facilitate the administrative procedures for the selection of sites and for the approval of the construction of new advanced nuclear power reactors designs. The plan recommends also carrying out an evaluation of the impact of the use of nuclear energy for electricity generation on the air quality.

As consequence of the adoption of the above plan, the nuclear industry indicated their readiness to identify specific sites for the construction of new nuclear power plants and to develop and put in the market nuclear advanced technologies. The government, at the same time, indicated their readiness to adopt new and easier licensing processes in order to facilitate the adoption of specific decisions on the construction of new nuclear power plants in the USA before 2010.

Independently of the decisions to be adopted regarding the future development of the nuclear energy sector in the USA and Canada, there is an understanding among all factors involved that most of the current nuclear power plants should continue to be operated in the future, particularly in the USA. This should be done with the maximum levels of security and with the appropriate management of the high-level nuclear waste in order to ensure the protection of the environment.

---

[22] According with IAEA information, the North American blackout was in fact only one of seven blackouts in a six-week period in 2003 that affected more than 120 million people in eight countries in several regions: Canada, Denmark, Finland, Italy, Malaysia, Sweden, UK and USA. In Sweden, in September, a nuclear power plant tripped (i.e. rapidly shut down), resulting in the loss of 1 200 MWe to the grid. Five minutes later a grid failure caused the shutdown of two units at another nuclear power plant with the loss of a further 1 800 MWe. To respond to this loss of 3 000 MWe (about 20% of Sweden's electricity consumption) the electric grid operators isolated the southern Sweden–eastern Denmark section of the grid, but the voltage eventually collapsed due to the insufficient power supply. At the time of the original reactor trip two high-voltage transmission lines and three links to neighboring countries were out of service for normal maintenance work and four nuclear units were off-line for annual overhauls. Their unavailability severely limited the options of the electric grid operators.

THE NUCLEAR POWER PROGRAMME IN THE UNITED STATES OF AMERICA (USA)

The USA is a pioneer of nuclear power development. With 104 nuclear power reactors in operation in 2009[23], the USA is the country with the highest number of nuclear power reactors in operation in the world and accounts for 19.7% of total electricity generation in the country. The evolution of nuclear power plants capacities in the USA during the period 1970-2007 and the number of number of nuclear power reactors in operation in this period are shown in Figures 28 and 29.

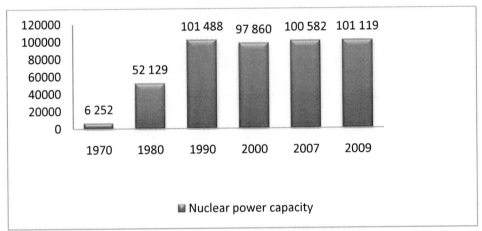

Source: CEA 8th Edition, 2008.

Figure 28. Evolution of nuclear power plants capacities in the USA during the period 1970-2007.

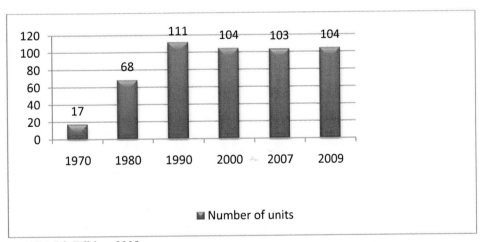

Source: CEA 8th Edition, 2008.

Figure 29. Evolution of nuclear power reactors in the USA during the period 1970-2007.

---

[23] In 2007, 103 licensed power reactors operated at 65 plant sites in 31 States (not including the Tennessee Valley Authority's [TVA's] Browns Ferry 1, which has not operated since 1985; TVA spent about US$1.8 billion to restart the operation of this reactor in 2008.

The US nuclear power industry grew to its present size following construction programmes initiated during the 1960s and 1970s when the use of nuclear energy was anticipated to be a low cost source of electricity. Increases in nuclear generating capacity during 1969-1996 made nuclear energy the second largest source of electricity generation in the USA, following coal. According with available public information, electricity production from US nuclear power plants is greater than from oil, natural gas and hydropower. Nuclear power plants generate more than half the electricity in six states. The 772 billion kilowatt-hours of nuclear electricity generated in the USA during 2002 was more than the nation's entire electrical output in 1963, when the first of today's large-scale commercial nuclear power reactors were ordered.

Existing nuclear power plants appear to hold a strong position in the ongoing restructuring of the electricity industry in the USA. In most cases, nuclear utilities have received favorable regulatory treatment of past construction costs, and average nuclear operating costs are currently estimated to be lower than those of competing fossil fuel technologies. Although eight US nuclear power reactors have permanently shut down since 1990, recent discussions within the power private industry could indicate greater industry interest in extending the operational life of several of the current nuclear power reactors in operation in the country and in the construction of new units, depending on the position that will be assumed by the US government on this last subject. Despite the shut downs, total US nuclear electrical output increased by more than one-third from 1990 to 2002, according to the Energy Information Administration. The increase resulted primarily from reduced downtime at the remaining plants, the startup of five new units and reactor modifications to boost capacity. (Holt and Behrens, 2003)

Better utilization of generating capacity has permitted nuclear power plants to maintain this relative position despite stopping the construction of new nuclear power reactors and the shut downs of several units for maintenance and refitting during the 1990s. It is important to note, however, that several of the nuclear power reactors that were permanently shut down during the 1990s were small or prototype units. The last unit was shut down during 1998[24].

The improved operation of nuclear power plants has helped drive down the cost of nuclear-generated electricity. Average operations and maintenance costs (including fuel but excluding capital costs) dropped steadily from a high of about US$3.5 cents/kWh in 1987 to below US$2 cents/kWh in 2001 (in 2001 dollars). By 2005, the average operating cost was US$1.7 cents/kWh. (Parke and Holt, 2007)

Since 1973, no energy generation company in the USA has been willing to order the construction of a new nuclear power plant[25]. Only preliminary steps towards purchasing and constructing a new nuclear power plants have been carried out by some of the generating companies until today in the absence of a promise of huge Federal subsidies to finance the construction of this type of power plants. The reason why no new nuclear power reactors has been built is not because of public opposition to the use of nuclear energy for electricity generation; not for the conditions to have a licensed geologic repository for the disposal of spent nuclear fuel, and certainly not because of the proliferation risks associated with commercial nuclear power plants. Rather, it is because new commercial nuclear power plants

---

[24] In the USA a total of 27 nuclear power reactors were shut down or dismantled during the period 1957- 1998.

[25] The only nuclear power reactor under construction since 1972 in the USA is the Watts Bar 2 unit with a net power capacity of 1 165 MWe.

were uneconomical in the USA until 2008 in comparison with other available energy sources. Now the situation has change due to the increase in the oil price in 2008, the raise in the price of natural gas and clean coal and the negative impact in the climate as a consequence of the increase emission of CO_2 to the atmosphere due to the use of fossil fuels for generating electricity.

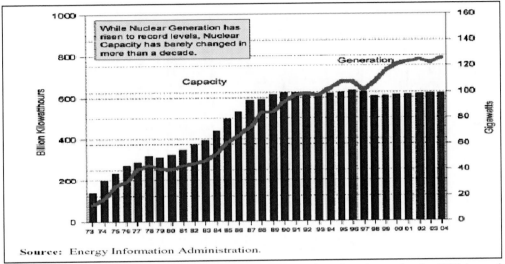

Note: Generation is read on the left scale (in billion kilowatt-hours) and capacity (in GW) is on the right.

Figure 30. Net nuclear generation vs. capacity during the period 1973-2004.

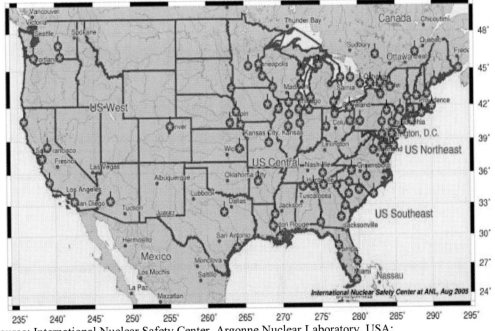

Source: International Nuclear Safety Center, Argonne Nuclear Laboratory, USA:

Figure 31. Location of the nuclear power plants in the USA.

It is important to stress that although there are in the USA the largest number of operating nuclear power reactors in the world in a single country, the number of cancelled projects, 138 is even larger.

As can be seen from Figure 31 almost all US nuclear power plants are located in the eastern part of the country. Only very few of them are located in the western part of the USA.

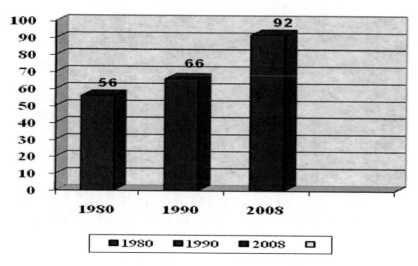

Source: IAEA and CEA 8th Edition, 2008.

Figure 32. Evolution of the capacity factor in the USA during the period 1980-2008.

The Energy Policy in the USA

After much preliminary debate the Energy Policy Act 2005 was approved by the US Congress without difficulties[26]. The Energy Policy Act included the following incentives for the nuclear power industry:

1) "production tax credit of US$1.8 cents/kWh from the first 6 000 MWe of new nuclear plants in their first 8 years of operation (same as for wind power on unlimited basis);
2) federal risk insurance of US$2 billion to cover regulatory delays in full-power operation of the first six advanced new plants;
3) rationalized tax on decommissioning funds (some reduced);
4) federal loan guarantees for advanced nuclear power reactors or other emission-free technologies up to 80% of the project cost;
5) the Price Anderson Act for nuclear liability protection extended for 20 years;
6) support for advanced nuclear technology".

---

[26] The vote was 74-26 in the Senate and 275-156 in the House of Representatives.

Table 10. Nuclear power reactors in operation and shut down in the USA

Station	Type	Capacity MWe	Operator[1]	Status	Reactor Supplier[2]	Construction Date	Criticality Date	Grid Date	Commercial Date	Shutdown Date
FARLEY-1	PWR	833	SOUTH	Operational	WEST	31-Oct-70	09-Aug-77	18-Aug-77	01-Dec-77	
FARLEY-2	PWR	842	SOUTH	Operational	WEST	31-Oct-70	05-May-81	25-May-81	30-Jul-81	
FITZPATRICK	BWR	840	ENTERGY	Operational	GE	01-Sept-68	17-Nov-74	01-Feb-75	28-Jul-75	
FORT CALHOUN-1	PWR	476	OPPD	Operational	CE	31-Jun-68	06-Aug-73	25-Aug-73	20-Jun-74	
GRAND GULF-1	BWR	1231	ENTERGY	Operational	GE	01-May-74	18-Aug-82	20-Oct-84	01-Jul-85	
H.B. ROBINSON-2	PWR	710	PROGRESS	Operational	WEST	01-Apr-67	20-Sept-70	26-Sept-70	07-Mar-71	
HATCH-1	BWR	836	SOUTH	Operational	GE	01-Sept-68	12-Sept-74	11-Nov-74	31-Dec-75	
HATCH-2	BWR	870	SOUTH	Operational	GE	31-Feb-72	04-Jul-78	22-Sept-78	05-Sept-79	
HOPE CREEK-1	BWR	1149	PSEG	Operational	GE	31-Mar-76	23-Jun-86	01-Aug-86	20-Dec-86	
INDIAN POINT-2	PWR	971	ENTERGY	Operational	WEST	31-Oct-66	22-May-73	26-Jun-73	15-Aug-74	
INDIAN POINT-3	PWR	984	ENTERGY	Operational	WEST	01-Nov-68	06-Apr-76	27-Apr-76	30-Aug-76	
KEWAUNEE	PWR	498	NUCMAN	Operational	WEST	01-Aug-68	07-Mar-74	08-Apr-74	16-Jun-74	
LASALLE-1	BWR	1130	EXELON	Operational	GE	01-Sept-73	21-Jun-82	04-Sept-82	01-Jan-84	
LASALLE-2	BWR	1130	EXELON	Operational	GE	31-Oct-73	12-Mar-84	20-Apr-84	19-Oct-84	
LIMERICK-1	BWR	1134	EXELON	Operational	GE	01-Apr-70	22-Dec-84	13-Apr-85	01-Feb-86	
LIMERICK-2	BWR	1134	EXELON	Operational	GE	01-Apr-70	12-Aug-89	01-Sept-89	08-Jan-90	
MCGUIRE-1	PWR	1100	DUKE	Operational	WEST	01-Apr-71	08-Aug-81	12-Sept-81	01-Dec-81	
MCGUIRE-2	PWR	1100	DUKE	Operational	WEST	01-Apr-71	03-May-83	23-May-83	01-Mar-84	
MILLSTONE-2	PWR	869	DOMINION	Operational	CE	01-Nov-69	17-Oct-75	09-Nov-75	26-Dec-75	
MILLSTONE-3	PWR	1136	DOMINION	Operational	WEST	01-May-74	23-Jan-86	12-Feb-86	23-Apr-86	
MONTICELLO	BWR	597	NUCMAN	Operational	GE	31-Jun-67	10-Dec-70	05-Mar-71	30-Jun-71	
NINE MILE POINT-1	BWR	621	CONSTELL	Operational	GE	01-Apr-65	05-Sep-69	09-Nov-69	01-Dec-69	
NINE MILE POINT-2	BWR	1135	CONSTELL	Operational	GE	01-Aug-75	23-May-87	08-Aug-87	11-Mar-88	
NORTH ANNA-1	PWR	925	DOMINION	Operational	WEST	31-Feb-71	05-Apr-78	17-Apr-78	06-Jun-78	
NORTH ANNA-2	PWR	917	DOMINION	Operational	WEST	01-Nov-70	12-Jun-80	25-Aug-80	14-Dec-80	
OCONEE-1	PWR	846	DUKE	Operational	B&W	01-Nov-67	19-Apr-73	06-May-73	15-Jul-73	
OCONEE-2	PWR	846	DUKE	Operational	B&W	01-Nov-67	1-Nov-73	05-Dec-73	09-Sept-74	
OCONEE-3	PWR	846	DUKE	Operational	B&W	01-Nov-67	05-Sept-74	18-Sept-74	16-Dec-74	
OYSTER CREEK	BWR	605	AMER	Operational	GE	01-Jan-64	03-May-69	23-Sept-69	01-Dec-69	
PALISADES	PWR	767	NUCMAN	Operational	CE	31-Feb-67	24-May-71	31-Dec-71	31-Dec-71	
PALO VERDE-1	PWR	1243	ANPP	Operational	CE	01-May-76	25-May-85	10-Jun-85	28-Jan-86	
PALO VERDE-2	PWR	1243	ANPP	Operational	CE	31-Jun-76	13-Apr-86	20-May-86	19-Sept-86	
PALO VERDE-3	PWR	1247	ANPP	Operational	CE	31-Jun-76	25-Oct-87	28-Nov-87	08-Jan-88	

Table 10. Continued

Station	Type	Capacity MWe	Operator[1]	Status	Reactor Supplier[2]	Construction Date	Criticality Date	Grid Date	Commercial Date	Shutdown Date
PEACH BOTTOM-2	BWR	1093	EXELON	Operational	GE	01-Jan-68	16-Sept-73	18-Feb-74	05-Jul-74	
PEACH BOTTOM-3	BWR	1093	EXELON	Operational	GE	01-Jan-68	07-Aug-74	01-Sept-74	23-Dec-74	
PERRY-1	BWR	1238	FIRSTENERGY	Operational	GE	01-Oct-74	06-Jun-86	19-Dec-86	13-Nov-87	
PILGRIM-1	BWR	667	ENTERGY	Operational	GE	01-Aug-68	16-Jun-72	19-Jul-72	01-Dec-72	
POINT BEACH-1	PWR	505	NUCMAN	Operational	WEST	01-Jul-67	02-Nov-70	06-Nov-70	21-Dec-70	
POINT BEACH-2	PWR	505	NUCMAN	Operational	WEST	01-Jul-68	30-May-72	02-Aug-72	01-Oct-72	
PRAIRIE ISLAND-1	PWR	525	NUCMAN	Operational	WEST	01-May-68	01-Dec-73	04-Dec-73	16-Dec-73	
PRAIRIE ISLAND-2	PWR	524	NUCMAN	Operational	WEST	01-May-69	17-Dec-74	21-Dec-74	21-Dec-74	
QUAD CITIES-1	BWR	762	EXELON	Operational	GE	01-Feb-67	18-Oct-71	12-Apr-72	18-Feb-73	
QUAD CITIES-2	BWR	855	EXELON	Operational	GE	01-Feb-67	26-Apr-72	23-May-72	10-Mar-73	
R.E. GINNA	PWR	498	RG3	Operational	WEST	01-Apr-66	08-Nov-69	02-Dec-69	01-Jul-70	
RIVER BEND-1	BWR	980	ENTERGY	Operational	GE	01-Mar-77	31-Oct-85	03-Dec-85	16-Jun-85	
SALEM-1	PWR	1111	PSEG	Operational	WEST	01-Jan-68	11-Dec-76	25-Dec-76	30-Jun-77	
SALEM-2	PWR	1110	PSEG	Operational	WEST	01-Jan-68	08-Aug-80	03-Jun-81	13-Oct-81	
SAN ONOFRE-2	PWR	1070	SCE	Operational	CE	01-Mar-74	26-Jul-82	23-Sept-82	08-Aug-83	
SAN ONOFRE-3	PWR	1080	SCE	Operational	CE	01-Mar-74	29-Aug-83	25-Sept-83	01-Apr-84	
SEABROOK-1	PWR	1161	FPL	Operational	WEST	01-Jul-76	13-Jun-89	29-May-90	19-Aug-90	
SEQUOYAH-1	PWR	1125	TVA	Operational	WEST	01-May-70	05-Jul-80	22-Jul-80	01-Jul-81	
SEQUOYAH-2	PWR	1125	TVA	Operational	WEST	01-May-70	05-Nov-81	23-Dec-81	01-Jun-82	
SHEARON HARRIS-1	PWR	900	PROGRESS	Operational	WEST	01-Jan-74	03-Jan-87	19-Jan-87	02-May-87	
SOUTH TEXAS-1	PWR	1264	STP	Operational	WEST	01-Sept-75	08-Mar-88	30-Mar-88	25-Aug-88	
SOUTH TEXAS-2	PWR	1265	STP	Operational	WEST	01-Sept-75	12-Mar-89	11-Apr-89	19-Jun-89	
ST LUCIE-1	PWR	839	FPL	Operational	CE	01-Jul-70	22-Apr-76	07-May-76	21-Dec-76	
ST LUCIE-2	PWR	839	FPL	Operational	CE	01-Jun-75	02-Jun-83	13-Jun-83	08-Aug-83	
SURRY-1	PWR	810	DOMINION	Operational	WEST	01-Jun-68	01-Jul-72	04-Jul-72	22-Dec-72	
SURRY-2	PWR	815	DOMINION	Operational	WEST	01-Jun-68	07-Mar-73	10-Mar-73	01-May-73	
SUSQUEHANNA-1	BWR	1105	PP&L	Operational	GE	01-Nov-73	10-Sept-82	16-Nov-82	08-Jun-83	
SUSQUEHANNA-2	BWR	1111	PP&L	Operational	GE	01-Nov-73	08-May-84	03-Jul-84	12-Feb-85	
THREE MILE ISLAND-1	PWR	816	AMEREN	Operational	B&W	01-May-68	05-Jun-74	19-Jun-74	02-Sept-74	

Station	Type	Capacity MWe	Operator[1]	Status	Reactor Supplier[2]	Construction Date	Criticality Date	Grid Date	Commercial Date	Shutdown Date
TURKEY POINT-3	PWR	693	FPL	Operational	WEST	01-Apr-67	20-Oct-72	02-Nov-72	14-Dec-72	
TURKEY POINT-4	PWR	693	FPL	Operational	WEST	01-Apr-67	11-Jun-73	21-Jun-73	07-Sept-73	
VERMONT YANKEE	BWR	505	ENTERGY	Operational	GE	01-Dec-67	24-Mar-72	20-Sept-72	30-Nov-72	
VIRGIL C. SUMMER-1	PWR	985	SCEG	Operational	WEST	01-Mar-73	22-Oct-82	15-Nov-82	01-Jan-84	
VOGTLE-1	PWR	1148	SOUTH	Operational	WEST	01-Aug-76	09-Mar-87	27-Mar-87	01-Jun-87	
VOGTLE-2	PWR	1149	SOUTH	Operational	WEST	01-Aug-76	28-Mar-89	10-Apr-89	20-May-89	
WATERFORD-3	PWR	1091	ENTERGY	Operational	CE	01-Nov-74	04-Mar-85	18-Mar-85	24-Sept-85	
WATTS BAR-1	PWR	1138	TVA	Operational	WEST	01-Dec-72	01-Jan-96	06-Feb-96	05-May-96	
WOLF CREEK	PWR	1170	WCLF	Operational	WEST	01-Jan-77	22-May-85	12-Jun-85	03-Sept-85	
MAINE YANKEE	PWR	860	MYAPC	Shut Down	CE	01-Oct-68	23-Oct-72	08-Nov-72	28-Dec-72	Aug.-97
MILLSTONE-1	BWR	641	DOMINION	Shut Down	GE	01-May-66	26-Oct-70	29-Nov-70	01-Mar-71	Jul.-98
HADDAM NECK	PWR	560	CYAPC	Shut Down	WEST	01-May-64	24-Jul-67	07-Aug-67	01-Jan-68	04-Dec-96
BIG ROCK POINT	BWR	67	CPC	Shut Down	GE	01-May-60	27-Sept-62	08-Dec-62	29-Mar-63	Aug.-97
ZION-1	PWR	1040	EXELON	Shut Down	WEST	01-Dec-68	19-Jun-73	28-Jun-73	31-Dec-73	Jan.-98
ZION-2	PWR	1040	EXELON	Shut Down	WEST	01-Dec-68	24-Dec-73	25-Dec-73	17-Sept-74	Jan.-98
PONUS	BWR	17	DOE/FRWE	Shut Down	GNE;FRWRA	01-Jan-60	01-Jan-64	14-Aug-54		01-Jun-68
CVTR	PHWR	17	CVPA	Shut Down	WEST	01-Jan-60	01-Mar-63	13-Dec-63		01-Jan-67
DRESDEN-1	BWR	197	EXELON	Shut Down	GE	01-May-56	15-Oct-59	15-Apr-60	04-Jul-60	31-Oct-78
ELK RIVER	BWR	22	RCPA	Shut Down	AC	01-Jan-59	01-Nov-62	24-Aug-63	01-Jul-64	01-Feb-68
ENRICO FERMI-1	FBR	65	DETED	Shut Down	UEC	01-Aug-56	23-Aug-63	05-Aug-66		29-Nov-72
FORT ST. VRAIN	HTGR	330	FSCC	Shut Down	GA	01-Sept-68	31-Jan-74	11-Dec-76	01-Jul-79	29-Aug-89
HUMBOLDT BAY	BWR	65	PGEC	Shut Down	GE	01-Nov-60	15-Feb-63	18-Apr-63	01-Aug-63	02-Jul-76
INDIAN POINT-1	PWR	257	CONED	Shut Down	B&W	01-May-56	02-Aug-62	16-Sept-62	01-Oct-62	31-Oct-74
LACROSSE	BWR	48	DPC	Shut Down	AC	01-Mar-63	11-Jul-67	26-Apr-68	07-Nov-69	30-Apr-87
PATHFINDER	BWR	59	NSP	Shut Down	AC	01-Jan-59	01-Jan-64	25-Jul-66		01-Oct-67
PEACH BOTTOM-1	HTGR	40	EXELON	Shut Down	GA	01-Feb-62	03-Mar-66	27-Jan-67	01-Jun-67	01-Nov-74
RANCHO SECO-1	PWR	873	SMUD	Shut Down	B&W	01-Apr-69	16-Sept-74	13-Oct-74	17-Apr-75	07-Jun-89
SAN ONOFRE-1	PWR	436	SCE	Shut Down	WEST	01-May-64	14-Jun-67	16-Jul-67	01-Jan-68	30-Nov-92
THREE MILE ISLAND-2	PWR	880	GPU	Shut Down	B&W	01-Nov-69	27-Mar-78	21-Apr-78	30-Dec-78	28-Mar-79
TROJAN	PWR	1095	PCRTGE	Shut Down	WEST	01-Feb-70	15-Dec-75	23-Dec-75	20-May-76	09-Nov-92
YANKEE NPS	PWR	167	YAEC	Shut Down	WEST	01-Nov-57	19-Aug-60	13-Nov-60	01-Jul-61	01-Oct-91

Source: IAEA.

It is important to note that the mentioned Act support the research activities for the development of Advanced High-Temperature Reactor and in the field of hydrogen. To support these research activities a significant amount of resources were allocated by the US government. A total of US$1.25 billion was authorized for an Advanced High-Temperature Reactor (next generation of nuclear power reactor) at the Idaho National Laboratory, capable of cogenerating hydrogen. Overall more than US$2 billion was provided for hydrogen demonstration projects. [99]

The Act also addressed climate change, requiring action on a national strategy to address the issue by 2007. In 2005, the USA emitted 5.9 billion tons of CO_2 to the atmosphere from energy use.

The USA energy policy regarding the use of nuclear energy for electricity generation is implemented through the US Department of Energy and involves tree important areas: a) research on new nuclear power reactor technologies; b) reinitiating nuclear power plant construction; and c) radioactive waste management.

One of the main elements of the US energy policy is "the development of standardized light water nuclear power reactors, which is the type of reactor more popular in the country, with the purpose to introduce it as soon as possible". The Department of Energy's Advanced Light Water Reactor Programme (ALWR) in the 1980s sought to create a standardized light water reactor at the earliest possible time. The implementation of this programme helped secure NRC certification for General Electric's ABWR and the Combustion Engineering's System 80+ Advanced Pressurized Water Reactor. The NRC gave final design approval to the ABWR and the System 80+ during the summer of 1994. Programmes initiated during the mid-1990s co-funded smaller (600 MWe) light-water reactors incorporating passive safety features. Westinghouse's AP-600 received design approval in 1998. [99]

Another main element of the US energy policy is a research programme of a new type of nuclear power reactor generation, the so-called "Generation IV", with new safety features, the reduction of nuclear waste and of proliferation risk, among others important innovative. The objective of this programme is to produce a Generation IV reactor by 2040.

Based on the adoption of the new US Energy Policy new bodies were established with the purpose to provide technical and scientific advice to the Department of Energy on nuclear energy matters. One of these bodies are the Nuclear Energy Research Advisory Committee (NERAC) and the International Nuclear Energy Research Initiative (I-NERI). I-NERI was established to serve as a key mechanism to set up bilateral agreements for international collaboration in developing Generation IV energy systems NERAC was established on 1 October 1998 to provide the Department of Energy and to the Office of Nuclear Energy, Science and Technology with independent advice on science and technical issues related to the Department of Energy's nuclear energy programme. NERAC reviews elements of the nuclear energy programme and provides advice and recommendations on long-range plans, priorities and strategies. NERAC also provides advice on national policy and scientific aspects on nuclear energy research as requested by the Secretary of Energy or the Director of Nuclear Energy. The Department of Energy created its Nuclear Energy Research Initiative (NERI) to address the technical and scientific issues affecting the future use of nuclear energy in the USA. NERI is expected to help preserve the nuclear science and engineering infrastructure within US universities, laboratories and industry; to advance the state of nuclear energy technology and to maintain a competitive position worldwide. [99]

Another important programmes adopted by the Department of Energy is the Nuclear Energy Plant Optimizer (NEPO) programme, with the purpose to improve performance in the operation of the current nuclear power plants in the USA. The primary areas of focus for the NEPO programme include plant ageing and optimization of electrical production. NEPO is also a public-private research and development partnership with equal or greater matching funds coming from industry. [99]

In the field of education and research the Department of Energy adopted the Nuclear Engineering Education Research (NEER) programme to support universities and professionals carrying out basic research in engineering programmes or in research reactors.

The Department of Energy's Office of Civilian Radioactive Waste Management (OCRWM) is the government office responsible for disposal of the nation's spent nuclear fuel and the high-level radioactive waste. The Department of Energy plans to store the spent nuclear waste in a deep geologic repository at Yucca Mountain, Nevada, but delays in the construction of this repository has delays the opening of this important site for the expansion of the US nuclear power programme in the future.

Evolution of the Nuclear Power Sector in the USA

Westinghouse designed the first fully commercial PWR type with a capacity of 250 MWe, in the Yankee Rowe nuclear power plant site. This unit was connected to the electric grid in 1960, started its commercial operation in 1961 and was shut down in October 1991.

Meanwhile, other type of nuclear power reactors, the BWR type, was developed by the Argonne National Laboratory and the first commercial nuclear power reactor, Dresden 1, with a capacity of 250 MWe designed by General Electric, was connected to the grid in 1960 and started up operation in the same year[27]. The Dresden 1 nuclear power reactor was shut down in October 1978. By the end of the 1960s, orders were being placed for PWR and BWR types of more than 1 000 MWe, and a major construction programme got under way. Since the first nuclear power reactor was constructed in 1957, until 2008 a total of 131 nuclear power reactors were constructed and entered into commercial operation. During the 1950s, a total of eight nuclear power reactors started their construction, two were connected to the grid and one entered into commercial operation. During the 1960s, a total of 62 nuclear power reactors started their construction, 20 were connected to the grid and 18 entered into commercial operation. During the 1970s, a total of 61 nuclear power reactors started their construction, 60 nuclear power reactors were connected to the electric grid and 60 units entered into commercial operation. During the 1980s, no nuclear power reactors were constructed, 45 units were connected to the grid and 47 units entered into commercial operations. During the 1990s, no nuclear power reactors were constructed, four units were connected to the electric grid and five units entered into commercial operations.

---

[27] A prototype of BWR, Vallecitos, ran from 1957 to 1963.

Table 11. Number of nuclear power reactors and the generation of electricity in the USA in 2008

	Number of nuclear units connected to the grid	Nuclear electricity generation (net TWh)	Nuclear percentage of total electricity supply (%)
United States of America	104	808.9	19.7

Source: WANO, 2008.

Source: Photograph courtesy of Ohio Citizen Action.

Figure33. Three Mile Island Nuclear power plant near Middletown, Pennsylvania.

As can be seen from the above figures, the peak in the construction of nuclear power reactors in the USA was, in1960s, with a total of 62 units initiating their construction. The peak of nuclear power reactors connected to the grid was reached in the 1970s with a total of 60 nuclear power reactors connected. The peak of nuclear power reactors entering into commercial operation was reached in the 1970s with 60 nuclear power reactors in this situation.

The PWRs and BWRs type of reactors remain practically the only types built commercially in the USA until today.

The development of the nuclear power programme in the USA suffered a major setback after the 1979 Three Mile Island nuclear accident, despite of the fact that no one was injured or exposed to harmful radiation as a consequence of the accident. Many orders and projects were cancelled or suspended and the nuclear construction industry went into the doldrums.

The last nuclear power reactor to be completed in the USA was Watts Bar 1, in 1996, and the construction license on a further four (Watts Bar 2, Bellefonte 1 and 2 and WNP 1) was recently extended, although there are no active construction activities on these sites. In October 2007, TVA announced that it had chosen the Bechtel group to complete the two thirds built Watts Bar 2 reactor for US$2.5 billion. Construction started in 1972, but was frozen in 1985 and abandoned in 1994. It is important to note that Watts Bar 1 was one of the most expensive nuclear power reactor ever built in the USA within its nuclear power programme. The completion of the mentioned unit took 23 years. Figure 34 shows the nuclear generation electricity in the USA from 1973 to 2002.

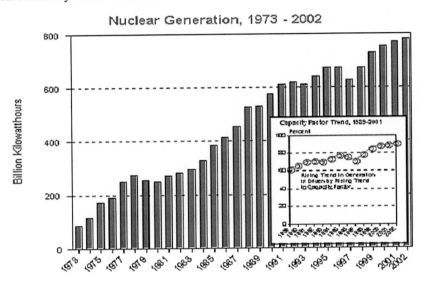

Source: IAEA.

Figure 34. Nuclear generation in the USA during the period 1978-2002.

Extension of the Lifetime of the Current Nuclear Power Reactors in Operation in the USA

The aging base of nuclear power plants for electricity generation and the increasing need for additional power in the USA, along with improved economic and safety performance of the majority of the nuclear power reactors currently in operation in the country, have push the government to accept the life extension of several of these reactors, particularly those that were built in the 1960s and 1970s[28]. At the same time, the US government has accepted, in principle, the construction of new nuclear power reactors in the coming years in order to satisfy a foresee increase in the energy demand. This position of the US government led many licensees to renew their operating licenses for an additional 20 years beyond their initial 40 year limits. Twenty-three nuclear power reactors have extended their operating licenses since 2000; this means 22% of the total nuclear power reactors currently in operation in the country. Applications to extend the licenses of at least 35 additional units were presented through 2006; this means 34% of the total nuclear power reactors currently in operation. It is expected that almost all operators of nuclear power plants eventually apply for operating license renewals of almost all operating nuclear power reactors in the USA, taking into

---

[28] Despite the failure of the US nuclear industry to build more nuclear power reactors the nuclear power industry remains highly successful in two main areas, increased output from existing nuclear power reactors and plant life extensions.

account the need to satisfy a growing demand of electricity in the coming years, the high level of oil price reached in 2007 and in 2008, the uncertainty on the level of the oil price in the coming years, and the increase in the price of natural gas and clean coal, among other factors.

In parallel to the extension of the operational licenses of an important number of nuclear power reactors [29], certain power companies are showing great interest in the construction of new nuclear power plants. Three utilities, Diminion Resources, Exelon and Entergy have applied during 2004 the so-called Early Site Permits (ESP). These applications will allow the above mentioned utilities to initiate nuclear power site clearances prior to commitments to build a nuclear power plant. Several other firms have indicated that they might be interested in requesting the initiation of the ESP process in the near future. Recently, the Detroit Edison company had selected the ESBWR reactor from General Electric-Hitachi for the possible construction of a new nuclear power reactor with a capacity of 1 520 MWe to be built in Michigan in the site of Fermi nuclear power plant. The license request was presented at the end of 2008[30].

Reduction of the Construction Time of Nuclear Power Plant in the USA

Progress has been made by nuclear supplier and utility in recent years in order to reduce the costs of building new nuclear power reactors by a reduction of the construction time. The NRC has also streamlined its licensing process for the construction of future nuclear power reactors with the same purpose. The goal of these changes is to shorten construction lead-times and to improve the economics of new nuclear power reactor technology. The US government goal is that these and other similar actions might encourage power industry to restart the construction of new nuclear power plants by the end of the decade.

However, it is important to stress the following: Investments in commercial nuclear power plants generating will only be forthcoming if investors are sure that the construction time will be reduced significantly and cost of producing electricity using nuclear energy will be lower than the costs associated with the use of alternative electric generation technologies. The renewed interest in the use nuclear energy for electricity generation has resulted, primarily, from higher prices for oil and natural gas occurred in 2007 and 2008, the decrease in oil proven reserves worldwide, a significantly improvement of the operation of existing nuclear power reactors, an increase in nuclear safety and uncertainty about future restrictions on coal emissions.

Until the recent price volatility, low fuel costs had helped gas-fired power plants dominate the market for new electric generation capacity since the late 1980s. Nuclear power's relatively stable costs and low air emissions may now appear more attractive, particularly combined with a substantial tax credit for nuclear generation and other incentives provided by the Energy Policy Act of 2005. However, large uncertainties about nuclear power plant construction costs still remain, along with doubts about progress on nuclear waste disposal and concerns about public opposition. All those problems helped cause the long

---

[29] According with the WANO information, since 1997 the NRC has approved 110 requests for extension capacity for the current operating nuclear power reactor in the USA, representing an increase of 4.7 GW in the nuclear capacity in the country.

[30] According with the US procedures for the construction of nuclear power plants, the presentation of a license request is not a commitment to initiate the construction of a plant.

cessation of US reactor orders and will need to be addressed before financing for new multibillion-dollar nuclear power plants is likely to be obtained. (Parker and Holt, 2007)

The main problems faced by the US nuclear industry related with the construction of new nuclear power reactors could be summarized in the following manner: Economic, problems in construction and opposition to them, which led to increased construction times and subsequently increased construction costs. Many utilities went bankrupt over nuclear power projects. The estimated cost of building a nuclear power plant rose from less than US$400 million in the 1970s to around US$4,000 million by the 1990s, while construction times doubled from 1970s to 1980s. These facts led the US business magazine *Forbes* in 1985 to describe the industry as "the largest managerial disaster in US business history, involving US$100 billion in wasted investments and cost overruns, exceeded in magnitude only by the Vietnam War and the then Savings and Loan crisis". (Schneider and Froggatt, 2007)

The Future of Fusion Technology in the USA

The nuclear fusion has been the power ideal for more than half century, but the problems that have being impeding until now the use of this type of energy for electricity production are not been solved in a satisfactory manner and for this reason, the use of fusion technology for the generation of electricity is not yet ready to be used commercially and will not so at least until 2050. According with several experts' opinions, it is expected that nuclear fusion will not be available for the production of electricity before 2050. The USA, that reduced significantly basic research in nuclear fusion some years ago, has now announced that the Laboratory Lawrence Livermore has begun fusion tests on May 2009. The tests will be extended until 2012 and the objective of it is to demonstrate that it is possible to generate thermonuclear energy. Why takes so long time to obtain specific results in this field. The answer is the following: The physics of the fusion is very difficult and the technology that there is to develop to prove the physical theoretical principles experimentally is also very complex and expensive, and all this demands long time", assures Mr. Diaz the Blonde, head of Investigation and Development of Lawrence Livermore National Laboratory, in California, USA.

According to Diaz of the Blonde, "we hoped that a power gain of the order of a factor of 10 takes place. But this is not sufficient, which causes that it is precise a mixed system fusion and fission, a concept that already formulated Andrei Sakharov. It is a very interesting alternative and it allows closing the cycle of the nuclear energy of a very safe form, since the part of the fission is not the normal one (it requires a critical mass of nuclear fuel neither uranium enrichment nor reprocessing of the radioactive waste), reason why the probability of a accident like the Chernobyl one is zero. The hybrids of fusion and fission can be a power alternative from 2025. These characteristics, plus the fact that in our concept the power gains are enormous and very high amounts of electricity of base without emitting CO_2 can be produced, allow us to think that this will be a very interesting thing in the mid-term". According to the calculations of Diaz of the Blonde, at least 10 more years are needed to construct the prototype of a commercial plant that generates energy using the fusion technology. However, there is a great expectation in using hybrids of fusion and fission technologies to produce energy and this could be ready in the second quarter of the 21th century. In the opinion of Diaz of the Blonde, the power model of the future must be mixed, with combinations of sources, including the improvement of the current renewable energies systems and the advance in the use of the fusion and the fission technologies. The scientist community is conscious of the distrusts that the nuclear energy provoke between the public opinion and, for this reason, the concept of fusion-fission by confinement has tremendous advantages and could allows to think about the possibility of expanding the nuclear energy in a safe form in the future reducing the volume of radioactive waste volume.

Management of the High-level Nuclear Waste

One of the problems that are limiting the construction of new nuclear power reactors in the USA is the management of nuclear waste, particularly the final disposal of the radioactive waste.

Highly radioactive spent nuclear fuel produced by nuclear power reactors poses a disposal problem that could be a significant factor in the consideration of new nuclear plant construction. The Nuclear Waste Policy Act of 1982 (NWPA, P.L. 97- 425) commits the federal government to providing for permanent disposal of spent nuclear fuel in return for a fee on nuclear power generation. The schedule for opening the planned national nuclear waste repository at Yucca Mountain, Nevada, has slipped far past NWPA's deadline of January 31,1998. DOE currently hopes to begin receiving waste at Yucca Mountain by 2017. In the meantime, more than 50 000 metric tons of spent fuel is being stored in pools of water or shielded casks at nuclear facility sites. NWPA limits the planned Yucca Mountain repository to the equivalent of 70 000 metric tons of spent fuel. (Parker and Holt, 2007)

It is important to single out that US nuclear power plants produce an average of 2 000 metric tons of spent fuel per year. Taking into account this average the Yucca Mountain limit is likely to be reached before any new nuclear power reactors begin coming on line. For this reason, the management of the spent nuclear fuel coming from any new nuclear power reactor that starts it operation after 2017 need to be thoroughly study before a decision is adopted on the construction of new nuclear power plants, because the Yucca Mountain repository cannot be used for final disposal of the new spent nuclear fuel coming from the operation of these nuclear power reactors.

Of course the storage of the spent nuclear fuel in the nuclear power reactor sites, within a period of 30 years after the reactors license expired, is one of the possibilities that the utilities have as an alternative now in order to proceed with the construction of new nuclear power plants. The NCR is estimating an operation license period of 70 years for the new types of nuclear power reactors. Therefore, NRC does not consider the lack of a permanent repository for spent fuel to be an obstacle to nuclear plant licensing [31]. (Parker and Holt, 2007)

In addition to what has been said in previous paragraphs, the reduction of the long construction time of a nuclear power reactors, as well as the reduction of the initial capital cost associated with the construction of this type of power plants, are two of the most

---

[31] However, it is important to bear in mind that at least six states, California, Connecticut, Kentucky, New Jersey, West Virginia and Wisconsin, have specific laws that link approval for new nuclear power plants to adequate waste disposal capacity and that the US Supreme Court has held that state authority over nuclear power plant construction is limited to economic considerations rather than safety, which is solely under NRC jurisdiction.

important factors that need to be solved, as soon as possible, in order to encourage investors to finance the construction of new nuclear power reactors in the USA.

New NRC Regulations

What the US authorities has done to reduce the construction time and to avoid a long waiting period for entry into operation of a nuclear power reactor in the future? Until 1989, licensing a new nuclear power facility involved a two-step process: 1) an NRC-issued construction permit that allowed an applicant to begin building a nuclear power reactor; and 2) an operating license that permitted the nuclear power plant to generate electricity for sale.

This procedure resulted in some cases in which completed or nearly completed nuclear power reactors awaited years to be granted operating licenses — delays that drove up the costs of the affected plants. In 1989, NRC issued regulations to streamline this process in three ways:

1) the Early Site Permit Programme (ESPR) allows utilities to get their proposed reactor sites approved by the NRC before a decision is made on whether or not to build the nuclear power plant;
2) Standard Design Certification (SDC) for advanced nuclear reactor designs allows vendors to get their designs approved by NRC for use in the USA, so utilities can then deploy them essentially "off the shelf";
3) Combined Construction and Operating License (COL) provides a "one-step" approval process, in which all licensing hearings for a proposed nuclear power plant are expected to be conducted before construction begins. The COL would then allow a completed plant to operate if inspections, tests, analyses and acceptance criteria (ITAAC) were met. This is intended to reduce the chances for regulatory delays after a plant is completed. (Parker and Holt, 2007)

After the license is issued, which could take three and a half years to be issued, the utility must decide whether to begin building the nuclear power plant or cancelled the project. Current projections of nuclear power plants construction schedules assume that a nuclear power plant can be built between four and seven years. At the end of the construction, the NRC verifies that the new nuclear power reactors meets all the necessary requirements established by the US competent authorities, with the purpose to allow the facility to start its operation as soon as possible. Overall, the process is anticipated that could take between 10 to 15 years to be concluded.

There were 27 license renewal applications presented in the USA in 2007. Twelve of these applications were under review in 2008. In general, as of January 2007, the NRC has renewed the operating license of 48 nuclear power reactors, representing 46.2% of the total nuclear power reactors operating in the USA. It is expected that NRC will renew the operating license of additional 85 nuclear power reactors current operating in the USA.

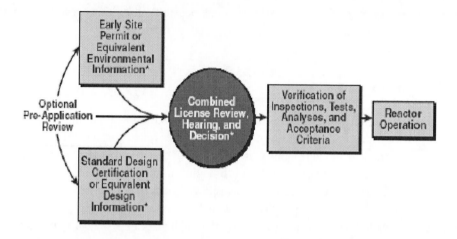

*A combined license application can reference an early site permit, a standard design certification, both, or neither. If an application does not reference an early site permit and/or a standard design certification, the applicant must provide an equivalent level of information in the combined license application.

Source: NRC.

Figure 35. Relationship between Combined Licenses, Early Site Permits and Standard Design Certification approved by NRC.

Finally, it is important to note that the NRC has authorized the power increase of two units of the North Anna nuclear power plant in the State of Virginia. The global power increase will be of approximately 23 MWe. The North Anna nuclear power plant has two PWR Westinghouse reactors that began operation in 1978 and 1980 respectively and whose operation license was extended by 20 years in 2003. At the same time, NRC has renovated for additional 20 years the operation license of unit 1 of the Three Mile Island nuclear power plant. The reactor, a PWR of 852 MWe that begun operation in 1974 and should be shut down definitively in 2014, is now authorized to operate up to 2034, after a process that included the revision of the environmental aspects and the operation safety of the unit. The company operating the Three Mile Island nuclear power plant has informed that, in 2009, two steam generators will be replaced during the stop programmed for the change of fuels. The unit is the fiftieth fifth American nuclear power reactor that received authorization to prolong their useful life. The NRC has under analysis the extension of the useful life of other 20 nuclear power reactors.

The Nuclear Industry in the USA

The US nuclear industry is one of the most advanced industries in the world nuclear sector. All necessary elements and components of a nuclear power plant can be produced in the country. Four companies supplying nuclear systems currently operate in the USA. These companies are the following:

a) Westinghouse Corporation, which built the majority of the PWRs.
b) Combustion Engineering (CE) and Babcock & Wilcox (B&W).
c) Westinghouse and CE, which are now part of Westinghouse BNFL while Areva NP now owns elements of B&W's nuclear technology.
d) General Electric which designed all presently operating BWRs in the USA.

There are at least two nuclear power reactor designs approved by the NRC for construction in the USA. These are the following: a) The Westinghouse System 80+ and AP 600; and b) General Electric and Hitachi ABWR designs of 1 300 - 1 500 MWe.

In the last case, it is important to single out that in September 2009 GE Nuclear Hitachi Energy informed that it has presented to the NRC the revised design of its new Economic Simplified Boiling Water Reactor (ESBWR). The presentation marks an important landmark in the advance in the process of NCR approval of the reactor of 1 520 MWe that two American companies have selected for the construction of two new nuclear power reactors in the coming years. In accordance with GE Hitachi, the new design improves significantly different aspects related with the safety of the reactor and it has economic advantages on the previous design.

There are currently eight nuclear reactor designs that are either undergoing certification or pre-certification procedures with the NRC. It is anticipated that other designs will join this process in the near future. The Westinghouse BNFL AP1000 design is currently undergoing final certification[32]. The following six nuclear power reactors designs are currently undergoing pre-certification:

1) the General Electric ESBWR reactor[33];
2) the Areva NP SWR-1000 reactor;
3) the General Atomics GT-MHR reactor;
4) the Atomic Energy of Canada ACR-700 advanced CANDU design reactor;
5) the Eskom PBMR reactor;
6) the Westinghouse BNFL IRIS reactor.

---

[32] Compared with reactors that produce the same amount of power, the BNFL AP 1000 reactor use 50% fewer valves, 35% fewer pumps, 70% less cabling and can be contained in a building almost half the size compare with other types of nuclear power reactors available in the market. However, the NRC has informed Westinghouse whose current owner is Toshiba that the building of the reactor AP1000 will require further analysis, new rehearsals and possibly design changes to be able to fulfill the requirements settled down by the NRC. Westinghouse has committed to make all the studies and necessary modifications to satisfy in a quick and definitive form the demands of the NRC, with the purpose of completing its goal of having the certifications required for 2011. According to their plans, the first AP 1000 reactor in the USA should be operative in 2016.

[33] The GE-ESBWR is a reactor built to withstand earthquakes, hurricanes, tornadoes and the impact of an aircraft collision. The ESBWR reactor, which has a design life of 60 years and can be built in three years only, according with the information provided by the supplier, is currently going through the US design certification process. Manufacturers are hopeful that the ESBWR reactor will be in commercial operation by 2015.

The US Uranium Resources and the Mining Industry

The USA ranks fourth in the world for known uranium resources, with reserves of 342 000 tU (reasonably assured plus inferred resources certified in 2005). Exploration expenditure reached US$50.3 million in 2007, which is more than doubled the 2006 level. The peak production was 16 800 tU reached in 1980, when there were over 250 uranium mines in operation in the country. This number abruptly dropped to 50 in 1984 when 5 700 tU was produced, and then there was steady decline until 2003, with most US uranium requirements for the operation of its nuclear power reactors being imported. By 2003, there were only two small mines in operations in the country producing a total of less than 1,000 tons of uranium per year. Most US production has been from New Mexico and Wyoming. Known resources are 167 000 tons of U-3O8 in Wyoming, 155,000 tons in New Mexico, 2,000 tons in Texas and around 50,000 tons in Utah, Colorado and Arizona, all to US$50/lb. [99]

The Public Opinion

US public opinion supports the use of nuclear energy for the production of electricity and this support is continuing an upward trend during the last years. In 2005, a National Survey was carried out by Bisconti Research, Inc., in which 1 000 adult's age18 and older participated. As result of this survey, around 70% of the public has a position in favor of using nuclear energy for electricity generation in the country in the future. In addition, the survey found that 85% of the participants agree that the current nuclear power plant license should be renewal.

According with the result of this survey, planning for the future has become a basic value in considerations about energy policy. Seventy seven percent agreed that electric companies should be prepared to build new nuclear power reactors, when needed in the future, and 71% would approve companies participating in federal site approval reviews for a possible construction of new units at nuclear power plant site nearest to where they live. Also, 69% said that if a new nuclear power reactor were needed to supply electricity, they would find it acceptable to add a new units at the nearest existing nuclear power plant. [142]

The Nuclear Energy Institute (NEI) has sponsored national polls for 23 years. The trends on some of the questions included in these survey cover two decades. In the period 1983-May 2005, the gap between those who favor and oppose the use of nuclear energy for electricity generation increased. Figure 36 shows this trend.

While the overall upward trend shows in Figure 36 is important, even more significant is the widening gap between those who strongly favor and strongly oppose the use of nuclear energy for electricity generation. Why this parameter is so important? The answer is very simple: Persons with strong opinions are most likely to take action one way or the other. In 2005, a total of 32% of Americans strongly favor nuclear energy compared with only 10% who are strongly opposed. In other words, one out of three strongly supports the use of nuclear energy for the production of electricity.

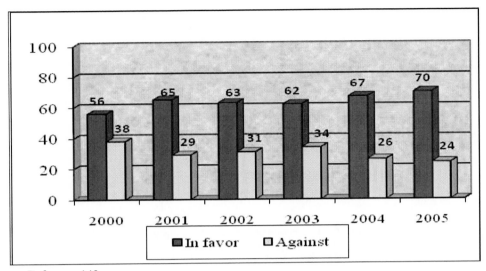

Source: Reference 142.

Figure 36. Percent of persons that are in favor and against the use of nuclear energy for electricity generation in the USA.

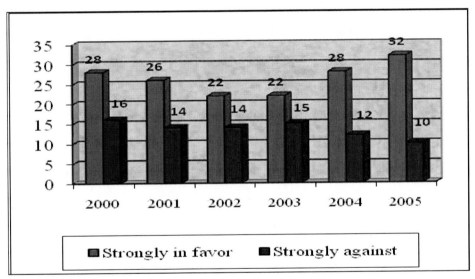

Source: Reference 142.

Figure 37. Percentage of persons in the USA that are strongly in favor and against of the use of nuclear energy for electricity generation.

Since 2003, there is a systematic increase in the number of persons in the USA that strongly support the use of nuclear energy for electricity generation. According with the data included in the figure below, this increment went up from 22% to 32% in the last two years, representing an increase of 10 points within this period.

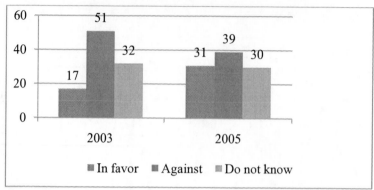

Source: Data reference 142.

Figure 38. Evolution of the position of the community on the use of nuclear energy for electricity production.

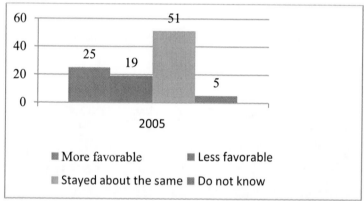

Source: Data reference 142.

Figure 38. Position of the community with regards the use of nuclear energy for electricity production.

The perception gap between actual and perceived public opinion may be narrowing now more people are aware that majority public opinion favors nuclear energy. Compared with two years ago, more people believe that the majority of their community favors the use of nuclear energy up from 17% to 31%, a 14 percentage-point increase". [142]

Among the American public, there is near consensus support for license renewal for nuclear power plants that continue to meet federal safety standards. There is also broad support for keeping the option to build more nuclear power reactors in the future and preparing now so that new reactors could be built, if needed, in the next decade. Fifty-eight percent agree with definitely building more nuclear power reactors. Opinions on this last measure vary according to perceived need. [142]

From what has been said in previous paragraphs, the US public opinion has generally been fairly positive, and has grown more so as people have had to think about security of energy supplies. In March 2006, a national survey revealed that 68% of people favoured the use of nuclear energy, while 86% believed nuclear would be important to meeting electricity needs in the years ahead. Some 73% would find a new reactor at the nearest nuclear power plant acceptable. In mid 2007, a survey of 1,150 people living within 16 km of nuclear power plants in the USA, but without any personal involvement with them, showed very strong support for the construction of new nuclear plants. Over 90% thought nuclear energy was important for future supply, 82% favoured it now, 77% said that new plants should definitely be built and 71% said they would accept the construction of a new nuclear power plant near them.

There was an overwhelmingly favourable view of local nuclear power plants, notably their safety. On nuclear waste, 71% said it was safe being stored at the plant site and 78% said the federal government should get on with developing the Yucca Mountain repository. Regarding reliable sources of information about nuclear energy, various nuclear power plant sources were rated 68% -74% compared with environmental groups 45% and anti-nuclear groups 22%. An April 2008, survey covering 1,000 people found that overall 82% said nuclear power will be important in meeting the nation's electricity needs in the years ahead. In a change since October 2007, most now put economic growth ahead of climate change and energy security as a prime concern, with air pollution trailing in a list of four. Public support for building new nuclear power plants strengthened three points to 78% since October.

A May 2008 survey covering 2,925 people carried out by Zogby showed 67% of Americans favoured building new nuclear power plants, with 46% registering strong support and 23% were opposed. Asked which kind of power plant they would prefer if it were built in their community, 43% said nuclear, 26% gas and 8% coal. [99]

In another nationwide survey, conducted in March 2009, covering 1,000 adults the results obtained was the following: Almost 70% of Americans favor the use of nuclear energy for electricity generation, with the number of Americans voicing strong support exceeding those strongly opposed by a margin of more than two to one[34]. The survey also showed that solid majorities of Americans are willing to see existing nuclear power plants expanded, believing that the USA should definitely build more nuclear power reactors in the future, and support federal incentives to build nuclear, wind and solar facilities to meet future electricity demand. This is the second consecutive survey which has found that 70% or more of Americans are in favor to use nuclear energy for the production of electricity and express their willingness to approve the construction of new nuclear power reactors at the nearest existing nuclear power plants. In September 2008, around 74% of respondents said they view nuclear energy favorably, with only 24% against.

Summing up can be stated that the majority of the American people favor the use of nuclear energy for electricity production as a real alternative to satisfy the foresee increase in the demand of electricity in the coming years.

Looking Forward

The future in the use of nuclear energy for electricity generation in the USA will depend on several factors. Among these factors are the following: 1) successfully dealing with nuclear waste issues, particularly high-level nuclear waste; 2) reduction of capital costs; 3) reduction in the construction time; 4) increase in nuclear safety; and 5) favorable government policies.

---

[34] The survey was conducted for the Nuclear Energy Institute by Bisconti Research Inc., in conjunction with GfK Roper. It has a margin of error of plus or minus 3% points.

The Department of Energy has recently initiated a Generation IV programme to develop innovative and new commercial nuclear power reactor designs by 2040. Why the development of new nuclear power reactors designs is so important? According with recent estimates, it is expected that the total US electricity consumption will increase another 20% to 30%, at least, over the next ten years. The power to satisfy the increase demand of energy has to come from somewhere. Neither solar, wind, hydroelectric nor other available alternatives energy sources in the market will provide the necessary additional energy supply to satisfy the foresee increase in energy demand in the USA in the future. The only technology that could decisively reduce the current level of US carbon emissions to the atmosphere in the near term and could satisfy the foresee increase in the energy demand is the nuclear technology that uses uranium as a fuel. Uranium is emission-free and is available in several countries located in stable political regions. It is time, due to economic, ecological and geopolitical reasons, that the US politicians adopt a decision to promote the use of nuclear energy for the generation of electricity in the country in the future.

At the same time, the North American company Babcock & Wilcox (BW) has presented recently a new modulate design of nuclear power reactor with a capacity power of 125 MW, that will be able to complement working in base to the eolic and solar power plants.

Up to now project of low nuclear power reactors had not been very successful because they went against the economy of scale and have to follow the same administrative process that the big nuclear power reactors to obtain the authorization of operation. However, the promoters of the new design of low nuclear power reactor believe now that their size makes that the evaluations of costs more precise and that it modulated design adapts better to the demands of the market, as for its use with other different energy sources.

A reactor of 125 MW can give electricity to at least 100,000 homes, and its power can enter in an electric grid without requiring special adjustments, like it happens in the case of reactors of 1 000 MW or more power. On the other hand, their construction is simple and it can be carried out, according to BW source, in their facilities located in USA and Canada.

The proposed reactor has passive security and an underground contention structure. A 5 year-old operative cycle has been planned without recharging of fuel. Their nucleus will have 59 fuel elements and the spent nuclear fuel will be stored in a pool for its 60 years of foreseen design life. The pressure vessel of the reactor will harbor the nucleus, the vapor generator, the coolant primary circuit and the mechanisms for the control of the bars. With this structure there will not be any penetrations of the primary system to the reactor vessel.

On the other hand, the electric company Energy Northwest, the owner of the Columbia nuclear power plant, has declared that there are studying the possibility to build a modulated nuclear power reactor of 45 MW to be produced by the company NuScale Power Inc., built in USA by the company Electric Boat that provide the nuclear power reactors for the nuclear submarines of the US fleet. The standard design would have 12 modules, with a total of 480 MW. Their construction today would be near to US$4,000 dollars for kW without further increase and it would be carried out from 36 to 42 months.

It is important to note that there are some initiatives that have been adopted by the US government to encourage the power industry to consider the construction of new nuclear power plants in the coming decades. Some of these initiatives have been described in previous paragraphs. Other initiatives have been adopted with the same purpose. One of these initiatives provided by the Energy Policy Act, is the adoption of the so-called Nuclear Production Tax Credit. Through this initiative a US$1.8 cents/kWh tax credit for up to 6 000

MW of new nuclear power capacity for the first eight years of operation, up to US$125 million annually per 1 000 MW, is offered by the US authorities to the utilities that request a license construction for a new nuclear power plant. An eligible nuclear power reactor under this initiative must be entered into service before January 1, 2021.

Another initiative to reduce cost and avoid unnecessary delays in the construction and in the beginning of the operation of new nuclear power plants, is the adoption of the so-called Regulatory Risk Insurance. Continuing concern over potential regulatory delays, despite the streamlined licensing system now available in the country described in previous paragraphs, prompted Congress to include an insurance system that would cover some of the principal and interest on debt and extra costs incurred in purchasing replacement power because of licensing delays. The first two new nuclear power reactors licensed by NRC that meet other criteria established by DOE could be reimbursed for all such costs, up to US$500 million apiece, whereas each of the next four newly licensed reactors could receive 50% reimbursement of up to US$250 million. (Parker and Holt, 2007)

Lastly, it s important to single out that the company Hyperion Power Generation of Denver, Colorado, has identified investors interesting in building small nuclear power reactors. The company has received already 70 intention letters and around US$100 million for this purpose. However, the company will have to wait four years before the final project reach the market. The US regulator still has to approve the design of this mini-reactor, process that could usually requires several years.

Summing up can be stated that the only real alternative the US government has to reduce the consumption of oil, gas and carbon for electricity generation in the coming years, satisfy the increase foresee in the energy demand and at the same time reduce the current high level of CO_2 emission to the atmosphere, is the construction of new nuclear power reactors.

THE NUCLEAR POWER PROGRAMME IN CANADA

The energy sector is an important part of Canada's economy converting the country in the world's sixth largest producer of electricity. Canada is also the world's largest producer and exporter of uranium. Canada has been a net exporter of most energy forms since 1969. Only in 2002, Canada exported energy to the USA for a value of C$50 billion[35].

Canada has one of the most diversified bases of electricity generation in the world, which includes hydroelectricity, natural gas, oil, coal, nuclear energy and renewables. The energy sector employs just fewer than 300,000 Canadians (or about 1.8% of the Canadian labor force) and accounts for about 6.2% of Canada's GDP[36]. The Canadian energy sector enjoys a strong presence in all primary energy commodities and strong electricity and energy efficiency industries. Electricity accounts for about 15% of domestic energy requirements. [15]

---

[35] The USA is by far Canada's largest customer (over 90% of Canada's energy exports). Virtually all of Canada's exports of oil, natural gas and electricity and 41% of uranium exports go to the USA.

[36] The information is until December 2003.

Canada had 18 nuclear power reactors in operation in 2009[37] with a net capacity of 12 652 MWe. The production of electricity using nuclear energy represented, in 2007, around 15% of the total Canada's electricity. The net load factor per type of all nuclear power reactors in operation in Canada in 2007 was 79.8%[38]. Canada shut down three nuclear power reactors, Gentilly 1 with a net capacity of 250 MWe in June 1977, Douglas Point with a net capacity of 206 MWe in May 1984 and NPD with a net capacity of 22 MWe in August 1987. Four additional units Bruce 1 and 2 and Pickering 2 and 3were in long-term shut down by Canada in 2007.

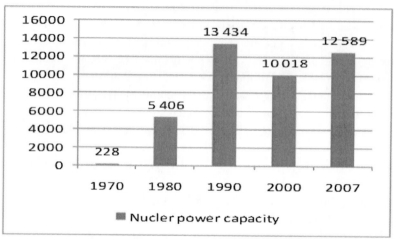

Source: CEA, 8th Edition, 2008.

Figure 40. Evolution of nuclear power plants capacities in Canada during the period 1970-2007.

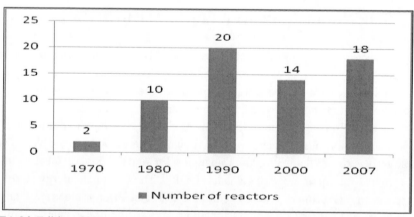

Source: CEA 8th Edition, 2008.

Figure 41. Evolution of nuclear power reactors in Canada during the period 1970-2007.

---

[37] The provinces in which these nuclear power reactors are located are the following: Ontario (16), Quebec (1) and New Brunswick (1).

[38] In Canada, there is only one type of nuclear power reactor in operation that was produced locally. It is the PHWR CANDU type.

Table 12. Canadian nuclear power data

	Canada	Ontario	New Brunswick	Québec
Total electricity generation (Growth %)	2.0	0.6	-10.0	3.6
Nuclear share of electricity generation (%)	12.8	50	21.0	2.5
Reactors in service	18	16	1	1
Net installed capacity (MW)	12 589	11 234	635	635

Sources: Natural Resources Canada, Statistics Canada and CEA, 8th Edition 2008.

The evolution of nuclear power plants capacities in Canada during the period 1970-2007 and the number of nuclear power reactors in operation in this period are shown in Figure 40.

Nuclear power reactors operating in Pickering nuclear power plant are shown in Figure 42. The location of all nuclear power plants operating in Canada is shown in Figure 43.

Source: http://www.icjt.org/npp/foto/747_2.jpg

Figure 42. Pickering nuclear power plant.

The Energy Policy in Canada

Canada's energy policy supports a variety of energy sources, including nuclear energy, in order to ensure a secure and sustainable energy supply for the country. The following are the three major areas of active federal energy policy development: a) conventional and renewable energies; b) nuclear energy; and c) environment.

In the case of nuclear energy, the Canadian government supports the use of nuclear energy for electricity generation as an important component of a diversified energy balance and as an effective manner to reduce CO_2 emissions to the atmosphere.

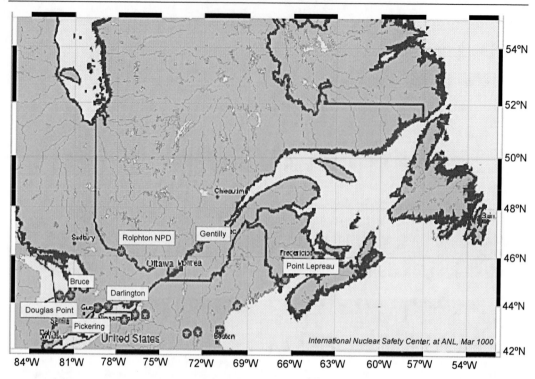

Source: International Nuclear Safety Center, At Argonne National Laboratory, USA.

Figure 43. Location of the nuclear power reactors in Canada.

The Atomic Energy of Canada Limited's (AECL)[39] is the federal government agency in charge of the promotion of the use of nuclear energy for electricity production in the country. AECL has both a public and commercial mandate. It has overall responsibility for Canada's nuclear research and development programmes as well as the Canadian reactor design (CANDU), engineering and marketing programme. Canada also has an indigenous nuclear power industry established around the CANDU reactor technology. Private sector firms, which undertake the manufacturing of CANDU components and the engineering and project management work for reactors both inside and outside of Canada, act as subcontractors to AECL. [15] The basic structure of the CANDU reactors is ahown in Figure 44.

There are now 44 CANDU nuclear power reactors in operation all over the world. The load factor of this type of reactor in 2007 was 77.5% (second place per type of reactor). The construction time for a CANDU-6 nuclear power reactor is between 4 and 5 years.

---

[39] In accordance with recent declarations made by the Minister of Natural Resources of Canada, is probable that the government decides the division in two parts of AECL. One part would be devoted to the sale nuclear power reactors and the other one to research and development of nuclear technology together with the production of radioisotopes. The government hired N.M. Rothshild & Sons for the elaboration of the restructuring plan of AECL. Although the complete report of the consultant company will be finished in 2010, the Minister was in favor of the division of the company and with the participation of private capitals.

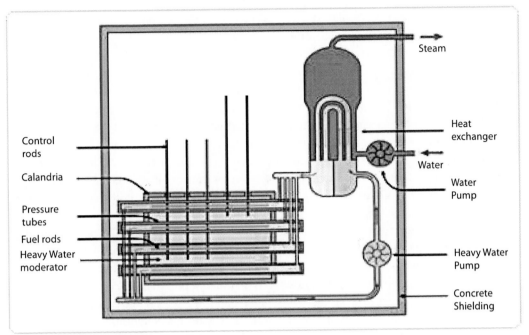

Source: Electrical Engineering Tutorials.

Figure 44. Pressurized Heavy Water Reactor (CANDU)[40].

One of the characteristics of the production of electricity in Canada is that this responsibility falls within the jurisdiction of the provinces, which at the same time, own the natural resources located in the provinces and are responsible for most aspects of regulation and energy sector development within their geographical boundaries. What is then the role of the federal government? The federal government is responsible for harmonizing energy policy at the national level, promoting regional economic development, frontier lands, offshore development, inter-provincial works (e.g. pipelines), international and inter-provincial trade. As a result of the division of the energy policy jurisdiction, Canada's electricity industry is organized along provincial lines. In most provinces the industry is highly integrated, with the bulk of the generation, transmission and distribution provided by a few dominant utilities. Although some of these utilities are privately owned, most are Crown corporations owned by the provinces. Among the major electric utilities, seven are provincially owned, seven are investor owned, two are municipally owned, and two are territorial Crown corporations. Provincial electric utilities own about 80% of Canada's total installed generating capacity and produced around 75% of total generated electricity. Traditionally, there have been three nuclear utilities in Canada: Ontario Power Generation (OPG), Hydro-Quebec and New Brunswick Power. A fourth, Bruce Power Inc. was added to the list in May 2001, when it leased the 8 nuclear power reactors at the Bruce generating station from OPG. [15] In addition to Bruce Power the three provincial utilities mentioned

---

[40] This nuclear power reactor is of a Canadian design, and it is known as CANDU reactor. These reactors are heavy-water-cooled and moderated PWR reactors. Instead of using a single large containment vessel as in a PWR system, the fuel is contained in hundreds of pressure tubes. These reactors are fuelled with natural uranium and are thermal neutron reactor designs. PHWR reactors can be refueled while at full-power, which makes them very efficient in their use of uranium (it allows for precise flux control in the core).

above, particularly OPG, have played a critical role in the development of Canada's nuclear power programme. These provincial utilities have worked closely with AECL in the design and construction of the CANDU nuclear power reactors currently in operation in their respective provinces.

Despite of the increase in oil price in 2007 and in 2008, the negative impact in the environment due to the increase use of fossil fuel, the ratification of the Kyoto Protocol and the positive and rich experience in the use of nuclear energy for electricity generation in the country, there are currently no firm plans in Canada to build additional nuclear power reactors. However, there is growing recognition that nuclear energy is an indispensable type of energy to meet future demand in the electricity sector, meet climate change and air quality commitments and, for this reason, should be taken into consideration in the configuration of the energy balance of the country for the coming years.

It is important to single out that, for the time being, the focus of the nuclear utilities operating nuclear power reactors is in the following two important areas: 1) servicing of existing nuclear power reactors; and 2) refurbishment of some of the nuclear power reactors operating in the country in order to increase their operational life.

However, for the next decade, there are better opportunities for the deployment of new nuclear power generating capacity in Canada. The reason for this conclusion is the following: AECL is currently working on the development of the 700 MW Advanced CANDU Reactor (ACR), and is aiming at reducing the capital cost associated with the construction of this type of reactor by up to 40%. The economics of the new nuclear power reactor has been ranked highly by international experts relative to other advanced reactors and has the potential to be cost competitive with other types of power generation. The ACR technology could provide an economic replacement for existing nuclear power reactors as they reach the end of their service lives, as well as for some new nuclear power plants in Canada and abroad. [16]

The Evolution of the Nuclear Energy Sector in CANADA

Canada has developed its own line of nuclear power reactors, the so-called CANDU reactor technology. Which are the contribution of the nuclear energy sector to the Canadian economy and the outcome of the investment made in this sector in the last 56 years? Nuclear energy contributes some C$5 billion per year to the Canadian economy and provides 20,000 direct jobs (2,000 in mining and uranium processing) and many more indirect jobs. The total nuclear electricity generated has a value of about C$3.7 billion per year and helps Canada minimize CO_2 emissions to the atmosphere from electric power generation.

About C$6 billion was invested in Canada to develop its nuclear power programme over the period 1952-2006. This investment has generated more than C$160 billion in GDP benefits to Canada from power production, research and development, CANDU exports, uranium, medical radioisotopes and professional services, according to AECL. A study by the Canadian Energy Research Institute found that the nuclear industry contributes about C$6 billion annually to Canada's GDP, while government research and development investment in it is about C$130 million. [18]

The Canadian's nuclear power programme is based on the use of heavy water natural uranium reactor system, which uses pressurized fuel channels instead of a pressure vessel, natural uranium instead of enriched uranium and heavy water as coolant/moderator instead of light water as coolant/moderator found in the PWR designs. There are three main reasons why Canada decided to proceed with the development and implementation of its own nuclear power programme using the technology developed by the national industry. These reasons are the following: a) because it had accumulated considerable experience in the heavy water natural uranium reactor system, which enabled Canada to make use of Canadian resources and technology; b) because in some regions of Canada, particularly Ontario, major hydro resources had been largely developed and fossil fuels would have to be imported; and c) because it had abundant supplies of uranium. [15]

The main milestones of the Canadian nuclear power programme are described briefly in the following paragraphs. In 1944, an engineering design team was brought together in Montreal, Quebec, to develop a heavy water moderated nuclear power reactor. The National Research Experimental reactor (NRX) was built at Chalk River, Ontario. The nuclear power reactor started up in 1947. It provided the basis for Canada's development of the very successful CANDU series of PHWR for power generation, and served as one of the most valuable research reactors in the world. [18]

In 1955, AECL, OPG and Canadian General Electric made a commitment to build the first small-scale prototype of a CANDU nuclear power reactor with a net capacity of 22 MWe at Rolphton, Ontario[41]. The nuclear power reactor started to be constructed in January 1958, was connected to the grid in June 1962, entered into commercial operation in October the same year and was shut down in August 1987. A larger prototype of nuclear power reactor with a net capacity of 200 MWe was constructed at Douglas Point, Ontario. This nuclear power reactor entered into commercial operation in January 1967 and was shut down in May 1984. These two units established the technological base for the development of Canada's nuclear power independent programme.

In June 1966, the construction of two 500 MWe nuclear power reactors at Pickering, Ontario began under a tri-partite agreement between OPG, AECL and the federal government. OPG later committed to build two more units to make an integrated 4-unit nuclear power plant; these units (Pickering A1 and A4) came into operation between July 1971 and June 1973.

In September 1966, the 250 MWe Gentilly 1 nuclear power reactor prototype started to be constructed. In April 1971, the reactor was connected to the grid up in the Quebec province and was shut down in June 1977. The reactor operated only six years.

The construction of six units in Bruce nuclear power plant (Units 3 to 8) started in July 1972. The construction of the last unit started in August 1979. The six units were connected to the grid during the period December 1977-March 1987. All units entered into commercial operation during the period February 1978-May 1987.

In July 1974, OPG decided to add 4 units at the Pickering nuclear power plant; the 4 units (Pickering 5 to 8) started to be constructed in the period November 1974-September 1976, were connected to the grid during the period December 1982-January 1986 and entered into commercial operation during the period May 1983-January 1985.

Four units at Darlington nuclear power plant were constructed during the period April 1982-July 1985. These units were connected to the grid during the period December1990-

---

[41] In Ontario, 16 commercial nuclear power reactors operate at three locations, providing 51% of the province's electricity in 2005.In 2009, these units generated 6 300 MW representing more than 50% of the total production of electricity of the provinces.

April 1993 and entered into commercial operation during the period November 1992-June 1993.

It is important to stress the following: The nuclear power reactors at Darlington, Ontario, provide the base design for the CANDU-9 design of around 900 MWe. It supplements the proven CANDU-6 of about 700 MWe, which has been an export success. The CANDU-9 has flexible fuel requirements ranging from natural uranium through slightly-enriched uranium, recovered uranium from reprocessing spent PWR fuel, mixed oxide (U and Pu) fuel, direct use of spent PWR fuel, to thorium. However, the design has been shelved but some of the innovation included in the CANDU-9 system, along with experience in building recent Korean and Chinese units, was then put back into the Enhanced CANDU-6 -built as twin units- with power increase to 750 MWe and flexible fuel options, plus 4.5 year construction and 60-year plant life (with mid-life pressure tube replacement). [18]

Finally, it is important to single that the modernization process and extension of the useful life of the reactor of Point Lepreau, in New Brunswick, Canada, has suffered a delay of about 16 months. The reactor that was taken out of service in March of 2008, should be again operative at the end of 2009. However, according with recent declarations of government and AECL representatives, the company in charge of the modernization of the works, the nuclear power reactor will not be in conditions of operating until February 2011. The delay is due mainly to holdup in assemblies to be carried out in the reactor.

Export of CANDU Nuclear Power Reactors

In the early 1970s, two CANDU-6 nuclear power reactors were successful sold to the following countries: Argentina in 1973 and South Korea in 1976. These two units went into service in the early 1980's. In 1979, an agreement was signed with Romania to build a multi-unit 650 MWe CANDU nuclear power plant at Cernavoda. The first nuclear power reactor, Cernavoda 1, was completed and went into commercial operation in December 1996. In 2002, an agreement was reached with Romania to complete the second unit at the Cernavoda nuclear power plant. The second unit went into commercial operation in October 2007.

In the 1990's, AECL sold three additional CANDU-6 units to South Korea. These units were Wolsond 3, 4 and 5, which went into commercial operation in the period July 1997 and October 1999.

In 1996, AECL entered the Chinese market by selling two CANDU-6 nuclear power reactors to the China National Nuclear Corporation. These units were Qinshan 3-1 and 3-2, which went into commercial operations in the period December 2002 and July 2003. In total, there are twelve CANDU-6 nuclear power reactors in operation or under construction outside Canada (see Table 14).

Three CANDU reactors are to be decommissioned. These nuclear power reactor are the following: a) NPD, a 25 MWe nuclear power reactor; b) Douglas Point, a 218 MWe nuclear power reactor; and c) Gentilly 1, a 266 MWe nuclear power reactor, all owned by AECL. All of these reactors are in a shutdown phase.

Table 13. Nuclear power reactors in operation in Canada

	MWe net	Status	Operator	First power*	Planned Close
Pickering 1	515	operating	Ontario Power Gen	1971/ 2005*	2022
Pickering 4	515	operating	Ontario Power Gen	1973/ 2003*	2018
Pickering 5	516	operating	Ontario Power Gen	1982	2014
Pickering 6	516	operating	Ontario Power Gen	1983	2015
Pickering 7	516	operating	Ontario Power Gen	1984	2016
Pickering 8	516	operating	Ontario Power Gen	1986	2017
Bruce 1**	750	laid up	Bruce Power	1977/2009-10	2035
Bruce 2**	750	laid up	Bruce Power	1976/2009/10	2035
Bruce 3	750	operating	Bruce Power	1977/ 2003*	2036
Bruce 4	750	operating	Bruce Power	1978/ 2003*	2036
Bruce 5	822	operating	Bruce Power	1984	
Bruce 6	822	operating	Bruce Power	1984	
Bruce 7	806	operating	Bruce Power	1986	
Bruce 8	795	operating	Bruce Power	1987	
Darlington 1	878	operating	Ontario Power Gen	1990	
Darlington 2	878	operating	Ontario Power Gen	1989	
Darlington 3	878	operating	Ontario Power Gen	1992	
Darlington 4	878	operating	Ontario Power Gen	1993	
Gentilly 2	635	operating	Hydro Quebec	1982	2011
Point Lepreau 1	635	operating	New Brunswick Power	1982	2034
Total operating (18)	12 652				

Source: IAEA data base and CEA, 8th Edition, 2008.

Table 14. Number of CANDU-6 reactors in operation or under construction outside Canada

Countries	Number of CANDU-6 reactors in operation or under construction outside Canada
Republic of Korea	4
China	2
India	2
Romania	2
Pakistan	1
Argentina	1
Total	12

Source: CEA, 8th Edition, 2008.

The Nuclear Industry in Canada

The Canadian nuclear industry covers all phases of the nuclear fuel cycle. The industry's activities are focused on the design, engineering, construction and servicing of CANDU nuclear power reactors in Canada and abroad; on fuel and component manufacturing and on the mining, milling, refining and conversion of uranium. The most significant members of the industry are AECL, the CNSC, provincial utilities and private sector firms involved in equipment manufacturing, engineering and the mix of private and government (both domestic and foreign) firms involved in uranium production. In addition, there are about 125 hospitals and universities across Canada performing isotope studies in research and/or nuclear medicine.

In Canada, there are over 150 companies that supply products and/or services to AECL and the utilities; 58% of these firms are located in Ontario, 14% in Alberta and 12% in Quebec. The remaining provinces have 16% of the suppliers to the nuclear industry; 56% of the nuclear industry supplier companies are in the manufacturing sector, 20% are in engineering and design and 16% in research and development. [15]

Despite of the achievement of the Canadian nuclear industry described in previous paragraphs, this do not means that there were no difficulties associated with the implementation of the nuclear power programme and with the operation of the nuclear power reactors in the country. Throughout their operational history, the Canadian reactors have been plagued by technical problems that led to construction cost over-runs and reduced annual capacity factors. In August 1997, Ontario Hydro announced that it would temporarily shut down its oldest seven units to allow a significant overhaul to be undertaken. At the time, it was the largest single shutdown in the international history of nuclear power, over 5 000 MWe of nuclear capacity, one third of Canada's nuclear plants capacity. The utility, Ontario Hydro, called for the "phased recovery" of its nuclear power reactors starting with extensive upgrades to the following operating units: Pickering, Bruce and Darlington. (Schneider and Froggatt, 2007)

The Production of Uranium in Canada

Canada is the world's leading producer of uranium[42]. Uranium exploration in Canada began in 1942 and was accelerate in 1947. In 1959, the country has 23 mines and 19 treatment plants in operation which allow Canada to export uranium with a value of C$300 million. In 2004, Canada produced 13 676 tons of uranium oxide concentrate, approximately 11.57 tons of uranium, representing around 30% of the total uranium world production. The value of the uranium production in 2004 was around C$800 million. Canada produces about one third of the world's uranium mine output[43].

In December 2007, the Areva group announced the creation of joint venture with the Japanese group Ourd-Canada and the Australian-Canadian company Denison-Mines with the purpose to develop the Midwest mine, which has an estimated capacity of 14 000 tons of

---

[42] As April 2008, the Canadian's proven uranium deposits amount to 297 000 tons.

[43] Canada and Australia are the main countries able to expand uranium production strongly as required to meet increases in world uranium demand.

uranium. Initial work on the project is expected to begin in 2009. Areva has around 69% stake in the joint venture. However, the French nuclear group Areva in charge of the work in one of these two mines decided, in December 2008, to delay developing the uranium mine in Canada. The operation was scheduled to begin in 2010, in light of recent falls in the price of uranium and an increase in exploitation costs in the region. The decision affected the Midwest mine, located in the Canadian province of Saskatchewan. Despite of this situation, experts expect that after 2011 the Canadian production of uranium will increase again as more new mines come into production.

Nuclear Safety Authority and the Licensing Process

On 31 May 2000, the Canadian Nuclear Safety Commission (CNSC) was created as the successor to the Atomic Energy Control Board (AECB), which had served as the regulator of Canada's nuclear industry for more than 50 years. The establishment of the CNSC followed the entry into force of the Nuclear Safety and Control (NSC) Act and its regulations. The CNSC's mission is "to regulate the use of nuclear energy and materials in the country with the purpose to protect health, safety, security and the environment and to respect Canada's international commitments in the field of the peaceful use of nuclear energy". Under the NSC Act, the CNSC's mandate involves four major areas: 1) "regulation of the development, production and use of nuclear energy in Canada; 2) regulation of the production, possession and use of nuclear substances, prescribed equipment and prescribed information; 3)implementation of measures respecting international control of the use of nuclear energy and substances, including measures respecting the non-proliferation of nuclear weapons; and 4) dissemination of scientific, technical and regulatory information concerning the activities of the CNSC and the effects on health and safety and the environment arising from the development and use of nuclear energy and nuclear substances".

What is the purpose of the Canadian regulatory system? The system is designed "to protect people and the environment from the risks associated with the development and use of nuclear energy and nuclear substances for peaceful purpose in different economic and social sectors". Companies, medical or academic institutions wishing to operate nuclear facilities or use nuclear substances for industrial, medical or academic purposes, must first obtain a license from the CNSC. It is a fundamental tenet of Canada's regulatory regime that licensees are primarily responsible for safety. The CNSC's role is to ensure that the applicants live up to their responsibility. The responsibility is, therefore, on the applicant or the holder of the license to justify the selection of a site, design, method of construction and mode of operation of a nuclear facility. When issuing a license, the CNSC must be satisfied that the companies have taken adequate measures to protect health and safety, the environment, security and to respect Canadian's international commitments and that the companies are qualified to carry out the licensed activities. Licensing matters for major nuclear facilities are carried out in public hearings by a seven-member tribunal. This is one of the most visible functions of the CNSC in the regulation of the nuclear industry.

The CNSC also controls the import and export of nuclear materials, nuclear technology and equipment that might be used to develop nuclear weapons (including so-called "dual use items"). CNSC staff inspects licensed activities, enforces compliance with regulations, and develops safety standards. [15]

Nuclear Fuel Cycle and Waste Management

The Nuclear Fuel Waste (NFW) Act requires "nuclear utilities to form a waste management organization whose mandate is to propose to the government of Canada approaches for the long-term management of nuclear fuel waste and to implement the approach that is selected by the government". The NFW Act also requires the utilities and AECL "to establish trust funds to finance the implementation of the selected long-term nuclear fuel waste management approach". Following the adoption of the NFW Act the Nuclear Waste Management Organization (NWMO) was established by the nuclear utilities in the fall of 2002.

In the case of low-level radioactive waste, the major nuclear utility in Canada is the OPG, which produces about 70% of the annual volume of this type of waste in their nuclear facilities. Due to the small level of volume of low level radioactive waste, there has been no pressing need in OPG for early disposal and, for this reason, the waste is being safely stored on an interim basis. However, it is important to single out that the utility fully recognized that, in the longer term, disposal is a necessary step in responsible waste management, so that future generations are not burdened with managing this waste. OPG is currently assessing possible options for the long-term management of low and intermediate level radioactive wastes. The year 2015 is considered an achievable target date for bringing a long-term management facility into service. The other major ongoing producer of low-level radioactive waste, AECL, had discussions with the CNSC to license a prototype below-ground concrete vault known as IRUS (Intrusion-Resistant Underground Structure) for relatively short-lived waste. The future application of IRUS technology is currently being reassessed by AECL. Until this, or another disposal facility is available, AECL will continue to store its on-going low-level radioactive waste in-ground and above-ground structures.

There are no facilities for the final disposal of high-level radioactive waste produced by the nuclear power reactors currently in operation in the country. All high-level radioactive waste is provisionally storage at the nuclear power plant sites.

The Public Opinion

Public opinion in Canada is shifting toward nuclear power. In 2008, around 70% of Canadians support refurbishment of existing nuclear power plants and more than 60% support the construction of new nuclear power reactors for the generation of electricity, with the purpose to satisfy the foresee increase in the energy demand in the coming years.

It is important to note that support of nuclear power dipped in 2005 but has since then climbed significantly. It is particularly high in the provinces that already are using nuclear power reactors for the production of electricity, though it has also grown in provinces such as Alberta where consumers realize the price of power is rising. Moreover, hard opposition to use nuclear energy for the production of electricity is, at an all-time, low in Canada. At the same time, the anti-nuclear lobby has been more effective in conveying its message than private industry, government or other pro-nuclear stakeholders. Whereas past concerns focused on the safety of nuclear power reactors, today's negative messaging targets the front and back-end of the generation cycle safety of uranium mining, high reactor construction costs, as well as nuclear waste management. Supporters of nuclear energy have not done an adequate job of addressing these concerns. [17]

In a survey carried out in October 2005 by the IAEA and another non-governmental organizations across 18 countries, including Canada, confirmed that respondents are most likely to choose the use of nuclear technology to treat human diseases such as cancer (39%) than for electricity generation (26%). The high preference for using nuclear technology in disease treatment is not surprising, considering the public's strong popular bias toward health and healthcare. In the case of Canada, 54% of the population supports the use of nuclear technology for the treatment of human diseases. Electricity generation is the second most preferred nuclear application. A total of 34% of the population consider that the use of nuclear energy for electricity population is safe and agree in constructing more nuclear power reactors, while 35% thinks that it is better to use the existing nuclear power plants as much as possible instead to build new ones. Around 22% of the Canadian has the opinion that the use of nuclear energy for electricity generation is very dangerous and all nuclear power reactors operating in the country should be closed.

However, it is important to note that with regards to the relationship between the use of nuclear energy for electricity generation and the impact on the climate, around 49% of the Canadian believe that for environmental reasons the use of nuclear energy for electricity production should not be expanded in the future, while 42% support the implementation of a programme for the expansion of nuclear power in the country in the coming years.

On this important issue most scientific believe the burning of fossil fuel, such as oil, coal or natural gas, is the main cause of climate change or global warming. Some say that nuclear power could play a role in protecting the world's climate, because it does not produce climate-changing or greenhouse gases. [36]

Looking Forward

Canada adopted a decision to extend the lifetime of the current nuclear power reactors in operation, instead of building new units in the near future.

A decision on refurbishing Pickering 5,6,7 and 8 (2 064 MWe) was made in 2008 and the work should be done over the period 2013-2016. Decisions on refurbishing Bruce 5,6,7 and 8 (3 400 MWe) will be undertaken about 2014 and refurbishing of Darlington units (3 524 MWe) will be made thereafter. Hydro Quebec is considering refurbishment of Gentilly 2 as an alternative to closing it about 2011.

For most of this coming decade, prospects for the construction of new nuclear power reactors in Canada are uncertain, even in the Ontario province, the major commercial utilization province of the CANDU system. The return to service of the remaining laid-up nuclear power reactors and the completion of gas-fired power units already under construction in the Ontario province should ensure adequate electricity supplies in the coming years. However, the construction of new nuclear power reactors in Ontario cannot be totally excluded in the future. The Minister of Energy of Ontario, George Smitherman, said that "the proposal of AECL to build new reactors for 2018 it was the only one that fulfills all the requirements settled down by the country but it was too much expensive". The position of minister of Energy of Ontario to reject the proposal of AECL is influenced by uncertainty that exists around the future of the company, because the Canadian government has plans to privatize AECL in the future. The Ontario provincial government wants that Ottawa, trough AECL, share part of the risk of the project, especially regarding the over cost that could be associated with the construction works of a new nuclear power reactor. Since the Ontario province government started to consider the construction of a new nuclear power reactor three years ago, the price of the construction of units has been duplicated increasing the concern of the Ontario province government about the final price of the project.

It is important to note that, in addition of the ECL offer, Westinghouse and the Areva also presented offers for the construction of two new nuclear power reactors but, according with the competent authorities in Ontario, their proposals were not adjusted to the requested. The country has budgeted C$26,000 million for the expansion and modernization of the Canadian nuclear power programme in the coming years with the purpose to be able to respond to the foresee increase in the demand of electricity.

While market prospects for the construction of new nuclear power reactors in the near to medium-term are not too promising, the refurbishment of existing units holds more promise. Hence, the refurbishment of existing nuclear power reactors would, at least in the medium-term, avoid the replacement of nuclear generating capacity with fossil-fuelled power plants. However, for the next decade, there are better opportunities for the deployment of new nuclear generating capacity in Canada. AECL is currently working on the development of the 700 MW Advanced CANDU reactors (ACR), and is aiming at reducing the capital cost to build a single unit by up to 40%. The ACR technology could provide an economic replacement for existing units as they reach the end of their service lives, as well as for some new nuclear power reactors in Canada and in other countries as well.

The new design of the CANDU system is undergoing pre-licensing assessment in the USA and Canada. New nuclear power reactors to be built in the future would likely be of the ACR design presently under development by AECL for domestic and international markets. The ACR aims to be cost-competitive with other systems of electricity generation, including natural gas. It also holds significant potential for use in Canada in hydrogen production. The modular construction of the new CANDU system means that AECL anticipates having major components built in USA shipyards, using a high degree of standardization of components. The ACR is designed to be built in pairs, and construction time is estimated at 44 months for the first unit, reducing to 36 months for the second and subsequent ones.

However, the current oil and gas price, the climate change, the need to reduce CO_2 emissions to the atmosphere, among other factors, could change this situation in favor of the use of nuclear energy for electricity generation in Canada in the coming years. Without the use of nuclear energy for electricity generation in Canada, the total GHG emissions gap would be between 20 to 35% higher in 2010 than currently forecast. At the same time, it is important to note that the use of nuclear energy for electricity generation in the country in the future is not exempt of problems and difficulties that represent a great challenge to the nuclear sector. Over the next two decades, nuclear energy in Canada will have to face major challenges in order to be able to compete with other technologies for generating electricity in an open and deregulated market environment. These challenges include: 1) the ability to develop a cost competitive ACR; 2) the ability to mobilize large capital investment for projects in an open market; 3) the sitting and licensing requirement for the construction of new nuclear reactors; 4) the price of fossil fuels; and 5) the development of mechanisms which will internalize the externalities related to the production of electricity from fossil fuels. [15]

The following are the planned and proposed nuclear power reactors to be constructed in Canada in the future.

Table 15. Planned and proposed Canadian nuclear power reactors

Utility	Site	Capacity	Type	Operation
Bruce Power	Bruce, Ont	4 x 1 000 approx	?	from 2014
OPG	Darlington, Ont	4 x 1 000 approx	?	from 2014
New Brunswick Power	Point Lepreau, NB	1 x 1 100	ACR-1000	
Bruce Power Alberta	Peace River, Alberta	1 x 1 100 or up to 4 x 1 000	?	from 2017

Source: Reference 18.

Chapter 4

THE CURRENT SITUATION AND PERSPECTIVES IN THE USE OF NUCLEAR ENERGY FOR ELECTRICITY GENERATION IN THE EUROPEAN REGION

The supply of energy will remain one of the major issues of the 21st century, especially in the European region, given its high dependency on energy imports. Energy demand continues to increase, raising concerns about supply, the economic competitiveness of different energy sources, and repercussions on the economic and social development and the environment. Consequently, consideration should be given to all these factors, to which others of special relevance may be added, such as liberalization of the energy markets, waste management and public acceptance of different technologies, all of which have a certain impact on the energy scene. [134]

It is important to single out that nuclear energy was an assured energy source for electricity generation in the European region in the 1960's. Between 1960's and 1990's, more than 190 nuclear power reactors were built in the European region increasing the share of nuclear energy in the energy balance of several European states. Two serious accidents, the first one at Three Mile Island in the USA in 1979 and the second one in 1986 in Chernobyl, Ukraine, the second one with important serious consequences for many countries in the European region, stop public support to the use of nuclear energy for electricity generation and led to a scaling back of the nuclear industry in the European region.

Due to different reasons, today's the debate about the use of nuclear energy for electricity generation started again in several countries of the European region. The first reason is the high price of oil reached during 2007 and 2008, and the tendency to a high oil price and an increment of the natural gas price. The second reason is the need to reduce CO_2 emissions to the atmosphere due to the commitments of the European states to the Kyoto Protocol. The third reason is the dependency of the Europe region, and particularly the EU, to the import of fossil fuel from politically unstable regions like the Middle East.

One of the main problems that the European region is facing is the electricity vulnerability. Europe electricity vulnerability depends on the following three main factors: The first factor is the margin of surplus capacity in relation to the peak power demand. In a public monopoly system, this margin is often comfortable, since one of the priorities of the monopoly is to prevent energy failure. It is certainly a costly strategy as reserve capacities are not used and it is the consumer who bears the cost. In an open market system, it is not necessarily in the companies' interest to have an overcapacity. Risks of energy failure are more likely, unless the regulator imposes a public service obligation either on the incumbent or on all the operators. The second factor is the interconnection rate between countries. Electricity interconnection with neighboring countries makes mutual assistance possible in periods of pressure on supply and demand and this limits the risks of failure. It is well known that the EC Directive recommends an interconnection rate of about 10% in terms of the installed electricity capacity of a country. The situation varies from one European country to another. The vulnerability of the "electricity peninsulas" (the Iberian, Italian and English peninsulas) is significantly higher than that of continental countries with many borders. The third factor is the net import rate, i.e. the percentage of electricity consumption, which is imported. About 60% of gas consumed within the EU crosses at least one border while cross-border electricity only amounts to 7% of the electricity consumed within the EU. Electricity long distance transport costs are high due to line losses and there is a maximum level of electricity imported that should be considered as politically acceptable. As electricity cannot be stored, dependence on imports is sometimes considered to imply a great risk and such dependence on a strategic good is unacceptable to some countries. [28]

During the consideration of the role that nuclear energy should have in the energy balance in the European region in the coming years, three main realities should be taken into account. These are the following: 1) expectations in the use of nuclear energy for electricity generation are rising again in several European states; 2) the answer to questions like 'is nuclear power economic" cannot be made using a single universal answer. As with just about everything else in life, the answer is 'it depends' — sometimes yes, sometimes no; and 3) whether the use of nuclear energy for electricity generation is more economics or not will depend on how cheap it is compared to other energy sources.

Certainly, the nuclear industry can influence this issue by bringing down costs, but there are factors outside the industry's control, such as the price of natural gas or of carbon credits that will also determine, for any particular investor, whether nuclear is a cost-effective option. (McDonald, 2008)

Currently, the region generates around 31% of its electricity from 197 nuclear power reactors currently in operation in 18 countries. According with some expert's opinion, and based on the commitments adopted by the European countries regarding the Kyoto Protocol, the above-mentioned proportion should be maintained or increased in order to meet the 2020 target, with an increase in the actual wattage generated to meet increasing power demand. The Kyoto Protocol requires industrialized countries to reduce their greenhouse gas emissions, most of which arise from burning fossil fuels such as oil and coal for electricity generation.

It is a fact that nuclear energy is already making a substantial contribution to an energy policy that is low carbon, cost-effective and that provides assured supply. At present, nuclear supplies almost one third of Europe's electricity, produces very low CO_2 emissions calculated over the entire fuel cycle (comparable to wind energy), and has a quasi 'indigenous' character, i.e., it can rely on a complete European nuclear fuel cycle. In addition, it contributes to the stabilization of electricity prices, owing to the favorable ratio of primary investment costs to fuel costs. (Blohm-Hieber, 2008)

Today a strong debate is happening among the oldest and most industrialized EU member states, which do not want slower growth, and are beginning to view nuclear energy as a real alternative for the production of electricity under the current energy crisis. The increase in oil price occurred in 2007 and 2008 changed the minds of people in the European region who turned against nuclear power after the Chernobyl nuclear accident.

Although less pronounced than in other parts of the world, energy and electricity consumption in the European region are expected to continue increasing over the foreseeable future, at least until 2030, and most likely beyond. At the same time, energy resources are becoming ever scarcer and more expensive, and excessive emissions of CO_2 to the atmosphere are driving the impending threat of climate change.

To satisfy the foresee growth in the demand of electricity in the coming years, the European energy balance should included all available energy technology, including nuclear energy. For this reason, nuclear energy should contribute to alleviate energy supply dependency, local air pollution and global climate change in the European region in the future. The EU's energy plan envisages the following three goals: a) 20% cut in EU's greenhouse gas (GHGs) emissions (or 30% as part of an international agreement); b) 20% energy share from renewable sources; and 3) 20% increase in energy efficiency. These targets are to be achieved by the year 2020. The ultimate goal of the plan is to limit the average global temperature rise to 2°C. (Blohm-Hieber, 2008)

In 2004, the nuclear share of electricity production in the EU reached 31%, which represents a 'non-emission' to the atmosphere of nearly 900 million tons of CO2 per year. This represents almost the quantity of carbon dioxide produced annually by the transport sector. Given these facts, it is very unlikely that the goal of a 20% CO2 emission reduction by 2020 can be achieved, if the EU energy mix does not include an important share of nuclear energy. In 2007, energy consumption accounts for 80% of Europe's GHGs emissions. In 2005, the nuclear share in the EU was 29.5%.

It is important to stress that unless Europe can make domestic energy more competitive in the next 20 to 30 years, around 60-70% of the EU energy requirements, compared with 50% today, will be met by reliance on imports from regions threatened by insecurity. [100] According with some experts' opinion, not only it is expected that this dependence would increase up to 60%-70% by 2030, but reliance on imports of gas is expected to increase from 57% to 84% by 2030. In the case of oil this increase would be from 82% to 93%. (Morales Pedraza, 2008)

The European region, particularly the EU, is consuming more and more energy and importing more and more energy products in order to satisfy it increase energy needs. External dependence for energy is constantly increasing and this situation is considered by many politicians and experts very danger from the political and economical point of view. If no measures are taken, in the next 20 to 30 years, 70% or more of the EU's energy requirements will be covered by energy imported from outside the European region. In 2002, almost 45% of oil imports come from the Middle East and 40% of natural gas from Russia[44]. Why the EU is increasing its energy import from these two places? One of the main reason is that the energy production within the EU is decreasing and there are not exploiting as much as they could their indigenous and renewable sources of energy. Instead of moving towards the stated objective of a 12% share for renewables in the European region energy balance, the

---

[44]The Middle East is a very unstable region and the supply of energy from Russia is not free of problems and difficulties, particularly the supply of gas through Ukraine. In the last years, two cuts of gas supplies happened affecting seriously the energy situation in the EU. The last cut happened in winter of 2009, when Russia reduced considerable the supply of gas through Ukraine.

region is stagnating around the 6% mark, which is half of the agreed objective. Despite of the efforts made during the last years by several European countries to increase the share of renewables energy in the energy balance mix, the outcome of these efforts are not enough and have no change significantly the current composition of the energy balance within the region as a whole. What is more worrisome is that there are no reason to believe that the current situation is going to be drastically change in the coming years, with the exception of the role that could be played by nuclear energy at least in several European countries.

During the consideration of the possible role to be played by nuclear energy in the European energy balance in the future, the following problems need to be thoroughly considered and favorable solved:

a) management of the radioactive waste, particularly high-level radioactive nuclear waste;
b) proliferation security;
c) operational safety of the nuclear power plants;
d) economic competitiveness;
e) public acceptance.

In several European countries, nuclear power is cost competitive with other forms of electricity generation, except where there is direct access to low-cost fossil fuels. One of the elements that make nuclear power competitive with other forms of energy is the fuel costs. It is, for nuclear power plants, a minor proportion of total generating costs; this proportion is around 20% of the total capital cost. In assessing the cost competitiveness of nuclear energy, decommissioning and waste disposal costs should be taken into account, which could increase the capital cost between 20% and 25%. If no actions are taken to solve the above-mentioned problems in the coming years, then the participation of nuclear energy for electricity generation in the European region will decline notably due to the following reasons: a) there are only 14 nuclear power reactors under construction in 5 countries. These countries are: Bulgaria (2), Finland (1), France (1), Ukraine (2) and the Russian Federation (8)[45]; b) the closure of several nuclear power reactors, due to the extinction of their exploitation license; and c) the implementation of the phase-out policy adopted by a group of European states, such as Belgium, Germany, Spain and Sweden.

Finland is one of the two European states that is constructing a third-generation PWW, designed by the French company Areva, the so called "EPR system". It was expected initially that the nuclear power reactor under construction would be connected to the electric grid in 2009 but delay in the construction of the unit has make impossible to respect this deadline. The other country that is constructing a third generation of nuclear power reactor is France. French state-owned power generating company, Électricité de France (EdF), is building, since 2007, an EPR system in the country. At the same time, in Eastern Europe the Bulgarian government awarded a contract for the construction of two units and Romania has restarted building a nuclear power reactor that was mothballed 15 years ago. Within the European

---

[45] Slovakian authorities has indicated that construction of Units 3 and 4 at Mochovce nuclear power plant was restarted at the end of 2007 and the Czech Republic's energy plan foresees the construction of two more nuclear power reactors by the end of the decade.

region, Russia has the largest plan for the construction of new nuclear power reactors in the coming years.

The Swiss parliament ended Switzerland's moratorium on building new nuclear power reactors and extended the operating lifetime of the country's five existing units. The British government has planned to build 10 nuclear power reactors in the coming years and the industry has requested the British's government to reduce regulatory and planning risks associated with the construction of nuclear power reactors in order to encourage the private industry to support the construction of new units in the country in the future.

At the same time, other European countries are thinking to restart the use of nuclear energy for electricity generation in the coming years, or are dropping previous plans to close all of their nuclear power plants in operation. For example, the Italian Industry Minister Claudio Scajola, country which closed its four nuclear power reactors after a 1987 referendum, has said that "the development of nuclear technologies remains an important element for Italy's energy policy". Following this statement the Italian government signed an agreement with the French company Areva and the Italian company ENA for the construction of four EPR systems in the coming years, reincorporating nuclear energy in the energy balance mix of the country. Sweden has dropped plans to close all its nuclear power plants by 2010, and Belgium's intention to start phasing out nuclear power in 2015 has run up against a finding, by the Federal Planning Bureau, that nuclear power is the best way for the country to meet its Kyoto Protocol commitments. Germany and Spain wish to gradually phase out all nuclear plants for safety reasons, and generate electricity from renewable sources instead, but both governments are under strong pressure from the private energy industry not to go ahead with their phase-out plans.

The three Baltic States, together with Poland, have agreed, in principle, to construct a nuclear power plant in Lithuania by 2015, and the country passed the necessary legislation in 2007 to make the construction of this plant possible. Turkey also passed new legislation to enable nuclear power plant construction. (McDonald, 2008)

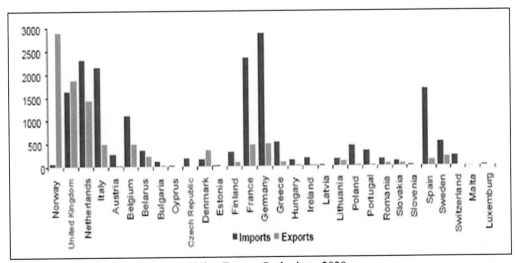

Source: Energy in Europe. European Union Energy Outlook, to 2020.

Figure 45. Imports and exports of oil products of EU-29 in 2004 (thousand barrels/day).

Until today, only three EU's countries have laws in force prohibiting the construction of nuclear power plants for electricity generation. These countries are Austria, Denmark and Ireland. There is no indication that these three countries are going to change their policy regarding the use of nuclear energy for electricity generation in the future.

The map with the location of all nuclear power plants in the European region is shown in Figure 46.

Source: International Nuclear Safety Center at Argonne National Laboratory, USA.

Figure 46. Map with the location of all nuclear power plants in the European region.

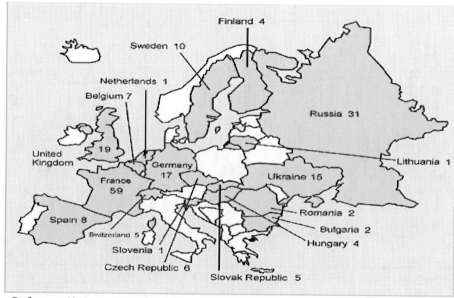

Source: Reference 49.

Figure 47. Number of nuclear power reactors per country in Europe.

Table 16. Nuclear share figures 1996-2006

Country or area	Nuclear share (%)											Nuclear electricity production (TWh)	Nuclear electricity production (TWh)
	1996	1997	1998	1999	2000	2001	2002	2003	2004	2005	2006	2005	2006
Belgium	57.2	60.0	55.2	57.7	55.3	58.0	57.3	55.5	55.1	55.6	54.4	45.3	44.3
Bulgaria	42.2	45.4	41.5	47.1	45.0	41.6	47.3	37.7	41.6	44.1	43.6	17.3	18.1
Czech Rep	20.0	19.3	20.5	20.8	26.7	19.8	24.5	31.1	31.2	30.5	31.5	23.2	24.5
Finland	28.1	30.4	27.4	33.0	30.0	30.6	29.8	27.3	26.6	32.9	28.0	22.3	22.0
France	77.4	78.2	75.8	75.0	76.4	77.1	78.0	77.7	78.1	78.5	78.1	430.9	428.7
Germany	30.3	30.6	28.3	31.2	34.5	30.5	29.9	28.1	32.1	31.0	31.8	154.6	158.7
Hungary	40.8	40.8	35.6	38.3	40.6	39.1	36.1	32.7	33.8	37.2	37.7	13.0	12.5
Lithuania	83.4	81.5	77.2	73.1	73.7	77.6	80.1	79.9	72.1	69.6	72.3	10.3	7.9
Netherlands	4.8	2.5	4.1	4.0	na	4.2	4.0	4.5	3.8	3.9	3.5	3.8	3.3
Romania	1.8	9.7	10.3	10.7	10.3	10.5	10.3	9.3	10.1	8.6	9.0	5.1	5.2
Russia	13.1	13.6	13.1	14.4	14.9	15.4	16.0	16.5	15.6	15.8	15.9	137.3	144.3
Slovakia	44.5	44.0	43.8	47.0	53.4	53.4	65.4	57.3	55.2	56.1	57.2	16.3	16.6
Slovenia	37.9	40.0	38.3	37.2	37.4	39.0	40.7	40.4	38.8	42.4	40.3	5.6	5.3
Spain	32.0	29.3	31.7	31.0	27.8	28.8	25.8	23.6	22.9	19.6	19.8	54.7	57.4
Sweden	52.4	46.2	45.7	46.8	39.0	43.9	45.7	49.6	51.8	46.7	48.0	70.0	65.0
Switzerland	44.5	40.6	41.1	36.0	38.2	36.0	39.5	39.7	40.0	32.1	37.4	22.1	26.4
UK	26.0	27.5	27.1	28.9	21.9	22.6	22.4	23.7	19.4	19.9	18.4	75.2	69.2
Ukraine	43.8	46.9	45.4	43.8	45.3	46.0	45.7	45.9	51.1	48.5	47.5	83.3	84.8

Legend: Data not yet available (na).

* estimates.

Source: WANO and IAEA.

Table 16 includes information on the nuclear share figures in the European region between 1996 and 2006.

The major reduction in the use of nuclear energy for electricity production it is expected to occur in the European region, dropping from 29.5% in 2005 to 12% in 2030[46], if no actions to stop this tendency is adopted by the European governments that are now using nuclear energy for electricity production, or are thinking to introduce the use of this type of energy for this purpose in the near future. The foresee reduction in the use of nuclear energy for electricity production in the European region in the coming years it is expected to be from 170 GWe to 74 GWe. This foresees reduction represents 56.4% of the region current total electricity production.

There are two main reasons why most European governments have distanced themselves in the past from the use of nuclear energy for electricity generation. One is political. The second is social. However, it is interesting to note the consequences of the inertia policy adopted by the energy industry, allowing governments to adopt positions, which are evidently not defensible in the long term from the economic point of view. A clear example of this position is the following: Some European governments adopted a energy policy to phase out the use of nuclear energy for electricity generation without any coherent strategy for the replacement of this type of energy for the generation of electricity. Such a position has only been possible because nuclear phase-out policy has largely meant retaining nuclear plants until their operational lifetimes are expired such that very little capacity has so far actually been lost. The approaching imminence of the closure of old nuclear power reactors and the realization that Europe's Kyoto obligations are looking increasingly difficult to meet are the reasons why there are now signs that some European governments are starting to reconsider. The long timescales required for the deployment of new nuclear power reactors means that the options for responding to changing demands are very limited. Some politicians would prefer to defer decisions of Europe's nuclear future until renewables have been given an opportunity to deliver. Unfortunately, by the time this becomes clear, it may well be too late to obtain the fullest advantage of nuclear generation. Even if renewables probe able to meet their target in Europe, they would barely match the contribution that nuclear energy is already making at present and further carbon savings will be needed. Faced with the extreme dangers of global climate change, the only sensible strategy is to keep the nuclear option open. (Hesketh, Worral and Weaver, 2004]

In the EU, the energy demand in 2002 was covered by oil in 41%, by gas in 22%, by coal in 16% (hard coal, lignite and peat), by nuclear power in 15% and by renewables in 6%. In 2005, the EU-27 satisfied its primary energy consumption with 36.7% oil, 24.6% gas, 17.7% coal, 14.2% nuclear, 6.7% renewables and 0.1% industrial waste. If nothing is done by the EU governments in the coming years, then, it is expected that the total energy picture in 2030 will continue to be dominated by fossil fuels: 38% by oil, 29% by gas, 19% by solid fuels. Renewables will be 8% and nuclear power only by 5.2%, this means a reduction in the nuclear share of 9%.

The electricity generating mix of the future must take into account, among other elements, the different economic, ecological, social and entrepreneurial interests that could exist within countries, putting aside any ideological factor. All available energy sources and technologies must be evaluated and meaningfully combined following due consideration of their advantages and disadvantages, their accessibility and level of reserves, their potential macroeconomic benefits, as well as the impact on the environment, among others. It is

---

[46] In another scenario this reduction could be even higher from 29% to 20%.

important to single out that the three main central goals of any energy policy and of the energy industry are always the following: 1) supply reliability; 2) cost efficiency; and 3) environmental soundness.

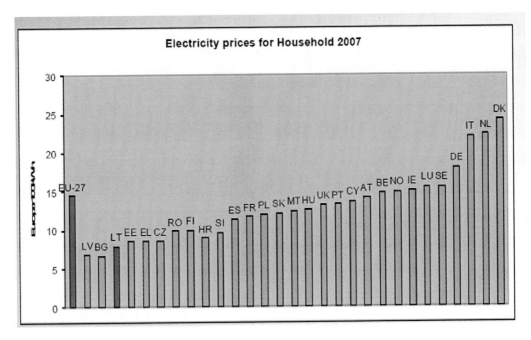

Source: European Commission, DG TREN, 2007.

Figure 48. Electricity for householder in 2007.

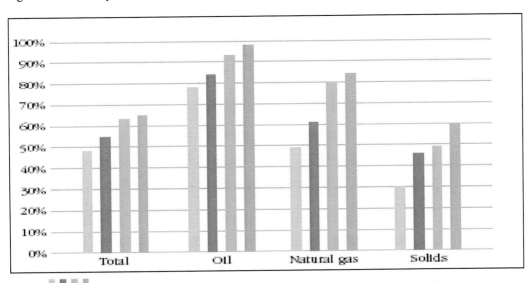

Source: European Commission DG, TREN, PRIMES.

Figure 49. Foreseen energy import dependence up to 2030.

Table 17. Reactors in operations and under construction in Europe (2008)

Country	Number of nuclear power reactors in operation	Net capacity MWe	Number of nuclear power reactors under construction	Net capacity MWe
Belgium	7	5 728	-	-
Bulgaria	2	1 906	2	1 900
Czech Republic	6	3 472	-	-
Finland	4	2 696	1	1 600
France	59	63 473	1	1 630
Germany	17	20 339	-	-
Hungary	4	1 826	-	-
Lithuania	1	1 185	-	-
Netherlands	1	485	-	-
Romania	2	1 310	-	-
Russian Federation	31	21 743	9	5 980
Slovakian Republic[47]	4	1 688	2	840
Slovenia	1	696	-	-
Spain	8	7 448	-	-
Sweden	10	9 014	-	-
Switzerland	5	3 220	-	-
Ukraine	15	13 168	2	1 900
United Kingdom	19	11 035	-	-
Total	196	170 432	17	13 850

Source: IAEA, WANO and CEA, 8th Edition, 2008.

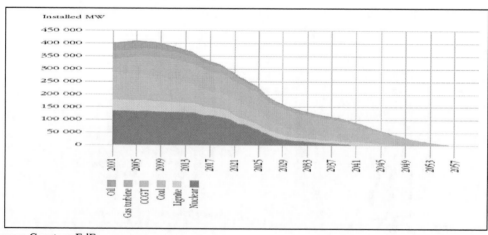

Source: Courtesy EdF.

Figure 50. Decline of installed fossil and nuclear power generation capacity (without renewal by new plants) in EU-15 per type of energy, showing a decrease of around 60% by 2030.

---

[47] In 2007, the Slovakian government decided to continue the construction of Units 3 and 4 at Mochovce.

Figure 50 shows the foresee decline of installed fossil and nuclear power generation capacity (without renewal by new plants) in the EU-15 per type of energy.

In Table 17, the number of nuclear power reactors in operation and the number of new units under construction by country in the European region are included.

According with the latest IAEA's information, the European region[48] has 196 operating nuclear power plants in 18 countries and 17 new units under construction in six countries. Overall, the nuclear power reactors in operation have a net capacity of 170 432 MWe. The 17 nuclear power reactors under construction will add 13 850 MWe to the existing capacity (see Table 17).

In Western Europe, capacity has remained relatively constant despite recent nuclear phase-out laws adopted in Belgium, Germany and Sweden, as well as the closure of old nuclear power reactors carried out in some other countries. The most advanced programme for the construction of new nuclear power capacity in Western Europe are in Finland and France, which are constructing two new reactors of third generation (EPR system). In Eastern Europe, only Bulgaria has two nuclear power reactors under construction, but some other countries, such as Romania, the Czech Republic and Slovakia, are planning to resumed or start the construction of new units in the near future.

The cut of the supply of Russian gas to Europe due to the dispute of prices with Ukraine in the last years has revived the interest in the use of nuclear energy for electricity generation and heating in several European countries, in order to reduce their energy dependence of Moscow. Slovakia and Bulgaria have announced their intention to restart the operation of old nuclear power reactors of Soviet design that were closed as previous condition for the entrance of these two countries to the EU due to shortest in the generation of electricity. These announcements have provoked a strong reaction from the EU Commission. The EU organ has declared openly that the reopening of these units would be illegal since the pact of adhesion to the EU requires the closing and dismantlement of them.

It is important to single out that the closure of the nuclear power reactors in Slovakia has reduced significantly the contribution of nuclear energy to the electric system of the country from 60.8% to 42%. In the case of Bulgaria, the nuclear contribution has decreased from 42% to 34%. Both countries with the closing of the old soviet design units have passed of being energy exporters to become energy importers, and this situation has a negative impact in their economies.

As for the other countries of the Eastern Europe, Hungary does not have plans for the construction of new nuclear power reactors, but has prolonged 20 years the useful life of its four units currently in operation, while Rumania that has two nuclear power reactors in operation is taking steps to build two new units in the coming years. The Czech Republic has placed nuclear energy as one of the priorities within its energy development plan for the future.

The Russian Federation continued its programme to extend licenses at 11 nuclear power reactors now in operation and for the construction of nine new units in the coming years. Ukraine, despite of the consequences of the Chernobyl nuclear accident, has two nuclear power reactors under construction with the purpose to add 1 900 MW to the electrical system.

---

[48] Including Russia and Ukraine. Belarus is excluded from Table 17.

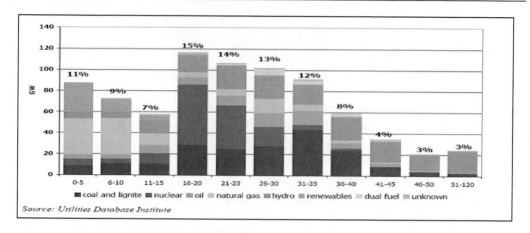

Figure 51. Breakdown of European power generation capacity by age (as of 31 December, 2004).

Considering Europe as a whole it is easy to reach the conclusion that France is the country with the highest number of nuclear power reactors in operation [59]. Other countries with an important number of nuclear power reactors in operation in the European region are the following: The Russian Federation (31), UK (19), Germany (17), Ukraine (15) and Sweden (10).

It is expected that during the next two decades, the European region might need between 200 GWe and 300 GWe of new generating capacity. Replacement of the existing energy infrastructure could double this amount. How to satisfy this foresee increase in the demand of energy in the European region without increasing the emission of CO_2, as well as other gas of green house effect to the atmosphere? To satisfy the foresee increase in the energy demand in the Europe region in the coming years could be a very difficult goal to be achieved, if the European countries decided to use mainly clean energy to reduce the emission of CO_2, as well as other gas of greenhouse effect, and if nuclear energy is excluded from the energy mix. Has been demonstrated that the use of nuclear energy for electricity generation is environmental clean, that the safety of the nuclear power reactors are very high and there is a safe and proven technology available for the disposal of the high-level nuclear waste produced by the operation of these reactors.

Energy saving, as an important element to reduce the consumption of energy in the European region, cannot make alone a significant contribution to the required extend, in spite of some estimation that 18% of the energy use could be saved using available technologies. The total estimated energy waste is the equivalent of the total energy needs of Austria, Denmark, the Netherlands, Finland and Greece combined.

THE FUTURE CONTRIBUTION OF NUCLEAR ENERGY TO THE ELECTRICITY GENERATION IN THE EUROPEAN REGION

When the future contribution of nuclear energy in the electricity production in the European region is analyzed, the following elements should be taken into consideration:

1) nuclear technology has reached its majority age and it is a safe and non-pollutant of the environment technology. In 2008, a total of 30 countries (31 countries if Taiwan is considered outside China) were using nuclear energy for electricity production, 18 of them in the European region;

2) the supply of fossil fuels could not be enough to support the growing energy needs of many European countries for the second half of this century;

3) a safety nuclear culture has been developed at world level, elevating the safety of the current nuclear power reactors in operation;

4) a new generation of nuclear power reactors will be available in the coming years with increase safety features and with less construction time;

5) the appropriate technologies exist for the safe treatment and for the final disposal of the high-level radioactive nuclear waste.

The acceptance by the public opinion of Finland and Sweden of the final disposal of high-level radioactive nuclear waste in their countries and the technological advances reached by the USA on this important subject, allows a little bit more optimistic view about the future of nuclear energy for electricity production in the European region. However, according with the IAEA Nuclear Technology Review of 2006, "no final high level waste repositories are expected to be in operation much before 2020".

It is important to note, when considering the future role to be played by nuclear energy in the energy balance in the European region, the following situation: Nearly 30% of Europe's generating capacity is now more than thirty years old. The breakdown of installed capacity by plant age reflects the technological history of Europe's electricity industry (see Figure 51). The oldest installations are hydroelectric. Following these in age are coal-fired power plants, most of which date from the 1960s to 1990s, and are now between 16 and 40 years old. In the 1970s, nuclear power started up, reaching a peak between 1980 and 1990, followed by a period during which development was halted. From the 1990s onwards, natural gas and renewables became more important. Renewable resources (mainly wind energy) have become especially popular over the last decade.

From Figure 51 can be clearly stated that coal and nuclear power plants account for more than 70% of all power plants that will be at least 30 years old in 2020. Replacement of more than 50% of the current electricity installations must be addressed from as early as 2010. The question of replacing capacity will first affect coal combustion and later nuclear energy. Specifically, replacing the coal capacity with cleaner forms of power generation will probably bring economic advantageous. Alternatives to replacement include termination of production, extension of operating licenses or radical renovation. [134]

THE EUROPEAN VISION OF CURRENT AND NEW NUCLEAR TECHNOLOGIES

The European vision for the short, medium and long-term development of nuclear energy technologies, with the aim of achieving a sustainable production of nuclear energy, can be summarized as follows:

f)   nuclear energy is a key element in Europe's future low carbon energy system;

g)   the perspective of an important development of nuclear energy in the world (nuclear market renaissance) is relying in Generation- III light water reactors, in which it is Europe's interest to maintain its present industrial leadership;

h)   the development of Generation IV fast-neutron reactors with closed fuel cycle, which requires technological breakthroughs. Such reactors could be deployed by the middle of the 21st century, to enhance significantly the sustainability of nuclear energy;

i)   Generation IV systems with closed fuel cycles to substantially minimize the volume, the radiotoxic content and thermal load of the residual high-level waste requiring geological disposal. As a consequence, the isolation time and repository volume can be reduced. [135]

FISSION POWER

The majority of the current nuclear power reactors in operation in the world use nuclear fission technology for the production of electricity. Nuclear fission energy is one of the highly technological sectors in which Europe has undisputedly acquired a world leadership. The renewal of a worldwide interest for nuclear fission technologies demonstrates a general recognition of the merits of this energy source. However, Europe's leadership in the world competition is now challenged by large-scale initiatives from the USA, Russia, China and India. Europe, which has the largest nuclear industry in the world, has continuously enforced a high safety level, while promoting fuel and system innovation thanks to its research programmes. In order to preserve this unique asset, it is imperative to strengthen the structure of EU research and development forces and its industrial community. [135]

In the field of fission technology, the position of several European countries is very clear: The technology is well development and, for this reason, the nuclear industry should take care of the design of more efficient and inherent safe nuclear power reactors. There are some progresses in this direction in the last years. It is expected that when decisions on the construction of new nuclear power reactors will be taken by a group of countries, particularly in the European region, several attractive new and more efficiency nuclear power reactor designs will be already available on the market using this type of technology.

FUSION POWER

By far, the largest financial contribution of Europe into research in the nuclear area is into fusion technology. The goal of European fusion research is to demonstrate the viability of this type of technology as a future energy option to meet the increase energy needs of a growing world population.

According with the EU Web page Fusion for Energy, "since 1957 a European Fusion Programme coordinated research has been established by successive EURATOM Framework Programmes. This has enabled the development of experiments that come ever closer to the conditions needed for an electricity-producing fusion reactor". However, after many years of large fusion research using different types of equipment, a conclusion was reached that a larger and more powerful device would be needed to create the conditions expected in a fusion reactor and to demonstrate its scientific and technical feasibility for the future production of energy.

Taking into account the scale of the challenge to create the necessary conditions expected in a fusion reactor to produce power, fusion has entered a new era of global cooperation with the creation of the project called "International Experimental Energy Reactor" (ITER) that brings together Europe and six other countries[49], representing over one half of the world's population. The purpose of these countries is to construct the largest ever fusion experiment in the south of France with the aim to dominate the fusion technology. The construction of this nuclear power reactor will enabled Europe to assume a leading position worldwide in the development of the fusion technology.

The construction costs of ITER are estimated at €5 billion to be spread over a period of ten years. A similar amount is foreseen for the twenty-year operation phase of ITER, which will follow the construction period. During the construction of the ITER device, 90% of the components by value will be contributed by the parties in kind, meaning that they will contribute the components themselves, rather than paying for them. Europe will contribute up to half of the construction costs as the host party and the other six parties will each contribute up to 10%, thus giving a 10% contingency within the present funding. (Westra, 2007)

However, the use of fusion reactors to produce electricity at experimental scale is not expected to be ready before 2020. At the same time, the fusion technology for the production of electricity is not expected to be commercially available before 2050, or even later. To reach this stage it is important to solve, in the coming years and in an acceptable manner, the main difficulties and technological problems that the use of this technology is now facing. This is something very difficult to predict at this stage. For this reason, the use of the fusion technology for electricity production is a long-term solution not only for the European countries interested in the use of nuclear energy for the generation of electricity but for other countries as well.

SAFETY OF NUCLEAR POWER REACTORS

One of the important issues that are limiting the wider use of nuclear energy for electricity production in the European region is the level of safety of several of the current type of nuclear power reactors in operations.

The EC started to discuss safety issues in 1975, when nuclear power programmes promoted by different European states had progressed and diverged along very different routes, including different legal frameworks, types and number of rectors, as well as national systems regulating them in force in several countries. (Morales Pedraza, 2008)

There were important differences in safety requirements for the safe operation of nuclear power reactors in Western and Eastern Europe. These differences make more difficult the search for a common acceptable solution to this sensitive problem. The level of safety to be applied to nuclear power reactors produced in Russia and other Eastern and Central Europe countries, was higher than the one that it was applied in the past. This situation forces the EU,

---

[49] The following countries are participating in the fusion experiment: EU (including Switzerland and represented by EURATOM), Japan, China, India, the Republic of Korea, the Russian Federation and the USA.

for the first time in the history, to start in the 1990s the process of carrying out an evaluation of nuclear safety in Eastern and Central Europe, particularly in those countries asking for their entry in the EU. The debates on nuclear safety in the context of the enlargement of the EU raised vital questions about what are the Western standards for nuclear safety. The objective to be reached within this debate is to ensure that all European states would have equivalent levels of nuclear safety. This by no means imply that the operator of nuclear power reactors should not retain the primary responsibility for the nuclear safety of the nuclear power reactors under the control of the national regulatory bodies, but to have minimum common nuclear safety standards that should be applied by all European countries. Over the past 20 years, reactor safety has improved significantly, both in and outside the European region. No early generation of Soviet (Chernobyl-type) Light Water Graphite Reactors (LWGRs) is in operation in the EU, and the present cohort of European reactors has had a good overall safety record. Since 1986, accident probabilities have decreased substantially because of improvements in reactor technology, peripheral equipment and operation practices. Most of the European reactors are equipped with confinement domes, ascertaining that, in the occurrence of an accident, radioactive material are not released to the external environment and the consequences associated to the accident are controlled to a minimum. Man-machine interactions in plant operation have been considerably perfected, and a better safety culture has been established through the creation of an international "early notification system", obliging operators to report any incident using the International Nuclear Event Scale (INES) of the IAEA. Continued efforts in maintaining and elaborating high safety standards are among the actions carried out by the operators of nuclear power plants for an expansion of nuclear power in Europe. (Morales Pedraza, 2008)

Opportunities exist for reactor safety enhancement through research and development on new nuclear power reactor designs. Innovative designs for nuclear power reactors that make greater use of passive-safety features and build on the construction and operation experience gained in today's nuclear power reactors already in operation. The EPR and the High Temperature Reactor (HTR) are examples of new types of reactor designs. EPRs are among the nuclear power reactor designs more likely to be used in the construction of new nuclear power reactors in the European region in the near future. The nuclear power reactors presently under construction in Finland and France are of this type – while for the longer run the HTRs type of nuclear power reactor may be added to existing nuclear capacity.

Furthermore, the EU is in the process of creating new directives to further improve nuclear power reactor operation safety and develop regulatory safety oversight on a European level. Among the issues that will be addressed in the new directives are the ascertaining of sufficient funds for the complete decommissioning of nuclear power plants, the exchanging of best operation practice for existing nuclear installations, the maintaining of high safety standards for nuclear power plants whose operation licenses are extended, and the providing of transparency for citizens related with the operation of nuclear power plants.

It is important to stress that the design of nuclear systems in Europe relies on the "defense in depth" principle. It consists in the prevention of accidents and the mitigation of their consequences and the protection of workers and populations against radiological hazards through the use of multiple barriers and safety systems. For the more recent reactor systems such as Generation III reactors, even extremely improbable accidents are taken into account. For example, the EPR was designed so that in the very unlikely event of a severe accident, radiological consequences would necessitate only very limited protective countermeasures in a relatively small area and for a limited time for the surrounding population. [135]

DECOMMISSIONING OF NUCLEAR POWER PLANTS AND OTHER NUCLEAR FACILITIES

In the majority of European countries with nuclear power plants in operation, responsibility for the funding of decommissioning and dismantling of nuclear facilities remains with the owner of the facility. The operator should maintain funds or financial guarantee for decommissioning and dismantling, as required by national legislation or operating licenses. The availability or not of a waste disposal facility is a significant factor in the selection of an appropriate decommissioning strategy. This is often a matter for national policy or government, and beyond the control of the operators of nuclear power plants. If no repository is available, high-level radioactive nuclear waste must be processed and stored in the nuclear power plant site until the appropriate repository is available. Ideally, the specifications for treatment and packaging of high-level radioactive nuclear waste will be consistent with the regulatory requirements for transport, storage and eventual disposal of dangerous wastes. These specifications define the radiological, mechanical, physical, chemical and biological properties of the waste and of any package. The construction and operation of a storage facility will involve costs that cannot be neglected. These costs are likely to be highly dependent on the type of facility, the type and quantity of waste arising from decommissioning, as typified by the difference in the characteristics and quantities of waste arising from light water and gas-cooled power reactors. Furthermore, if such storage areas are located and remain on the site of the nuclear facility being decommissioned, they will prevent the full release of the site from nuclear regulatory control and block its availability for unrestricted use. In particular cases where the volume of decommissioning waste is large, as in the case of graphite-moderated power reactors for example, this situation may discourage immediate dismantling and encourage instead the option of safe enclosure until a waste disposal facility is available. In some cases, national or centralized storage facilities are built which allow the operator to carry out the decommissioning up to the release of the site for unrestricted reuse. [120]

One of the important issues that need to be in the mind of government authorities and the nuclear industry in charge of the operation of a nuclear power plants is to keep good records and precise information about of all nuclear facilities. Relevant knowledge and technical information about nuclear facilities is of prime importance for their future safe decontamination and dismantling. Conservation of such knowledge is a key element that needs to be included in any decommissioning strategy.

In some countries of the European region a great experience has been accumulated regarding the implementation of decommissioning strategies during the shut-down process of different nuclear facilities. In Germany, for example, 19 nuclear power reactors, 31 research reactors and critical assemblies, as well as nine fuel cycle facilities have been permanently shut down. Two of the nuclear power reactors, 21 of the research reactors and critical assemblies and four of the fuel cycle facilities have now been decommissioned, and the sites of the two nuclear power reactors restored and released from regulatory control. The other nuclear power reactors are currently in safe enclosure or are being dismantled, and their sites will be returned to a condition suitable for unrestricted reuse. In Belgium, about half of the cells in a major fuel reprocessing plant have been emptied and decontaminated and the other half are currently in the process of being brought to the same state. A small prototype PWR is in the process of being dismantled and decontaminated, and laboratories used previously for nuclear research and development have been decontaminated, released from radiological control and are now being used for conventional research. [131]

Decommissioning programmes implemented in France and in the UK, although more extensive in scale, are broadly similar in principle to the one implemented in Belgium. In France, 11 nuclear power reactors have been shut down until 2007. In the case of the UK, 26 nuclear power reactors were also shut down during the period 1950-2007. In Italy, four nuclear power reactors were shut down after the Chernobyl nuclear accident. In the case of the Netherlands, one nuclear power reactor were shut down in the same period as well as three nuclear power reactors in Sweden. In Eastern Europe, the following nuclear power reactors were shut down during the period 1950-2007: four in Bulgaria; three in Slovakia; five in Russia and four in Ukraine.

Lastly, it is important to note that other European countries with relative young nuclear power programmes, like Finland, the Czech Republic and Hungary, for example, have no decommissioning programmes underway. In these countries there are no plans for the adoption of a decommissioning strategy in the following years.

Main Elements of the EU Energy Policy

The main elements of the EU Energy Policy are the following: a) a 30% reduction in greenhouse gas emissions by developed countries by 2020 compared to 1990. In addition, 2050 global GHG emissions must be reduced by up to 50% compared to 1990, implying reductions in industrialized countries of 60-80% by 2050; and b) an EU commitment now to achieve, in any event, at least a 20% reduction of greenhouse gases by 2020 compared to 1990. [2]

It is important to note that the application of the new EU Energy Policy should reduce the CO_2 emissions in some 2 150 million of tons by 2030, that is to say, 16% below previously forecasts. This level of emission is equivalent to the CO_2 current emissions of Germany, UK, France and Italy all together. (Morales Pedraza, 2008)

Meeting the EU's commitment to act now on greenhouse gases should be at the center of the new EU Energy Policy for three reasons: 1) CO_2 emissions from energy make up 80% of EU-GHG emissions; reducing emissions means using less energy and using more clean locally produced energy; 2) limiting the EU's growing exposure to increased volatility and prices for oil and gas; and 3) potentially bringing about a more competitive EU energy market, stimulating innovation technology and jobs. (Morales Pedraza, 2008)

The EU Energy Policy should be prepared based on the realities of the different countries and should contemplate, among others, the following elements:

1) the increase in the saving and in the efficiency of the use of the different energy sources now available in the region. Since the 1970s, the European countries have

been guaranteeing an economic growth sustained without significant increments in the energy consumption. This achievement is the result not only of a more rational use of the energy but also to a more efficient use of it. In the 1970s and 1980s, the growth of the European countries was, as average of 2.4% annually, being the increment of the energy requirements of only 0.4%;

2) relevant technological advance that can be able to revolutionize the conversion, transport and storage of energy, as well as to elevate it efficiency;

3) the impact in the environment of the use of the different energy sources;

4) an increase in the resources allocated for research and development of the different renewable energy sources and for the discovery of new and even more efficient energy sources;

5) an increase of the international cooperation between the European countries and other countries, with the purpose to join effort in the use of the renewable energy sources now available in the world in the most efficient manner. The search for new more efficient energy sources should also be a subject of international cooperation;

6) an increase of the investments in each one of the links of the energy supplies chain. (Morales Pedraza, 2008)

Nuclear energy development programmes in each of the EU countries should take into account the above mentioned EU Energy Policy.

MAIN ELEMENTS OF THE EU ENERGY STRATEGY

Developing an EU Energy Strategy is being a long-term challenge for the EC. This needs a clear but flexible framework: clear in that it represents a common approach endorsed at the highest level; flexible in that it needs periodic updating.

During the preparation of any future EU Energy Strategy the following opinions and factors, among others, needs to be considered: a) government representatives; b) legislative representatives; c) managers; d) investors; e) suppliers; f) consumers; g) energy sources available; h) energy demands; and i) foresee energy price.

Six priority areas have been identified and included in the current EU Energy Strategy. These priorities areas are the following:

1) energy for growth and jobs in Europe: completing the internal European electricity and gas markets;

2) an Internal Energy Market that guarantees security of supply: solidarity between member states.

3) tackling security and competitiveness of energy supply: towards a sustainable, efficient and diverse energy mix lives;

4) an integrated approach: to tackling climate change;

5) encouraging innovation: to strategic European energy technology plan;

6)   towards to coherent external energy policy[50]. (Morales Pedraza, 2008)

The above-mentioned EU Energy Strategy contains several important elements that all EU countries should take into account during the elaboration of their own energy strategy for the introduction or for the expansion of the use of nuclear energy for electricity generation in the coming years.

THE PUBLIC OPINION

In October of 2001, a survey was carried out within the EU that shows that the majority of people would support the nuclear energy option for the generation of electricity, if and only if problems related with the final disposal of high-level radioactive nuclear waste are properly solved. Only in Austria, the numbers of people that are against the use of nuclear energy for electricity production outnumber those who agree. When citizens from EU-25 member states were asked what national governments should focus on in order to reduce its energy dependency, only 12% answered that first the use of nuclear energy should be further developed.

Between February and June 2005, the EC carried out a survey of nuclear energy waste and public acceptance of nuclear power in the EU. The survey highlights the following important aspects. For example, people who consider themselves well informed clearly show a better acceptance in all phases of nuclear waste. However, it should also be noted that only 25% of the citizens of the EU consider themselves well informed. The situation is especially negative, in terms of the opinions of women and young people aged between 15 and 24. The vast majority do not want further delays in setting up national strategies for high-level radioactive waste. They clearly want to be involved in the decision-making process and in the selection of the disposal sites. Harmonized strategies and management policies for radioactive waste are needed for the whole of the EU. Environmental non-governmental organizations are considered the most trustworthy sources of information, followed by independent scientists and the authorities. Far less trusted are national agencies responsible for nuclear waste. Across the EU, 37% of the people surveyed were in favor of nuclear energy, while 55% were against it. [134]

It is important to note that due to the increasing focus on climate change, particularly in the mass media, a shift towards a more positive perception of nuclear power seems to be taking place within the European region. Given the different approaches to nuclear power, European countries can be classified as follows: a) those using nuclear power; b) those using nuclear power, but with phase-out policies such as Germany, Sweden and Belgium; and c) those not (yet) using nuclear power (e.g. Italy and Poland), but with plans for the introduction of a nuclear power programme in the future; and d) those that have no plans for the use of nuclear energy for the production of electricity and have no intention to adopt such plan in the future.

The perception of the public opinion regarding the use of nuclear energy for the production of electricity in group (a) is not homogeneous. According with reference 33, it

---

[50] More information on the EU Energy Strategy can be found in reference 66.

ranges from very negative (Croatia, Russia) to very positive (Finland, Czech Republic, Romania and Bulgaria). Over the last 25 years in Finland, a positive perception of nuclear power has increased from only 25% in 1982 to 50% in 2005. In the same period, the negative perception of nuclear power has dropped from almost 40% in 1982 to 20% in 2005. Thus, Finland has seen a complete shift in the public opinion and perception of nuclear energy over the past 25 years. In the UK, there is also a trend towards a better public attitude towards nuclear energy. Polls conducted in December 2005, showed that 41% of those interviewed were in favor of new nuclear power plants, whereas a year earlier, the share was only 35%. [134]

The share of nuclear power in the overall power production does not seem to play any role for public acceptance in some European countries. For example, in Spain and Switzerland, the main issues of concern are nuclear waste and its final disposal. At the same time, the political process for building new nuclear power plants is extremely complicated in several European countries.

In the case of group (b), there is also a clear trend towards higher acceptance of the use of nuclear energy for electricity generation in several countries included in this group. The high acceptance of nuclear energy has significantly risen in Sweden in the last years, despite the adoption of nuclear phase-out policy by the government. In polls conducted in 2005, a total of 83% of those interviewed either wanted to keep the country's nuclear power reactor in operation or replace them with new ones. In the polls carried out in 2006, a total of 85% wanted to keep the countries' ten nuclear power reactors operational or support the construction of new ones. In Germany, polls show unclear results, 54% of those interviewed believe that, despite the phase-out policy, nuclear energy should continue to play a role for a long time, but only 22% want nuclear energy to secure the German electricity demand for the next 20 or 30 years.

In group (c) countries included in this group viewed nuclear energy rather differently. Due to its attempt of reducing power production from coal, Poland public opinion is very positive towards the use of nuclear energy for electricity production. In Italy, a shift in the public opinion has also taken place in recent years. In polls carried recently, 54% of those interviewed think that Italy should build new nuclear power plants instead of importing electricity produced in French using nuclear energy. At the same time, 70% of the Italian people interviewed think that it does not make much sense for Italy not to have nuclear power plants on its own territory, while being surrounded by countries with nuclear power plants in operation.

It is important to note as the impacts of climate change and the vulnerability of the European economy to foreign fuel imports become more evident, it is likely that the gradual shift in public opinion of the last decade will further develop towards less skepticism or in favor of nuclear energy. (Van der Zwaan, 2006)

Summing up can be stated that the public attitude to the use of nuclear energy for electricity production has had a major influence on the use of this type of energy for this purpose in most of the European countries. In some specific cases, the negative reaction of the public opinion to the use of nuclear energy for the generation of electricity has prevented utilities to enter into commercial operation or to force government to adopt a nuclear phase out policy. Public opposition banned nuclear power in Denmark and Ireland, prevented the operation of an already built nuclear power plant in Austria, led to a short-term phase out in Italy and to long-term phase out in Belgium, Germany and Sweden. The public opinion regarding the use of nuclear energy for electricity generation varies widely amongst countries within the European region. The highest acceptance of the use of this type of energy for electricity production seems to be in the Nordic and Eastern European countries, while the lowest is in the non-nuclear and Southern countries. Spain is a noteworthy It has several operating nuclear power reactors with a very low acceptance. In Figure 53a map with the result of the public opinion poll carried out by the EC in 25 countries of the European region is shown.

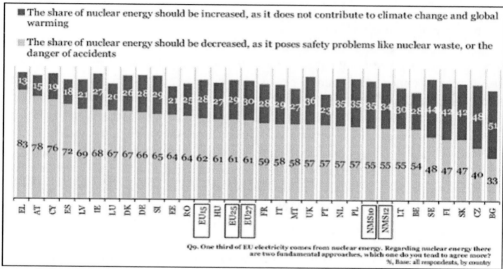

Source: Gallup, Attitudes on issues related to EU Energy Policy, EC, DG TREN, April 2007.

Figure 52. Public opinion within the EU on nuclear power.

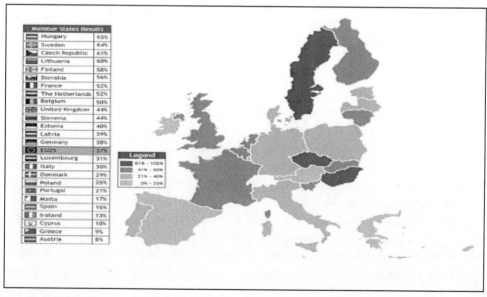

Source: Special Eurobarometer, Radioactive Waste, commissioned by DG TREN, EC, September 2006.

Figure 53. Map showing the level of support of nuclear power programme in EU-25.

LOOKING FORWARD

The future of nuclear power within the European region depends of the country you are considering. There are several groups of countries that can be identified. These groups are the following:

a) countries using nuclear power and with plans to expand the use of this type of energy for electricity production in the coming years: Bulgaria, Finland, France, Hungary, Lithuania, Romania, the Russian Federation, Slovakia, Slovenia, Switzerland, UK and Ukraine;

b) countries using nuclear power but without approved plans to expand the use of this type of energy for electricity production in the coming years: Belgium, Germany, the Netherlands, Spain and Sweden;

c) countries not using nuclear power but with plans to introduce the use of this type of energy for electricity production in the coming years: Italy and Poland;

d) countries not using nuclear power and without plans to expand the use of this type of energy for electricity production in the coming years: Albania, Austria, Bosnia and Herzegovina, Croatia, Cyprus, Denmark, Estonia, Federal Republic of Yugoslavia, Greece, Holy See, Iceland, Ireland, Island, Latvia, Liechtenstein, Leetonia, Luxembourg, Monaco, Montenegro, Serbia and the former Yugoslav Republic of Macedonia.

THE NUCLEAR POWER PROGRAMME IN FRANCE

In 2007, French total net electricity generation was 545 TWh. The total electricity consumption was 482 TWh, this means 7 700 kWh per person. The net nuclear electricity generation was 418.6 TWh France has the highest proportion of nuclear power production of any country in the world. In 2007, the 59 nuclear power reactors operating in the country produced 76.85% of the total electricity generation of France and 47% of all nuclear electricity in the EU. The nuclear capacity installed is 63 473 MWe.

Fifty-eight of the total nuclear power reactors operating in the country are PWR. This amount represents 98.3% of the total nuclear power reactors operating in France. Out of this total, thirty four units have 900 MWe of capacity; twenty units have 1 300 MWe of capacity, and four units 1 450 MWe of capacity. All of these nuclear power reactors were constructed by the French manufacturer Framatome. France has one fast breeder reactor (Phoenix) operated by EdF with a capacity of 230 MWe. The total nuclear capacity installed in France is of over 63 GWe[51]. In 2007, the net load factor of the French PWR reactors was 75.9%. France has 14 nuclear research reactors of various type and sizes all constructed between 1959 and 1980.

In 2007, the annual electrical power production and the nuclear power production in France are shown in Table 18.

---

[51] France's nuclear power programme has cost some FF400 billion in 1993 currency, excluding interest during construction. (33)

Table 18. Annual electric power production and nuclear power production in France in 2008

Total power production	Nuclear power production
549 TWh	418. 3 TWh

Source: IAEA PRIS, 2009.

It is important to single out that a few years ago France was a net electricity importer. Due to the introduction of its nuclear power programme France is now a net electricity exporter. Over the last decade, France has exported between 60 to 70 billion kWh net each year to neighboring countries and EdF expects exports to continue at the level of 65-70 TWh per year in the future. The nuclear power production exported to other countries, such as Italy, Portugal, Belgium and the UK reached almost 20% of the total electricity generated by this type of energy. At the same time, France has exported its PWR reactor technology to Belgium, South Africa, South Korea and China. There are two 900 MWe French reactors operating at Koeberg, near Cape Town in South Africa, two units at Ulchin in South Korea and four units at Daya Bay and Lingao in China, near Hong Kong. (Schneider, 2008)

It is important to stress the following: France has achieved a high degree of standardization in several of its nuclear power reactors currently in operation. The level of standardization already achieved in the construction of the French nuclear power reactors is the highest in the world.

The evolution of the nuclear power capacities in France during the period 1970-2007 and the number of nuclear power reactors in operation in this period are shown in Figures 54 and 55.

In 2006, EdF sales revenue was €58.9 billion and debt had fallen to €14.9 billion. The cost of nuclear-generated electricity fell by 7% from 1998 to 2001 and is now about €3 cents/kWh, which is very competitive in Europe. The back-end costs (reprocessing, wastes disposal, etc) are fairly small when compared to the total kWh cost, typically about 5%. It gains for exporting electricity is over €3 billion per year. The location of the nuclear power plants operating in France in shown in Figure 56.

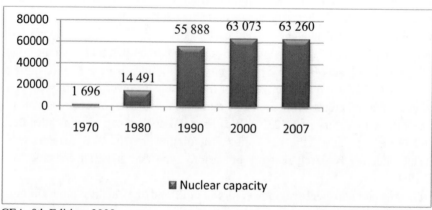

Source: CEA, 8th Edition, 2008.

Figure 54. Evolution of the nuclear power capacities in France during the period 1970-2007.

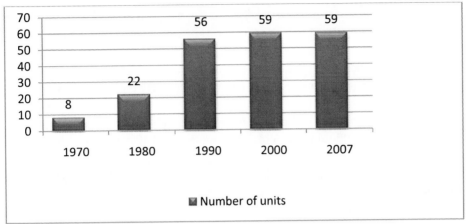

Source: CEA, 8th Edition, 2008.

Figure 55. Number of nuclear power reactors in operation during the period 1970-2007.

Source: International Nuclear Safety Center at Argonne National Laboratory, USA.

Figure 56. Location of the nuclear power plants in France.

The Energy Policy in France

After the end of the World War II, France's economic and social development relied mainly on the deployment of energy intensive industries. Until the 1970s, domestic coal and hydropower resources were the main energy sources used to satisfy the increasing energy needs of the country. However, French domestic fossil fuel resources were limited and its extraction very costly and, for this reason, the country had to rely heavily on imports for its energy supply. By 1973, imports were covering more than 75% of national energy consumption, compared to 38% in 1960. After the 1970s oil crisis, the country was in need of better energy independence. At that time, implementation of a large nuclear power programme became a major element of France's energy policy, including also energy saving measures, efficiency improvement and research and development in the field of renewable energies. The share of nuclear power in primary energy supply increased from less than 2% in the late 1970s to about one third in the mid 1990s and reached 42% in 2003. The main macro-economic impacts of France's energy policy are: a) drastic improvement in the energy trade balance; b) stabilization of domestic energy prices at a rather low level; c) increase competitiveness of French companies on international markets; and d) deployment of a nuclear industry sector covering nuclear power reactor construction and the whole of the fuel cycle.

Increased awareness of environmental constraints reflects in the French energy mix, aiming to reduce the negative impacts of energy production on health and environment. In this regard, substitution of nuclear power to fossil fuel for electricity generation resulted in a drastic reduction of atmospheric emissions from the energy sector. In 1999, a parliamentary debate reaffirmed the three following main objectives of the French energy policy: a) security of supply; b) respect for the environment (especially greenhouse gases); and c) proper attention to radioactive waste management. However, taking into account the present energy situation the main objectives of the French energy policy were change. The new objectives of the French energy policy are now the following: a) to optimize the utilization of existing equipment, i.e., power plants and fuel cycle facilities; b) design and implement a policy with regard to final disposal of high level radioactive waste; and c) develop the next generation of reactors improving the use of natural uranium and minimizing waste production. [32]

It is important to single out the main macro-economic impacts of currently France's energy policy. These are the following: a) drastic improvement in the energy trade balance of the country; b) stabilization of domestic energy prices at a rather low level; c) increased competitiveness of French companies on international energy markets; and d) develop of a nuclear industry sector, including nuclear power reactor construction and all components of the nuclear fuel cycle.

After the 1970s oil crisis, the country was in need of better energy independence. For this reason, and based on the lack of sufficient fossil fuel sources available in the country, a large nuclear power programme became a major element of France's energy policy, in addition to energy saving, efficiency improvement and research and development in the field of renewable energies. As a result of the implementation of the French energy policy, the share of nuclear power in the total electricity production in France increased from 3.4% in 1970 to about 23.5% in 1980. In 2007, the share of nuclear power in the total electricity production in France went up to 76.85% being the largest share in the world. In 2008, the nuclear share reached 76.18%.

However, and despite of the achievement reached by France in the use of nuclear energy for the generation of electricity, it is important to note that in the longer term, energy-policy makers should pay particular attention to the development of new nuclear technologies, including research in high temperature gas-cooled reactors, super-critical water-cooled reactors and designs using liquid sodium or lead (or lead alloys) as a coolant with the objective to increase further the level of safety of the new nuclear power reactors designs.

The Evolution of the Nuclear Energy Sector in France

During the post World War II reconstruction period, France's economic and social development relied mainly on the deployment of energy intensive industries. The rapidly increasing energy needs were partly met by domestic coal and hydropower resources. However, due to limited and costly French domestic fossil fuel resources the country had to rely heavily on imports for its energy supply.

In 1974, just after the first oil shock, the French government decided to expand rapidly the country's nuclear power capacity with the purpose to increase its energy independence. This decision was taken bearing in mind the substantial heavy engineering expertise France had at that time but with few indigenous energy resources available in the country for the production of electricity. For this reason, the use of nuclear energy for electricity generation was the only choice that could be adopted at that time by the French's government with the purpose not only to increase the production of electricity to satisfy the foresee increase in the electricity demand but to minimize, as much as possible, energy imports for this purpose and in order to achieve greater energy security.

As a result of the 1974 decision, France has now a substantial level of energy independence and is one of the countries in the European region with the lowest cost in the production of electricity. It also has an extremely low level of CO_2 emissions per capita from electricity generation, since over 90% of its electricity is generating by nuclear power reactors or by hydro. Table 19 and figure shows the French national production of electricity using different types of energy from 1950 to 2007.

Historically, the development of nuclear power programme in France passes through four phases.

Table 19. French national production of electricity during the period 1950-2007

Year	Fossil fuels (TWh)	Hydro (TWh)	Nuclear (TWh)	Others (TWh)	Total (TWh)
1950	17	16	-	-	33
1955	24	26	-	-	50
1960	32	41	-	-	72
1965	54	46	1	-	101
1970	79	57	5	-	141
1975	101	60	17	-	179
1980	119	70	58	-	247
1985	52	64	213	-	329
1990	45	57	298	-	400
1995	37	76	359	-	471
2000	50	72	395	-	517
2005	59	56	430	4	549
2007	55	63	419	7.9	545

Source: Nuclear power plants in the world, edition 2008, CEA.

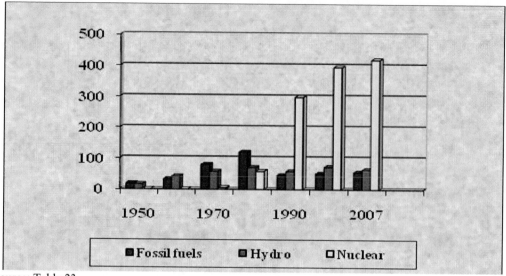

Source: Table 23.

Figure 57. French national production of electricity during the period 1950-2007.

Phase I: During the 1960s, France started the promotion of its indigenous nuclear power reactor design based on the Westinghouse nuclear power reactor design. These were mainly natural uranium-gas cooled and fast breeder's reactors. During this period six units started to be constructed, six were connected to the electric grid, included three nuclear power reactors started to be constructed in the 1950s, and seven entered into commercial operations. All of these reactors were already shut down. International developments in the nuclear industry led in the late 1960s to the recognition that the French nuclear power reactor designs were not competitive with light water reactors. For this reason, the French industry took, in 1969, the decision to build light water reactors under license, whilst restructuring the domestic nuclear industry to improve competitiveness.

Source: Photograph courtesy of www.ecolo.org/photos/npp/.

Figure 58. Nuclear power plant at Flamanville.

Phase II: From 1974 to 1981, emphasis was put on adaptation of the Westinghouse nuclear power reactor design for the development of a French nuclear power reactor. The oil crisis of the 1970s accelerated the implementation of the French nuclear power programme. The capacity of French nuclear power reactors increased from 900 MWe to 1 300 MWe and later to 1 450 MWe. France developed and implemented, in parallel with its nuclear power programme, a strong domestic nuclear fuel cycle industry, built upon the infrastructure originally established by CEA. During the 1970s, a total of 39 nuclear power reactors started to be constructed, six units were connected to the electric grid and five units entered into commercial operations.

Phase III: In 1981, Framatome terminated its license agreement with Westinghouse and negotiated a new agreement. The purpose of this new agreement was to give France a greater autonomy in the development of its nuclear power programme. Based on this agreement, Framatome developed a wide range of servicing expertise and capabilities in nuclear power reactor operation and maintenance services. In the same year, France adopted its energy policy to a lower–economic growth, together with the occurrence of over-capacity in the national electricity supply system. The achievement of the 1 450 MWe N4 model was the landmark for a totally autonomous French nuclear power reactor design. During the 1980s, 18 nuclear power reactors started to be constructed, 41 were connected to the electric grid and 43 units entered into commercial operation. In the 1990s, no new nuclear power reactors started to be constructed, but 10 units were connected to the electric grid and six units entered into commercial operation.

Phase IV: In 2000, Framatome merged its nuclear activities with German company Siemens. As result of this merge, the Areva group was established. Framatome holds 66% of the Areva company and Siemens 34%. During the 2000s, no new nuclear power reactors started to be constructed or connected to the electric grid and four units entered into commercial operation.

From the above paragraphs the following can be concluded: The peak in the construction of nuclear power reactors was achieved in the 1970s, when 39 units started to be constructed. The highest number of nuclear power reactors connected to the electric grid was achieved in the 1980s, when 41 units were connected to the electric grid. The highest number of nuclear power reactors entered into commercial operations was achieved in the 1980s, when 43 units started its operations. In the following figures some nuclear power plants currently in operation in France are shown.

In 2003, the 1 600 MW EPR nuclear power reactor, designed by Framatome based on the French nuclear power reactor N4 and the German Konvoi nuclear power reactor type, was ordered by a consortium of Finnish industrial companies in order to expand the nuclear generation electricity in the country. It is the first model of EPR reactors to be built in the European region in several years. In May 2006, the EdF board approved construction of a new 1 650 MWe EPR unit at Flamanville, Normandy, alongside two existing 1 300 MWe units. After considerable preparatory work first concrete was poured on schedule in December 2007 and construction is expected to take 54 months.

In France, a total of 11 nuclear power reactors were shut-down until 2007. The list of nuclear power reactors in operation in France is shown in Table 20.

Table 20. Nuclear power reactors in operation in France

Reactor	MWe net	Commercial operation
Blayais 1-4	910	12/81, 2/83, 11/83, 10/83
Bugey 2-3	910	3/79, 3/79
Bugey 4-5	880	7/79-1/80
Chinon B 1-4	905	2/84, 8/84, 3/87, 4/88
Cruas 1-4	915	4/84, 4/85, 9/84, 2/85
Dampierre 1-4	890	9/80, 2/81, 5/81, 11/81
Fessenheim 1-2	880	1/78, 4/78
Gravelines 1-4	910	11/80, 12/80, 6/81, 10/81
Gravelines 5-6	910	1/85, 10/85
Saint-Laurent B 1-2	915	8/83, 8/83
Tricastin 1-4	915	12/80, 12/80, 5/81, 11/81
Belleville 1 & 2	1 310	6/88, 1/89
Cattenom 1-4	1 300	4/87, 2/88, 2/91, 1/92
Flamanville 1-2	1 330	12/86, 3/87
Golfech 1-2	1 310	2/91, 3/94
Nogent 1-2	1310	2/88, 5/89
Paluel 1-4	1 330	12/85, 12/85, 2/86, 6/86
Penly 1-2	1 330	12/90, 11/92
Saint-Alban 1-2	1 335	5/86, 3/87
Chooz B 1-2	1 500	5/2000, 9/20001999
Civaux 1-2	1 495	1/2002, 4/2002
Phenix	130	7/74
Total 59	63 260	

Source: CEA, 8th Edition, 2008.

Until 2007, eleven experimental and nuclear power reactors are being decommissioned in France, eight of them first-generation gas-cooled graphite-moderated types and six were very similar to the UK Magnox reactor type. The other three include the 1 200 MWe Super Phenix fast reactor[52], the 1966 prototype 305 MWe PWR at Chooz, and an experimental 70 MWe GCHWR at Brennilis, which ran during the period 1967-85. A license was issued for dismantling Brennilis, in 2006, and for Chooz A, in 2007. The total net nuclear capacity retired from the grid was 3 951 MWe.

---

[52] France abandoned its fast breeder reactor programme in1998 when the only industrial scale plutonium fuelled breeder in the world, the 1 200 MW Superphénix in Creys-Malville, was officially shut down permanently. Started up in 1986 it produced some electricity only in six of the twelve years it was officially in operation. Its lifetime load factor was less than 7%. Plagued by technical problems and a long list of incidents, the cost of the adventure was estimated by the French Court of Auditors at FRF60 billion (close to €9.15 billion) in 1996. However, the estimate included only FRF5 billion (€0.760 billion) for decommissioning. That figure alone had increased to over €2 billion by 2003. At a lifetime power generation of some 8.3 TWh, Superphénix has produced the kWh at about €1.35 (to be compared with the French feed-in tariff of €0.55 per building integrated solar kWh). (Thomas, 2005)

Table 21. Nuclear power reactors under construction in France

	Type	MWe net	Construction start	Grid connection	Commercial operation
Flamanville 3	EPR	1 620	12/07	12/11	5/12

Source: WANO.

To carry out decommissioning activities EdF puts aside €0.14 cents/kWh for decommissioning and at the end of 2004 it carried provisions of €9.9 billion for this purpose. By 2010, it will have fully funded the eventual decommissioning of its nuclear power plants (from 2035). Early in 2006, it held €25 billion segregated for this purpose, and is on track for €35 billion in 2010. [33]

Finally, until 31 December 2007, France exported nine nuclear power reactors to the following four countries: South Africa (2), Belgium (3), China (2) and the Republic of Korea (2), with a total net capacity of 8 529 MWe. All of these reactors are still connected to the grid in these countries.

Nuclear Safety Authority and the Licensing Process

It is important to single out that nuclear legislation in France has been developed in line with technological advances and growth in the atomic energy field. Therefore, many of the enactments governing nuclear activities in the country can be found in the general French legislation on environmental protection, water supply, atmospheric pollution, public health and labor. [32] The main French legislations in the nuclear field are the following: a) Act No. 68-493 (30 October 1968), setting special rules as to third party liability in the field of nuclear energy; b) 19 July 1952 Act, now embodied in the Public Health Code, specifying licensing requirements for the use of radioisotopes; c) Act No. 80-572 of 25 July 1980 on the protection and control of nuclear materials; and d) Act No. 91-1381 concerning research on radioactive waste management.

French nuclear legislation began to be developed in 1945 when the Ordinance No. 45-2563 of 18 October 1945 was adopted by the French government. This development had several landmarks: a) in 1963, a system for licensing and controlling major nuclear installations was introduced, setting government responsibility in matters of population and occupational safety (Decree of 11 December 1963); b) in 1973, this system was expanded to cover the development of the nuclear power programme and to better define the role of government authorities; and c) in 1966, with the adoption of the decree of 20 June 1966, including the Euratom Directives as part of the French radiation protection regulations.

In the course of the 1980's, the enactments setting up the CEA were amended so as to strengthen its inter-ministerial status. At the same time, a tripartite Board of Administration including staff representatives was created. The main task of CEA was laid down in September 1992 by the government. According with the government decision, "CEA should concentrate its activities in developing the control of atom uses for purposes of energy, health, defense and industry, while remaining attentive to the requests made by its industrial and research partners". More specifically, the inter-ministerial committee of 1 June 1999 requested CEA "to strengthen long-term research on future reactors capable of reducing, and even eliminate the production of long-lived radioactive waste. In addition, CEA was given a particular responsibility for research and development on alternative and renewable energies. [32]

The French authorities involved in the licensing procedure for the setting up of large nuclear installations are: a) the Minister for Industry; and b) the Minister for Ecology and Sustainable Development.

In the beginning of 2002, the General Directorate for Nuclear Safety and Radioprotection (DGSNR) was created with the purpose to carry out measurements or analytical work in order to determine the level of radioactivity or ionizing radiation that might become hazardous to health in various environmental situations for individuals as well as for the population as a whole. It also co-ordinates and defines controls for the radiation protection of workers and is involved in the safety plans to be put in action in case of radioactive incident. DGSNR reports to the Ministers for Industry, Health and Ecology and Sustainable Development and is mainly responsible for:

a) studying problems raised by site selection;
b) establishing the procedures for licensing large nuclear installations (licenses for setting up, commissioning, disposal, etc.);
c) organizing and directing the control of these installations;
d) drafting general technical regulations and following their implementation;
e) establishing plans in the event of an accident occurring in a large nuclear installation;
f) proposing and organizing public information on nuclear safety.

The licensing procedure in force in France regarding the authorization for the establishment of nuclear installations is governed by Decree No. 63-1128 of 11 December 1963. This decree lays down the technical requirements and other formalities which the operator of nuclear installations must comply with. For nuclear power reactors, for instance, there are generally two stages: 1) Fuel loading and commissioning tests; and 2) entry into operation. In all cases, the final decision needs to be approved by the Ministers for Industry and for Ecology and Sustainable Development. The consent of the Minister for Health is also requested.

It is important to note that, in 2006, the new Nuclear Safety Authority (Autorite de Surete Nucleaire, ASN)[53], became the regulatory authority responsible for nuclear safety and radiological protection in France. ASN took these functions from the DGSNR. ASN reports to the Ministers of Environment, Industry and Health. However, its major licensing decisions still to be approved by the French government.

Nuclear Fuel Cycle and Waste Management

France, as a nuclear weapons state, has the full control of the nuclear fuel cycle. The following are the main information regarding nuclear installations and activities related with

---

[53] ASN is an independent body within the French's government structure.

the French nuclear fuel cycle. France uses some 12 400 tons of uranium oxide concentrate (10 500 tons of U) per year to ensure the operation of its nuclear power plants. Much of this comes from Canada, Niger, Australia, Kazakhstan and Russia, mostly under long-term contracts. Beyond this, it is self-sufficient and has conversion, enrichment, uranium fuel fabrication and MOX fuel fabrication plants operational, together with reprocessing and a waste management programme. Most fuel cycle activities are carried out by Areva NC. Uranium concentrates are converted to hexafluoride at the Comurhex Pierrelatte plant in the Rhone Valley, which commenced operation in 1959. In May 2007, Areva NC announced plans for a new conversion project - Comurhex II - with facilities at Malvesi and Tricastin to strengthen its global position in the front end of the fuel cycle. The €610 million nuclear facility will have a capacity of 15 000 tons of uranium per year from 2012, with scope for increase to 21 000 tons of uranium per year later on. Enrichment of uranium takes place at the 1978 Eurodif plant at Tricastin nearby, with 10.8 million SWU capacity (enough to supply some 81 000 MWe of generating capacity, about one third more than France's total generating capacity). [33]

In 2003, Areva brought 50% stake in Urenco's Enrichment Technology Company (ETC). The deal enables Areva to use Urenco/ETC technology to replace its Eurodif gas diffusion enrichment plant at Tricastin with a new enrichment plant. The agreement was signed in mid 2006, and the construction license for the new enrichment plant was approved by ASN in February 2007. The €3 billion two-unit plant has a nominal annual capacity of 7.5 million SWU and will be built and operated by Areva NC subsidiary Societe d'Enrichissement du Tricastin (SET). The nuclear installation is expected to reach full capacity in 2014. The second unit is expected to be fully operational in 2016. When fully operational, in 2018, the whole plant will free up some 3 000 MWe of Tricastin nuclear power plant's capacity for the French electric grid - over 20 billion kWh per year (€800 million per year).

According to the Law n° 91-1381 on Research and Management of Nuclear Wastes of 1991, the government should define three complementary lines of research aimed at finding a solution for managing high-level and long-lived intermediate-level radioactive waste over the very long term: partitioning and transmutation of long-lived radioactive elements, reversible and irreversible deep geological disposal and conditioning and long-term near-surface storage.

The Public Opinion

The French nuclear programme began at the second half of 1950s. The beginning of this programme caused widespread large-scale and occasionally violent demonstrations, leading to the death of a protester on at least one occasion. The magnitude of these demonstrations increase in the 1970s and the burst of nuclear power plants construction were ended in the middle of the 1980s. In the 1990s, only one nuclear power plant initiated its construction. Since then the relatively slow build rate of additional nuclear power reactors has diminished the focus for public concern and the number and magnitude of the demonstrations against the use of nuclear energy for the generation of electricity in the country. However, this not means that the opposition to nuclear power in France does not remain strong.

From time to time demonstrations against the use of nuclear energy for electricity production have been organized in the country. In 2006, for example, Greenpeace activists surrounded and chained themselves to a truck transporting plutonium from the La Hague, a nuclear recycling facility located in this region, to an undisclosed burial site. This action stopped the transfer of these fissionable materials over French town.

French environmentalists say the economic benefits of nuclear energy are far outweighed by the dangers of it's by products, particularly for the high-level radioactive wastes it generates, which has to be storage in special facilities for long period. At the same time, French environmentalist also say that nuclear facilities are potential targets for terrorist groups putting in danger the live of the people living near the facilities.

In 2003, a poll had shown that 67% of people thought that environmental protection was the single most important energy policy goal. However, 58% thought that nuclear power caused climate change while only 46% thought that coal burning did so. The debate was to prepare the way for defining the energy mix for the next 30 years in the context of sustainable development at a European and at global levels. But the French government and the powerful nuclear lobby are firm in their position regarding the use of nuclear energy for electricity generation in the future in order to maintain the current energy independence of France.

In October 2007, President Sarkozy promised "that France was not going to replace its existing nuclear power plants in its entirety and, for this reason, in the future no more than 60% of France's electricity will be produced using nuclear energy", well down on the level reached in 2008 that was 76.16%. The statement of the President was made with the purpose to reduce the tension surrounded the decision of the French government to continue using nuclear energy for the production of electricity in the future and to replace the current nuclear power reactors with a new generation of reactors in the coming years.

A poll of 20,790 respondents was conducted between July 15 and November 4, 2008 by WorldPublicOpinion.org, a collaborative research project involving research centers from around the world and managed by the Programme on International Policy Attitudes (PIPA) at the University of Maryland. As a result of this poll carried out in France 4 in 10 (40%) agree with placing less emphasis on building new nuclear power reactors in the future while 26% say there should be more nuclear power reactors built to face the foresee increase in the demand of energy.

Despite of the French demonstrations carried out against the use of nuclear energy for electricity generation in the country in the past, and the result of the abovementioned polls, it is important to note that now the majority of the French public opinion is either in favor of using nuclear energy for electricity generation or at least is not firmly against the use of this type of energy for this purpose in the future.

Looking Forward

After years of inactivity in the construction of nuclear power reactors in the country, in May 2006, the EdF Board approved the construction of a 1 650 MWe EPR unit at Flamanville, Normandy. This is going to be the second EPR nuclear power reactor to be built in Europe. The construction works started on December 2007 and it is expected to take around four and half years (54 months) to be finished. It is also expected that the nuclear power reactor begins commercial operation in 2012. The overnight capital cost is expected to be €3.3 billion, and power from it €4.6 c/kWh - about the same as from new combined cycle gas turbine at current gas prices and with no carbon emission charge. In January 2007, EdF ordered the main nuclear part of the reactor from Areva. The turbine-generator section was ordered in 2006 from Alstom.

The President of France has confirmed recently that EdF "will build in Penley, near Dieppe, in the north of France, a second EPR unit with a capacity of 1 650 MWe". The nuclear power plant of Penley already has in operation two PWR units with a capacity of 1 300 MWe each. The group GdF-Suez, the great French energy group, will have a minority participation in the construction of the new nuclear power reactor. The Italian company Enel was also invited to be part of the company that will participate in the construction of the new nuclear power reactor. It expected that the construction activities begin in 2012 and that the nuclear power reactor enters in operation in 2017. France's second new-generation Areva EPR is to have substantial equity from GdF-Suez, whose nuclear base is in Belgium. With oil giant Total it will take a one third share in the 1 650 MWe EPR to be built at Penly. Italy's Enel is expected to take up its right to 12.5%, so the split is likely to be EdF 54.2%, GdF-Suez 25%, Enel 12.5% and Total 8.3%, making it an unusually international project for France. GdF-Suez said it was "a major step forward in the implementation of its nuclear strategy", though through Electrabel it already operates all seven Belgian nuclear power reactors, and has shares of two Chooz units in France. It also bought a 5% share in Areva's new enrichment plant in France. Areva, GdF-Suez and Total are together bidding to build a pair of EPRs in Abu Dhabi. It is important to note that Total is the first major oil company to invest in nuclear power in recent years.

In August 2005, EdF announced that it plans to replace its 58 present reactors with EPR nuclear reactors from 2020, at the rate of about one 1 650 MWe unit per year. It would require 40 of these types of reactors to reach present capacity. This will be confirmed about 2015 on the basis of experience with the initial EPR unit at Flamanville. EdF's development strategy selected the nuclear replacement option on the basis of nuclear economic performance, for the stability of its costs and out of respect for environmental constraints[54]. [33]

However, in January 2006, the President of France announced that the CEA "was to embark upon designing a prototype of the new Generation IV system to be operating in 2020, bringing forward the timeline for this by some five years. This decision could delay the ambitious French nuclear power programme mentioned above. France has been pursuing the following three Generation IV system technologies: a) Gas-Cooled Fast Reactor; b) Sodium-Cooled Fast Reactor; and c) Very High Temperature Reactor (gas-cooled). While Areva has been working on the last two types of Generation IV systems, the main interest in the Very High Temperature Reactors has been in the USA, as well as South Africa and China. CEA interest in the fast reactors is on the basis that they will produce less waste and will better exploit uranium resources, including the 220 000 tons of depleted uranium and some reprocessed uranium stockpiled in France.

In December 2006, the government's Atomic Energy Committee decided to proceed with a Generation IV Sodium-Cooled Fast Reactor prototype whose design features are to be decided by 2012 and the start up aimed for 2020. A new generation of Sodium-Cooled Fast Reactor with innovations intended to improve the competitiveness and the safety of this reactor type is the reference approach for this prototype. A Gas-Cooled Fast Reactor design is to be developed in parallel as an alternative option. The prototype will also have the mission

---

[54] Use of other designs such as Westinghouse's AP 1000 of GE's ASBWR is possible.

of demonstrating advanced recycling modes intended to improve the ultimate high-level and long-lived waste to be disposed of. The objective is to have one type of competitive fast reactor technology ready for industrial deployment in France and for export after 2035-2040. The prototype, possibly built near Phenix at Marcoule, will be 250 MWe to 800 MWe and is expected to cost about €1.5 to €2 billion. [33]

Electricite de France SA said, in November 2008, that "it will invest up to €50 billion with its partners over the next 12 years to build next-generation of nuclear power reactors in Europe, the USA and China". It also said that "its share of the investment would run to between €12 billion to €20 billion by 2020". EdF confirmed in a statement that "the first of its new EPR nuclear power reactor would be operational in 2012". That will be followed by a Chinese EPR in 2013, a USA unit in 2016 and a British nuclear power reactor at the end of 2017. EdF expects 140 GWe of new nuclear capacity to be built around the world by 2020, on top of the 370 GWe already in operation. Most of the new capacity is going to be built in China, India and other Asian countries.

THE NUCLEAR POWER PROGRAMME IN LITHUANIA

Lithuania has only one nuclear power plant in operation with two RMBK-1500 Soviet reactors[55]. The country depends to a large extent on the Ignalina nuclear power plant, which generated 64.4% of the total electricity of the country in 2007 and 72.89% in 2008. The net load factor of the Ignalina nuclear power reactors in 2007 was 87.4%. Lithuania export electricity to Latvia, Estonia, Belarus and Kaliningrad.

It is important to single out that the Ignalina nuclear power plant does not have a containment, which would contain accidentally released of radioactive materials and, for this reason, the risk of operating this type of nuclear power plants cannot be reduced, even with the introduction of additional safety measures with the purpose to increase the safety operation of the plant. The design lifetime of the two operating nuclear power reactors is 30 years. The information related with the nuclear power reactors in Lithuania is shown in Table 22.

The evolution of the nuclear power capacities in Lithuania during the period 1970-2007 is shown in the following figures.

Table 22. Nuclear power reactors in Lithuania

Station	Type	Net Capacity	Construction date	Criticality date	Connection to the grid date	Commercial date	Shutdown date
Ignalina-1	LWGR	1185	May-77	Oct-83	Dec-83	May-84	Dec-04
Ignalina-2	LWGR	1185	01-Jan-78	Dec-86	Aug-87	Aug-87	in 2009

Source: CEA, 8th Edition, 2008.

---

[55] The capacity of the two nuclear power reactors was down rated to about 1 300 MWe for safety reasons.

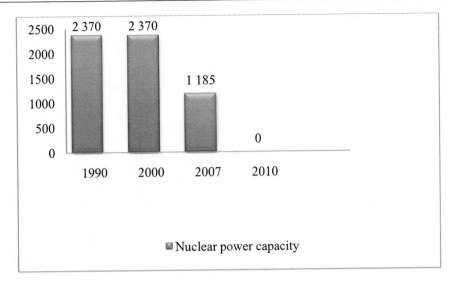

Source: CEA, 8th Edition, 2008.

Figure 59. Evolution of the nuclear power capacities in Lithuania during the period 1970-2007.

Table 23. Annual electrical power production for 2008

Total power production (including nuclear)	Nuclear power production
12 539.7 GWhe	9 140.04 GWhe

Source: IAEA PRIS, 2009.

As can be easily see in Figure 59, the closure of Unit 1 of the Ignalina nuclear power plant reduce the country's nuclear capacity from 2 370 MWe to 1 185 MWe. The shutdown of Unit 2 at the end of 2009 will reduce to zero the country's nuclear capacity in 2010.

Tables 23-24 include the annual electrical power production for 2007 in Lithuania and the total production of electricity using nuclear energy, as well as other information regarding the use of nuclear energy for electricity generation in the country.

The location of the Ignalina nuclear power plant is shown Figure 60.

Table 24. Nuclear share during the period 1995-2007

Year	1995	1996	1997	1998	1999	2000	2001	2002	2003	2007
Nuclear	11.8	12.7	12.0	13.6	9.86	8.42	11.36	14.1	15.5	9.1
Total	13.9	15.2	14.9	17.6	13.5	11.4	14.7	17.7	19.5	14.1
Nuclear (%)	84.9	83.4	80.9	76.9	72.9	73.7	77.3	79.7	79.5	64.4

Source: IAEA country information and CEA, 8th Edition, 2008.

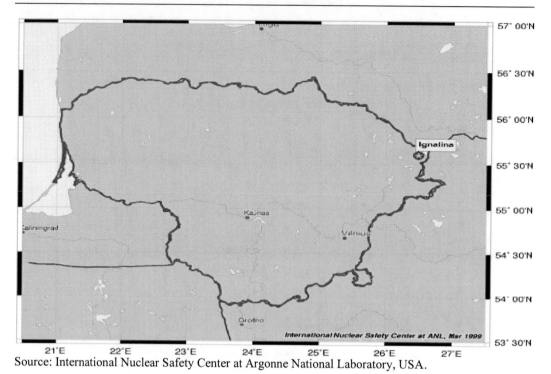

Source: International Nuclear Safety Center at Argonne National Laboratory, USA.

Figure 60. The location of the Ignalina nuclear power plant.

As can be easily see in Figure 61, the participation of nuclear energy in the production of electricity in Lithuania is decreasing since 2003 from 79.5% to 64.4% in 2007.

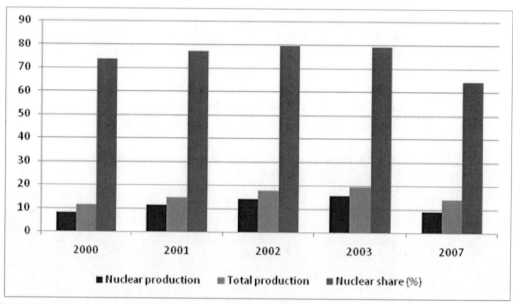

Source: IAEA.

Figure 61. Nuclear power and total electricity generation during the period 2000-2007.

The Lithuanian Energy Policy and Strategy

The Law on Energy adopted by the Lithuanian government and the Siemas obliges to a revision of the Lithuania's National Energy Strategy every five years. The first National Energy Strategy was approved by the government of Lithuania in 1994. Five years later, on 5 October 1999, the Seimas (Parliament) approved the second National Energy Strategy. The revision of the second strategy was scheduled to be carried out in 2004. However, the resolution of Lithuania to join the EU and the related pre-accession processes required an approval of a revised strategy two years earlier than anticipated. According with the Lithuanian National Energy Strategy adopted in 2002, the following strategic objectives were set:

1) to ensure a reliable and secure energy supply at lowest cost and with minimum environmental pollution, as well as constantly enhancing the operational efficiency of the energy sector;

2) to liberalize electricity and natural gas sectors by opening the market, in accordance with the requirements of EU directives;

3) to privatize energy enterprises, subject to privatization, in the natural gas transmission and distribution and power sector, as well as to continue privatization of oil refining and transportation enterprises;

4) within the terms agreed with the EU, to develop and start performing a set of measures facilitating the implementation of the EU environmental directives in the energy sector, as well as to ensure compliance with nuclear safety requirements;

5) to ensure that 90-day stocks of crude oil and petroleum products are built up by 2010, according to the agreed schedule;

6) to prepare for the decommissioning of the nuclear power reactors of the Ignalina nuclear power plant, the disposal of radioactive waste and the long-term storage of spent nuclear fuel;

7) to integrate the Lithuanian energy systems into the energy systems of the EU within the next 10 years;

8) to further develop regional co-operation and collaboration with a view to creating a common Baltic electricity market within the next five years;

9) to pursue an active policy of integration into the Western and Central European electricity markets and to ensure that conditions conforming to the Energy Charter, EU legislation and practices are applied to the transit of energy resources through Lithuania;

10) to increase the efficiency of district heating systems;

11) to achieve that the share of the electricity generated in the combined heat and power operation mode would account for at least 35% in the electricity generation balance at the end of the period;

12) to strive for a share of renewable energy resources of up to 12% in the total primary energy balance by 2010;

13) to improve energy sector management, i.e. strengthen institutions in the sector, improve the skills and knowledge of specialists of those institutions. [67]

As mentioned in point 6 above, the second National Energy Strategy includes plans for the decommissioning of the Ignalina nuclear power plant. The first unit was shut down in December 2004 and the shutdown of Unit 2 was scheduled to be carried at the end of 2009. The costs of immediate decommissioning of the Ignalina nuclear power plant were estimated as €1,134 billion (excluding inflation and pay increase for the staff of the nuclear power plant).

In Protocol 4 of the Accession Treaty of Lithuania to the EU stated the following: "European Community commits to provide Lithuania with €285 million for the period 2004-2006 and to provide adequate additional Community assistance to the decommissioning effort beyond 2006. Calculated preliminary cost for the implementation of decommissioning projects of Ignalina nuclear power plant measures related to its safe operation, upgrading and security supply projects require about €1 billion for the period 2007-2013. The EC committed to provide €815 million assistance for this period.

However, and despite of the decision of the Lithuanian government to shut down the Ignalina nuclear power plant, the intention of the Lithuanian authorities is not to renounce to the use of nuclear energy for electricity production, but to stop using the Ignalina nuclear power plant for this purpose, due mainly for safety reason. For this grounds, the Lithuanian National Energy Strategy states that, "with a view to remaining a nuclear energy state in the future and generating electricity in nuclear power plants complying with modern safety requirements, Lithuania will legally, financially and politically support investments in the construction of a new unit or reactor with the use of the existing infrastructure at the Ignalina nuclear power plant". [67]

In Lithuania the decision-making process for building nuclear facilities is rather complex and requires, besides the normal environmental impact assessment procedure necessary for major power plant projects, the approval of the decision-in-principle by the Parliament. The Nuclear Energy Act of 1987 defines the procedures required for new nuclear power plants irrespective of private or state ownership. The same is true what comes to other nuclear facilities, such as the waste management facilities and to decommissioning.

With the construction of a modern nuclear power plant, Lithuania would not fall dependent on energy imports after the decommissioning of the Ignalina nuclear power plant. The start of operation of a new nuclear power plant is foresee for 2015 and is expected to increase the level of electricity generation output in 2016.

The Evolution of the Nuclear Power Sector in Lithuania

The decision to build a nuclear power plant in the Baltic region for electricity supply to all Baltic States, Belarus and Kaliningrad was made by the former government of the Soviet Union at the beginning of the 1970s. After the Lithuanian government approved the proposal, the Lithuanian authorities started to search for the appropriate site for the nuclear power plant. The shore of Drukšiai lake near the borders of Lithuania, Latvia and Belarus was selected. The Ignalina nuclear power plant was initially conceived to host several RBMK 1500 MW reactors[56]. Construction of the first unit commenced in May 1977; the second unit followed in

---

[56] It is important to note that the RMBK 1500 is the most advanced version of the RMBK nuclear power reactors design series built by the former Soviet Union.

January 1978, and the third unit in 1985. The first unit was connected to the electric grid in December 1983 and entered into commercial operation in August 1978; the second unit was connected to the grid in August 1987 and entered into commercial operation in the same year. In August 1988, the former USSR Council of Ministers suspended the construction of the third unit. In November 1993, the Lithuanian government decided to abandon the construction of Unit 3 and, in 1996, started dismantling the existing structure of this unit. The first unit was shut down in December 2004. The net load factor of Unit 2 in 2007 was 87.4%.

Impact of the Decommissioning of the Ignalina Nuclear Power Plant

There should be no doubt that the closure of the Ignalina nuclear power plant, which provided 72.86% of the total electricity of the country in 2008, would have a great economic impact in the economy of Lithuania. According with Lithuania National Energy Strategy, as a result of the closure of Unit 2 at the Ignalina nuclear power plant, the average electricity generation cost will increase by 3 Lithuanian cent/kWh approximately. In calculating the average electricity generation cost, all the components of the operational costs of the nuclear power plants (expenses for fuel, repairs, wages, etc.), investments in the construction of new units or modernization of the existing ones, as well as the costs of the management and disposal of new radioactive waste and spent nuclear fuel have been taken into account. [67]

The closure of Unit 2 of the Ignalina nuclear power plant will have a greater negative impact on the national economy of the country than the closure of Unit 1. There should be no doubt that the closure of Units 1 and 2 will have direct socio-economic, environmental, energy-related and other consequences, including impact on the security of energy supply. The solution of problems resulting from these consequences will require a lot of time. In addressing the issue of the closure of the Ignalina nuclear power plant, the consequences of this closure, including the maintenance of the reactors of the Ignalina nuclear power plant after their shutdown, also problems concerning the dismantling of the reactors, radioactive waste management, compensation for socio-economic consequences in the region, modernization of electricity capacities and construction of alternative electricity generating sources, as well as environmental problems related to the decommissioning of the Ignalina nuclear power plant should be considered. The decommissioning of Units 1 and 2 of the Ignalina nuclear power plant will entail technical decommissioning costs. The structure of a near-surface repository for low and intermediate-level short-lived radioactive waste will be finally chosen only after decisions on waste management and disposal strategies are taken. Costs for spent nuclear fuel disposal will be adjusted in accordance with the chosen waste management strategy. [67]

It is important to note that with the closure of the Ignalina nuclear power plant by the end of 2009, primary energy demand in the basic scenario would increase only by approximately 30% during the period until 2020. The closure of the Ignalina nuclear power plant will have, without any doubt, negative consequences for the energy sector of the country. These consequences are, among others, the following:

1) the energy sector will be affected: After the closure of the Ignalina nuclear power plant the electricity production will decrease. This situation combined with economic

growth and a rise in power demand, will result in a negative power balance and energy shortages;

2) the environment will be badly affected: The portion of electricity generated by the Ignalina nuclear power plant will have to be replaced using another type of fossil fuel. It is expected that the production of the Ignalina nuclear power plant will be replaced primarily by electricity produced at gas-fired power plants. This replacement will significantly increase CO_2 emissions;

3) security of supply will be affected: Lithuania import natural gas and oil from Russia. After the closure f the Ignalina nuclear power plant, the main primary energy source in the country will be natural gas. This implies that Lithuania will become heavily dependent upon Russia for natural gas imports and, with doubt, this dependency will poses both an economic and a political threat to the country;

4) social consequences: One of the negative impact of the closure of the Ignalina nuclear power plant is in the field of manpower because their employees will lose their jobs and due to the characteristics of their jobs the majority of them will have a limited chance of switching to another kind of work or of moving to another country to find work elsewhere. Furthermore, it will exacerbate the Russian-speaking minority issue, since many of the workers who will lose their jobs originally came from Russia, Ukraine and Byelorussia in the 1970s to work at Ignalina nuclear power plant and most of them cannot talk Lithuanian. So far, the country has been the only Baltic State not to be confronted with the issue of the Russian-speaking minority. The closing of the Ignalina nuclear power plant could, therefore, create ethnic tension between Lithuanian and Russian-speaking minority.

The closure of the Ignalina nuclear power plant will also have positive consequences, i.e. the amount of radioactive waste in Lithuania will stop increasing, thereby reducing the costs of the storage of radioactive waste and spent nuclear fuel and this will created a better environment for competition in the energy sector.

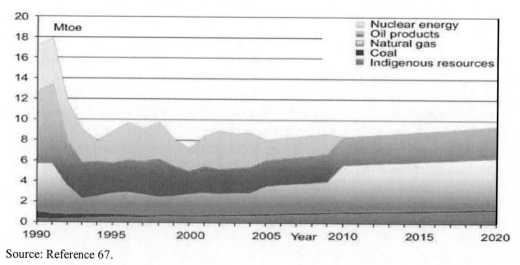

Source: Reference 67.

Figure 62. Forecast of primary energy demand in Lithuania.

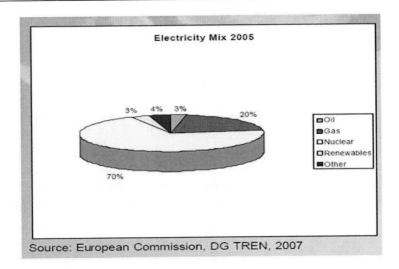

Figure 63. Electricity mix in Lithuania in 2005.

On the basis of the available information and the technical economic analysis carried out by the Lithuanian authorities, it can be stated that upon the closure of both units of the Ignalina nuclear power plant, the following measures will be necessary in order to ensure the least costs of the development and operation of power and district heating systems, as well as higher reliability of electricity supply: 1) modernization of the Lithuanian power plants, the major electricity source and of the Vilnius and Kaunas power plants: installation of new burners, modern control and management equipment and flue gas cleaning equipment. Renovation of the Kaunas Hydro Power Plant; 2) construction of new power plants in Klaipëda, Diauliai and Panevëþys, a combined cycle gas turbine condensing power plant and additional power plants in other cities, if necessary; and 3) reconstruction of the existing boiler-houses: Installation of gas turbines and generators or small power plants using indigenous fuel, provided that their installation would be economically feasible taking into account the local conditions and that they could compete with renewed large power plants.

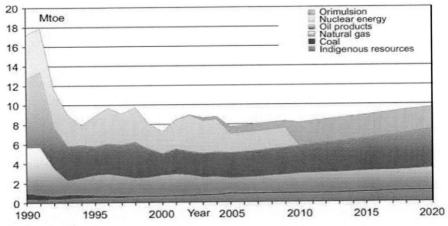

Source: Reference 67.

Figure 64. Forecast of primary energy demand (in case of modernization of Lithuanian power plants).

The strategy for the development of the Lithuanian power sector is based on the continuity and development of the use of nuclear energy for the production of electricity in the future.

Lack of Electricity Connections with others Countries within the EU

One of the problems that Lithuania is currently facing is the lack of electricity connections with others countries within the EU. The Lithuanian electricity and gas networks have no direct connections to the energy systems of Western Europe, thus, they are dependent on a single supplier of natural gas. To solve this situation, a decision to construct a high-voltage grid to allow the connection between Lithuania, the Nordic countries and Poland is targeted to be finalized by 2012. The strategy aims at integrating the electric energy market of the Baltic States into the markets of EU states, to establish in the country a common regional natural gas storage facility, as well as a common liquefied natural gas import terminal by 2012.

Nuclear Safety Authority and the Licensing Process

After the collapse of the Soviet Union, the Lithuanian government has to assume all safety responsibility associated with the operation of the Ignalina nuclear power plant as well as all safety responsibilities of other nuclear facilities and of the radioactive waste management. For this reason, in October 1991, the government established the State Nuclear Power Safety Inspectorate (VATESI) as the national regulatory authority with responsibility for the functions of safety and control of nuclear installations and the supervision of accounting for nuclear materials. One year later, in October 1992, the first statute of VATESI was approved, regulating its activities and determining the basic objectives, functions and rights of the inspections. The new statute of VATESI was approved in July 2002. VATESI now operated under the Ministry for the Economy.

Pursuant to its statute, VATESI "is responsible for the state regulation of nuclear safety at the Ignalina nuclear power plant as well as for the safety of other nuclear facilities and of the radioactive waste management". The duties of VATESI in its capacity as national nuclear regulatory authority include: 1) drafting and, under the authority of the government, approving safety standards and rules for the design, construction and operation of nuclear facilities, storage of nuclear and radioactive materials and waste disposal; 2) ensuring adherence to the requirements set out in licenses and safety rules through assessment of the safety of nuclear facilities; 3) establishing the system of accounting for and control of nuclear materials; 4) issuing licenses for the acquisition, possession and transportation of nuclear materials and the storage and disposal of radioactive waste.

Nuclear Fuel Cycle and Waste Management

It is important to single out that Lithuania has no nuclear fuel fabrication industry and, for this reason, all nuclear fuel used to ensure the operation of the Ignalina nuclear power plant was supplied by Russia. Originally, spent nuclear fuel from Ignalina nuclear power plant was managed by Soviet agencies in charge of reprocessing and final disposal activities. However, with the disintegration of the Soviet Union, Lithuania was obliged to find other solutions. Now Ignalina nuclear power plant itself manages its operational radioactive waste. There are facilities for processing and storage of liquid and solid waste in the country. In November 1993, Lithuanian government approved temporary storage of spent fuel in containers for fifty years until solutions for final conditioning and disposal are found. Interim dry storage for spent nuclear fuel was built and commissioned in 1999 on the site of Ignalina nuclear power plant. A new Interim Spent Fuel Storage (ISFS) was built on Ignalina nuclear power plant site as a pre-decommissioning project. The Ministry of Economy established, in July 2001, the Radioactive Waste Management Agency (RATA), to implement the management and final disposal of all transferred to it radioactive waste, generated by the Ignalina nuclear power plant during the operation and decommissioning process, and radioactive waste from small producers (hospitals, industry, research institutions, etc.). One of the main task of RATA will be to construct and operate the repositories for short-lived and long-lived radioactive waste. [56]

In February 2002, the government approved the Radioactive Waste Management Strategy for 2002-2009 and a three year programme to modernize the management and storage of solid short-lived and long-lived radioactive waste of Ignalina nuclear power plant. The purpose of the programme was to perform investigations and to prepare recommendations on implementation of a near surface repository for low and intermediate-level short-lived radioactive waste. Three possible sites has been identify in the Ignalina region.

The Public Opinion

There is still a public debate about the benefits and threat of the possible future construction of a new nuclear power plant in Lithuania. In the opinion of some experts, companies planning the construction of new nuclear power plants should take into consideration all other options, including energy efficiency and use of local resources, before the decision to go ahead with the project is finally adopted by the competent national authorities. They should be noted that the use of nuclear energy for the production of electricity is not as cheap as compared to other sources of energy as it is often argued, emphasized EU Energy Commissioner A. Piebalgs speaking at a conference in Riga. Some experts, say t A. Piebalgs are open "to consider the possible role of nuclear energy for the electricity generation like never before". He insisted that "decisions on building new plants should not be taken in a hurry". He also pointed out that "future growth of energy consumption tends to be overestimated by companies who are serving their own needs". His opinion regarding the price of nuclear energy differed from the one presented in Vilnius in the feasibility study that claimed it would be cheaper than other options. "The only nuclear energy that is cheap – is the one that is coming from the power plants that have been paid for. It's not cheaper se. Costs of managing nuclear waste should be taken into consideration too. I am not against such projects, but all options should be considered. We know that big power companies not always have the same interest as consumers", Commissioner said. "Right now certain lobby groups are working very actively on promoting the construction of a new nuclear power plant. Some parts of the feasibility study presented to the public barely reflect the reality", I must say, "lobbyists are so busy praising the project, because in the end it is them who are going to benefit from those contracts we will have to sign", - he added.

Those in favor of nuclear energy keep pointing out that Lithuania together with other Baltic States depends on Russian energy resources, and this is a situation that they would like to change in the coming years with the purpose to reduce significantly this dependency. EU Energy Commissioner A. Piebalgs insisted "that more emphasis should be put on finding ways to explore local resources and investments should be made in energy efficiency means. According to him, "this is the only way Europe can meet energy-related future challenges". If nuclear is to be chosen, he went on, "there would be no way back. When you go for a big infrastructure project, you should evaluate all other options, because afterwards it will be the decisive factor in all your lives. In all the countries where there's a decentralized energy supply, possibilities for energy efficiency and local resources should be looked at first of all, especially if the governments are behind it".

The estimates are that Lithuania may need one or two 800-1 600 MW nuclear power reactors while there is a demand of some 5 000 MW in the Baltic region. Energy experts point out, that a new 350 MW power line will connect Finland with Estonia. In addition to that, there are plans to connect Lithuania with Sweden and Poland with 1 000 MW lines each. This high-voltage (400 kV) 1000 MW DC Power Bridge line will cost around €250-€300 million and will improve transmission capacity between Lithuania and Poland. The line will be built by 2015 and will be half funded by the EC. Some experts say that would be enough to meet the energy demand of the region, while others draw the attention that although Ignalina nuclear power plant goes out of operation at the end of 2009, new connections could become operable no earlier than 2010-2012. Those in favor of the construction of a new nuclear power plant insist that it would provide Lithuania with cheaper source of energy.

In the mean time, financial analyst V. Katkus admits, "that Lithuanian energy experts do not share a common opinion regarding the new nuclear power reactor. They disagree whether thermal plants should be developed further or a new nuclear power plant should be built. Estimates are that after the year 2010 some 53% of electricity will be generated in thermal plants in the Baltics. Here the argument about dependency on Russian gas comes in. Even if Russia will demonstrate a good political will, the possibility of technical failure should not be left out completely. We have to balance security and economics".

In the other hands, Latvian greens did not welcome the news from neighboring Lithuania about plans to build a new nuclear power reactor. According to A.Ozola-Matula, the leader of the Latvian green movement, "there is a lack of public discussion in Latvia regarding government's plans to participate in this project". She says "NGO's are planning to initiate a regional public awareness campaign. Not only we are concerned; other NGOs and other local groups are too. There is a lack of transparency on dealing with this important matter. All we know is that there was a feasibility study presented in Vilnius, but "Latvenergo" discussed possibility to take part in this project behind the closed doors. We think that such discussions should be more open, so that we also know how much we will be paying for the energy in the future. We doubt that it will be cheaper, especially if they plan to invest a billion," – activist wondered.

Summing up can be stated that though the media sometimes develops anti-nuclear arguments in its articles or broadcasts that target the nuclear industry, public opinion has, in general, a positive attitude towards nuclear energy for the production of electricity. According to energy experts, J. Gylys and S. Ziedelis, "a majority of the Lithuanians have nothing against new, modern and safe western-type nuclear power plants, if the electricity produced there is cheaper".

Without any doubt, Lithuanians have one of the most positive attitude towards the use of nuclear energy for electricity generation within the EU. This position, in the opinion of energy experts, is largely attributing to their insecurity over energy supplies, particularly from Russia. In general, 64% of Lithuanian views the use of nuclear energy for the production of electricity favorably. However, it is important to single out that there is a prevailing sense of a security situation in which certain domestic developments and regional instability are seen as the biggest and most immediate threats to the Lithuanian people. When asked what the most dangerous threats to Lithuania are, 45% of the respondents pointed to internal dangers. Residents of Lithuania were given a list of factors that might influence security and stability in their country. Among them the situation of the Ignalina nuclear power plant was included with a rate of 10.2%.

Looking Forward

Several studies and research projects concerning the analysis and forecasting of possible future changes in power balance in Lithuania and in the entire Baltic region have been carried out bearing in mind the closure of Unit 2 of Ignalina nuclear power plant at the end of 2009. The results can be summarized as follows: 1) power reserves are decreasing and the energy balance has become negative in most countries, particularly those neighboring on Lithuania; 2) the power balance in the Lithuanian energy system will become negative sometime between 2010 and 2020 and new bigger power generating capacity is needed, if the balance is to remain positive. New nuclear power plants or combined-cycle gas turbine power plants could achieve this purpose; 3) Lithuania's energy supply system once the Ignalina nuclear power plant is finally shut down will not comply with the main security requirements and will be extremely vulnerable due to lack of diversification of primary energy sources and energy supply routes. Security of energy supply could be substantially improved, if new modern nuclear power plant were built.

These well-known arguments were further reinforced by the important changes in the gas market that occurred in January 2006. A sudden jump in gas prices from GASPROM (about 40% on average for Lithuania) and interruptions of the gas supply from Russia to Ukraine and Georgia demonstrated that problem of security of energy supply have becomes much more important than other considerations.

According with the declaration adopted by the Presidents of the three Baltic States, the following actions should be carried out by these states in the future in order to increase the security of energy supply:

1) support the development of a common European energy policy as a guarantee to the security of supply at the Community level;

2) the energy security problem that affects the Baltic States should be addressed at EU level. In order to integrate the Baltic States into the EU energy market, it is necessary to define appropriate measures that would diminish the existing fragmentation of the EU energy market;

3) there is a need to integrate the EU's energy, external relations and security policies. A harmonized EU external energy policy should be established vis-à-vis third countries and organizations, notably with Russia and the OPEC countries. The EU should speak to suppliers of energy in one strong voice. For example, the EU and international instruments, such as G-8, World Trade Organization, Energy Charter Treaty, among others, should be effectively employed to ensure the transparency of energy supply and the liberalization of energy markets;

4) call for the development of an EU mechanism that prepares for and ensures solidarity and assistance to a country facing difficulties following damage to its essential infrastructure or disruptions in energy supply. This includes enhancing Europe's gas stocks, *inter alia* utilizing the vast potential storage capacity of the Baltic States to ensure against short-term supply disruptions to the EU;

5) state that while being in favor of a common EU energy policy, the necessity to maintain national sovereignty over the choice of primary energy sources and structure of energy mix is paramount;

6) call the EC and the EU member states to develop an action plan of immediate measures aimed at enhancing EU energy security. Such an action plan should be approved by the Council on the basis of the above mentioned assessment of the EC.

In a communiqué adopted by the prime ministers of the three Baltic States, they declared "their collective action plan for launching concrete short-term activities to promote energy security in the region". The Communiqué outlines the aims of the action plan as follows:

1) to attempt to broaden the Baltic energy market until 2009 and to harmonize standards in the Baltic electricity market consistent with those applied in the Nordic countries' electricity market (Nordpool);

2) to support the construction of electric grid interconnections between Baltic States and the rest of EU on the basis of full cooperation;

3) to promote an initiative to build a new nuclear power plant in Lithuania;

4) to invite state-owned energy companies in the three Baltic States to invest in the design and construction of a new nuclear power plant in Lithuania, on the basis of agreed terms and conditions applicable to each party involved;

5) to follow the principle of consensus for all involved parties when inviting other companies to participate in the project;

6) to examine possibilities to erect terminals for liquid gas and to develop gas storage capacities in the Baltic regions;

7) to examine the general conditions governing the importing of electricity to the Baltic States from states not included in the European economic space and the possibility of parties involved in the new nuclear power plant construction project in Lithuania to sign long-term contracts for the buying-selling of electricity.

Lithuania is in favor of an efficient, liberal and competitive EU electricity market. Energy security should be implemented as widely as possible on an EU scale, with the EC entrusted with greater responsibility for the coordination of energy projects. It is expected that the negative effect on the country's energy balance as well as the diversity of primary energies due to the premature decommissioning of the Ignalina nuclear power plant will be compensated for by the rapid integration of Lithuania's electricity transmission network and other infrastructural facilities into the EU energy networks.

Finally, it is important to single out that the new government in Lithuania is questioning the need to construct a new nuclear power plant in the country and the new president is ordering a deep study on such project. The president is asking that the study to be carried out include the possibility to use other type of energy for the production of electricity as well. This position of the new Lithuanian government regarding the future role of nuclear energy in the energy balance of the country, could delay the adoption of a final decision whether to construct a new nuclear power plant in the country in the near future.

THE NUCLEAR POWER PROGRAMME IN SLOVAKIA

Electricity consumption in Slovakia has been fairly steady since 1990. Generating capacity in 2004 was 7.7 GWe and from this capacity 34% is nuclear. In 2005, a total of 30.8 billion kWh was produced, and from this total around 55% was produced by nuclear power plants. Up to the end of 2006, Slovakia was a net exporter of electricity exporting some 2 billion kWh per year to other countries.

In 2008, four PWR systems were in operation in Slovakia. The total nuclear power capacity in the country in 2009 was 1 688 MWe. The production of electricity by the nuclear power reactors in operation in Slovakia in 2008 was 15 453.4 GWe. In 2008, the share of nuclear energy in the total electricity generation in Slovakia was 56.42%. The net load factor of the nuclear power reactors in operation in the country in 2007 was 79.5%.

In February 2005, Slovakia's economics minister authorized the sale of 66% of Slovenské Elektrárne (SE), the country's nuclear operator, to Italy's Enel S.p.A. for €840 million. SE, which operates five Russian designed WWERs, is the largest electricity generator in Slovakia and the second largest in Central and Eastern Europe. The company's reactors generated more than 15 000 GWh in 2004, supplying over half of the country's power. With respect to the Slovakian power system it is important to be aware of the following: The Slovakian power system is in a strategic position in the heart of Europe, with good power connections to both Eastern and Western European markets. When the last nuclear unit began commercial operation, Slovakia became both self-sufficient in electricity supply and a power-exporting country. The government remains strongly committed to the future of nuclear energy, which also has good public support. (Kovan, 2005)

The Energy Policy in Slovakia

The main document defining targets, directions and framework of power development in the country, is the Power Policy of the Slovak Republic, approved by the Slovak government decree No. 5 dated 12 January 2000. This policy document defines the framework for new orientation of the power sector in the country. The document has the following three main pillars: 1) preparation for the integration into internal markets of the EU; 2) security in power supplies; and 3) sustainable development.

The main target was the transformation of the power sector into a compatible one that is able and prepared to be incorporated into a united European market. The power sector transformation is conditioned by meeting the following basic measures: a) restructuring and privatization of power utilities; b) establishment of independent regulatory authority; c) making energy prices more realistic for all categories of consumers; and d) completion and approval of legislation adapting power sector.

The main objectives and actions contained in the Slovakia Energy Policy are the following:

1) satisfy the energy needs of the society in a reliable, safe, most effective and ecologically acceptable way. Harmonization of the Slovak Republic legislation with the legislation of the EU.
2) liberalization of the electricity and natural gas market;
3) fulfillment of the international agreement in the fields of ecology, nuclear safety, investments and energy trade[57];
4) reduce the energy intensity to the level in the EU member countries;
5) build up the storage capacities up to the volume of 90-days emergency oil stock and oil products stock (until 2010);
6) strengthen the strategic position of the Slovak Republic in the area of transit of strategic energy supplies, through development of gas and crude oil pipeline system;
7) resolve the concept of the final sections of the radioactive fuel cycle in nuclear power plants;
8) shut down the second Bohunice V1 reactor some-time in 2008;
9) increase the share of renewable energy sources in the coverage of consumption of primary energy resource.

The intentions of the Slovakia's Power Policy are the following:
a) create competitive power sector;
b) establish conditions for stakeholders to enter electricity grid and to create competitive environment;
c) minimize involvement of the state in the direct control of the sector;
d) ensure non-discriminating and transparent conditions for all subjects participating in the generation, transmission, distribution and sale of power;
e) make possible a gradual liberalization of power market for legitimate customers.

---

[57] These international agreements are, among others, the following: Kyoto Protocol, Nuclear Safety Treaty, Supplementary Agreement to Energy Charter Treaty (ECT), Protocol on Energy Efficiency and Ecology Aspects of the ECT, United Nations Framework Convention on Climate Change (UNFCCC) and Aarhus Convention.

According with the IAEA sources, the strategic goal of the Slovakia's Energy Policy is to ensure fuel and energy for all consumers. The energy shall be: 1) produced with the lowest costs and impacts on the environment; 2) transported to the consumer safely and reliably and in the quality required; 3) used in the field of generation, transport and consumption as effective as possible.

The Evolution of the Nuclear Power Sector in Slovakia

Nuclear energy development in Slovakia started in 1956 when the country was part of the former Czechoslovakia Socialist Republic (CSSR). In this year, an intergovernmental agreement between the former USSR and CSSR for the construction of an industrial-research nuclear power plant on the territory of CSSR was signed. In 1957, an investment enterprise named Nuclear Power Plant A-1 was established by the decision of the Governmental Committee for Nuclear Energy and of the Authority for Nuclear Power Management of the CSSR.

One year later, in January 1958, started the construction of the first nuclear power reactor in Bohunice[58]. This first nuclear power reactor was HWGCR unit built by Skoda with a net power capacity of 110 MWe. The unit started to be constructed in January 1958, was connected to the electric grid in October 1972 and entered into commercial operation in December 1972. The nuclear power reactor was shut down in May 1979. Why the unit was shut down definitively after only 7 years of operation? The following events force the early closure of this unit: a) A serious incident at the reactor occurred 1976; and b) Another severe accident occurred in 1977 during reactor refueling. To avoid any further accident in the reactor the CSSR government decided, in 1978, the decommissioning of the reactor and, in 1999, adopted the decision No. 137/1999, approving stage 1 of the decommissioning plan scheduled till 2007.

In 1972, started the construction of two WWER 440/230 soviet design units built by Atomenergoexport of Russia and Skoda from CSSR, with a power capacity of 408 MWe each. The nuclear power reactors were Bohunice 1 and 2 and were connected to the electric grid in December 1978 and March 1980 respectively. Bohunice 1 entered into commercial operation in April 1980 and was shut down in December 2006. Bohunice 2 entered into commercial operation in January 1981.

In 1976, construction started on two type of WWER 440/213 soviet design units built by Skoda and designed by Energoproject with a power net capacity of 408 MWe each. These two nuclear power reactors (Bohunice 3 and 4) were connected to the electric grid in August 1984 and August 1985 respectively. Bohunice 3 entered into commercial operation in February 1985 and Bohunice 4 in December 1985. Regarding the safety of the Bohunice Units 3 and 4 it is important to single out the following: Since 1990, significant improvements have raised the safety level of Bohunice 3 and 4 to that of Western European nuclear power reactors of the same age. The condition of the pressure vessels indicates that annealing will not be necessary. The modernization programme implemented during the period 1999-2008 includes, for example, the installation of in-service diagnostic systems, the

---

[58] Since 1958, a total of nine nuclear power reactors were constructed, two of them were closed in 1977 and the other in 2006, two others units were not concluded. Five units are currently in operation in the country.

renovation of instrumentation and control systems, the improvement of electrical systems and fire and seismic upgrading, with a view to extending operational life to 40 years (until 2025). (Kovan, 2005)

In 1981, the Czechoslovakian government decided the construction of four nuclear power reactors at Mochovce site, using WWER 440/213 soviet design. The net power capacity of each of the first two units was 405 MWe and Skoda was responsible for their construction. Units 1 and 2 were connected to the electric grid in July 1998 and December 1999 respectively. Unit 1 entered into commercial operation in October 1998 and Unit 2 in April 2004. It is important to stress that Mochovce nuclear power plants ranks among the world's leading plants in terms of operational safety, according to WANO performance indicators. Work on Units 3 and 4 started in 1986 and halted in 1992 when Units 1 and 2 were 90% and 75% complete, and Units 3 and 4 were between 30% and 40 complete. Work on the first two units was restarted in 1995. Unable to get funding for the project on acceptable terms from the European Bank for Reconstruction and Development, the Slovak government arranged financing from Czech, French, German, Russian, and Slovak banks. The project, implemented by a consortium of Framatome, Siemens and Russian suppliers, made Mochovce the first Soviet designed reactor to meet international safety standards. Preservation work has been carried out on Units 3 and 4 since 1992. While the government remains supportive of nuclear power and wants the two units to be completed, it has said it would not help finance the project. (Kovan, 2005)

The technology use in the construction of Mochovce 1 and 2 is based on a robust design, low unit power, low power density and high volumes of water in the cooling circuits, which ensure a wide and stable operational range for the plant, with large time margins available for operator corrective actions. This implies high performances of the plant in accident prevention and, consequently, an inherently high level of nuclear safety. [65]

After reviews of their safety, phase 1 upgrading of Bohnuice 1 and 2 was undertaken during the period 1991-95, and phase 2 - intended to achieve Western European standards - was carried out up to 2000. In 2001, Slovakia relicensed Bohunice 1 and 2 units for another decade though following the upgrades that were carried out in these two units their operating life was expected to run until 2015.

As a precondition for Slovak to entry into the EU in 2004, the Slovak government committed to an early closing of Bohunice Units 1 and 2. The reason to an early shut down of these two units was a number of safety deficiencies that this early soviet model reactor has. The original date specified for closing these two units was 2000, though subsequently 2006 and 2008 were agreed in relation to EU accession. Both units were subject, since 1991, of a major refurbishment, including replacement of the emergency core cooling systems and modernizing the control systems. The cost of the refurbishment work Bohunice 1 was around US$300 million.

The nuclear power reactors currently in operation in Slovakia were producing electricity at half the average cost for all Slovak energy sources. For this reason, the early closure of these nuclear power reactors before Mochovce Units 3 and 4 enter into commercial operation leave the country short of power.

The total production of electricity by nuclear energy in 2007 is shown in Table 26.

Table 25. Nuclear power reactors in operation in Slovakia

Reactors	Model V-PWR	First power	Announced closure
Bohunice 3	V-213	1984	2025
Bohunice 4	V-213	1985	2025
Mochovce 1	V-213	1998	-
Mochovce 2	V-213	1999	-
Total (4)			

Source: CEA, 8th Edition, 2008.

In Slovakia, three nuclear power reactors were definitively shut-down. These reactors were A1 Bohunice in May 1979 and Bohunice 1 and 2 in December 2006 and 2008 respectively. The location of the nuclear power plants in Slovakia is shown in Figure 67.

Table 26. The total production of electricity for 2008

Total power production (including nuclear)	Nuclear power production
27 389 GWhe	15 453.4 GWhe

Source: IAEA PRIS, 2009.

Source: Photograph courtesy of the IAEA.

Figure 65. Mochovce nuclear power plant.

Source: Photograph courtesy of IAEA.

Figure 66. Bohunice nuclear power plant.

Source: Mochovce 3 and 4 basic facts, new clear power, Enel, Slovakia, 2008.

Figure 67. Location of the nuclear power plants in Slovakia.

Nuclear Safety Authority and the Licensing Process

The Nuclear Regulatory Authority (ÚJD) of the Slovak Republic was established on 1 January 1993, after the division of the former CSSR. Its initial rights result from the Law No. 2/1993 of the Slovak National Council now replaced by the Act No. 575/2001, which is in force since 1 January 2002. The ÚJD is an independent state regulatory authority reporting directly to the government and its chairman is appointed by the cabinet. The ÚJD executes the state regulation over: a) nuclear safety of nuclear installations, including the regulation over the management of nuclear waste, spent nuclear fuel and other phases of nuclear fuel cycle; b) nuclear materials, including their control and recording; and c) quality of selected equipment and instrumentation.

The ÚJD ensures: 1)review of intentions how to use nuclear energy in respect on nuclear safety; 2) evaluation and inspection of emergency plans; and 3) fulfillment of the commitments of the Slovak Republic resulting from international agreements in the field of nuclear safety and nuclear safeguards.

According to the law No. 541/2004, the ÚJD performs: a) routine inspections by site inspectors; b) special inspections by nuclear safety inspectors; and c) team inspections.

The licensing procedures in Slovakia have three main levels. These levels are the following: 1) selection of construction site; 2) commencement of construction; and 3) permanent operation.

With regards to the selection of the construction site, the regional construction offices issue decisions for the selection of construction site of nuclear installations based on the approval of the ÚJD, Ministry of Health and of other offices and organizations of state authorities. ÚJD, since 1st December 2004, is the special government authority for final decision for the construction, operation and decommissioning of nuclear installations. Prior to issuing a license for permanent operation, the ÚJD performs inspections in line with the approved programmes of active and non-active tests and issues approvals for fuel loading, physical start-up, power start-up and trial operation. The basic mandatory condition for issuing any approval related to nuclear safety is to develop and submit a Safety Analysis Report and other prescribed safety documentation and to comply with the conditions from previous approval proceedings and with the decisions of the ÚJD. As regards approvals, responsibilities of this authority are specified in the law No. 50/1976 (Construction Law), in Act No. 541/2004 and in regulations of the Ministry for Environment Nos. 453/2000 and 55/2001. The licensee is responsible for the safety of its nuclear installation. [122]

On June 14, 2001, the Act No. 276/2001 on the regulation in network industries and on amendments in certain other laws was approved by the Slovak National Council. The bill specifies: a) establishment, authority and activities of the Office for Regulation of Network Industries; b) object and conditions of the state regulation in network industries; and c) conditions for the execution of regulated activities and rights and obligations of regulated subjects.

Nuclear Fuel Cycle and Waste Management

The current policy of radioactive waste management in Slovakia is included in the Resolution No. 190/94 of the Slovak government. This policy can be characterized as follows:

1) basic solidification methods of liquid radioactive waste, radioactive sludge and exhausted ion-exchanging resins into a form suitable for final disposal, included cementation and bitumenation;

2) the volume of solid radioactive waste will be minimized by applying compaction and incineration;

3) the treated liquid or solid radioactive waste is then grouted by active mixture of concrete and concentrate into fiber-reinforced concrete containers. These containers are suitable for transportation as well as for long-term storage and final disposal;

4) for treatment of intermediate level waste and radioactive waste with high contents of trans-uranium, specific liquid radioactive waste produced during the storage of spent fuel at nuclear power plant A1 that is in form of sludge and chromic, is necessary to apply a vitrification method;

5) low-active soil and concrete rubble shall be arranged into layers on supervised stock-piles.

6) the low-activity metal waste shall be treated by applying fragmentation and decontamination and cleaned material can be than released into the environment;

7) for the treatment of metal radioactive waste, that cannot be released into the environment a melting unit shall be installed and used for its conditioning;

8) institutional radioactive waste shall be treated and conditioned into a form acceptable for final disposal. Waste of open source character will be conditioned by applying standard methods. The disused sealed sources shall be conditioned into a form suitable for centralized long-term storage or disposal;

9) the conditioned radioactive waste produced during the operation and decommissioning of nuclear power plants and the conditioned institutional radioactive waste that meet the acceptance criteria shall be disposed of in the National Repository Mochovce;

10) the waste which is not acceptable for the National Repository Mochovce shall be stored at the power plant site. An integral storage shall be installed at Bohunice to allow storing of radioactive waste that is not acceptable for National Radwaste Repository;

11) an integral storage for disposal of the radioactive waste, which does not meet the criteria for disposal in near surface repository do not represent the final solution. Therefore, for disposal of mentioned "non acceptable" waste a deep geological repository shall be built.

The Slovakian Policy in the field of spent fuel is the following: the spent fuel should be disposed of without reprocessing. For this reason, a near surface facility designed for the disposal of solid and solidified low and intermediate-level radioactive waste was built in

Mochovce and Bohunice nuclear power plant sites. The final disposal of the spent nuclear fuel produced by these two nuclear power plants is expected to be in deep underground geological repository to be built in the future in a site to be selected.

The basic policy of spent nuclear fuel and radioactive waste management has been established by the Resolutions No. 930/1992 and No. 190/1994 approved by the Slovak government in 1992 and 1994 respectively. In 2001, the Slovak government in his Resolution No. 5/2001 accepted the proposal on the schedule of economical and material solution on the management of spent nuclear fuel and decommissioning process of nuclear facilities, and submitted a policy of decommissioning of nuclear facilities and management of spent nuclear fuel, according to the Act on Environmental Impact Assessment, by the end of 2007.

It is important to note that the operation of nuclear power reactors in Slovakia adopts an open fuel cycle since the reactors WWER-440 are not licensed to utilize MOX fuel. Discharged spent fuel is stored for three years in spent fuel pools of the main generation building. Further long-term spent nuclear fuel storage (40 to 50 years after its removal from the reactor), which is required prior to its final disposal in a repository, will be assured in separate interim storage facilities at Bohunice and Mochovce sites. Reprocessing of spent nuclear fuel from Bohunice and Mochovce nuclear power plants is not included into the concept of spent nuclear fuel management. An interim spent nuclear fuel wet storage facility (ISFS-SE-VYZ) has been in operation at Bohunice site since 1987. ISFS-SE-VYZ has been already reconstructed in order to increase its storage capacity. A project to enhance its seismic resistance and to improve its safety and cooling system was accomplished in 1999. The storage facility is re-licensed for extension of storage period to 50 years[59]. [122]

The National Radwaste Repository, located near the Mochovce nuclear power plant, is a near surface facility designed for the disposal of solid and solidified low and intermediate-level radioactive waste. Commercial operation of the National Radioactive Waste Repository started on October 1999, based on the Regulatory Authority Decision No.335/99. Other radioactive waste storage facility is located in the Bohunice nuclear power plant site. It is expected that in 2015 the repository located at Mochovce nuclear power plant site will be full and, for this reason, it will be necessary to build a new similar storage facility at the same site. A project of spent nuclear fuel storage facility at Mochovce site (ISFS-EMO) was approved and is currently in preparation. According with available information, the facility will probably be based on the dry storage technology. A long-term spent fuel storage facility is expected to cost about €100 million. The first phase of decommissioning the A1 reactor was completed in 2007. Preparation for decommissioning the two Bohunice units will begin in 2012 and the work will be carried out in 13 years at an estimated cost of about €500 million.

The final disposal of the spent nuclear fuel produced by the two nuclear power plants currently in operation in Slovakia is expected to be in deep underground geological repository to be built in a site to be selected in the future. Activities on the selection of an adequate site for the final disposal of spent nuclear fuel are being carried out by the national competent authorities, but no final decision has been taken on this sensitive issue. At the same time, other possibilities for the management of the spent nuclear fuel have been considered. One of

---

[59] It is important to note that the enlargement of the storage capacity using new compact cask containers will represents an increase from original 5 040 pcs to 14 112 pcs of fuel assemblies, which will be sufficient to storage all spent nuclear fuel from Bohunice nuclear power reactors produced for their whole operation period and for Mochovce nuclear power reactors until 2015.

these possibilities is the transportation of the spent nuclear fuel into foreign countries for final disposal or reprocessing without importing the products back into Slovakia. Another possibility that have been considered is an international or regional solution on the final spent nuclear fuel disposal. However, this kind of solution are not been approved and for the time being could take some years before a solution can be found acceptable to all parties involved. Several proposals have been presented by different countries and the IAEA and other international organizations are studying them thoroughly, but until now there is no clear indication that one or more of them could represent a real breakthrough. It is important to note that by the end on 2004 all Slovak WWER 440 units used 9 300 fuel assemblies; from this amount approximately 700 assemblies were exported to the Russian Federation, another 6 800 pieces are stored in wet interim storage facility located at Bohunice site and the rest of about 1 800 spent nuclear fuel assemblies are cooled down and stored in pools adjacent to the reactors. [122]

Lastly, it is important to stress that the treatment and storage of radioactive wastes produced in the Slovak Republic is performed in compliance with the legislative and procedures developed by ÚJD Operation of all treatment and storage radioactive waste in nuclear facilities is licensed and regularly supervised by ÚJD.

With regards to the nuclear fuel used in the nuclear power plants currently in operation in Slovakia, it is important to single out the following: All nuclear fuel used to ensure the operation of five nuclear power reactors in Slovakia has been fabricated, first, in the former Soviet Union and later on in the Russian Federation. The Russian nuclear fuel supplier provides completed fuel assemblies, including nuclear material, its conversion and enrichment. In 2003, SE concluded new fuel contract with Russian supplier for delivery of fresh nuclear fuel for Bohunice Units 3 and 4 and Mochovce Units 1 and 2 for the period 2005-2010. The supplied fuel was of new generation with the purpose to achieve better efficiency and lower annual consumption of nuclear materials.

The Public Opinion

In recent pool carried out by the independent agency GfK in 2007, over two-third of the Slovak population and more than 87% of the population living in 10 km area around the Mochovce nuclear power plant are in favor of the completion of Units 3 and 4. The main opposition related with the use of nuclear energy for electricity production in Slovakia comes from neighboring Austria. This country has adopted a law prohibiting the use of nuclear energy for electricity production and is one of the few EU countries that are not using this type of energy for this purpose and has no plans to use nuclear energy for the generation of electricity in the future.

According with press information, however, the completion of the Units 3 and 4 at the Mochovce nuclear power plant meet certain public opinion opposition. Apart from traditionally anti-nuclear oriented Austrian neighbor, objections were raised also by Green Fraction at the European Parliament. They said that "Mochovce design and its equipment are not based on accident and earthquake safety considerations". In addition, they claim that "Mochovce construction is far too advanced to be retrofitted to comply with international safety standards". At the same time, ÚJD chairperson Marta Ziakova says "she cannot imagine conditions under which Mochovce 3 and 4 building permit could be taken away. The construction is under permanent surveillance of Slovakian nuclear authorities", she said. The two units will not be built according to the original plan and will be equipped by modern systems in line with the current international safety regulations. Numerous evaluations of foreign institutions, including the IAEA and WANO, confirmed "that the safety of Slovak reactors is at the level of operators in Western Europe", said the Mochovce operator spokesman Juraj Kopriva.

Looking Forward

At the end of October 1995, the Slovak Republic and Russia signed an agreement for the conclusion of Units 3 and 4 of Mochovce nuclear power plant. In addition, a loan of US$150 million for completing Unit 3 will be provided by Russia as well as the supply fuel for the plant and will reprocess the spent nuclear fuel. According to Slovak Prime Minister, "between US$22.9 and US$25.8 million will be used to work on Units 3 and 4". The cost of completing Units 3 and 4 is estimated at US$1.14 billion. In February 2007, the Slovakian authorities announced that "it would proceed with the construction of Units 3 and 4 at Mochovce nuclear power plant later in the year. The Italian company Enel had agreed to invest around €1.8 billion with this purpose. It is expected that Units 3 and 4 will entry in operation in 2012 and 2013 respectively. With the entry into operation of these two units, the nuclear capacity of the country will be increased in 840 MWe. The Slovakian authorities have already invested €576 million in these two units. It is important to stress that Units 3 and 4 at Bohunice nuclear power plants were subject to an upgrade programme in order to improve seismic resistance, cooling and instrument and control systems.

However, and despite of what have been said in the previous paragraphs, it is important to stress the following: The extent of structural improvements of Mochovce Units 3 and 4 is limited because, in 1993, when construction was stopped, already 70% of civil construction was completed and 30% of the equipment had been supplied, among them the reactor vessel and steam generators, which were partially installed, have been mothballed and are stored at the plant. The structural limitations mainly concern the confinement structures and the possibility of physical separation of safety systems. Neither the reactor building nor the bubble condenser are resistant against external events such as an air plane crash or a missile attack.

In the other hand, it is important to single out that Units 3 and 4 does not have a containment, only a confinement with a bubble condenser to limit the pressure from large pipe ruptures. The tests showed the functioning of the bubble condenser for design basis accidents and some severe accidents, but not all. Seismic design has to be proven in relation to the earthquake risk at the site. Seismic design is a weak point of all WWER 440/213 units. Moreover, seismic evaluation and seismic design have evolved fundamentally in recent years. It is unclear whether Mochovce Units 3 and 4 can meet the latest seismic design standards of the IAEA.

In June 2009, the Czech company Skoda JS and the Slovak company SE signed in Bratislava a contract to conclude the construction of Units 3 and 4 of the Mochovce nuclear power plant. The contract was signed for US$517 million for the supply of important equipment, including part of the instrumentation and control system. The heavy equipment for these two units was manufactured by Skoda and surrendered to the nuclear power plant in the 1980s, but the project was stopped. The plan adopted foresees that these two units begin to operate in 2012 and 2013, respectively.

Despite of the difficulties in the operation of the first nuclear power reactor at Bohunice and the strong opposition of the Austrian authorities to the use of nuclear energy for electricity generation in Slovakia, the Slovakian government is considering the possibility of building a fifth unit at Mochovce nuclear power plant and also new units at Bohunice nuclear power plant, or in a new site located in the eastern part of the country (Kecerovce).

At the same time, it is important to single out that Slovakia and the Czech Republic have created a mixed company to build a new nuclear power plant at Bohunice, with the purpose to replace the two Slovak units that were stopped as condition to the access of the country to the EU. After a feasibility study carried by the competent authorities of both countries it was decided that the construction of this new plant could be carried out in 2013. This agreement also opens the possibility to enlarge Temelin nuclear power plant in the Czech Republic, as it was foreseen initially, with the construction of other two nuclear power reactors of WWER-1000 type similar to the ones already operating in this site.

Summing up can be stated that the government's list of priority power projects for the future shows 1 200 MWe for Bohunice for commissioning by 2025 at a cost of €3 billion, the possible construction of a 1 200 MWe nuclear power plant at Kecerovce after the closure of Bohunice Units 3 and 4 in 2025 at a cost of €3.5 billion. Preparation for decommissioning Units 3 and 4 will begin in 2012. It is expected that this work will be done in 13 years at an estimated cost of about €500 million.

THE NUCLEAR POWER PROGRAMME IN BELGIUM

One of the main characteristic of the country is that it has no gas, no uranium, no oil and very limited hydro resources. For this reason, it is fair to say that since 1993, when the last Belgian coal mine was closed, and as far as energy stocks are concerned, Belgium is effectively energy dependent for almost 100%. Only the currently small amount of renewable energy sources used in the country reduces the import dependency to somewhat less than 100%. Based on what has been said before can be stated that Belgium is highly dependent on foreign countries for its energy supply and, therefore, it has to integrate its energy policy into a larger framework on the international level. Working towards this goal implies finding a dependable energy supply on viable economic conditions that also sustains environmental quality. [8]

Belgium has seven PWR units in operation in 2009 with a total capacity of 5 728 MWe. Nuclear production of electricity was of 43.4 TWh in 2007, representing 54% of the total production of electricity in that year and 43.35 TWh in 2008, representing 50.76% of the total for that year. The net load factor of the nuclear power reactors in operation in the country in 2007 was 89.9%.

The composition of the (gross) Belgian electricity generation fuel mix up to 2005 is shown in Table 27.

Table 27. Composition of the (gross) Belgian electricity-generation fuel mix up to 2005

GWh

	1998	1999	2000	2001	2002	2003	2004	2005	% in 2005
Nuclear	46,165	49,017	48,157	46,349	47,360	47,379	47,312	47,596	54.7
Hydro, pumping, wind	1,508	1,502	1,713	1,676	1,546	1,404	1,736	1,831	2.1
Combustible renewables and waste	1,062	1,208	1,219	1,458	1,655	1,609	1,760	2,250	2.6
Coal	14,187	9,939	12,916	9,936	10,029	9,638	9,147	8,199	9.4
Gas	17,739	21,820	19,091	18,608	20,499	23,579	23,812	25,409	29.2
Liquid combustible	2,580	1,035	797	1,665	972	1,007	1,675	1,740	2.0
Total	83,241	84,521	83,893	79,692	82,061	84,616	85,442	87,025	100.0

Source: FPS Economy, DG Energy (D'haeseleer, 2007).

From Table 27 can be easily confirm that the nuclear sector is by large the most important sector in the production of electricity in the country. For this reason, the implementation of the phase-out policy adopted by the Belgian government without having a clear and competitive energy alternative will have a negative impact in the electricity price, in the commitment of the government regarding the Kyoto Protocol, will increase the need to import more electricity from other countries and the dependency of Belgium from energy external source for the production of electricity. The dependency of the country of external energy source for the production of electricity is clearly shown in Table 36. The dependency of the country from external energy source for the production of electricity increase from 11 645 GWh to 18 719 GWh, this means an increase of 161% during the period 2000- 2006. The implementation of the phase-out policy could increase this dependency significantly.

However, it is important to note that the balance of electricity exchanges does not mean that Belgium is not able to provide its own electricity generation most of the time, although it is becoming more difficult to cover its own peak. The margin has decreased dramatically over the last years due to the increased demand, on the one hand, and the almost halted investments in dispatchable power plants, on the other hand. The exchanges are related to the prices in neighboring countries versus those in Belgium and a better functioning of the global French-Belgian-Dutch electricity market due to enhanced transmission capacity and a change of the capacity allocation on the French-Belgian border.

Table 28. Balance of electricity exchanges 2000-2006

GWh	2000	2001	2002	2003	2004	2005	2006
Import	11 645	15 818	16 658	14 664	14 567	14 328	18 719
Export	7 319	6 712	9 070	8 254	6 789	8,024	8,696
Balance	4 326	9 106	7 588	6 410	7 778	6 304	10 023

Source: FPS Economy, DG Energy + Elia (D'haeseleer, 2007).

Table 29. Electric energy exchanges between Belgium, the Netherlands, France and Luxemburg in 2004, 2005 and 2006

(GWh)	2004	2005	2006	Evolution 2005-2006
France				%
Import	7,591.0	6,750.3	10,636.2	57.57%
Export	1,179.7	2,220.6	1,981.1	-10.79%
The Netherlands				
Import	4,630.1	5,073.8	5,603.6	10.44%
Export	4,052.9	4,430.1	5,017.8	13.27%
Luxemburg				
Import	2,380.8	2,366.4	2,0478.8	4.75%
Export	1,576.7	1,373.2	1,696.9	23.57%

Source: ELIA, 2004, 2005 and 2006 (D'haeseleer, 2007).

The Energy Policy in Belgium

The Belgian energy policy was prepared on the basis of a balanced mixture of contributing elements. These elements are the following. First, if important post-Kyoto carbon-reduction limits are pursued, energy savings will have to be an important component of the Belgian's energy policy. Then, a diversity of primary energy sources and conversion technologies should be opted for, with a cost-effective integration of renewables, whereby the cost effectiveness is best geared by carbon prices rather than absolute objectives. Given the existing constraints and the costs reported, taking into account all hypotheses and uncertainties involved, and based on the combination of scientific, technical and economic arguments, can be conclude, that in case the nuclear phase out is implemented, the expected post-Kyoto constraint is expected to be extremely expensive and strongly perturbing for the Belgium economic industry. Even after having incurred a major part of the very high costs, the risks of not satisfying a reliable energy provision under the assumed constraints, are indeed very large. (D'haeseleer, 2007)

The current Belgian's energy policy focus on the reduction of greenhouse gases in order to reach the agree quota foreseen by the Kyoto Protocol, which is in the case of Belgium, a reduction of 7.5% from the 1990's level. In order to fulfill its international obligations the government has taken different measures, such as a reduction of taxes to promote the use of clean' energy, energy saving in industry, transport and households, as well as the promotion of the construction of large wind farms offshore with a capacity of nearly 6-10% of the national electricity generation.

However, and spite of all measures adopted and the goodwill of the government, it will be very difficult for Belgium to achieve the Kyoto Protocol goals and the existence of the nuclear phase-out law is not helping meeting this target, even when taking into account that the nuclear phase-out law in Belgium will begin to be implemented in 2015, after the first Kyoto Protocol commitment period ended. After having utilized the other solution paths, such as energy savings and renewable energy to a maximum reasonable extent, substantial relief of this extremely heavy task to reduce domestic CO_2 emissions can be further obtained if carbon capture and storage (CCS) would be available, or if nuclear power were allowed to continue operation beyond 2015 and 2025. (D'haeseleer, 2007)

The Nuclear Phase-out Law

It is important to single out that the circumstances under which the nuclear phase-out law was adopted in 2003 have indeed changed significantly in the last years; now the urgency for climate-change actions is becoming more apparent and the era of very cheap oil and gas prices is almost certainly behind us. This new situation forces a reconsideration of the overall Belgian energy policy, including the use of nuclear energy for electricity generation in the future.

By the Federal Act of 31 January 2003, the Belgium authorities decided to abandon the use of fissile nuclear energy for industrial electricity production. This was done by prohibiting the building of new nuclear power plants and by limiting the operational period of the existing ones to 40 years. The Act does not affect the operation of research facilities, and does not rule out the use of the fusion technology for the generation of electricity in the future. The phase-out law can only be overridden by new legislation or by a government decision based on a recommendation from the regulator, if Belgium's security of supply would be threatened by closing the plants. [8]

Despite of the phase-out law adopted by the Belgium authorities, power upgrades of the existing nuclear power reactors through steam generator replacement, turbine refurbishments, etc. are authorized. However, in this case a license adaptation application should be submitted to the competent national authorities for approval.

Although this is not the intention to make an in-depth analysis of the phase-out law in this section it is, however, instructive to make some comments on this important issue for the energy sector of the country and to single out the context in which it was enacted. The Belgian nuclear phase-out law was officially published in the Belgian Official Journal on February 28, 2003. It is important to note that in Belgium there was no predefined technical lifetime for the operation of a nuclear power plant. There was only an economic lifetime, always subject to stringent safety regulations, and a political lifetime, if the authorities decide that a nuclear power plant must be closed. The political character of the phase-out law is evident from the timing.

In 1999, the government appointed the Ampere Commission with the mandate to prepare a report on electricity demand and options for meeting it in the 21st century. After three months of the establishment of the Ampere Commission, the new governing coalition formed after the elections agreed in its coalition agreement and, later on in the governmental declaration, that a nuclear phase law out should be implemented, resulting in a shutdown of the Belgian nuclear power plants after 40 years of operation. The conclusions of the Ampere Commission adopted in 2000 were that "nuclear power was important to Belgium and recommended its further development". However, the conclusion of the Ampere Commission to change the phase-out policy was not considered by the new government coalition.

A careful examination of the explanatory memorandum and the phase-out law shows that many arguments used to justify the phase-out policy do not stand up to serious scientific scrutiny. Even stronger, particular arguments made by the Ampere Commission have been taken out of context and have been made improper use of it in order to justify the end. In any case, the most worrisome element of the phase-out law is that the consequences of the implementation of this law have not been fully investigated. Taking into account the present circumstances, the overall situation on the international energy scene has changed dramatically: The substantially risen fossil fuel prices in the last years and the unstable geopolitical situation, have a severe impact on the security of supply and accelerate climate change. This added to the fact that a careful evaluation at the time would have shown that a nuclear phase out would already have been extremely challenging, makes it almost impossible now to keep the closure calendar as foreseen in the law. (D'haeseleer, 2007)

A summary of the cost associated to the implementation of the nuclear phase-out law adopted by the Belgium government is the following: One of the main reasons why a nuclear phase-out is so expensive are the expected imposed GHG reduction obligations that Belgium will have to face by 2030. In addition:

1) Belgium gives up a cheap way to reduce CO_2 emissions domestically;
2) phasing out 6 000 MW of cheap base load capacity will lead to an increase in electricity prices;
3) allowing nuclear power plants to continue would allow the state to negotiate a concession fee;
4) Giving up nuclear power increases the country import dependency; this reduced security of supply has a cost;
5) by postponing decommissioning of the nuclear power reactors, the decommissioning fund will grow substantially;
6) although not really an actual cost, letting a future government negotiate with nuclear power plant owners by using the "carrot" of a nuclear operation extension, can keep certain elements of the energy system under the control of the Belgian authorities. (D'haeseleer, 2007)

Summing up can be stated that considering the major challenges faced by the Belgian energy economy, it must be concluded that, especially in the light of the very stringent GHG-reduction efforts expected, an actual implementation of the Belgian nuclear phase-out law turns out to be expensive, as too much opportunity will be missed and Belgium will has to pay a substantial amount for the premature closure of its nuclear power plants.

The following figure show that postponing the decommissioning of the nuclear power plants in Belgium from 40 to 60 years will not only increase the decommissioning fund but make this activity cheaper.

The pink figure shows the costs for an operation period of 40 years and the dark-blue the same but considering a period of 60 years. In total, the area under the pink curve, represents the amount needed for decommissioning after 40 years, whereas the area under the dark-blue curve represents the amount needed for decommissioning after 60 years[60]. From the mentioned figure can be concluded that there is an important saving by postponing in 20 years the decommissioning of the nuclear power plants currently in operation in Belgium, and this important element should be taken into consideration by the current Belgian's government.

According with IAEA sources, the electricity production activity is totally liberalized in Belgium. Three categories of electricity producers can be distinguished:

1) *Electricity companies:* In 2002, they cover 97.9% of the domestic production. The most important producers are the private company ELECTRABEL and the public company SPE. Some smaller units (mainly co-generation and renewables) are owned by distributors or newly founded companies.

2) *Auto-producers:* They generate electricity themselves to cover their own needs. They are mainly active in the chemistry and metallurgy sector and represent 1.5 % of the total electricity production.

3) *Autonomous producers:* They are mainly active in the service sector and produce electricity as a complementary activity (e.g. in the framework of waste incineration) for resale to a third party. They are mainly active in the service sector and represent only 0.6% of the total electricity production. [74]

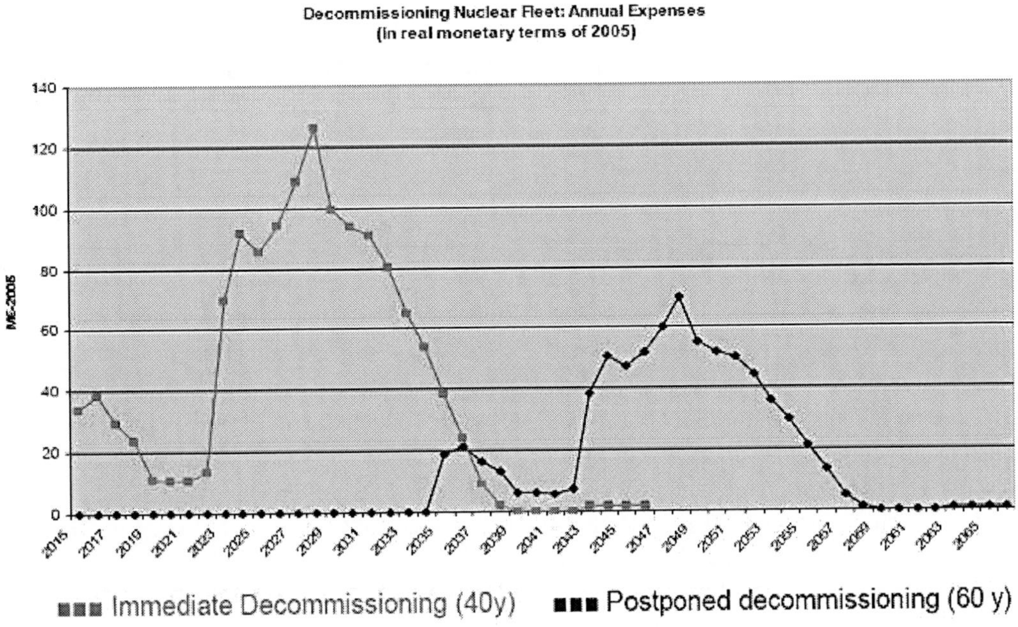

Decommissioning Nuclear Fleet: Annual Expenses
(In real monetary terms of 2005)

■■■ Immediate Decommissioning (40y)   ■■■ Postponed decommissioning (60 y)

Source: D'haeseleer, 2007.

Figure 68. Cost of the decommissioning activities in 40 and 60 years.

---

[60] It is important to single out that the Figure 68 is based on numbers available in early 2004. Using an annual discount factor of 3% in real terms, this cost figure has been shifted to the right by 20 years, but properly discounted. (D'haeseleer, 2007)

The Evolution of the Nuclear Power Sector in Belgium

Since the 1980s, nuclear power overtakes coal. In 2008, nuclear power provided 50.76% of Belgium's gross electricity production, fourth only to France, Lithuania and Slovakia in relative importance within the EU[61].

The historical evolution of the nuclear energy sector in Belgium is described in detail in IAEA documents[62]. A summary of this evolution is included in the following paragraphs.

Nuclear activities in Belgium began during World War II, when started uranium production in its mines in Africa. In 1957, Belgian engineers took part in the commissioning of the first commercial nuclear power plant in the USA, with the purpose to gain the necessary experience in the use of this type of energy for electricity production and to train their engineers with the purpose to build a nuclear power plant in the country in the future. In the same year, Belgium authorities started the construction of the BR3 PWR prototype in Mol with a power capacity of 11 MW. The nuclear power reactor was connected to the grid in October 1962 and entered into commercial operation in the same month and year. The reactor was shut down in June 1987 after produced 855.3 GWh. This was the first nuclear power reactor imported from the USA.

In 1966, the Franco-Belgian nuclear power plant (Chooz A) with a power capacity of 305 MWe was commissioned. The nuclear power reactor was decommissioned in 1991.

In 1968, the order to build two PWR units in Doel (Doel 1 and 2) was issue. The construction of the first nuclear power reactor was initiated in July 1969. Doel 1 was connected to the grid in August 1974 and entered into commercial operation in February 1975. The construction of the second unit started in September 1971. Doel 2 was connected to the grid in August 1975 and entered into commercial operation in December 1975. In 2007, the net load factor of Doel 1 was 88.2% and of Doel 2 was 91.8%. These two units have a power capacity of 392 MWe and 433 MWe respectively and were constructed by Loop Westinghouse.

In 1969, the order to build a nuclear power reactor in Tihange (Tihange 1) was approved. The power capacity of the first unit was 962 MWe. The construction work started in June 1970; the unit was connected to the grid in March 1975 and entered into commercial operation in October of the same year. In 1974, the Belgium government approved the construction of two additional nuclear power reactors in Doel (Doel 3 and 4), and two more en Tihange (Tihange 2 and 3). Unit 3 in Doel was connected to the grid in June 1982 and entered into commercial operation in October 1982. Unit 2 of Tihange was connected to the grid in October 1982 and entered into commercial operation in June 1983. The power capacity of these two units are 1 006 MWe and 1 008 MWe respectively. Unit 4 in Doel and Unit 3 in Tihange were connected to grid in April 1985 and in September of the same year respectively and both units entered into commercial operation in 1985. One of the reactors (Doel 4) has a power capacity of 1 008-MWe and was constructed by Loop Framatome; the second has a capacity of 1 015 MWe and was constructed by the same firm. The total nuclear net capacity available in the country in 2007 was 5 824 MWe. In 2008, this figure was 5 728 MWe, representing a reduction of 1.1%.

---

[61] See Figure 5.
[62] See Reference 8.

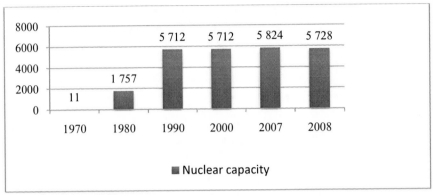

Source: CEA, 8th Edition, 2008.

Figure 69. Evolution of the nuclear power capacity in the country during the period 1970-2007.

Source: Photograph courtesy of nuclear power plants in the world (http://www.icjt.org/npp/foto/784_1.jpg).

Figure 70. Tihange nuclear power plant.

Table 30. Belgian nuclear power reactors in operation

Reactors	Model	First power	40-year license
Doel 1	PWR	1974	2014
Doel 2	PWR	1975	2015
Doel 3	PWR	1982	2022
Doel 4	PWR	1985	2025
Tihange 1	PWR	1975	2015
Tihange 2	PWR	1982	2022
Tihange 3	PWR	1985	2025
Total (7)			

Source: CEA, 8th Edition, 2008.

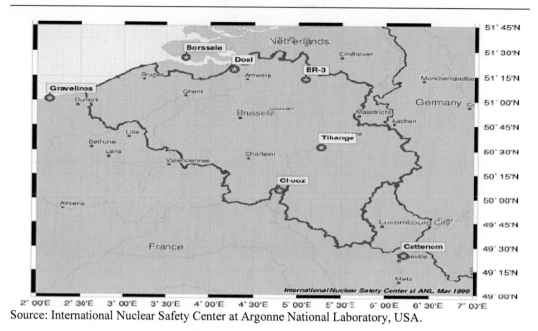

Source: International Nuclear Safety Center at Argonne National Laboratory, USA.

Figure 71. Location of the nuclear power plants in Belgium.

The evolution of the nuclear net power capacity in the country during the period 1970-2008 is shown in Figure 69.

The Tihange nuclear power plant is shown in Figure 70.

The construction of an 8th nuclear power reactor with a power capacity of 1 400 MWe was indefinitely postponed by the government in 1988.

The location of the Belgium nuclear power plants are shown in Figure 71.

Nuclear Safety Authority and the Licensing Process

The licensing process for the construction and operation of nuclear installations in Belgium are under the authority of the Minister of Labor and the Minister of Interior, according with the Royal decree of August 7, 1995, which have the guardianship over the Federal Agency for Nuclear Control (FANC). This government office is responsible for promulgating and enforcing regulations designed to protect the employees of the nuclear power plants and the population against the hazards of ionizing radiations. The Agency is assisted in technical matters by a Scientific Council of experts and representatives from various authorities responsible for nuclear safety, acting in an advisory role. State agencies, such as the Association Vinçotte Nuclear (AVN), carry out official acceptance procedures for nuclear installations prior to commissioning and exercise supervision over these installations during operation. Final authorization for nuclear plant commissioning rests with the King. [8]

The main steps in the Belgian licensing procedure for the construction and operation of nuclear installations are the following. The first step in the process is filing of an application. The request for the license is then sent to the Director General of the FANC, together with the following relevant information: a) characteristics of the installation; b) planned safety measures; c) an Environmental Impact Assessment; and d) a study of the premises and the demographic, geological, meteorological characteristics of the area of the installation, among others. The request has to contain also a preliminary safety report and a report describing the incidences of the environment.

Another stage in the process is to consult the Scientific Council with the purpose to provide a preliminary advice about the request. The application presented is sent to the applicant. At this stage, the EC is also consulted, if necessary, according to article 37 of the Euratom Treaty, as well as all the municipalities in a radius of 5 km around the installation, who will inform their population on this matter and to the province involved. After the advice of the municipalities, the province and of the EC has been received, the file is submitted to the Scientific Council once more, which then gives its definitive advice. [8]

Other stage in the process is the preparation, by the Minister of Labor and the Minister of Interior, a Royal Decree to be submitted to King for approval. This Royal Decree gives the construction and operation license authorization. It contains the conditions to be respected during the construction of the nuclear power plant. The Decree stipulates, among other things, the content of the safety report. After the construction of the nuclear installation, and before the start of the operation the state authority proceeds with the acceptance of the nuclear installation. This acceptance must establish the conformity of the nuclear installation with the general regulation, the stipulations of the construction and operation license and the safety report. If the acceptance is favorable, the Minister of Interior proposes to the King to confirm the construction and operation license, which are granted for an unlimited period.

Nuclear Fuel Cycle and Waste Management

It is important to single out that Belgium has no natural uranium that can be mined economically for its use in the Belgium nuclear power plants. In the past, there was a limited production of about 40 tons per year from imported phosphates. This has been terminated because of economic reasons while uranium prices were low. Synatom, which is responsible for all aspects of the nuclear fuel cycle, secures the uranium supply for the nuclear power reactors in operation in the country through medium-and long term contracts with exporters from Australia, Canada, Russia and central and southern Africa. The uranium fuel-element fabrication plant in Dessel has a production capacity of 400 tons of uranium per year, which is more than enough to meet the country's needs. The mixed oxide fuel (MOX) factory of Belgonucléaire, also in Dessel, has been shut down in 2006. (D'haeseleer, 2007)

It is important to single out that the Act of 11 April 2003 entrusted Synatom with the following four missions: a) "the definition of a decommissioning strategy in close collaboration with the nuclear operator and the development of scenarios for the treatment of spent nuclear fuel and nuclear waste; b) the evaluation of the related future expenses; c) the calculation of the necessary nuclear provisions; and d) the management of the funds corresponding to these provisions".

In compliance with the Act, a number of actions have been taken by Synatom, including an agreement with the Belgian state and Electrabel, modifying the structure of the share ownership of Synatom and defining the solvency requirements and conditions for loans granted by Synatom to Electrabel and SPE.

The national agency for radioactive waste and fissile materials management (ONDRAF/NIRAS) is responsible for the safe management of all radioactive materials in the country, including transport, treatment, conditioning, storage and disposal. Its main facility is at the Mol-Dessel site, run by its subsidiary Belgoprocess.

In June 2006, the government decided that low-level and short-lived intermediate-level wastes would be disposed of in a surface repository at Dessel. The municipality of Mol had also been considered and expressed a willingness for the facility to be there. Research on deep geological disposal of intermediate and high-level wastes is underway and focused on the clays at Mol. During the period 1980- 1984, the Hades underground research laboratory was constructed 225 m deep in the Boom clay. Funding of waste management and decommissioning is through Synatom and borne by the waste producers, notably the power companies. [74]

The Public Opinion

The Belgium government adopted a decision, in 1999, to phase out nuclear power altogether started in 2015. According with this decision, the closure of all nuclear power reactors should conclude in 2025. However, the Ampere Commission in its report presented a recommendation to the Belgian's government "to keep the nuclear option open by maintaining the scientific and technological potential needed to ensure optimal conditions for safety and performance, by preserving the national know-how on nuclear energy and by participating in mostly private-sector research and development on future reactor types" (see the Ampere Commission report).

As in the other European countries, in Belgium the current debates regarding the use of nuclear energy for electricity generation covers different issues such as the climate change, energy supply security, the safety of the nuclear power reactors in operation, the management of the high-level radioactive nuclear waste, among others. As the impacts of climate change and the vulnerability of the Belgium economy to foreign fuel imports become more evident, it is likely that the gradual shift in public opinion will further develop towards less skepticism or even in favor of the use of nuclear energy for electricity generation.

Of course, any severe nuclear incident related to these aspects, such as the use by terrorists of a simple atomic bomb or radiological device, or another major civil reactor accident, will likewise imply a major setback for the popularity of nuclear energy in Belgium.

Looking Forward

In July 1999, the new Belgian government announces the closure of all Belgian nuclear power plants when they reach their 40-years lifetime. At the same time, a moratorium on reprocessing of nuclear spent fuel was adopted.

In December 2001, an agreement was reached between the Belgian government and the electricity sector on financing the dismantling of old nuclear power reactors at the Mol site, and on the management of the provisions for spent nuclear fuel disposal and dismantling of the Belgian nuclear power plants at the end of their lifetime[63].

---

[63] It is important to stress that Belgian companies supplied about 80% of the systems and equipment for the country's nuclear power programme.

In January 2003, the Belgian Senate approved the legislation on phasing out Belgian nuclear power plants no later than 40 years from the date on which operations started. This decision will lead to a decrease in nuclear electricity generation and an increase of the share of fossil fuelled power plants, especially through the construction of new combined-cycle gas turbine units. However, the law adopted by the Senate does not affect the operation of nuclear research facilities, and does not rule out the use of the fusion technology for electricity generation in the future, when this new nuclear technology for the production of electricity is available. It is important to note that the phase-out law adopted by the Senate can only be overridden by new legislation adopted by this organ, or by a government decision based on a recommendation from the regulator, if Belgium's security of energy supply would be threatened by the closure of the nuclear power plants.

In 2007, the Commission on Energy Policy 2030 set up by the government to study the energy policy adopted by the government, reach the conclusion that "a fundamental review of energy policy was required and, in particular, that nuclear power should be utilized long-term in order to meet CO_2 reduction commitments, enhance energy security and maintain economic stability". It also concludes that "the 2003 phase-out decision should be reconsidered as it would double the price of electricity, deny Belgium a cheap way of meeting the country's CO_2 emission reduction targets and increase import dependency. Instead, the operating lives of the seven nuclear power reactors should be extended". According with this report, giving up nuclear power increases Belgian's import dependency, roughly from about 65% to an overall 90-95%. This leads to two extra cost issues: a) the reduced security of supply, with sometimes instantaneous power dependency of more than 95%, making Belgium more vulnerable to interruptions of continued delivery, especially of gas, and this leads to an increased cost; and b) whereas nuclear power is characterized by a rather cheap fuel cost, maintenance and part of the construction cost can be considered as domestic costs, gas dependence must be imported, leading to a higher burden on our financial balance. (D'haeseleer, 2007)

Finally, it is important to single out the following. In 2009, the Belgian government deferred his plans of closing all of its nuclear power plants for another ten years, but the generating utilities should pay for this postponement. The Belgian government, at the same time, decided to authorize the functioning of its first three nuclear power reactors, Doel I and II and Tihange I, until 2025, ten years longer than those under the current law. It is important to note that this law, which was adopted in 2003, provides also the withdraw of the rest of the nuclear power reactors when they reach the 40 years service. The law also prohibited the construction of new nuclear power reactors. The new decision reached after a proposal by the Belgian Minister of Energy, Paul Magnette, based on the technical report GEMIX commissioned by the Executive to meet best energy distribution and to achieve security of supply with stable prices during the period 2020-2030. Nuclear energy produced, in 2008, represent 53,76% of the total electricity consumed in the country. The study indicates that "if current nuclear power reactors are closed, then this action would jeopardize the security of supply and would not reduce CO_2 emissions". The report mention the possibility for a later extension of the functioning of the three units ten years more; this mean until 2035 and the other four units by 20 additional years of operation. The government will request the operator, the Tractebel company, an annually financial contribution to the state budget between €215 and €245 million during the period 2010-2014, and the commitment to invest in improvements in energy efficiency, capture and storage of CO_2 and the use of renewable energy. The exact amount will be negotiated with a Monitoring Committee recently created on the basis of the energy price evolution in the market. GDF-Suez, owner of Tractebel, declared that is taking notes of the government's communication and confirmed its desire to complete the negotiations, as soon as possible, although in principle is not accepting the concept of the proposed contribution. The new fund will be used by the government to develop these abovementioned plans.

THE NUCLEAR POWER PROGRAMME IN SWEDEN

Electricity production In Sweden started in the 1880s when the first generating plants based on hydroelectric power were built. They were small and intended to supply power to industries and communities in the close vicinity. Hundreds of small hydroelectric power plants were constructed during that period. During many years hydroelectric power was the main energy sources for electricity generation in the country. Today around 90% of the electricity produced in the country come from hydro power and nuclear power with both producing about equal amount. Sweden has the highest per-capita use of nuclear power in the world, almost 8 000 nuclear kWh per year. This is somewhat more than in France, and three times as much as in the USA. (Wikdahl, 2008)

Sweden was the first of the Scandinavian countries to have a commercial nuclear power plant. The construction work of the first unit, with a net capacity of 467 MWe, started in August 1966 (Oskarshamn 1)[64]. In April 2009, Sweden has 10 nuclear power reactors in operation, with a net capacity of 9 014 MWe. In 2007, Sweden produced 61 300 TWh of electricity using nuclear energy. In 2008, the production was 61 335.9 TWh. The net load factor of the BWRs system in operation in the country, in 2007, was 82.2%. In the case of the PWRs system this load factor was 79.2%.

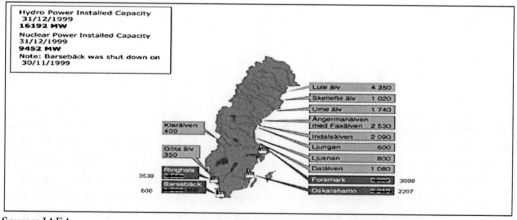

Source: IAEA.

Figure 72. Location of the nuclear power plants in Sweden.

---

[64] It is important to note that the first PHWR nuclear power reactor with a net capacity of 10 MWe, Agesta, was initiated its construction in December 1957, was connected to the grid in May 1964 and was shut down in June 1974.

Sweden's energy requirement is covered both by imported energy, primarily oil, coal, natural gas and nuclear fuel, and by domestic energy in the form of hydropower, wood and peat plus waste products from the forestry industry (bark and liquors). Originally, all energy was domestic, primarily wood and hydropower. However, during the 19th century, coal began to be imported and came to play an important role until the beginning of World War II, when oil and hydropower together became the base of the energy supply. [129]

The Energy Policy in Sweden

The main objective in the long and short term of the Swedish Energy Policy is "to ensure reliable supplies of electricity and other forms of energy carriers at prices that are competitive with those of other countries. It is intended to create the right conditions for cost-efficient Swedish energy supply and efficient energy use with minimum adverse effect on health, the environment or climate. Extension of cooperation in the fields of energy, the environment and climate around the Baltic is also an important objective".

The climate change is one of the major concern for the Swedish' government and a number of measures have been taken to reduce the emission of CO_2 and other contaminated gases to the atmosphere. During the spring of 2000, the Climate Committee published its proposal for a Swedish climate strategy, in which it suggested "a national target for Sweden involving reducing the emission of greenhouse gases by 2% between 2008 and 2012, relative to the 1990 level. To achieve this objective, the Committee suggests "a programme of work at both national and international levels". One of the elements of the international work is that Sweden should push for the introduction of European trade in emission rights of greenhouse gases. At the national level, work includes information campaigns linked to demonstration projects and investment subsidies. Some proposals also involve tightening up existing regulations. [129]

The Phase-out Policy

The Swedish's Energy Policy was affected by The Three Mile Island nuclear accident in the USA. As a consequence of this accident, the Swedish's government adopted a decision "to call a public referendum in Sweden, to remove the issue from the election campaign late in 1979. The 1980 referendum canvassed three options for phasing out nuclear energy, but none for maintaining it. A clear majority of voters favored running the existing plants and those under construction as long as they contributed economically in effect to the end of their normal operating lives (assumed then to be 25 years)". [107] It is important to single out that the Swedish Parliament decided to embargo further expansion of the use of nuclear energy for the production of electricity in the county in the future and to start the decommissioning of the 10 nuclear power reactors currently under operation by 2010, if new energy sources were available and can replace them technically and economically. The use of this new type of energy source should be also an acceptable alternative from the environment point of view.

In addition to the Three Mile Island nuclear accident, the Chernobyl nuclear disaster, which was initially detected at a Swedish nuclear power plant, created additional pressure on the Swedish's government to shut down all nuclear power reactors operating in the country.

In 1988, the government decided to begin in 1995 the phase-out of all nuclear power reactors, but this decision was overturned in 1991 due to strong opposition of the trade unions to implement it. Three years later the government appointed an Energy Commission consisting principally of backbench politicians, which reported at the end of 1995 that a complete phase-out of nuclear power by 2010 would be economically and environmentally impossible. However, it said that one unit might be shut down by 1998. This gave rise to intense political maneuvering among the main political parties, all of them minority, with varied attitudes to industrial, nuclear and environmental issues. The Social Democrats ruled a minority government but with any one of the other parties they were able to get a majority in parliament. Early in 1997, an agreement was forged between the Social Democrats and two of the other parties, which resulted in a decision to close Barseback Units 1 and 2, both 600 MWe BWRs constructed by ASEA-Atom and commissioned in 1975 and 1977 respectively. They were closed in 1999 and 2005 respectively. The positive aspect of the decision to close Barseback Units 1 and 2 is that the other nuclear power reactors gained a reprieve beyond 2010, and will be able to run for about 40 years, this means during the period 2012-2025. [94]

It is important to note that due to uncertainties about energy prices and the security of energy supply, in 1965 a decision was adopted by the Sweden government to supplement the production of electricity using nuclear energy. Today most of Sweden's electricity is produced by the following three type of energy: a) hydropower; b) nuclear power; and c) conventional thermal power plants, but in this last case accounting for only about 5%. Oil-fired cold condensing power plants and gas turbines are used today, primarily, as reserve capacity during dry years due to low precipitation.

The Evolution of the Nuclear Power Sector in Sweden

Sweden now has 10 nuclear power reactors in operation providing, in 2007, a total of 61.3 TWh or 46% of the total net electricity production of the country (140 TWh). The total nuclear capacity installed is 9 014 MWe. In 2008, the nuclear production was 61 335.8 TWh representing 42% of the total (145.9 TWh) The evolution of the nuclear capacity in the country and the evolution of the nuclear power reactors operating in the country during the period 1970-2007 are shown in Figures 73 and 74.

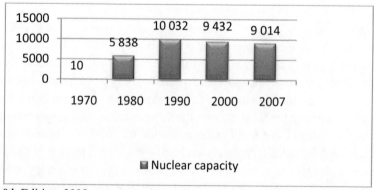

Source: CEA, 8th Edition, 2008.

Figure 73. Evolution of the nuclear capacity in the country during the period 1970-2007.

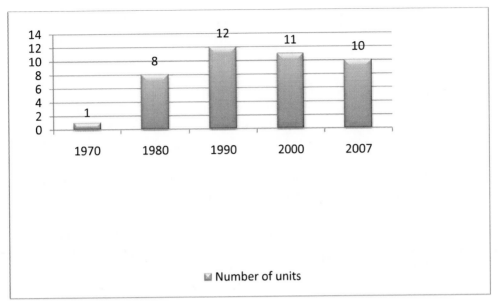

Source: CEA, 8th Edition, 2008.

Figure 74. Number of units in operation in the country during the period 1970-2007.

The nuclear power reactors currently in operation in Sweden are shown in Table 31.

Table 32 includes information on the outputs of all Swedish nuclear power reactors currently in operation. From the table below can be easily seen that the original nuclear power capacity of 8 840 MWe was increase up to 9 014 MWe, this means an increase of 7.9%. However, it is planned to increase this level even further up to 10 465 MWe in the following years, representing a further increase of 18.3%.

Table 31. Nuclear power reactors in operation in Sweden

Reactor	Type	Net MWe	Construction date	Connected to the electric grid	Commercial operation
Oskarshamn 1	BWR	467	August 1966	August 1971	February 1972
Oskarshamn 2	BWR	598	September 1969	October 1974	January 1975
Oskarshamn 3	BWR	1 150	May 1980	March 1985	August 1985
Ringhals 1	BWR	855	February 1969	October 1974	January 1976
Ringhals 2	PWR	867	October 1970	August 1974	May 1975
Ringhals 3	PWR	985	September 1972	September 1980	September 1981
Ringhals 4	PWR	935	November 1973	June 1982	November 1983
Forsmark 1	BWR	987	June 1973	June 1980	December 1980
Forsmark 2	BWR	1 000	January 1975	January 1981	July 1981
Forsmark 3	BWR	1 170	January 1979	March 1985	August 1985
Total 10		9 014			

Source: CEA, 8th Edition, 2008.

The historical evolution of the use of nuclear energy for electricity generation in Sweden is briefly described in the following paragraphs. The first interest in the use of nuclear energy for different purpose, including the generation of electricity from the Sweden government was shown in 1947, when AB Atomenergi was constituted as a research organization. Up to 1955, the programme was orientated towards basic research and concentrated on a small natural uranium/heavy water reactor. In 1956, an official ad-hoc commission proposed the implementation of a research and development programme, with the purpose to introduce the use of nuclear energy for electricity generation and heat production programme based on natural uranium and heavy water. AB Atomenergi was proposed to be responsible of this programme. [129] The legislation in the nuclear field started with a general Atomic Energy Act in 1956, followed by a general Radiation Protection Act in 1958.

In 1957, the Sweden's government took a decision to build a small nuclear power reactor for production of heat and electricity. The name was Ågesta and the construction of this small nuclear power reactor started in December 1957. The unit entered into commercial operation to produce 65 MWe in May 1964. Part of the production, 55 MWe, was used for heating a suburban of Stockholm and 10 MWe for electricity production. The reactor was shut down in June 1974.

In 1960, an act about emergency planning in case of a nuclear accident was introduced by the Sweden's government and, in 1968, the Third Party Liability Act was adopted.

In 1963, the construction of a heavy water nuclear power reactor Marviken began, with an electric output of 140 MWe using saturated steam or an electric output of 200 MWe using superheated steam. The project was stopped in 1970 only a few months before the fuel loading. The reactor was never used for power production.

In 1959, the Atomic Power Consortium, AKK, asked for a governmental approval for the construction of a BWR unit, based on a design from General Electric, with a net capacity of 60 MWe. However, the project was never realized due to different considerations. The information regarding the construction date, connection to the grid and entered into commercial operation of the nuclear power reactors currently in operation in Sweden is included in Table 39.

Table 32. Increase in the nuclear capacity of the Swedish's nuclear power reactors

Reactor	Initial power Capacity MWe	Power capacity in 2007. MWe	Planned power capacity MWe
Oskarshamn 1	460	467	487
Oskarshamn 2	580	598	840
Oskarshamn 3	1 100	1 150	1 450
Ringhals 1	750	855	880
Ringhals 2	785	867	920
Ringhals 3	915	985	1 110
Ringhals 4	915	935	1 160
Forsmark 1	900	987	1 134
Forsmark 2	900	1 000	1 134
Forsmark 3	1 100	1 170	1 360
Total	8 840	9 014	10 465

Source: IAEA and CEA, 8th Edition, 2008.

In November 1999, after 28 years of the beginning of the construction of Barsebäck 1, the unit became the first nuclear power reactor to be definitively shut down as result of the application of the Nuclear Power Decommissioning Act of January 1998. The Act allows the government, within a specified framework, to decide that the right to operate a nuclear power plant will cease to apply at a certain point in time. Such a decision infers the right to compensation by the state for losses incurred. The Barsebäck 2 unit was definitively closed in May 2005. To compensate the closure of Barsebäck 1 and 2 the Sweden's government decided that the other nuclear power reactors gained a reprieve to operate beyond 2010, and will be able to run for about 40 years. Compensation for the premature closure of Barsebäck 1 cost the Swedish taxpayers €593 million plus a payment for operating Barsebäck 2 on its own. The compensation for Barsebäck 2 was agreed at €583 million. The closure of Barsebäck 1 and 2 will have a substantial impact on the environment. Barsebäck's output of approximately 4 TW/h per year will primarily be replaced by imports from coal-fired power plants operating in Denmark and Germany.

Until 2008, four power reactors - Agesta, Marviken (never operated) and Barseback-1 and 2 are being decommissioned, along with three nuclear research reactors (R1, R2 and R2O at Studsvik). The research reactor R1 has now been dismantled.

In 1977, the Stipulation Act was approved and, in 1980, the Sweden's Parliament passed an act on public control of the safety work at the nuclear power plants. In 1981, an act on the financing of future costs for spent nuclear fuel was also approved by the Sweden's Parliament. Taking into account the different acts on nuclear power approved so far by the Sweden's Parliament, in 1984, the whole system of acts was revised. The Atomic Energy Act, the Stipulation Act, the Act on Public Control and part of the Financing Act were combined in one single act named Act on Nuclear Activities.

Sweden was the first country in the world in penalizing the use of nuclear energy for the production of electricity. According to the Financing Act from 1981, the nuclear utilities have to pay a fee per produced kW/h to a state fund. The fund shall cover all future costs for handling and final storage of all waste and for decommissioning of all the facilities. In the late 1990s, the Sweden's government imposed a capacity tax on nuclear power at SEK5 514 per MW thermal per month (€0.30- €0.32 cents) potentially produced. In January 2006, the tax was almost doubled to SEK10,200 per MWt (about €0.6 cents/kWh). In 2007, it was proposed to increase it further to SEK12,684 per MWt from 2008 - total SEK4 billion (€435 million, meaning about €0.67/kWh). [94]

Why the uses of nuclear energy for the production of electricity expand so fast in Sweden after the Second World War? The reason was a yearly increase in the power consumption of 7% during several years and the impossibility to satisfy this increase using hydropower. Neither the Sweden's government nor the utilities wanted to use oil fired power plants to satisfy this increase because this will increase further the dependence on oil imports. For this reason, the only available option that the Sweden's government had, at that time, in their hands to increase the electricity generation and satisfy the demand of energy was the use of nuclear energy for this purpose.

However, just before the general election in 1976, nuclear power became a main political matter due the nuclear waste issue. The leader of the Centre Party became Prime Minister with the government consisting of a non-socialistic coalition. The new government arranged a huge investigation of the risks and economies of nuclear power compared to other energy sources in an official Ad-hoc Energy Commission. In 1977, a unique act about the nuclear waste was accepted by the Sweden Parliament. According to the new waste act, called the Stipulation Act, the utilities would not be allowed to load fuel into a new nuclear power reactor (and several were in the pipe line) before it had been shown that it was possible to arrange a final storage of the waste in an absolutely safe way. Before Parliament accepted the act a remark was added in the minutes saying that the word "absolutely" should not be interpreted in a "draconian" way. [129]

Nuclear Safety Authority and the Licensing Process

The Nuclear Safety Authority in Sweden (SKI) is under the Ministry of Environment. SKI's Board is appointed by the government and consists of politicians and experts, and is chaired by the Director General of SKI. SKI has three advisory committees: a) the Reactor Safety Committee; b) the Safeguards Committee; and c) the Research Committee. SKI is responsible for supervising the implementation of the Act on Nuclear Activities in Sweden. SKI has the following overall tasks:

a) "ensuring that Swedish nuclear installations have adequate defense-in-depth methods that prevent serious incidents or accidents originating from technology, organization or competence. In addition, the dispersion of nuclear substances must also be prevented or limited if an accident should occur;

b) adequately protecting nuclear installations and nuclear substances under Swedish jurisdiction against terrorist activities, sabotage or theft;

c) ensuring that the Swedish government, in co-operation with the competent international safeguards agencies, is provided with adequate information on and control over nuclear substances and nuclear technology, which are held, used and traded, and which come under Swedish jurisdiction. This must be done to ensure that such substances or technology will not be used in any way contrary to Swedish legislation and Sweden's international obligations in the area of non-proliferation;

d) carrying out the final disposal of spent nuclear fuel and nuclear waste in such a way that any possible leakage of radioactive substances should, within various periods of time, be expected to remain under tolerable levels. Future generations should not be exposed to greater health and environmental risks than are tolerated by society today;

e) ensuring that the nuclear industry carries out comprehensive and appropriate research and development programmes to achieve the safe handling and final disposal of spent nuclear fuel. Methods must also be developed for decommissioning and dismantling nuclear installations, and sufficient funds should be set aside for such future expenses;

f) keeping decision-makers and the general public well informed about nuclear risks and safety, and about supervision and the final storage of spent nuclear fuel and nuclear waste;

g) actively contributing to developing and strengthening efforts in the areas of international nuclear safety and non-proliferation, particularly within the framework of the EU. As a member of the EU, Sweden should actively work for increased and effective environmental measures in neighboring countries, especially in the Baltic area and in central and Eastern Europe.

Also, SKI shall:

a) provide a clear definition of safety requirements;
b) control compliance with requirements by supervision focusing on processes influencing safety;
c) report and inform. SKI shall issue regular reports on the safety status of the nuclear power plants and the quality of licensee safety work and, in general, implement active public information services with regard to events and circumstances within its area of regulatory responsibilities;
d) initiate improvements in safety and non-proliferation whenever justified by operating experience, research and development;
e) maintain and develop competence. SKI shall promote maintenance and development of competence for safety at licensees and nationally;
f) maintain emergency preparedness at SKI. SKI shall be prepared to advise emergency management authorities in order to limit detrimental health effects and social consequences in case of radioactive releases or situations where there is a threat of such releases".

It is important to note that by law, anyone conducting nuclear activities has the full and undivided responsibility for taking the necessary steps to ensure safety, to meet non-proliferation requirements and to ensure that spent nuclear fuel and nuclear waste are managed and disposed of safely. SKI should monitor the way in which licensees meet these responsibilities by establishing a clear, independent picture of the safety situation at the facilities and of the quality of the licensees' safety work. [129]

Nuclear Fuel Cycle and Waste Management

To overcome problems associated with the management of nuclear waste and, in particular the final disposal of nuclear waste, the power industry started a very wide-ranging technical development work in order to identify acceptable methods for final storage of spent nuclear fuel. The result was KBS-3, a project that was presented for the first time in 1983. A major programme of research has been carried out since then, with the most recent report being submitted to the reactor safety authorities in the autumn of 2007. The waste project includes a central intermediate storage for spent fuel in operation since 1986, a final repository for operational waste in operation since 1988 and a 500 m deep final repository for spent nuclear fuel to be built in the future. Two possible sites for a final repository have been investigated in detail. Both are in the vicinity of existing nuclear power plants: one in Öskarshamn and the other at Forsmark. (Wikdahl, 2008)

In June 2009, and following the advice of SKB, the government has decided to choose Östhammar as the location of the geologic repository for the final disposal of the Swedish spent nuclear fuels produced by the nuclear power plants currently in operation. According to

Claes Thegerström, consultant of SKB, "the reasons for the election of Östhamma, is that the granite of the underground in this location is more stable and has less water that the others location selected". The adoption of this decision puts an end to years of preparations and of investigation works, as well as of uncertainty for the politicians and for the population of Öskarshamn and Forsmark communities. SKB plans to request the construction authorization next year and expects that the government takes a decision in 2013, year in which they would like to begin the construction works. It is expected that construction works conclude in 2023. The spent nuclear fuel will be placed in steel capsules and copper and will be buried to about 500 meters deep in the granite. SKB had built in Öskarshamn a demonstration facility for the encapsulation that will be enlarged as necessary. Another facility will also be built in this location for the assembly and confirmation of the capsules that will be transported to Östhammar later by ship.

It is important to single out that when the selection of the site for the deep final depository being approved by the Swedish Parliament, the deep repository will only be put into operation by depositing approximately 5-10% of the existing spent nuclear fuel. After an evaluation and a renewed licensing process, the repository will be put in full scale operation. All the costs for managing and disposing of Sweden's nuclear waste shall be paid by the owners of the nuclear power plants. This also applies to the costs of decommissioning the nuclear power plants and disposing of the decommissioning waste. They pay a fee set by the government to a state fund administered by the SKI to cover waste management and decommissioning (€0.21 cents/kWh).

With respect to the supply of uranium for the normal operation of the Swedish nuclear power reactors the situation is the following: Swedish utilities import all their need of uranium and enrichment services. The Swedish utilities buy part of their fuel elements from abroad. The spent nuclear fuel from all the Swedish nuclear power plants is transported by boat to the central interim storage CLAB. The facility started operation in 1985 and is situated close to the Oskarshamn nuclear power plant site. It has been substantially expanded during the last years. [129]

In Sweden, the handling of low nuclear waste is finally deposed of at local dumps and some of it incinerated at Studsvik. All other waste from nuclear power reactor operation is transported to the final repository for radioactive operational waste (SFR), in operation since 1988. SFR is located close to the Forsmark nuclear power plant site. Most of the waste from decommissioning of the nuclear power reactors will be disposed at SFR. SFR has 63 000 cubic meter capacity and receives about 1 000 cubic meters per year.

The Public Opinion

A week after the Three Miles Island nuclear accident, all the political parties agreed to arrange a referendum about the future of nuclear power in Sweden. A special law was adopted forbidding the start of the operation of all new nuclear power reactors until the referendum is carried out. The referendum was carried out in March 1980 and some months later the Sweden's Parliament decided, in accordance with the result of the referendum, "to allow the start of the operation of all the nuclear power reactors, which were ready to operate or the continuation of the construction of the new nuclear power reactors". It was also decided "that nuclear power would be phased out by 2010, if new energy sources were available at that time and could be introduced in such a way that it would not affect the social welfare programme and the employment in the heavy industry".

The Chernobyl nuclear accident resulted in a new political debate about the future of the Swedish nuclear power programme. Sweden Parliament decided, in 1988, "that the phase out of nuclear power reactors would start within the period 1995 to 1996, with two units to be closed". However, after the decision to phase out nuclear power plants in the country, the industry and the labor unions started an intensive debate, based on official reports prepared by the government, singling out that the total cost of an early phasing out (after 25 years operation instead of 40 years, which is the assumed technical life time of the Swedish nuclear power reactors) would cost the society more than SEK200 billion. The price of energy for the electricity intensive industry (paper and steel) would double and this increase in the energy price will force an estimate between 50,000 to 100,000 persons to lose their jobs. Based on this estimates the Sweden Parliament decided, in 1991, not to start the phase out of nuclear power reactors by 1995.

Taking into account that, in 2007, nuclear power accounts for 46% of the total electricity produced in Sweden, the shutdowns of all nuclear power plants could trigger record price increases in almost all sectors of the Sweden's economy. However, the Swedish government's energy agency said "the nation's electricity supply was not currently at great risk because it can rely more on hydropower during the summer months". Form many experts this is a very optimist statement.

It is important to stress that in the last four years, and in the light of concerns about climate change, the public opinion in Sweden adopted a more positive attitude in favor of the use of nuclear energy for electricity generation. In April 2004, total of 17% supported a nuclear phase-out, 27% favored continued operation of all the country's nuclear power plants, 32% favored this plus their replacement in due course, and 21% wanted to further develop nuclear power in Sweden. The total support for maintaining or increasing nuclear power thus was 80% as the government tried to negotiate a phase out. This total support had risen to 83% in March 2005, with a similar proportion saying that limiting greenhouse gas emissions should be the top environmental priority. With slightly different questions, total support for maintaining or developing nuclear power was 79% in June 2006 and fluctuated around this level to November 2007 when it was 77%. A self-assessed 18% (26% of men, 11% of women) said, in November 2007, that they had become more positive towards nuclear power in the light of concerns about climate change, while 7% (4% of men, 10% of women) said that they had become more negative. This may be related to 14% who thought that nuclear power was a source of CO_2 with a large impact on the environment (8% of men and 21% of women). [94]

Looking Forward

After the early closure of Barseback-1 and 2 nuclear power reactors, the government is working with the utilities to expand nuclear capacity to replace the 1 200 MWe lost. The actions adopted are the following: 1) Ringhals 3 was subject to a major uprate. Steam generator was replaced in 2007. Early in 2008, it was operating at 985 MWe net, which is a little less than anticipated, but a further 5% uprate is expected in the coming months; 2) the older BWR Unit 1, a 15 MWe uprate was completed in 2007, with another 15 MWe to follow in the future; and 3) Ringhals 4 had a 30 MWe uprate to 935 MWe following replacement of its low pressure turbines in 2007. Exchange of high pressure turbines and steam generators in 2011 and other work to be carried out in this unit in the future is expected to yield a further 240 MWe. The total uprate for Ringhals nuclear power plant is likely to be more than 400 MWe costing €225 million to be carried out during the 2008-2010. In 2004, low pressure turbines were replaced in Unit 3 of the Ringhals nuclear power plant, giving a 30 MWe uprate. The same is being done for Units 1 and 2.

In 2005, Sweden authorities approved a uprate of the Oskarshamn 3 to 1 450 MWe net and this was confirmed by the government in January 2006. The US$450 million project involves turbine upgrade in 2008, as well as a reactor upgrade. These modifications will extend the nuclear power plant's life to 60 years. It is important to note that the operation of the unit has been authorized only for a period of one year, when the unit will be stopped for refueling. In that moment, the owner of the nuclear power plant should requests another authorization to operate the unit definitively.

Regarding the future of the nuclear power programme in Sweden it is important to note the following. In February 2009, the Swedish government announced his intention of annulling the prohibition that was imposed in 1980s to build more nuclear power reactors or to prolong the life of those already in operation. In the last years, through their state electricity company, Vattenfall, the government had followed an intelligent policy that combined strategic vision of acquiring an external power capacity with a commercial orientation that took to the company to acquire important generation assets in the north of Europe. The use of renewable energy for electricity generation was promoted by the government as well as a reduction of the CO2 emissions to the atmosphere and an increase in the security supply of energy.

The First Minister Fredrik Reinfeldt took the decision that the current nuclear power reactors in operation should be substituted by new ones and to considerer the possibilities to construct new nuclear power reactors in the future, if necessary. The decision was adopted taking into account first recent problems of gas supply in the Baltic countries due to the dispute between Russia and Ukraine that put in evidence the strong energy dependence of the whole region, and second due to the wish of Sweden to be a world leader in the reduction of emissions of polluting gases emissions to the atmosphere. Without any doubt, Sweden is one of the countries with more sensibility for the nature, and its environmental politicians are a reference for environmentalist groups. The Swedish decision of 1980s prohibiting the construction of nuclear power plant was exhibited by these groups as a model to follow by other countries. Thirty years later the conservative coalition that is in power in the country rectifies this decision and affirms that the nuclear power reactors in operation should be replaced by new nuclear power reactors with the purpose to reduce the emissions of polluting gases and fulfill the commitments of and obligations adopted by the Sweden's government with the Kyoto Protocol. There are several factors that make this decision tremendously important. The most significant factor is that this rectification was made thanks to a change in the public opinion in the country. Now around 83% of the Sweden's public opinion supports the nuclear power option. Another factor is that the Swedish society is, according to different studies carried out, the best informed on nuclear energy issues.

However, there is also an important element that needs to be taking into account. This element is the following: Sweden, the country with more public companies in Europe, will adopt the Finnish position on nuclear matters, that is to say, the construction of new nuclear power reactors should be a private initiative promoted by the big industrial consumers. In other words, the Sweden's government will not subsidize the construction of new nuclear power reactors in the future and this is something that could make more difficult the construction of these types of power reactors in the country in the future.

THE NUCLEAR POWER PROGRAMME IN UKRAINE

In the case of Ukraine, the main energy sources available in the country are uranium and coal. According with IAEA source, "the confirmed coal reserves in the country could cover the Ukraine's energy needs for 200-300 years". The country imports mostly all of their needs in oil and gas from Russia. A large share of primary energy supply in Ukraine comes from the country's uranium and substantial coal resources. After the breakdown of the Soviet Union in 1991, the country's economy collapsed and its electricity consumption declined dramatically from 296 billion kWh in 1990 to 170 billion kWh in 2000, all the decrease being from coal and gas plants. Total electricity production in 2007 amounted to 195 billion kWh, 84.3 billion kWh were produced by nuclear power plants and with 9 billion kWh net exports. [97] The total capacity of Ukraine is over 52 GWe. Ukraine has 15 nuclear power reactors in operation in 2009 at five nuclear power plants with a net nuclear capacity of 13 168 MWe. The government are constructing two new units and have plan for the construction of 22 new nuclear power reactors in the future with a net capacity of 30 800 MWe. In 2007, the distribution of power was the following: around 47.4% came from coal and gas (approximately 20% from gas), 48% from nuclear (in 2008 was 47.40%) and 4,6% from hydro.

It is expected a major increase in electricity demand to 307 billion kWh per year by 2020 and 420 billion kWh by 2030. Despite of the Chernobyl nuclear accident and the shutdown of four of its units, the policy of the government is to continue supplying approximately half of its energy demand from nuclear power plants. To achieve this goal it will require having more than 20 GWe of nuclear capacity in 2030.

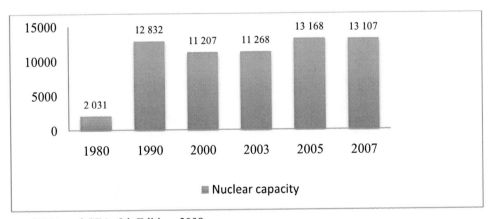

Source: WNA and CEA, 8th Edition, 2008.

Figure 75. The evolution of the net nuclear capacity during the period 2003-2007.

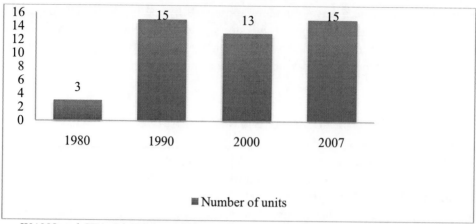

Source: WANO and CEA, 8th Edition, 2008.

Figure 76. The evolution of the number of nuclear power reactors in operation during the period 2003-
2007.

The nuclear net capacity installed in the country increased from 11 268 MWe in 2003 to
13 168 MWe in 2005 to 13 107 MWe in 2007 and to 13 168 MWe in 2008. This increase was
due to addition of two new WWER-1000 nuclear power reactors, Khmelnitski 2 and Rovno 4,
which entered into commercial operation in December 2005 and in April 2006 respectively.
The cost of completing these two nuclear power reactors was estimated to be around
US$1,236 (US$621 for the Khmelnitski 2 and US$642 for Rovne 4). The net load factor of
the nuclear power reactor in operation in the country in 2007 was 76%.

The evolution of the net nuclear capacity installed in the country during the period 2003
to 2007 is shown in Figure 75.

All nuclear power reactors installed in Ukraine are Russian WWER systems of three
different generations. Out of the 15 nuclear power reactors currently in operation in the
country, two are 440 MWe V-213 models of second generation, two are the larger WWER 1
000, V-302 and V- 338 second generation, and eleven are WWER-1 000, V-320 third
generation.

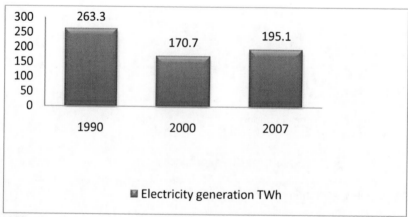

Source: WANO.

Figure 77. The evolution of the electricity generation I Ukraine during the period 1990-2007.

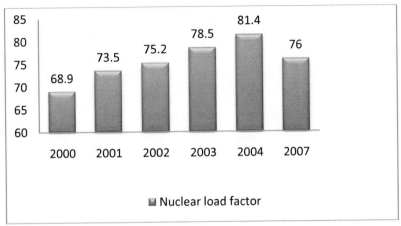

Source: WANO and CEA, 8th Edition, 2008.

Figure 78. Evolution of the net load factor during the period 2000-2007.

Figure 77 shows the evolution of the generation of electricity from 1990 to 2007. From the figure can be concluded the following: During the period 1990-2000 the electricity generation in the country felt from 263.3 TWh in 1990 to 170.7 TWh in 2000, this means a decrease of 65%. From 2000, the generation of electricity shows a modest but continuing increase from 170.7 TWh to 195.1 TWh in 2007, this means an increase of 114%.

From Figure 78 can be concluded the following: During the period 2000-2004 the net load factor of the nuclear power reactors in operation in Ukraine increased in a systematic manner from 68.9% to 81.4%, this means an increase of 12.5%. However, during the period 2004-2007, the net load factor decrease from 81.4% to 76%; this means a decrease of 5.4%.

To increase the participation of nuclear energy in the energy balance of the country in the future with the purpose to satisfy the foresee increase in the energy demand, the Ukrainian Parliament approved, in 1996, an energy policy and its top priorities, which were included in the National Energy Programme for the period 1996-2010 (NEP). The programme approved includes the following main tasks:

a) solution of the problems related to a stable and reliable supply of fuel and energy resources in order to meet Ukraine's demand;
b) decrease in foreign economic dependence of fuel and energy resource;
c) wide and large-scale implementation of energy saving technologies;
d) expansion of utilization of non-conventional energy sources;
e) decrease in harmful effects on the environment of the fuel-energy complex's sectors;
f) expansion of the share of extra budgetary sources for programme financing, due to the budgetary deficit and because of the development of market relations;
g) development of domestic basis for machinery building to meet the fuel-energy complex's needs.

According to the NEP, the following are the main measures and activities to be promoted related with the development of electric power in Ukraine:

1) implementation of energy saving measures;
2) orientation towards a Ukrainian fossil fuel base (coal): the refurbishment of a fossil fuel balance focused on increasing the share of coal in the electric power production and on decreasing the natural gas and crude oil utilization;
3) development of the nuclear industry in the future, taking into account the present deficiency in fossil fuel in Ukraine. The country is planning to commission two nuclear power reactors with a high priority at Khmelnitsky and Rovno nuclear power plant sites and two units with a medium priority at Khmelnitsky nuclear power plant site;
4) the commissioning of new hydropower capacities is foreseen at Dnestr and Kanev nuclear power plant sites, taking into account the deficiency of the hydropower plants whose capacity can be changed;
5) a primary task comprises the technical upgrading and rehabilitation of thermal power plants in order to extend their plant life to an additional period of 15 to 20 years and to improve their environmental and economic conditions.

The locations of the nuclear power plants in Ukraine are the following: South Ukraine, Zaporizhzhe Rivne (or Rovno), Khmelnitsky and Chernobyl.

Source: Courtesy of Wikimedia Commons (Eric Gaba).

Figure 79. Location of the nuclear power plants in Ukraine.

Table 33. Nuclear power reactors operating, under construction and shut down in Ukraine

Station	Type	Net capacity MWe	Status	Construction date	Electric grid connection	Commercial date	Shut down or planned date
Khmelnitski-1	WWER V-320	950	Operational	01-11-81	31-12-87	13-08-88	2018 2032
Khmelnitski-2	WWER V-320	950	Operational	01-02-85	08-04	12-05	2035 2050
Rovno-1	WWER V-213	381	Operational	01-08-73	31-12-80	21-09-81	2011 2026
Rovno-2	WWER V-213	376	Operational	01-10-73	30-12-81	30-07-82	2012 2027
Rovno-3	WWER V-320	950	Operational	01-02-80	21-12-86	16-05-87	2017 2032
Rovno-4	WWER V-320	950	Operational	01-08-86	10-10-04	04-06	2035 2050
South Ukraine-1	WWER V-302	950	Operational	01-03-77	31-12-82	18-10-83	2012 2027
South Ukraine-2	WWER V-338	950	Operational	01-11-79	06-01-85	06-04-85	2015 2030
South Ukraine-3	WWER V-320	950	Operational	01-02-85	20-09-89	29-12-89	2019 2034
Zaporozhe-1	WWER V-320	950	Operational	01-04-80	12-84	25-12-85	2015 2030
Zaporozhe-2	WWER V-320	950	Operational	01-01-81	22-07-85	15-02-86	2016 2031
Zaporozhe-3	WWER V-320	950	Operational	01-04-82	10-12-86	05-03-87	2017 2032
Zaporozhe-4	WWER V-320	950	Operational	01-04-83	18-12-87	14-04-88	2018 2033
Zaporozhe-5	WWER V-320	950	Operational	01-11-85	14-08-89	27-10-89	2019 2034
Zaporozhe-6	WWER V-320	950	Operational	01-06-86	19-10-95	16-09-96	2026 2041
Khmelnitski-3	WWER	950	Under const.	01-03-86	-	-	-
Khmelnitski-	WWER	950	Under const.	01-0287	-	-	-
Chernobyl-1	LWGR	725	Shut down	01-03-70	26-09-77	27-05-78	30-11-96
Chernobyl-2	LWGR	925	Shut down	01-02-73	21-12-78	28-05-79	30-10-91
Chernobyl-3	LWGR	925	Shut down	01-03-76	03-12-81	08-06-82	15-12-00
Chernobyl-4	LWGR	925	Shut down	01-04-79	22-12-83	26-03-84	26-04-86
Total 21 (15 operational)		18 507					

Source: IAEA and CEA, 8th Edition, 2008.

The Evolution of the Nuclear Power Sector in Ukraine

Nuclear energy development started in Ukraine in 1970 with the construction of the Chernobyl nuclear power plant. During the period 1970-1980, nuclear power rapidly grew in Ukraine. The information regarding the construction date, the date in which the different units were connected to the electric grid and the date when the units entered into commercial operation are included in Table 33.

It is important to note that Ukraine has been building nuclear power plants since 1973, and for most of them, the construction works started before the Chernobyl nuclear accident occurred. After this accident, eight nuclear power reactors were connected to the electric grid increasing the net nuclear capacity in 8 550 MWe. After the accident at the 4th unit at Chernobyl, the Supreme Soviet of Ukraine adopted, on 2 August 1990, a moratorium to build new nuclear power reactors in the country. The construction work at Unit 6 at Zaporozhe nuclear power plant site was interrupted and the construction of four new WWER systems at Khmelnitsky and Rovno nuclear power plant sites was also halted.

This situation raised a number of social problems for the nuclear power sector, which were added to the traditional technical and economic ones, that is safe, reliable and competitive operation of nuclear power plants. Some of the problems to be solved associated with the Chernobyl nuclear accident were the following: 1) undermined confidence in nuclear electricity generation; 2) massive out-flow of highly skilled and experienced personnel due to the affected prestige of their profession and, subsequently, declined level of living; and 3) re-orientation of nuclear and building industrial enterprises because of the lack of orders, creation of new jobs for personnel. (Kovryzhkin, 2004)

In the second part of 1991, the collapse of the Soviet Union affected the structure of the nuclear energy sector in the country, resulting in the separation of its enterprises and loss of the centralized management system. Bearing in mind this situation, the Cabinet of Minister took the decision to make the nuclear power plant managers personally responsible for the safe operation of the nuclear power plants operating in the country. In order to create the state management system, to ensure the safe operation of the nuclear power plants in the country, the Ukrainian State Committee on Nuclear Power Utilization was established (Goskomatom) by the Decree of the Cabinet of Ministers of Ukraine on 16 January 1993. On 6 May 1997, a Ministry of Energy of Ukraine was established, in accordance with a Decree by the President, on the basis of the former Ministry of Energy and Electrification, as well as on the basis of the State Committee on Nuclear Power Utilization. This new ministry includes the State Department for Nuclear Power that was entrusted with the functions of the state authority responsible for nuclear power sector administration. On 14 April 2000, the Ministry of Fuel and Power Industry of Ukraine (Mintopenergo) was established by the Decree of the President of Ukraine #598/2000, on the basis of the Ministry of Coal Industry of Ukraine; Ministry of Power Industry of Ukraine; State Department of Ukraine on Electric Power Issues; State Department of Ukraine on Oil, Gas and Oil-refining Industry; State Department of Ukraine on Nuclear Power. The main task of Mintopenergo is the State management of the fuel energy complex. [138]

Source: Photograph courtesy of Wikimedia Commons (Elena Filatova).

Figure 80. Chernobyl nuclear power plant with the sarcophagus.

The National Nuclear Energy Generating Company (Energoatom, NNEGC) was established, in accordance with the government's Decree dated 17 October 1996. This company has been set up with the aim of improving the electricity supply to the national economy and the population, the nuclear power plants operation, the competition ability under market conditions and modifying the structure of nuclear energy management, in accordance to the requirements of the acting legislation. In May 1997, NNEGC joined the WANO-Moscow Centre as an associate member.

In December 2006, the Ukraine government took the decision to form Ukratomprom with the mandate to development a national nuclear fuel cycle.

The Chernobyl Nuclear Accident

The Chernobyl nuclear accident is the worst nuclear accident ever occurred in a nuclear power plant. The accident at Chernobyl Unit 4 resulted from a combination of design and technical deficiencies with a grave operator error. According with WANO report, on 25 April prior to a routine shutdown, the reactor crew at Chernobyl 4 began preparing for a test to determine how long turbines would spin and supply power to the main circulating pumps following a loss of main electrical power supply. This test had been carried out at Chernobyl the previous year, but the power from the turbine ran down too rapidly, so new voltage regulator designs were to be tested.

A series of operator actions, including the disabling of automatic shutdown mechanisms, preceded the attempted test early on 26 April. By the time that the operator moved to shut down the reactor, the unit was in an extremely unstable condition. A peculiarity of the design of the control rods caused a dramatic power surge as they were inserted into the reactor.

The interaction of very hot fuel with the cooling water led to fuel fragmentation along with rapid steam production and an increase in pressure. The design characteristics of the unit were such that substantial damage to even three or four fuel assemblies can - and did - result in the destruction of the reactor. The overpressure caused the 1 000 tons cover plate of the reactor to become partially detached, rupturing the fuel channels and jamming all the control rods, which by that time were only halfway down. Intense steam generation then spread throughout the whole core (fed by water dumped into the core due to the rupture of the emergency cooling circuit) causing a steam explosion and releasing fission products to the atmosphere. About two to three seconds later, a second explosion threw out fragments from the fuel channels and hot graphite. There is some dispute among experts about the character of this second explosion, but it is likely to have been caused by the production of hydrogen from zirconium-steam reactions. Two workers died as a result of these explosions. The graphite (about a quarter of the 1 200 tons of it was estimated to have been ejected) and fuel became incandescent and started a number of fires, causing the main release of radioactivity into the environment. A total of about 14 EBq of radioactivity was released, over half of it being from biologically-inert noble gases.

However, in January 1993, the IAEA issued a revised analysis of the Chernobyl nuclear accident, attributing the main root cause to the reactor's design and not to operator error[65]. In 2005, the IAEA and the World Health Organization (WHO) reported that "only 56 people had died directly from the incident, mainly accident workers[66]. They estimated another 4,000 deaths among workers and local residents". However, according to unofficial statistics not yet confirmed at least 15,000 people could have died as a direct result of the explosion.

After the Chernobyl nuclear accident the pressure of the international community to close nuclear power plants in operation in many countries increased significantly, independently of the type of nuclear power reactors used. In 1995, a Memorandum of Understanding was signed between the governments of the G-7 countries, the EC and Ukraine, agreeing with the closure of the Chernobyl nuclear power reactors. Based on this Memorandum, Unit 4 was shut down in June 1984, Unit 2 in October 1991 after a huge fire in the unit, Unit 1 in November 1996 and Unit 3 in December 2000.

Following the nuclear accident, Unit 4 was encased in a giant concrete sarcophagus (See Figure 99), constructed above the destroyed reactor by hundreds of thousands of soldiers and civilian, including nuclear experts, to prevent further leakage of radioactive material. However, it is important to note that the sarcophagus built in 1986 is considered to be unstable and could collapse in the future. To avoid that this happens, in December 2000, the USA promised to contribute the largest G-7 amount to repair the sarcophagus. A waste management facility began construction in 2001 for the treatment of fuel and other wastes from decommissioned Units 1, 2 and 3. A stabilizing steel structure was extended in December 2006 to spread some of the load on the walls damaged by the explosion.

The Memorandum stipulated the agreement on international financial aid to Ukraine to support Chernobyl decommissioning, power sector restructuring, completion of Khmelnitsky 2 and Rovno 4 units, thermal and hydro plant rehabilitation, construction of a pumped storage

---

[65] The IAEA in its 1986 analysis had cited the operators' actions as the principal cause of the accident.

[66] According with WANO source, the accident destroyed the Chernobyl 4 unit, killing 30 operators and firemen within three months and several further deaths later. One person was killed immediately and a second died in hospital soon after as a result of injuries received. Another person is reported to have died at the time from a coronary thrombosis. Acute radiation syndrome (ARS) was originally diagnosed in 237 people on-site and involved with the clean-up and it was later confirmed in 134 cases. Of these, 28 people died as a result of ARS within a few weeks of the accident. Nineteen more subsequently died between 1987 and 2004 but their deaths cannot necessarily be attributed to radiation exposure.

plant, and to support energy efficiency projects, in accordance with Ukraine's energy sector strategy. In 2000, the European Bank for Reconstruction and Development (EBRD) approved by an 89% vote apart from abstentions, a US$215 million loan towards completion of Khmelnitsky 2 and Rovno 4 units. The EBRD funding, was a modest part (14.5%) of the US$1,480 million estimated to be required for the decommissioning of the Chernobyl nuclear power plant. Conditions on the loan included safety enhancement of all Ukraine nuclear power reactors, independence for the country's nuclear regulator and electricity market reform.

Following approval of the EBRD loan, the EC approved a US$585 million loan to Energoatom. The EC said that approval of this Euratom funding a few days before the permanent closure of Chernobyl gives a clear sign of the Commission's commitment to nuclear safety, as well as to the deepening of EU relations with Ukraine. It will finance the completion, modernization and commissioning of two third-generation nuclear units. The EC pointed out that it and the EBRD had concluded that the project met all safety, environmental, economic and financial criteria. [97]

It is important to single out the following: Russia offered US$225 million credit for the purchase of equipment for Khmelnitsky 2 and Rovno 4 units and for fuel. A loan of US$44 million for completion of both units was approved by Russia in 2002. The arrangement will cover also goods and services from Russia. It followed signing of a US$144 million agreement in June, including about US$100 million for purchase of fuel. However, the promised loans of US$215 million and the Euratom's US$585 million were deferred late in 2001 because the government had baulked at doubling the wholesale price of power to US$2.5 cent/kWh as required by EBRD. At the same time, Ukraine rejected almost all approved Russian loans. The Ukrainian government estimates that for the completion, site works and upgrades for Khmelnitsky 2 and Rovno 4 at US$621 million and US$642 million respectively.

In August 2004, the Ukrainian President said that Western governments had failed to honor their 1995 undertakings to assist his country in exchange for closing the Chernobyl nuclear power plant, particularly in relation with the work of Khmelnitsky 2 and Rovno 4 completion, electric grid infrastructure and a pumped storage hydro plant.

Nuclear Safety Authority and the Licensing Process

Taking into account the Chernobyl nuclear accident, the safety of the nuclear power reactors in operation in Ukraine became the major concern by the government, the parliament and the public opinion. For this reason, the President of Ukraine, upon submission of the Cabinet of Ministers, took a decision "to establish the State Nuclear Regulatory Committee as a central executive authority with the special status on the basis of the Department of Nuclear Regulation and Main State Inspectorate of the Ministry of Environment Protection and Natural Resources". In compliance with Decree #1303 of the President of Ukraine "On State Nuclear and Radiation Safety Regulation", dated 5 December 2000, the newly established authority is entrusted with the following functions: 1) set criteria, requirements and conditions of safety in nuclear energy utilization; 2) issue permits and licenses for carrying our activities in this field; 3) exercise state supervision over compliance with the legislation, norms, regulations and standards on nuclear and radiation safety; and 4) exercise other functions of national regulatory body of nuclear and radiation safety as defined in the Convention on Nuclear Safety and Joint Convention on the Spent Fuel Management and on the Safety of Radioactive Waste Management.

The tasks, functions and mandates of the regulatory authority are defined in details in the Statute on the State Nuclear Regulatory Committee of Ukraine, which was approved by the Decree No.155 of the President of Ukraine dated March 6, 2001. This Statute entrusts "the State Nuclear Regulatory Committee with the function of the central authority and contact point which is responsible for physical protection of nuclear material in compliance with the Convention on Physical Protection of Nuclear Material, and with the functions of the competent authority and contact point in charge of transmission and receipt of notifications in case of nuclear accident in accordance with the Convention on Early Notification of a Nuclear Accident. It is also responsible for organization and performance of researches in the field of nuclear and radiation safety assurance, and for co-ordination of activities of the executive authorities as regards their performance of nuclear and radiation safety regulation". The Statute provides for the mandates required for execution of the regulatory activity, in particular "the right to access the territory of facilities and to obtain necessary information, mandates to apply financial sanctions and to limit or stop activities involving violations of safety conditions, mandates dealing with the orders to carry out activities required for maintaining the safety level, right to hear the reports of the officials and to pass the materials to law-enforcement authorities".

In addition to the establishment of a regulatory authority, a system of technical support organizations was also established. It includes the following organizations: a) State Scientific and Technical Centre of Nuclear and Radiation Safety, an independent scientific and technical and expert organization which provides support for the regulatory activities; and b) affiliates of the State Scientific and Technical Centre of Nuclear and Radiation Safety in Kharkiv, Odesa and Slavutich, which perform the functions of specific scientific and technical support to the regulatory authority, according to their specialization.

Due to the different decisions adopted by the Ukraine government in the field of safety can be stated that now there is a clear separation between the functions of the regulatory authority and those of other body or organization concerned with the promotion or utilization of nuclear energy in the country. In carrying out its activity the regulatory authority is independent of any central bodies of the executive authority responsible for the nuclear energy utilization. This provision is secured by Article 23 of the Law of Ukraine "On Nuclear Energy Utilization and Radiation Safety". NNEGC Energoatom, as a license holder, bears full responsibility for the radiological protection and safety of nuclear installations irrespective of the activity and responsibility of suppliers or state nuclear and radiation safety regulatory authorities. In line with responsibilities laid by the Ukrainian law on the operating body, NNEGC Energoatom shall: a) assure nuclear and radiation safety; b) develop and take measures on safety improvement of nuclear installations. For instance, based on the previous analysis and assessment of the safety level of nuclear power plants, long-term measures for upgrading and safety improvement of Ukraine's nuclear power plants were developed; and c) provide for radiological protection of personnel, public and environment. The new radiation safety norms with stricter requirements were put in effect in 1998. The operating body has developed and is implementing the programme on bringing its activity in conformity with these requirements.

Today, NNEGC Energoatom holds licenses by the regulatory authority for activities as follows: a) construction of nuclear installations - 2 licenses (Unit 4 of Rovno nuclear power plant and Unit 2 of Khmelnitski nuclear power plant); and b) design of nuclear facility (spent fuel storage facility at Chernobyl nuclear power plant). [138]

Nuclear Fuel Cycle and Waste Management

The development of the nuclear industry of Ukraine is going on under the following four main programmes:

1) comprehensive programme for nuclear fuel cycle establishment;
2) programme concerning integration of Ukrainian zirconium production into the nuclear fuel for the WWER-1000 system;
3) programme dealing with the spent nuclear fuel management. This programme concerns establishment of on-site (near-site) storage for long-term (up to 5 years) dry storage of spent nuclear fuel at all nuclear power plants site in Ukraine, organization of spent nuclear fuel containers manufacture in Ukraine and the establishment of central storage for long-term dry storage of the spent nuclear fuel. The problem of the reliable and safe radioactive waste storage also has to be solved;
4) the programme of equipment manufacture and technology mastering for the nuclear power plants of Ukraine. [138]

For the normal functioning of a nuclear power plant, it is indispensable to supply it with fresh nuclear fuel. For this reason, a comprehensive programme for the nuclear fuel cycle creation in Ukraine was developed some time ago. However, it did not materialize, due to different political and financial reasons. One of these reasons is the existence of the former Soviet Union which was, until the beginning of the 1990s, the responsible for the supply of fresh nuclear fuel to all Ukraine nuclear power plants.

However, the December 2006 decision to form Ukratomprom revived intentions to build a fuel fabrication plant. Russia has since offered to transfer fuel fabrication technology. At the same time, Ukraine has been seeking cooperation with other countries, which experience in the nuclear fuel cycle as a part of its effort to increase its supply of low-cost nuclear electricity and to reduce its imports of natural gas and other energy sources from Russia.

It is important to stress that, for the time being, there is no intention of the Ukraine authorities to close the nuclear fuel cycle, although this possibility remains under consideration.

With respect to the production of uranium, which is one of the main energy source available in the country, the situation is the following: The uranium ore mining and uranium concentrate production in Ukraine is performed by the Vostochny Uranium Ore Mining and Processing Enterprise (VostGOK). The main product of the enterprise is natural uranium concentrate. The commercial and industrial activity of the VostGOK on the natural uranium mining and processing is aimed at implementing the Comprehensive Programme of Nuclear Fuel Cycle Creation in Ukraine. The programme on the development of the nuclear energy's raw material sources allows for covering the nuclear power plants demand for uranium and is focused on two main areas: a) maintaining of the existing production capacities; and b) increasing of the output up to the level of the nuclear energy demand for natural uranium. [138] Ukrainian uranium concentrate and zirconium alloy produced by Ukraine are sent to Russia for fuel fabrication. The nuclear fuel produced from these Ukrainian components by TVEL in Russia then returns to Ukraine for its use in the nuclear power plants currently in operation in the country.

It is important to single out that the country depends also on Russia to provide other nuclear fuel cycle services, notably enrichment. To reduce the dependency from Russia, and in order to diversify nuclear fuel supplies, Energoatom started implementation of the Ukraine Nuclear Fuel Qualification Project (UNFQP). The project assumes the use of US-manufactured fuel in the WWER-1000 system following the selection of Westinghouse as a vendor on a tender basis. In 2005, South Ukraine's third unit was the country's first to use the six lead test assemblies supplied by Westinghouse, which were placed into the reactor core together with Russian fuel for a period of pilot operation. A reload batch of 42 fuel assemblies will be provided by Westinghouse in mid 2009 for a three-year period of commercial operation at the unit with regular monitoring and reporting. [97] However, it is important to note that, in the opinion of some experts, the use of Westinghouse assemblies in Russian-made reactors will considerably increase the risk of an accident. This fear has an objective basis. Finland has recently decided to continue buying Russian fuel for its Russian-built reactors and declined Westinghouse's offer because the use of American fuel at the Temelin nuclear power plant in the Czech Republic nearly caused an accident. The management of the Paks nuclear power plant in Hungary entrusted the cleaning of fuel assemblies at its second block to the French-German company Framatome ANP. The use of an "alien" technology resulted in the malfunction of 30 fuel assemblies and almost caused an accident. The Hungarian authorities called on Russian specialists for help in this cleaning process, who managed to remedy the situation only three and a half years later. The IAEA, which investigated the fuel-cleaning incident at Paks, rated it at Level 3 (serious incident), according with the International Nuclear Event Scale (INES).

At the same time, the Ukrainian government approved, on December 2008, a draft agreement with Russia and Kazakhstan on the involvement of the country in the work of an international uranium enrichment facility in East Siberia proposes by Russia to the IAEA, with the purpose to reinforce the international character of the enrichment activities. Plans for a nuclear center in Angarsk, 5 100 km from Moscow, were proposed by Russia in early 2007 as a means of allowing countries, including Iran, to develop civilian nuclear power without having to enrich their own uranium. The establishment of this type of international centers will reduce the fear over nuclear weapons proliferation. The draft agreement stipulates Ukraine's to acquire a 10% stake in the enterprise, in which Russia's state-controlled nuclear power corporation Rosenergoatom holds a 51% controlling stake and Kazakhstan's Kazatomprom has 10%. The planned network of uranium enrichment centers, which would also be responsible for the disposal of nuclear waste, will work under the supervision of the IAEA. Since this center would be a backup or reserve mechanism, it would be designed in a way not to disrupt the existing commercial market in nuclear fuels.

With respect to the final disposal of spent nuclear fuel the situation in Ukraine is the following: There is no full radioactive waste management strategy adopted until now by the Ukraine government. Pending any decision on this issue, the policy adopted by the government is to storage the used fuel for at least for 50 years. To settle the spent nuclear fuel and radioactive waste management problems, the Cabinet of Ministers adopted Decree No.542, dated 5 April 1999, according to which a new Comprehensive Programme on Radioactive Waste Management was established. Following the Order No. 7, issued by the Minister of Fuel and Energy on 13 January 2000, the Comprehensive Programme of Management of NPP[67] Spent Nuclear Fuel was established. In accordance with the adopted decisions, and with the financial assistance of the Nuclear Safety Account, jointly established by the G-7, the European Commission and some other countries, a storage facility of spent nuclear fuel is being constructed on the Chernobyl nuclear power plant site, which the purpose to storage used fuel from decommissioned reactors at Chernobyl nuclear power plant. A new dry storage facility is under construction in this site. The design of the storage facility allows for safe storage of 25 000 spent nuclear fuel assemblies and additionally 3,000 used absorbers in a dry-type storage facility. After commissioning a new storage, the existing one will be decommissioned. Activities on the first stage of spent nuclear fuel storage were completed at Zaporizhzhe nuclear power plant site based on the method of dry storage in reinforced concrete casks. The Nuclear Regulatory Committee completed assessment of the Storage Safety Analysis Report and granted the license to the Operating Body-Utility. [138]

Used fuel is mostly stored on site though some WWER-440 fuel is again being sent to Russia for reprocessing. At Zaporozhe nuclear power plant site, a long-term dry storage facility for spent nuclear fuel has operated since 2001, but other WWER-1000 spent nuclear fuel is sent to Russia for storage. A centralized dry storage facility for spent nuclear fuel is proposed for construction in the government's new energy strategy, to operate from 2010. In December 2005, the Ukrainian government signed a US$150 million agreement with the US-based Holtec International to implement the Central Spent Fuel Storage Project for Ukraine's WWER systems and, in April 2007, Energoatom and Holtec signed the contract to proceed with this project. In September 2007, Holtec International and the Ukrainian government signed a contract to complete the placement of Chernobyl's used nuclear fuel in dry storage systems. Removing the radioactive fuel from the three undamaged Chernobyl units is essential to the start of decommissioning of the Chernobyl nuclear power plant. The project is estimated to be worth €200 million over 52 months. There is full endorsement from the Assembly of Donors, who provides funding for Chernobyl remediation and decommissioning. Holtec also won a tender conducted by the State Specialized Enterprise Chernobyl Nuclear power plant to develop a storage system design for the damaged used fuel in dry storage. From 2011, high-level wastes from reprocessing Ukrainian fuel will be returned from Russia to Ukraine and will go to the central dry storage facility. [97]

The Public Opinion

The sociological service of the Razumkov Center conducted two national opinion polls among 2,010 and 2,008 respondents over 18 years of age between April 23 and April 28 and between May 27 and June 2[68]. The outcome of the poll shows that the Chernobyl consequences and construction of new nuclear power reactors are not the most serious societal concerns in comparison with low incomes, unemployment, and crime. The Chernobyl

---

[67] Nuclear power plant.
[68] For more information about these polls see reference 117.

problems rated only fourth among the six gravest concerns (these worried only 14.1% of respondents). The construction of new nuclear facilities worried twice fewer respondents 7.4%. At the same time, respondents over 40, who remember the Chernobyl disaster very well, are concerned more deeply than those aged under 40: 15% versus 12.8%. This problem appears a lot more serious to residents of the areas where nuclear facilities are located than residents of other parts of the country: 17% percent versus 13.6% on the Chernobyl issue; 12.5% versus 6.5% on the issue of construction of new nuclear power plants. [117]

Another important result of the poll is that prospects for nuclear power generation appear more realistic to residents of the areas where nuclear power plants are located than to respondents in other parts of Ukraine: 36% versus 26.3% respectively. On the whole, Ukrainians believes that the nuclear industry is a positive factor in achieving the country's energy independence (39.3% vs. 30% of those who do not). This opinion is prevalent among younger respondents (44%) and residents of nuclear-free areas (39.4% vs. 28.3%). In the areas where nuclear facilities are located the opinions on this issue split almost equally: 39.4% vs. 40.4%.

Despite of the Chernobyl nuclear accident, how the accident was handled by the soviet nuclear authorities, the negative impact of the accident on the environment and the population living in the surrounding area of the Chernobyl nuclear power plant, only 24.6% of Ukrainians consider Ukrainian nuclear power plants extremely dangerous, and 40.3% appraise them as substantially dangerous. On the other hand, very few respondents (3.5%) believe that Ukrainian nuclear power plants are absolutely safe; 24.1% of respondents assess them as relatively safe and 7.5% are undecided. It is important to single out that there is a notable difference between the assessments of nuclear safety levels by residents of areas where nuclear facilities are located and residents living outside these areas. The former appear to be more rational than the latter: a mere 0.3% of residents living near of nuclear industry areas believe that Ukrainian nuclear power plants are absolutely safe versus 4.1% of respondents living in areas where no nuclear facilities are located.

Regarding the necessity to build new nuclear power reactors the replies were the following: 54.9% of respondents consider that there is no need to construct new nuclear power plant in the country, while 26.8% are in favor and 18.3% was undecided. One of the outcome of the poll reflects that there are more opponents to such construction plans among residents of areas where nuclear power plants are located and among older respondents: 60.6% vs. 53.9% and 58.5% vs. 49.5% respectively.

Regarding the plan of the government to construct 11 new nuclear power reactors by 2030[69], the idea was supported by 19.9% of respondents while 57.2% were against; 9.6% were indifferent, and 13.3% were undecided[70]. The plan of the government to construct 11 new nuclear power plants in the coming years was rejected by activists promoting the following action: "NO to new reactors, YES to energy conservation!" near the Cabinet of Ministers of Ukraine in Kiev. The plans were announced in May 2005 and later confirmed at parliamentary hearings on the development of Ukraine's energy sector. "We are not calling

---

[69] Nine more new nuclear power reactors are proposing to be constructed after 2030.

[70] One of the possible reasons for a high opposition to the construction of new nuclear power reactors in the country in the future could be a lack of information of the Ukrainian's population about the government's plan to build new nuclear power reactors. The poll reflects that 91.1% of the population does not know were these new nuclear power reactors will be located. In other words, the Ukrainian citizens are insufficiently informed about the government's plans to develop the national nuclear power industry.

for the immediate closure of nuclear power plants in Ukraine but we do oppose new construction and life extensions," said Yevgeny Kolishevsky, executive director of the Voice of Nature NGO. "We strongly demand that the new government of Ukraine includes a set of concrete measures aimed at increasing energy efficiency in all sectors of the Ukrainian economy into the Programme of Energy Sector Development of Ukraine for the period running to 2030. Yury Urbansky, from the National Ecological Centre of Ukraine and CEE Bankwatch Network, said, "We have serious concerns about the safety of Ukraine's nuclear power plants. The implementation of the state programme of a safety upgrade for existing nuclear power has failed, the parliament ratified loans provided by the EBRD and Euroatom for the modernization of Khmelnitsky 2 and Rivne 4 commissioned in 2004 with a one year delay, problems connected with radioactive waste management continue to go unsolved and there are no signs for progress in the near future. A government ignorant of the issues of nuclear safety has no right to even think about plans for nuclear expansion!" (Saprykin, 2005)

A new World Public Opinion poll of 21 nations founds very strong support for the government requiring utilities to use more alternative energy, such as wind and solar, and requiring businesses to use energy more efficiently, even if these steps increase the costs of energy and other products. Fewer than half of the nations polled favor putting more emphasis on nuclear energy or on coal or oil. In the case of Ukraine, 56% were in favor to use alternative energy source for electricity production and 49% put more emphasis in increasing of the efficiency in the generation of electricity than in the use of nuclear energy for the same purpose.

In a poll of 20,790 respondents conducted between July 15 and November 4, 2008 by WorldPublicOpinion.org, a collaborative research project involving research centers from around the world and managed by the Programme on International Policy Attitudes at the University of Maryland, found that nearly half (49%) of the respondents in Ukraine say the building of nuclear power plants should receive less emphasis, while only 9% believe they should receive greater emphasis. This position of the public opinion in Ukraine is influence by the fear of having a second nuclear accident in the country.

Looking Forward

A 1993 deadline to close two of the Chernobyl nuclear power reactors grew closer, Ukrainian authorities expressed their concern that they may have acted too quickly with their decision to close the first two units of the Chernobyl nuclear power plant. In April 1993, the chairman of Ukraine's Parliamentary Standing Committee on Basic Industrial Development said "that the Committee intended to ask the Ukrainian Cabinet of Ministers to lift the moratorium on new nuclear power plant construction". The Commission for Nuclear Policy held a public hearing in May 1993 on the issue of lifting the moratorium and extending operation of the Chernobyl nuclear power plant. After postponing a decision in the summer, the Ukrainian Parliament voted in October to continue operating the Chernobyl nuclear power plant and to lift the moratorium on new plant construction. Parliament cited Ukraine's energy shortage as the reason. The vote cleared the way for completion of three partly-built WWER-1000 units- Zaporozhye 6[71], Rovno 4 and Khmelnstkiy 2. [71]

---

[71] The construction of this nuclear power reactors finished in September 1996.

In February 1994, the President of Ukraine issued a directive calling for the completion, by 1999, of five WWER-1000s units under construction: Zaporozhe 6, Rovno 4 and Khmelnitsky 2, 3 and 4. In the case of the Khmelnitsky 2, the financial resources need to complete the construction of the unit was estimated at US$257 million. For Rovno 4, the financial resources to complete the construction work of this unit was estimated at US$267 million. The Ukrainian's Energy Minister approved a contract for the continuation of the construction at the Khmelnitsky's nuclear power plant. The contract was given to a Russian company AtomStroyExport. The estimated cost of the work is of about US$4,000 million and it will be financed in 85% with a Russian loan. The remaining 15% will be financed with funds provide by Ukraine.

At the end of 1995, Zaporozhe Unit 6 was connected to the grid making Zaporozhe the largest nuclear power plant in Europe, with a net capacity of 5 700 MWe and the third in the world[72]. In August and October 2004, Khmelnitsky 2 and Rovno 4 respectively were connected to the electric grid, bringing their long and interrupted construction to an end and adding 1 900 MWe to replace that lost by closure of Chernobyl 1 and 3 in 1996 and 2003 respectively.

It is important to stress that all nuclear power reactors in Ukraine have already reached 50% of their designed service life. During the period from 2013 till 2020, eight WWER-1000 and two WWER-440 units now in operation should be shut down. To compensate any possible loss of power due to the closure of any of the mentioned units, and to take care of the foresee increase in the consumption of electricity in the country in the coming years, Ukraine authorities has taken a decision to build up to 20 new nuclear power reactors and extending the life of 13 of the existing PWRs systems currently in operation. These measures will ensure that nuclear power continues to represent about 50% of electricity generated in Ukraine in the near future. The main driving force behind the support for nuclear power is the desire of the Ukraine authorities to reduce dependence on Russian oil and gas.

THE NUCLEAR POWER PROGRAMME IN BULGARIA

Bulgaria has very few domestic energy resources that can be used for the generation of electricity and, for this reason, since 1956 the government has favored the use of nuclear energy for this purpose. Proven oil and gas reserves for the country have declined for a number of years and, in 2004, were only about 5 million tons of oil equivalents. In fact, it was less than six months normal hydrocarbon consumption for Bulgaria at that time. Hydropower potential is also limited since most of Bulgaria's rivers are small and the only large river, the Danube, has a small drop in altitude where it forms Bulgaria's northern border with Romania. Largely because of this constraint, hydro capacity accounts in 2004 for about 17.6% of the country's total installed generating capacity and an even smaller percentage of generation; thermal power is 50%, and nuclear power is 28.4% of the country's total installed generating capacity. [14]

It is important to note that electricity consumption in Bulgaria has been growing slowly since 1980, but it has been a significant exporter of power to other countries. In 2005,

---

[72] The second largest nuclear power plant operating n Europe is Gravelines, near Dunkerque in France, with six PWR nuclear power reactors with a total net capacity of 5 460 MWe.

Bulgaria's National Electricity Company (NEK) produced 44 billion kWh and exported 7 billion of kWh to Greece, Turkey, Serbia and Macedonia. However, with the closure of two older nuclear power units at the end of 2006, there are no significant surpluses of power that can be used for exports.

The energy sector is an important section of the Bulgarian industry. Its structure and development are based predominantly on imported energy sources and the used of domestic low-quality lignite coal. Despite of its limited energy resources the energy sector is well balanced. The electricity production in the country is based on the use of: a) solid fuel; b) nuclear power; c) natural gas; d) hydro resources; and e) renewables. The share of the imported energy resources referred to the general import of raw materials, other materials, investment and consumer goods is about 18%. From these imported energy resources only 35.4% are used in the energy sector for the production of electricity and heat power. The nuclear share is 37.6% of the imported energy resources.

Bulgaria has two nuclear power reactors in operation, Kozloduy 1 and 2, with a power net capacity of 1 906 MWe, producing 14.7 TWh in 2007. This production represents, in that year, 32% of the total electricity production of the country. In 2008, this share was almost the same (32.92%). Bulgaria is now constructing two nuclear power reactors, Belene 1 and 2, with a total net capacity of 1 900 MWe, and the enter into commercial operation of these two units will increase significantly the nuclear share in the future.

The Energy Policy and the National Energy Strategy in Bulgaria

After the collapse of the Soviet Union, the Bulgarian's government adopted a new energy policy and strategy in order to increase the efficiency of its energy sector with the purpose to satisfy the foresee increase in the demand of electricity. This new energy policy was adopted bearing in mind the new political and economic situation of the country and the plans to become member of the EU in the future.

The energy strategy of Bulgaria is based on the national priorities and corresponds to the new lasting positive political and economic trends in the country, as well as to the requirements of the European guidelines, the principles of market mechanisms and the government's programme. It is determined by the requirements for ensuring sustainable economic growth and raising the living standard of the Bulgarian population. The strategy has been developed in conformity with the natural and geographic factors determining the inherent role of the country in the region, and the optimal mix of energy resources used in accordance with the specific conditions. According with the IAEA[73], the main objectives guiding the energy development are:

a) through competitive energy to competitive economy;
b) continuous and safe coverage of the national energy needs with minimum public cost;
c) providing energy independence for the country;
d) reduction in specific energy intensity per GDP unit in economy;
e) ecologically oriented development;

---

[73] See reference 14.

f) establishment of competitive internal energy market;
g) integration of the Bulgarian energy system and energy market with that of the EU;
h) maintaining the nuclear safety at an acceptable level;
i) utilization of local renewable energy sources.

However, the establishment of an up-to-date and market-oriented energy sector cannot be achieved only with the adoption of the above objectives. It is important to take into account a series of prerequisites that have been missing up to this date, namely: a) normalization of energy prices in line with the justified full economic costs and phasing out of the subsidies for generators; b) financial recovering and establishment of energy companies operating on a commercial basis; c) properly functioning of regulatory authorities and mechanisms; d) market rules and structures; and e) appropriate legal framework.

Concurrent actions should be undertaken, mainly in the following areas:

1) financial restructuring;
2) establishment of financially viable commercialized companies;
3) institutional changes;
4) enhancement of the role, autonomy and influence of the regulatory body (SERC);
5) commercial restructuring;
6) transition from administration to regulation and introduction of clear regulatory rules for the market players;
7) deregulation;
8) introduction of clear and sustainable market rules and a clear schedule for the opening of the internal and external market to competition, including delegation to SERC of the powers to enforce market rules;
9) legal changes;
10) discussion and adoption of a new energy law which would ensure a legal framework for the successful implementation of the above areas of the reform;
11) privatization;
12) transfer of ownership aiming to attract investments and to bring the management practice in line with up-to-date standards.

At the same time, it is important that the government adopt a set of principles based on which its energy policy should rest. The following are those principles:

a) introduction of market relations, based on cost-reflective tariffs and free contracting;
b) the active role of the state in the creation of a clear and stable legal and regulatory framework for investments, commercial activity and protection of public interests;
c) creation of a legal, regulatory and market environment prior to the implementation of new large-scale investment and privatization projects;
d) pro-active energy efficiency policy as a means for improving the competitiveness of the economy, security of energy supply and environmental protection;

e) efficient social protection through shifting government subsidies from the producer to the consumer, through energy efficiency measures and introduction of socially-oriented tariffs;

f) positioning of Bulgaria as a reliable country for the provision of future transit of oil, natural gas and electric power and as a dispatching and market centre in the region.

In the future, the Bulgarian's government energy policy will rely on the following two main energy sources: a) nuclear energy; and b) local lignite coal.

The Evolution of the Nuclear Power Sector in Bulgaria

Since 1956, the Bulgarian government decided to introduce the use of nuclear energy for electricity generation in the country, with the purpose to satisfy the foresee increase in the electricity demand. Ten years later, in 1966, an agreement was signed with the former Soviet Union for the construction of several nuclear power reactors providing the basis of the country's nuclear power programme. In the absence of Bulgarian own safety and regulatory bodies at that time, these functions was assigned to Soviet authorities using their own standards.

The plan included the construction of six nuclear power reactors. The first pair of nuclear power reactors was of the WWER-440/230 type. The second pair was WWER-440/230 model but with many of the much-improved safety features of the WWER-440/213 system. The third pair was the larger WWER-1000/320 model. All these units were part of the Kozloduy nuclear power plant, located close to the Danube River in the border with Romania. The construction of the first unit at Kozloduy nuclear power plant commenced on October 1969. The innovative approach used for the construction of the plant gives possibility to significantly decrease in the construction time of the Kozloduy nuclear power plant in complete compliance with the construction regulations.

Source: Photograph courtesy of Wikipedia (courtesy Ivailo Borisov).

Figure 81. Kozloduy nuclear power plant..

The capacities on Kozloduy nuclear power plant site were commissioned at three stages. The first stage was completed with the construction of Units 1 and 2. These units were connected to the electric grid in July 1974 and in September 1975 and entered into commercial operation in October 1974 and in November 1975 respectively. Both nuclear power reactors were shut down in December 2002. In parallel with the construction of Units 1 and 2, in October 1976, the construction of Units 3 and 4 started. These units were connected to the electric grid in December 1980 and May 1982 and entered into commercial operation in January 1981 and in June 1982 respectively. The construction of Units 3 and 4 represented the second stage of development of the Kozloduy nuclear power plant site. Both units were shut down in December 2006. The commissioning of Units 5 and 6, the third stage of development of the Kozloduy nuclear power plant site, was completed in December 1988 and 1993 respectively. At the end of the third stage, the Kozloduy nuclear power plant site comprises six PWRs of Soviet design. The total power net capacity installed was 3 538 MWe.

Until May 2008, four of these units were shut-down[74] and Units 5 and 6 were upgraded and modernized under a programme that was implemented until 2006. The purpose of this programme was to expand the operational live of these last two units, the only ones currently in operation in the Kozloduy nuclear power plant. The cost of the safety improvements extended to Units 5 and 6 were more than €120 million per year during the period 1998-2002.

After the commissioning of Unit 6, in December 1993, the imports of energy gradually decreased and exports increased. As a result of that process, after 1997 Bulgaria became one of the leading exporters of electric power in the region. At the same time, domestic demand was reduced and more efficient energy utilization was achieved. Units 5 and 6 have already received major updates to their original design, and Kozloduy nuclear power plants management sees no technical obstacles for their safe operation until 2010/2012, which coincides with the design lifetime of the units. It is important to note that during the period 1993-2003, the Kozloduy nuclear power plant has been providing between 40 to 47% of the average annual electricity produced in Bulgaria. The total production of electricity by nuclear energy in 2008 is shown in Table 34.

The foresee increase in the demand of electricity until 2020 and how this demand will be satisfied by the different energy source available in the country are shown in Figure 82.

According with figure 101, the new demand of electricity until 2020 will be satisfied by the construction of new nuclear and thermal power plants.

A second site for the construction of a new nuclear power plant with four or six large units was chosen in the area of Belene, also near the Danube River in the border with Romania. Site works started in 1980 and construction of the first WWER-1000/320 unit started in 1987. This was partly built (40%, with 80% delivery of equipment) but was aborted in 1991 due to lack of funds and pressure from environmental groups. The go-ahead for completing the nuclear power reactor at Belene followed a series of public discussions on the results of an environmental impact assessment and a feasibility study for the project. Those discussions indicated very strong political and public support for the project at the local and national levels. The project involves completing the partially built Unit 1 and building a second more advanced unit. The project has an estimated cost of €2 to €3 billion. (Kovan, 2005)

---

[74] The EC pledged €200 million to decommission all four of Kozloduy's WWER-440 reactors, if closed by the end of 2006.

Table 34. Annual electricity power production for 2008

Total power production (including nuclear)	Nuclear power production
44 787 GWhe	14 742 GWhe

Source: IAEA PRIS, 2009.

On April 8, 2005, the Bulgarian Council of Ministers approved the conclusion of the construction of two 1 000 MWe units with PWRs in Belene. These new nuclear power capacities guarantee not only the country's electricity supplies, but the future development of the Bulgarian nuclear power industry. At the same time, they will create significant social and economic benefits, thus providing new employment opportunities and securing the long term supply of cost-effective electric power.

The following were the main elements used by the Bulgarian authorities to support the development of the nuclear sector in the country in the future: 1) the nuclear electricity cost in the country is lower than the cost of other major energy alternatives except hydropower, including spent fuel and radioactive management costs and decommissioning costs; 2) low fuel cost make the cost of nuclear energy for the generation of electricity more stable, predictable and less sensitive to swings in fuel prices; 3) any productivity increases is translated directly into profit in deregulated electricity market, in contrast to previous market operations under regulation; 4) nuclear fuel has the advantage of being a highly concentrated source of energy, cheap and easily transportable, located in more politically stable regions; and 5) nuclear power emits virtually no greenhouse gases.

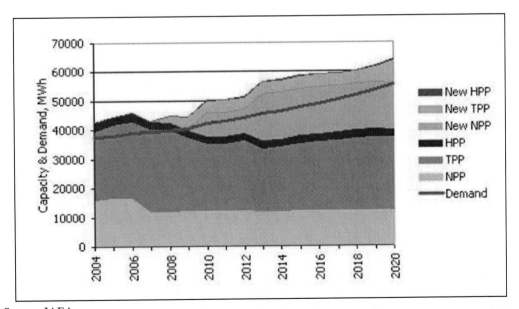

Source: IAEA.

Figure 82. Foresee increase in the demand of electricity until 2020 and how this demand will be satisfied.

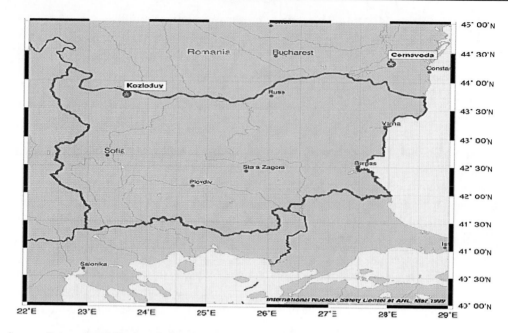

Source: International Nuclear Safety Center at Argonne National Laboratory, USA.

Figure 83. Location of the Kozloduy nuclear power plant.

The information about the Bulgarian nuclear power reactors currently in operation, shut down and under construction until December 2008 is shown in Tables 35-37.

Table 35. Nuclear power reactors currently in operation in Bulgaria

Reactors	Model	Net MWe	First power	Entry into operation	Date of closure
Kozloduy 5	V-320	953	1987	September 1988	2010/2012
Kozloduy 6	V-320	953	1991	December 1993	2010/2012
Total operating reactors: 2		1 906			

Sources. IAEA and CEA, 8th Edition, 2008.

Table 36. Nuclear power reactors shut down in Bulgaria

Reactors	Net MWe	First power	Entry into operation	Shut down
Kozloduy 1	408	1974	October 1974	December 2002
Kozloduy 2	408	1975	November 1975	December 2002
Kozloduy 3	408	1980	January 1981	December 2006
Kozloduy 4	408	1982	June 1982	December 2006
Total: 4	1 632			

Source: CEA, 8th Edition, 2008.

Table 37. Nuclear power reactor under construction

Reactors	Net MWe	Initial construction date
Belene 1	950	January 1987
Belene 2	950	March 1987
Total: 2	1 900	

Source: CEA, 8th Edition, 2008 and WANO data base.

Nuclear Safety Authority and the Licensing Process

The National Regulatory Authority in the field of safe and peaceful use of nuclear energy in Bulgaria is the Nuclear Regulatory Agency (NRA). The legal framework within NRA operates in the country is provided for in the Act on the Safe Use of Nuclear Energy (ASUNE) in force from July 2002. According to Article 4 (1) of the Act, "state regulation of the safe use of nuclear energy and ionizing radiation, the safety of radioactive waste management and the safety of spent fuel management is implemented by the Chairman of the Nuclear Regulatory Agency". Article 4 specifies: "the Chairman is an independent specialized authority of the executive power, shall be designated by a decision of the Council of Ministers, shall be appointed by the Prime Minister for a mandate of five years and may be selected for one more term of office (mandate)". The functions of the NRA are effectively separated from those of the bodies and organizations involved in promotion or use of nuclear technology in the country. Pursuant to Article 5 of the ASUNE, "the NRA shall have the following powers:

1) grant, amend, supplement, renew, suspend and revoke licenses and permits;
2) control the fulfillment of safety requirements and standards, as well as the conditions of licenses and permits granted;
3) issue and withdraw individual licenses;
4) undertake enforcement measures and impose administrative penalties;
5) assign (contract) nuclear safety and radiation protection related external expertise, researches and studies;
6) implement the interactions with other competent authorities of the executive power vested with some regulatory and control functions and propose to the Council of Ministers measures for co-ordination of the activities;
7) implement the international cooperation of the Republic of Bulgaria in the field;
8) provide individuals, legal persons and state agencies with objective information referring to nuclear safety and radiation protection;
9) submit annually to the Council of Ministers a report on the status of nuclear safety and radiation protection, as well as on the operation of the Agency;
10) organize and co-ordinate the drafting process and submit to the Council of Ministers the reports for implementation of country obligations under the Convention on

Nuclear Safety and the Joint Convention on the Safety of Spent Fuel Management and on the Safety of Radioactive Waste Management;

11) organize and co-ordinate implementation of the obligations of the Republic of Bulgaria arising from the Agreement Between the People's Republic of Bulgaria and the IAEA for the Application of the Safeguards in Connection with the Treaty on the Non-proliferation of Nuclear Weapons, as well as from the Additional Protocol to the Agreement;

12) perform the functions of a competent authority and a contact point for notification of an accident and for provision of assistance according to the Convention on Early Notification of a Nuclear Accident and the Convention on Assistance in the Case of a Nuclear Accident or Radiological Emergency;

13) develop and submit for adoption to the Council of Ministers regulations for the application of the ASUNE;

14) exercise other powers as may be entrusted thereto by the national legislation".

The main legal provisions for the licensing of nuclear installations in Bulgaria are outlined in the ASUNE. The Act specifies "the conditions, the order, terms and time limits for issuance of licenses and permits. The NRA Chairman based on a written application by the applicant shall issue licenses and permits for utilization of nuclear energy. According to ASUNE Article 20 (1), "a license shall be issued for a term of validity not exceeding ten years".

The adoption of a new nuclear law was an indispensable step in the process of enhances the nuclear sector in Bulgaria after the collapse of the Soviet Union and to ensure the safe development of the nuclear power sector. According to the ASUNE, "the NRA Chairman has no power to issue regulations". The NRA Chairman develops and submits regulations for the application of the Act to the Council of Ministers for adoption (Article 5, item 14 of the ASUNE), through the Deputy Prime Minister. The ASUNE specifies areas, which have to be regulated by the regulations and submitted to the Council of Ministers on a motion of the NRA Chairman or other state authority. After entry into force of ASUNE, the NRA initiated a large-scale legislative programme for development of a comprehensive set of regulations in the area of the safe use of nuclear energy and ionizing radiation, safe management of radioactive waste and spent nuclear fuel, physical protection, emergency planning and preparedness, etc.

The regulatory practice of the NRA has been build up and developed during a long time by taking into account of the legislative requirements, Agency's own experience and the good international practices. The NRA continuously improves its regulatory practice by conducting self-assessment, inviting well known international experts as management consultants, inviting independent external reviews, as well as by making use of the vast potential of the IAEA and the leading regulators of the world scale. [14]

Nuclear Fuel Cycle and Waste Management

The nuclear fuel cycle in Bulgaria includes uranium purchase, conversion, enrichment, fabrication, interim storage, spent fuel transportation, reprocessing and the disposal of high radioactive nuclear waste. These services are provided by Russia on the basis of an agreement adopted between Bulgaria and Russia and following long term commercial contracts for fuel supply and spent fuel reprocessing.

Bulgaria is now developing a new strategy for spent fuel management and radioactive waste. Two options of spent nuclear fuel management were analyzed: a) spent nuclear fuel reprocessing; and b) spent nuclear fuel disposal. The leveled fuel cycle cost was used as a criterion for decision making. It is important to note that the result of this analysis indicate that there is a small cost difference between the prompt reprocessing option compared with the long-term storage and direct disposal alternative. Based on best estimate data, the reference cases show a difference of less than 20% of the total nuclear fuel cycle cost taking into account all spent nuclear fuel accumulated at the Kozloduy nuclear power plant site, the cost of the direct disposal option being lower. In light of the underlying cost uncertainties, this small cost difference between the reprocessing and direct disposal options is considered to be insignificant, and in any event, represents a negligible difference in overall generating cost terms. It is likely that considerations of national energy strategy including reactor type, environmental impact, balance of payments and public acceptability will play a more important role in deciding a fuel cycle policy than the small economic difference identified. [14]

The new strategy target adopted by the Bulgarian authorities is to minimize the impact on the population and environment of the radioactive wastes and spent nuclear fuel stored at the Kozloduy nuclear power plant site, in pools situated near the reactors constructed in the 1990s. It is situated in a separate building on the territory of the Kozloduy nuclear power plant site, nearby Units 3 and 4. According to the design, the facility is to be filled in 10 years and the assemblies can be stored in it for a period of 30 years. After 3-5 years storage in the near reactor pools, the spent nuclear fuel is transported to the facility. The facility received its license operation in 2001.

During the operation of a nuclear power plant liquid and solid radioactive wastes are generated some of them very danger for the population and the environment. Compared to the unit of produced energy, the radioactive waste quantities generated by the operation of a nuclear power plant are over 10 000 times lower than the wastes produced by the operation of coal-fired power plant, but several time more danger. The generally accepted principles for radioactive waste management define that the radioactive waste produce by a nuclear power plant has to be collected, treated, conditioned and stored in a way that provides protection of human health and the environment now and in future with the purpose not to be a burden to the future generations.

For the management of the radioactive waste generated at Kozloduy nuclear power plant site, a facility was constructed for treatment, conditioning and storage of low and intermediate-level liquid and solid radioactive wastes. According with the IAEA data, in recent years the nuclear power plant generated annually an average of about 350 m^3 liquid radioactive waste, 200 m^3 conditioned solid radioactive waste and 20 m^3 low and intermediate ion exchange resins.

The commissioning of this nuclear facility gave a permanent solution of the issue for reliable storage of radioactive waste and is also a significant contribution to environment protection. As a result of the commissioning of this facility and the programme applied during the last years to minimize the radioactive wastes, the speed of treatment and conditioning of radioactive waste for long term storage has increased. The spent nuclear fuel is stored at the plant site under conditions which provide safety for the environment and population. After storage in special at reactor spent nuclear fuel pools, the fuel is removed to a specially constructed Wet Spent Fuel Storage Facility. The capacity of the facility allows storage of all spent nuclear fuel assemblies being discharged now and for the future years until the commissioning of a new facility. The conditions created for safe storage of spent nuclear fuel at the plant site, together with the fact that part of the fuel is transported for reprocessing and long term storage in Russia, provide a mid-term solution of the spent nuclear fuel safe management issue. [14]

With respect to the treatment of nuclear waste it is important to single the following: Iberdrola Engineering and Construction company from Spain has begun the works for the construction in Bulgaria of a plant for the treatment of radioactive waste using an innovative technology (plasma) that reduces the volume of the waste when it is subject of a temperatures up to 12 000° C. The Spanish company declared, in an official statement made in June 2009, that the company "will have 80% of the consortium formed with the participation of the Belgian company Belgoprocess, with great experience in the treatment of residuals, and has allocated around €30 million to initiate the operation of the joint company". The plant will be built in the site of the Kozloduy nuclear power plant and the construction works will be carried out in a period of four years. The main objective of the new facility is the reduction of the volume of radioactive waste storage at the Kozloduy nuclear power plant site, as well as to storage the waste generated during the decommissioning of the four units already shut down.

The Public Opinion

The Chernobyl nuclear accident made nuclear safety a sensitive political issue in Bulgaria. After this accident public opinion, now a much more significant factor for policy makers, had turned strongly against the nuclear industry. However, in the middle of the 1990s, people opinion regarding the use of nuclear energy for electricity generation in the country turnout to be more positive. Around 70% of the people have the opinion that Bulgaria should continue to use nuclear energy for the production of electricity while only11% are against. These recent data confirmed the outcome of previous studies carried out during 2004 and 2006.

A second nuclear power plant was started at Belene, to add six 1 000 MW power reactors by the end of the Tenth Five-Year Plan. But the construction of the Belene nuclear power plant was halted in 1989 by public opposition and disclosure that both Kozloduy and Belene were located in earthquake-prone regions. Long-term plans for the use of nuclear power for heat generation were also shelved at that time. In the case of the Belene nuclear power plant, project members of the Bulgarian Academy of Science organized a strong protest against the construction of this plant, based on the result of a scientific research study carried out in 1990 concerning the seismogenic nature of the specific area in which the plant is going to be built. During years the construction of the Belene nuclear power plant was halted. However, and despite the initial opposition to the construction of the Belene nuclear power plant, the Bulgarian government revived plans for building two units at the Belene site in 2004 and awarded a construction contract to the Russian company Atomstroyexport, which had put forward a bid in cooperation with the French/German Areva NP.

Since then the Bulgarian government has been appealing to French, German, Italian, Swiss and US banking groups, asking for investing and financial support to carry out the construction of the Belene nuclear power plant without success, due to the constant and persistent protest of the international antinuclear and environmental movements against the construction of the Belene nuclear power plant.

Summing up can be stated that there is some opposition to the construction of the Belene nuclear power plant not only in the Belene region itself, but also across the borders in Romania and Greece. In these two countries municipality leaders and civil organizations representing hundreds of thousands of citizens have declared their opposition to the project. Citizens from the F.Y.R. of Macedonia have also accused the Bulgarian's government to violate the Espoo Convention on cross-boundary impacts in force for both countries. However, it is important to single out that in the latest polls carried out in Bulgaria in 2007, the construction of the Belene nuclear power plant was supported by 73% of the people, while 13% were against.

In the case of the closure of four nuclear power reactors at the Kozloduy nuclear power plant it is important to stress the following. The decision adopted by the Bulgarian's government to shut down the nuclear power plant was the reason to a political struggle in Bulgaria. The Bulgarian Socialist Party (BSP), garnered 518,000 signatures to protest against the decommissioning, and a powerful nuclear engineering lobby have united to put pressure on the government to drop its commitment to the EU to decommission four reactors. (Van der Zwaan, 2006)

Pro-nuclear groups say that the commitment to close the Kozloduy nuclear power plant was adopted by the Bulgarian' government under pressure of the EC during the accession negotiations for the entry of the country in the EU. The closure of four units in Kozloduy nuclear power plant was adopted without considering the improvement in the safety of the nuclear power reactors carried by the Bulgarian's authorities in the 1990s, in close cooperation with the IAEA, WANO and the EU. The design of the nuclear power reactors was one of the main elements that were considered for the closure of the four units.

The first safety review was undertaken by the IAEA in 1991. Large-scale renovations of Units 1–4 at Kozloduy were implemented. The work was undertaken in close consultation with the IAEA, WANO and the EU to improve safety and bring the units closer to international norms. From 1998 to 2002, a more thorough modernization was undertaken in line with IAEA safety criteria to bring the units into conformity with current world standards. This was approved by the Bulgarian Nuclear Regulatory Agency, but only fully implemented on Units 3 and 4. An upgrade and modernization program for Units 5 and 6 was extended to 2006, but there is no great concern about the safety of these units, which conform to international standards. (Kovan, 2005)

Lastly, it is important to stress that in an opinion poll carried out in February 2004, around 46% of respondents said keeping the reactors was more important than joining the EU, while only 30% took the opposite view. According to the Alpha Research polling agency, 66% of the plant's supporters believed preserving it would keep electricity prices low and 53% felt the reactors would ensure Bulgaria remained strong in the international energy market. Some 41% opposed foreign experts making any decisions on the country's national interests. (Shkodrova, 2004)

Looking Forward

According with the decision adopted in October 2006 by the Bulgarian authorities, Russia firm Atomstroyexport was chosen to build the Belene nuclear power plant. The plant will comprise two 1 000-MWe AES-92 WWER third generation nuclear power reactors. The contract was signed in January 2008 for €3.9 billion, excluding the cost of the supply of the first cores. However, it is expected that the cost of the construction of the two units will be around €6- €7 billion. The new units will be similar to those being built by Atomstroyexport in China and India. The first unit is expected to be on line in 2014 or 2015. Instrumentation and control for the Belene nuclear power plant will be supplied by Areva.

The Minister of Economy and Energy of Bulgaria has announced that the German electric company RWE will contribute to the financing and the operation of the new nuclear power plant that the country has begun to build in Belene. RWE will acquire, for a value of US$1,753 million, 49% of the actions of the Belene Power Company (BPC). RWE has also offered loans for US$1,170 million for purchase of equipment and other supplies. The Belgian generating company Electrabel, property of the French GDF Suez, will acquire part of the actions bought by RWE, maintaining the condition that the 51% of the total of the actions of BPC are in hands of the Bulgarian State.

On the basis of the above information can be stated that the use of nuclear energy for the generation of electricity will continue to be one of the main type of energy included in the energy mix in Bulgaria in the coming years.

THE NUCLEAR POWER PROGRAMME IN SLOVENIA

Slovenia has rather limited energy reserves and, for this reason, should imports almost all its oil and gas to be used for the electricity generation and heating- essentially from Algeria and Russia. Lignite is its main source of energy of the country. The proven and recoverable reserves of low quality brown coal and lignite amount to 190 million tons. Oil reserves are very scarce and the proven oil reserves is of 850 000 tons. The annual oil exploitation is of 2 500 tons. The estimated hydro reserves of Slovenia are up to 9 TWh per year, out of which 3.5 TWh are already exploited. The country is connected to two gas pipelines from Algeria and Russia respectively. [123] According with WANO source, electricity consumption in Slovenia per capita is about 6 000 kWh per year.

Slovenia has one nuclear power plant in operation located in Krsko near the Croacian border, with a net power capacity of 696 MWe. The reactor is of the PWR type. The Krsko nuclear power plant produced, in 2007, around 5.9 TWh, which represents 42% of the total electricity produced by the country. In 2008, the nuclear share was 41.71%, a little bit less than in 2007. The net load factor of the nuclear power reactor in operation in the country, in 2007, was 93%. To satisfy the foresee increase in the electricity demand of the country in the coming years, the Slovenian's government is considering the construction of one additional nuclear power reactor with a net capacity of 1 000 MWe in 2017.

The evolution of the nuclear capacity in the country during the period 1970-2007 is shown in Figure 84.

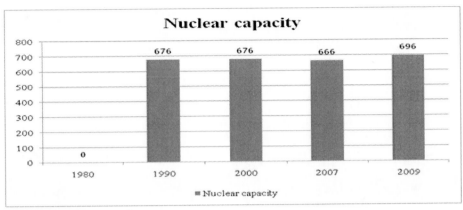

Source: CEA, 8th Edition, 2008.

Figure 84. Evolution of the nuclear power capacity in the country during the period 1970-2007.

In 2006, the participation of nuclear energy in the production of electricity was 40.3%. Two years later, in 2008, the production of electricity by nuclear energy sources in Slovenia increased up to 41.71%. The total production of electricity in Slovenia using nuclear energy, in 2007, is shown in Table 38.

The location of the Krsko nuclear power plant is shown in Figure 85.

Table 38. Annual electrical power production in Slovenia for 2008

Total power production (including nuclear)	Nuclear power production
14 316.6 GWhe	5 972 GWhe

Source: IAEA, WANO and CEA, 8th Edition, 2008.

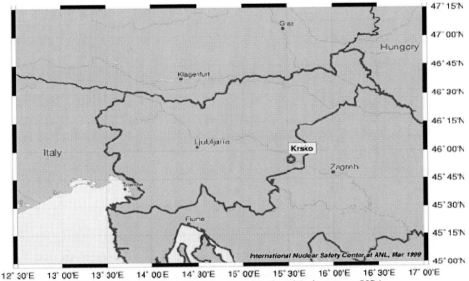

Source: International Nuclear Safety Center at Argonne National Laboratory, USA.

Figure 85. Location of the Krsko nuclear power plant..

Source: Photograph courtesy of Wikipedia (Sl-Ziga).

Figure 86. Krsko nuclear power plant.

Table 39. Information regarding the Krsko nuclear power plant

	Construction date	Criticality date	Connection to the electric grid date	Commercial date
Krsko	March 1975	September 1981	October 1981	January 1983

Source: CEA, 8th Edition, 2008.

Table 39 includes additional information regarding the Krsko nuclear power plant.

The Energy Policy in Slovenia

The government of Slovenia adopted, in January 1996, a resolution describing its energy policy objectives and main priorities for the development of its energy system in the country for the coming years. The resolution adopted was titled "Resolution on the Strategy of Energy Use and Supply of Slovenia". Three years later, in September 1999, the government adopted a new energy law to provide the legal framework for the development of the nuclear energy sector in the country. In 2003, the Slovenian government endorsed the National Energy Plan (NEP). The time horizon of NEP is 2000-2020. The adopted economic scenario foresees that Slovenia's will grow 4% annually. After 2009, it will stabilize around 2.2% per annum until 2020. According to the economic scenario described in NEP, as the main driving force for energy demand, total primary energy demand will grow in the period 2000-2020 by 1.1% annually. Electricity is foreseen to grow 1.5% annually. NEP sets the guidelines for the energy policy of the country taking into account the three main objectives of the energy sector: 1) security of energy supply; 2) competitive energy prices; and 3) sustainable development and mitigation of environmental impacts of energy generation and consumption.

The main goals of the Slovenian's Energy Policy could be summarized in the following activities:

a) to maintain the present availability of energy sources;
b) to further improvements of technical reliability of energy networks and quality of supply;
c) to implement measures for the efficient use of energy;
d) to match electricity at the level of final consumption with international standards;
e) to promote the opening of electricity and natural gas market;
f) to promote the efficient use of energy;
g) to increase the share of renewable sources of energy in the primary energy demand;
h) to maintain the long-term exploitation of lignite;
i) to keep the nuclear option for electricity generation[75];
j) to enable the 90 days reserves of liquid fuels.

The Evolution of the Nuclear Power Sector in Slovenia

The Krsko nuclear power plant was the first built in the former Yugoslavia. The plant is an important electrical power supply source for both Slovenia and Croatia.

The evolution of the introduction of nuclear energy for electricity generation in Slovenia is briefly described in the following paragraphs. In November 1973, US firm Westinghouse was selected to build Yugoslavia's first nuclear power plant at Krsko, in Slovenia, 30 miles West of Zagreb, Croatia. The construction works started in March 1975[76]. In September 1981, test production of electricity started and October the unit was connected to the electric grid. In January 1983, the Krsko nuclear power plant begins operation at full capacity. The net load factor of the Krsko nuclear power plant in 2007 was 90.3%. Assuming a normal 40-year life expectancy, the Krsko nuclear power plant can be expected to remain in operation until the year 2023. However, the government is considering extending its lifetime.

However, it is important to single out that the operation of the Krsko nuclear power plant was not exempt of difficulties. Several problems have been attributed to the operation of the plant. One of these problems was related with the issue of radioactive waste disposal. Radioactive fuel is now stored in cooling pools inside the plant, but this capacity was already exhausted. In addition, medium and low radioactive waste is being compressed and stored also at the site. To solve this problem, the spent nuclear fuel poll has recently been successfully re-racked to provide sufficient capacity for plant lifetime and even for possible lifetime extension. After re-racking the original 828 positions for spent fuel assemblies in the poll were increased to 1,694 positions. The discharge rate is around 36 fuel assemblies per fuel cycle. [123] In April 2005, the Krsko nuclear power plant shuts down twice: once automatically because of a capacity reduction during a test of turbine valves, a second time because of a minor glitch in the ventilation system of the non-nuclear portion of the plant.

---

[75] In 2004, a resolution was adopted by the government stating that the nuclear option must be kept open. It includes also a project of building a second nuclear power reactor at Krsko nuclear power plant as a realistic option in order to satisfy the foresee increase in the electricity demand until 2020.

[76] According to Yugoslav media, the country's industry built 68% of the Krsko nuclear power plant.

After the necessary inspections made by the Slovenia's competent authorities to ensure the safety operation of the plant, the nuclear power reactor was immediately reconnected to the electric grid.

Nuclear Safety Authority and the Licensing Process

The Slovenian nuclear authority is the Slovenian Nuclear Safety Administration (SNSA), which is part of the Ministry of Environment, Spatial Planning and Energy. The activities of SNSA cover five main areas: a) nuclear safety; b) radiological safety; c) nuclear and radioactive materials; d) inspection control; and e) legal and international cooperation.

The SNSA authority performs the following specialized technical and developmental administrative tasks:

a) nuclear and radiation safety;
b) radioactive waste management;
c) carrying out practices involving radiation and use of radiation sources, except in medicine or veterinary medicine;
d) protection of people and environment against ionizing radiation;
e) physical protection of nuclear materials and facilities;
f) non-proliferation of nuclear materials and safeguards;
g) import, export and transit of nuclear and radioactive materials and radioactive waste;
h) radiation monitoring;
i) liability for nuclear damage.

In accordance with the recommendations made by the IAEA, the SNSA should not promote the use of nuclear energy for the production of electricity in the country, limiting to its activities to act as an independent licensing office within the Ministry of Environment, Spatial Planning and Energy, which is the government agency in charge of the energy sector.

The major nuclear facilities supervised by the SNSA are the Krsko nuclear power plant, the TRIGA Mark II research reactor of 250 kW thermal power operates within the Reactor Centre of the Jozef Stefan Institute, the interim storage of low and medium radioactive waste at the Reactor Centre site operated by the Agency for Radioactive Waste Management and the closed uranium mine Zirovski Vrh.

The SNSA Division of Nuclear Safety deals with licenses and with analyses, which are used to support the licensing by performing and/or reviewing the safety analysis. The SNSA Division of Radiation Safety verifies radiation safety, except in the field of human and veterinary medicine, and is responsible for radiation dosimetry control and radiation monitoring. The SNSA Division of Nuclear and Radioactive Materials deals with trade, transport and treatment of such materials. In carrying out its activities the Division shares responsibility in the field of physical protection of nuclear power plants and nuclear materials with the Ministry of the Interior. It also deals with the treatment, temporary storage and disposal of radioactive waste and participates in the selection of sites for nuclear facilities, especially those destined for radioactive waste. Finally, it is responsible for safeguards and illicit trafficking issue. [123]

The SNSA Legal and International Cooperation Division are involved with licensing procedures and the preparation of legislation on nuclear and radiation safety, among other activities. During the implementation of its activities the Division should maintain close coordination and cooperation with international organizations such as the IAEA, OECD/NEA, EU, among others, and with foreign regulatory authorities of other countries in the field of nuclear and radiation safety, particularly with those countries within bilateral agreements in force as well.

The SNSA Division of Inspection Control supervises license-holders in fulfilling the safety requirements contained in the laws, regulations and in their licenses. Inspections may be done one at a time, or may form part of an overall plan of inspections. To increase their efficiency, inspections may be unannounced. Regular inspections in Krsko nuclear power plant are carried out on a weekly basis. [123]

The main national laws and regulations in nuclear power is the Act on Protection against Ionizing Radiation and Nuclear Safety adopted by the Parliament of Slovenia on 11 July 2002 and entered into force on 1 October 2002. The new act is adjusted to the EU legislation in the field of radiation and nuclear safety and to international agreements already ratified or signed by the Republic of Slovenia. The act includes the main principles in the field of nuclear and radiation safety and the provisions on:

a) practices involving ionizing radiation (reporting an intention, a permit to carry out practices involving radiation and a permit to use a radiation source);

b) protection of people against ionizing radiation (principles, justification, dose limits, protection of exposed workers and medical exposure);

c) radiation and nuclear safety (the classification of facilities, use of land, construction and carrying out of construction and mining activities, trial and actual operation of radiation and nuclear facilities, radioactive contamination, radioactive waste and spent fuel management, import, export and transit of nuclear and radioactive substances and radioactive waste and intervention measures);

d) issue, renewal, modification, withdrawal or expiry of a license;

e) physical protection of nuclear facilities and nuclear substances;

f) non-proliferation of nuclear weapons and safeguards;

g) monitoring radioactivity in the environment;

h) removal of the consequences of an emergency event;

i) report on protection against radiation and on nuclear safety, records containing information on radiation sources and practices involving radiation;

j) financing of protection against ionizing radiation and of nuclear safety (costs incurred by the users and public expenses) and compensation for the limited use of land due to a nuclear facility;

k) inspection, penal provisions and transitional and final provisions.

Nuclear Fuel Cycle and Waste Management

The Krško nuclear power plant is the main producer of all nuclear waste in Slovenia, particularly high radioactive nuclear waste. The contribution of other producers is relatively small. The nuclear waste produced by the nuclear power plant is stored at the Krško site and the facility used for this purpose is also operated by the nuclear power plant operator. With the waste volume reduction and improvements in waste treatment the facility, in which low and intermediate level waste is stored, still meets the requirements established by the government, but a long-term solution is needed. Similarly to the low and intermediate levels of waste, storage for spent nuclear fuel is located at the Krsko nuclear power plant and managed by the plant operator.

It is important to note that the disposal solutions for low and intermediate levels of waste or spent nuclear fuel are available neither in Slovenia nor in Croatia. In both countries, the site selection processes for the low and intermediate levels of waste repositories were initiated in early nineties and developed to different stages but so far none of the countries succeeded in confirming the site. Simultaneously with the site-selection process, the conceptual design of a repository for low and intermediate levels of waste is being prepared and the performance and safety assessment of the disposal facility is being developed. On the other hand, the final solution for spent nuclear fuel remains undefined. The only document treating the long-term management of spent nuclear fuel is "The Strategy for Long-term Spent Fuel Management", which was adopted by the Slovenian government in 1996. The strategy for long-term spent nuclear fuel management analyses different possibilities of long-term management and possible final solutions for spent nuclear fuel. The preparation of the strategy was strongly influenced by the small quantities of spent nuclear fuel generated in Slovenia, the expected phase-out of nuclear energy and at that time still unresolved question of ownership of the Krško nuclear power plant. [123]

There are two options under consideration for the management of the spent nuclear fuel of the Krsko nuclear power plant. One is reprocessing and the second is direct disposal. Between the two options the strategy proposes direct disposal. Taking into account that the nuclear power plant lifetime expires only in 2023, the decision on a final solution for the disposal for spent nuclear fuel need not be taken before the end of operation of the nuclear power plant lifetime. The strategy recommends adopting a decision no later than 2020. In accordance with the requirements of the decommissioning plan for the nuclear power plant adopted in 1996, the repository facility will be needed not before 2050.

The 1996, strategy for long-term management of used fuel recommends direct disposal of the spent nuclear fuel produced by the Krsko nuclear power plant, but leaves open the possibility of a later decision to reprocess or the export of the spent nuclear fuel to another country, or to accept a multinational solution of the spent nuclear fuel, particularly for assist countries with small nuclear power programmes in the management of its spent nuclear fuel.

The Public Opinion

A survey on the production of clean electricity in Slovenia was carried out by the agency Ninamedia between 14 and 15 April 2005. Around 300 persons were inquired. The results of the survey can be summarized as follows: If people had possibility to choice electrical energy from different energy sources the most of examinees (86%) would decide for natural sources, 5.5% would choose fossil energy sources and only 4% nuclear energy.

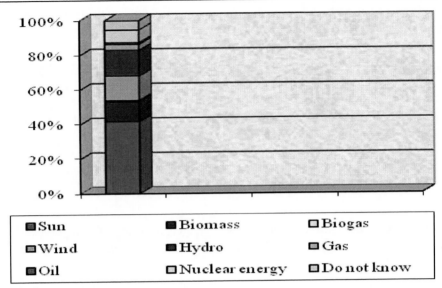

Source: Reference 127.

Figure 87. The participation of different types of energy in the energy balance in Slovenia.

Around 60.3% of examinees are prepared to pay a higher price for electrical energy from renewable energy sources. A quarter of them are willing to pay a price which is higher just for 2%; a total of 37% are willing to pay till 5% higher price, and 23.3% till 10%.

Most of examinees considering that the strongest trend of growth in Slovenia in the field of green electricity production in the next 5 years will have hydropower (28%), follow by solar energy (22%), wind energy (19.7%), biomass energy (13.0%), biogas energy (4.7%) and geothermal energy (2.7%).

The electrical energy source of the future is, for most of examinees, sun (41.7%), hydro (15%), wind (14%) and biomass (10%). Nuclear power is the energy source for the future for only 7.3% of examinees. [127]

Despite of the fact that there is no active strong opposition to the use of nuclear energy for electricity generation in the country, only a small part of the population, according with this survey, are thinking in the use of this type of energy for the production of electricity in the future. More than 70% supports the use of alternative energy source for the generation of electricity in Slovenia in the coming years.

Looking Forward

In Slovenia, a second nuclear power plant is under the consideration by the Slovenian authorities. However, before a decision is taken regarding the construction of a second nuclear power plant in the country, three issues should be clearly solved. These are the following: a) social acceptance of the use of nuclear energy for electricity generation; b) safe functioning of the nuclear power plants; and c) safe disposal of medium and low level radioactive waste.

At the same time, and taking into account the unresolved dispute about the ownership of the Krsko nuclear power plant, the Croatian parliamentary parties give consent to ratify a Croatian-Slovene agreement for the dismantlement of the Krsko nuclear power plant in the future. The plant, scheduled to operate until 2023, would undergo dismantlement procedures from 2017 until 2042. In March 2005, a joint Slovene-Croatian commission agrees on the establishment of a programme for the dismantlement of the Krsko nuclear power plant and the storage of low and medium radioactive waste. The Slovenian government has appropriated €115 million for the implementation of the programme.

THE NUCLEAR POWER PROGRAMME IN HUNGARY

According with the IAEA, the main energy sources in Hungary is hard coal, lignite, oil and gas. The total reserves of hard coal are estimated to be around 714 million of tons, but only around 100 million of tons are economically recoverable. Reserves of lower-rank coal are much larger. Coal mining has been in a critical situation for a long time, due to difficult geological conditions, poor heating value and low quality of the Hungarian lignites. As result of this situation, there is a declining annual production of coal. Hungary produced about 20-22 million tons of coal, including 5-6 million tons of poor quality lignite. However, the low average calorific value, the high sulphur contents and the high production costs limit both the extent of recovery and the specific field's utilization.

Hungary is a producer of crude oil. The present production of approximately 2 million tons per year covers less than a quarter of the national demand and is decreasing. Oil reserves are put at 58 million of tons. About 6 billion m^3 per year of natural gases are produced in Hungary, covering roughly half of the total demand. Gas production is also decreasing. Gas reserves are estimated to be about 113 billion m^3.

Hungary has four WWER units of Soviet design with a total net capacity of 1 826 MWe, and located in Paks nuclear power plant site, about 5 km south of the town of Paks, on the right bank of the Danube River. In 2007, the nuclear net electricity generation was 13.9 TWh representing 37% of the total electricity generation in the country that year, which was a little bit over 35TWh. In 2008, the nuclear share was 37.15%. The net load factor of the nuclear power reactors in operation in the country, in 2007, was 87.2%. Nuclear power provides electricity to the Hungarian's population at the lowest cost considering all energy options available in the country.

The Energy Policy and Strategic Goals in Hungary

The Hungarian authorities adopted, in 1994, a new energy policy with the purpose to take into account the new political and economical situation existent in the country after the collapse of the Soviet Union and the disappearances of the European socialist group. The main strategic goals of the Hungarian energy policy are the following:

Table 40. Annual electrical power production for 2008

Total power production (including nuclear)	Nuclear power production
37 598 GWhe	13 968 GWhe

Source: IAEA PRIS, 2009.

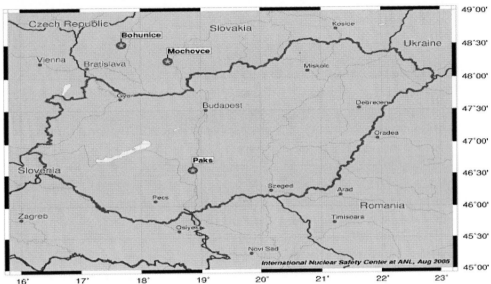

Source: International Nuclear Safety Center at Argonne National Laboratory, USA.

Figure 88. The location of the Paks nuclear power plant.

Source: Photograph courtesy of International Nuclear Safety Programme.

Figure 89. Paks nuclear power plant.

1) exemption of the unilateral import dependency of the country;
2) restraint of the environmental damages of energetic;
3) helping of thrift and more efficient consuming;
4) extension of social acceptance system;
5) flexible development system;
6) creating of the development sources (capital investment).

The major Hungarian energy policy directives approved by the parliament are the following:

1) maintaining and increasing energy supply stability;
2) increasing energy efficiency and the role of energy conservation thereby improving the competitiveness of the Hungarian economy;
3) establishing a market conforming organizational, economic and legal background;
4) enforcing environmental protection aspects;
5) promoting European integration in the energy sector.

Since its approval in 1993, certain parts of the directives and strategic objectives have already been implemented while others are currently under way. These objectives are the following: a) increase of storage capacity for oil; b) the construction of a new gas transmission pipeline between Austria and Hungary; c) the plan for building secondary reserve capacity for the connection of the Hungarian electrical grid with the electric energy system of Western Europe, among others.

The Evolution of the Nuclear Power Sector in Hungary

The use of nuclear energy started in Hungary in 1959 with the construction of the first Hungarian research reactor at Csilleberc on the outskirts of Budapest. The research reactor of Soviet design was refurbished by Hungarian experts after 30 years of operation, and put into operation again by the Nuclear Energy Research Institute in 1993.

In 1966, it was decided to build the first nuclear power plant in Hungary with four WWER-440/213 Soviet reactors model. The construction work started in August 1974 for the first two reactors and in October 1979 for the other two. Paks 1 was connected to the electric grid in December 1982 and entered into commercial operation in August 1983. Paks 2 was connected to the electric grid in September 1984 and entered into commercial operation in November 1984. Paks 3 was connected to the electric grid in September 1986 and entered into commercial operation in December of the same year. Paks 4 was connected to the electric grid in August 1987 and entered into commercial operation in November of the same year.

In addition to the four nuclear power reactors already in operation in Paks, two new units Paks 5 and 6 using WWER-1000 Soviet type reactors with a net capacity of 950 MWe each were planned but later cancelled due to decreased power demand in early 1990s. However, the feasibility of building these two nuclear power reactors is again under consideration by Paks and Atomstroyexport due to a foresee increase in the demand of electricity in the coming years. In 1998, Paks proposed building a further 600-700 MWe of capacity using a

Westinghouse AP600 model, or an AECL CANDU-6, or an Atomstroyexport/ Siemens WWER-640, but this proposal was rejected by the Hungarian national utility because it did not fit government policy at that time.

The design lifetime of the four nuclear power reactors operating n Hungary was initially planned for 30 years, but in November 2005 the Hungarian Parliament endorsed plans to extend the four units lifetimes by 20 years (to 2032-2037). In the past five years, extensive work has been done to prepare the case for extending the operating license. An expert team has prepared a detailed assessment of the plant status, the aging and lifetime prognosis of plant structures, systems and components, and defined the renovation that is needed. The assessment showed that the continuation of operation for another 20 years is feasible. Paks also plans to increase each unit's electrical power to 500 MWe. The four units of Paks nuclear power plant are equipped with all engineered safety systems that could be considered as similar to the Western PWRs system constructed in the same period, including confinement of special pressure suppression system. For this reason, can be stated that Paks nuclear power plant was the first of the former Soviet design reactors to be upgraded to meet modern safety standards. It was also the first nuclear power plant to undergo peer review by the WANO. (Kovan, 2005)

However, and despite of the safety measures adopted to improve the safety operation of the Paks nuclear power reactors, a nuclear incident occurred in April 2003 in the plant that force the Hungarian nuclear authorities, with the support of the IAEA, to introduce additional measures to increase the safety culture at all levels in the plant. What happens? During a fuel cleaning operation, some fuel elements in a cleaning tank were severely damaged when the fuel overheated. While the cause of the incident was soon established, assessments by the Hungarian safety authority and the IAEA pointed to some underlying problems in safety culture. A number of measures were adopted by the operator of the nuclear power plant with the purpose to strengthen the safety culture at all levels of the organization, from top management down.

In order to improve the safety of the Paks nuclear power reactors, between 1996 and 2002, a programme of safety improvements was initiated based largely on the results of the Advanced General and New Evaluation of Safety (AGNES) project, which reevaluated the safety of the plant according to Western standards using up-to-date assessment tools. AGNES revealed no dramatic new conditions to challenge the fundamental safety of the units. The subsequent US$300 million modernization programme, using the plant's own internal resources, included the following: a) renovation of the reactor protection system; b) improvements in seismic safety; c) replacement of instrumentation and control cable penetrations and cable using loss of- coolant accident–resistant; d) enhancement of fire safety; e) implementation of steam generator leak management measures; f) modification of the primary overpressure protection system; g) addition of emergency iodine filter systems; and h) implementation of measures to reduce human error potential.

Due to the conservative design of the Paks nuclear power reactors, there are several safety merits, which have been proved by the outstanding operational records of the plant. Nevertheless, a series of specific modifications were introduced in all units of the Paks nuclear power plant, with the purpose to increase by 20 MWe the nominal power capacity of the nuclear power reactors. The present total gross power capacity per unit is 470 MWe for Paks 2 and 3 and 500 MWe for Paks 1 and 4, increasing the total gross power capacity of the nuclear power plant up to 1 940 MWe (1 826 MWe net). Despite of the initial increase in the power capacity of the nuclear power reactors mentioned above, the Hungarian authorities has plans to increase even further the power capacity of Paks 2 and 3 to 500 MWe in the coming years. It is estimated that the increase in the power capacity of these two units will cost around US$900 million and will be carried out until 2030.

It is important to single out that despite of the intensive use of nuclear energy for electricity generation and for other peaceful purposes in the country, there are no nuclear suppliers in Hungary. The main components of the Paks nuclear power plant were made in Russia and in the Czech Republic.

Table 41 shows the evolution and the status of the introduction of the nuclear power programme in Hungary.

The evolution of the nuclear capacity of the Paks nuclear power plant during the period 1990-2007 is shown in Figure 90.

Table 41. Status of the Paks nuclear power plant

Station	Type	Net capacity MWe	Construction date	Grid connection date	Commercial date
Paks-1	WWER	470	01 Aug. 1974	28 Dec. 1982	10 Aug. 1983
Paks-2	WWER	443	01 Aug 1974	06 Sept. 1984	14 Nov. 1984
Paks-3	WWER	443	01 Oct. 1979	28 Sept. 1986	01 Dec. 1986
Paks-4	WWER	470	01 Oct. 1979	16 Aug. 1987	01 Nov. 1987
Total 4		1 826			

Sources: IAEA, PRIS.

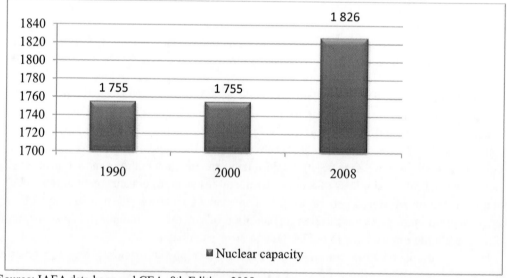

Source: IAEA data base and CEA, 8th Edition, 2008.

Figure 90. The evolution of the nuclear capacity in Hungary in the period 1990-2007.

Nuclear Safety Authority and the Licensing Process

It is important to note the following: There are two types of licensing procedures in force in Hungary. Nuclear safety licenses are required for nuclear facilities for the following phases: selection of the site, construction, commissioning, operation, modification and decommissioning of the nuclear facility. Nuclear safety licenses for nuclear equipment and devices are required for their fabrication, import, installation, commissioning and operation, modification or reconstruction, reparation and removal. The licensing of a nuclear facility is performed step by step. It means that the whole process of licensing consists of subsequent licensing procedures according to the mentioned phases. The sequence of phases is fixed and the licensing procedure of a phase may not be started before completing the previous one. The licensing procedures are similar for different nuclear facilities, but every type of nuclear facility requires a set of special directives. [43]

The Act on Atomic Energy defines the role of the Hungarian Atomic Energy Commission and the Hungarian Atomic Energy Authority (HAEA). HAEA is the nuclear safety regulatory body of the country and the main counterpart for IAEA. HAEA coordinates or performs the particular regulatory tasks necessary to ensure the safe application of nuclear energy. It is an independent administrative body operating through its Nuclear Safety Directorate, under the supervision of the government. The Authority's scope of responsibility covers not only nuclear safety licensing of nuclear facilities and of their systems, structures and components, but also safety assessment and review of Licensee's reports as well as inspection, including enforcement actions, like withdrawing operational license and it may impose penalties on the licensee.

Licensing of nuclear installations falls within the regulatory activities of the HAEA, and in this framework and strictly guided by the relevant legal regulations, the HAEA authorizes the licensee to perform certain activities. In all cases, this authorization shall be based on safety analysis. The guarantee of nuclear safety and the fulfillment of the regulations ensuring the safety and of the particular requirements during the licensed activity shall be demonstrated in the safety analysis.

In accordance with the Act on Atomic Energy, the tasks of the HAEA with regard to licensing of nuclear installations are as follows: a) "nuclear safety licensing. This is a prerequisite for establishing, installing, extending, commissioning, operating, modifying, and decommissioning a nuclear facility; b) regulatory licensing of (building) structures for nuclear facilities; c) nuclear safety and engineering radiation protection licensing of nuclear equipment for activities related to the design, manufacture, assembly (installation), commissioning, operation, modification (repair), importing, decommissioning; d) licensing of transportation of radioactive materials in accordance with the laws on transport of dangerous materials; e) approval of packaging of radioactive materials; and f) preliminary licensing of nuclear export and import.

For certain facilities, beyond the regulatory licensing procedure, the Act on Atomic Energy prescribes higher approval as well. In order to take over the ownership of an existing nuclear facility and/or to take possession of any title, preliminary government approval in principle is required. In order to commence preparatory activities with a view to establishing a new nuclear facility or a radioactive waste storage (disposal) facility, as well as to extend an existing nuclear power plant with any additional unit containing one or more nuclear reactors, Parliament's preliminary approval in principle is required.

Nuclear Fuel Cycle and Waste Management

There are three areas in Hungary where uranium are located, but only one of them, located in the Mecsek Mountain, has been exploited in the past. The uranium resources in Hungary are the following: 20 000 metric tons of exploitable uranium resources and 10 000 metric tons of additional reserves. The uranium ore was processed and converted to yellowcake at Mecsek and then shipped to Russia. However, in recent years Hungary has stopped the production of uranium ore.

Fuel cycle services were guaranteed by the former Soviet Union when Hungary purchased Soviet design reactors, including the fabrication of fuel assemblies, the shipping of the fabricated fuel assemblies to Hungary, and the return of spent nuclear fuel to the former Soviet Union. Hungary does not have other fuel cycle capabilities such as fuel conversion, enrichment, and fabrication. There are no reprocessing capabilities in Hungary, and no plans to develop any. Hungarian spent nuclear fuel has been reprocessed in Russia and the recovered plutonium does not have to be returned to Hungary. The government has at present no plans for recycling plutonium as fuel. In 1995, a new fuel management strategy was adopted in a attempt to diversify the purchasing of nuclear fuel. The new strategy includes the creation of the conditions for purchasing nuclear fuel from second supplier and the preparatory work for the use of a new type of Russian fuel assembly with profiled enrichment to allow change over to four year's fuel cycle. The first Russian produced profiled fuel, with a mean enrichment of 3.82%, was loaded into Unit 3 during 2000. [43]

With respect to spent nuclear fuel the Hungary policy can be divided in two period. Since the beginning of the operation of the Paks nuclear power plant, the Hungarian policy on spent nuclear fuel storage was to store the spent nuclear fuel produced by the operation of the four nuclear power reactors in pools at the Paks nuclear power plant site for five years and, after that period, transport the spent nuclear fuel to the former Soviet Union under an existing agreement. However, in spring 1992 the Russian Parliament passed an environmental law forbidding the import of foreign spent nuclear fuel affecting all transfer of spent nuclear fuel to the former Soviet Union, but despite of this resolution, Hungary was able to continuing sending a few spent nuclear fuel shipments up until February 1995.

Taking into account the situation with the shipment of the spent nuclear fuel to the former Soviet Union, the Hungarian Atomic Energy Commission issued a license on 4 February 1995 to the Hungarian Power Companies Ltd. for the construction of a spent nuclear fuel interim dry storage facility at the Paks site. The GEC Alsthom's modular dry storage technology was chosen. In an agreement with the local government authorities, Paks nuclear power plant guaranteed that no spent nuclear fuel would be placed in the dry storage facility as long as the Russians accepted the spent nuclear fuel. The dry storage facility is designed to hold spent nuclear fuel for 50 years. According to the Hungarian-Soviet Inter-Governmental Agreement on Co-operation on the construction of the Paks nuclear power plant, concluded on 28 December 1966, and the Protocol concluded on 1 April 1994 attached to such Agreement, the Soviet and now Russian party undertakes to accept spent nuclear fuel elements from the Paks nuclear power plant in such a manner that the radioactive waste materials and other by-products arising from the treatment of such fuel is not returned to Hungary. Until 1992, return of the spent nuclear fuel elements was conducted without problems, under conditions which were very favorable for Hungary but which, nevertheless, deviated from normal international practice. Following the collapse of the Soviet Union, however, this method of returning waste became less and less reliable. For this reason, and in the interests of ensuring operation of the Paks nuclear power plant, it became necessary to find an interim solution (50 years) for the storage of spent nuclear fuel units. [43]

With respect to the management of nuclear waste the situation in Hungary is the following: Radioactive waste is produced not only by the operation of the nuclear power reactors but also by medical and industrial nuclear activities. The strategy on low and intermediate nuclear waste disposal in is that these wastes should be buried in cemented form in steel drums in a shallow-ground disposal site, maintained for 600 years. Since 1986, low and intermediate nuclear waste from the Paks nuclear power plant has been stored at Paks site, due to public opposition to its continued burial at the existing disposal sites at Puspokszilagy and Ofalu.

In the past, activities related to the disposal of radioactive waste were conducted within the framework of the state budget for waste not originating from the power generation industry. As of 1 January 1998, the Atomic Energy Act established the Central Nuclear Financial Fund based on the payments of parties using nuclear energy. The goal of this Fund is to attend to the storage of radioactive waste, interim storage and final disposal of spent nuclear fuel and to finance the decommissioning (dismantling) of nuclear facilities. Pursuant to the Act on Atomic Energy, an authority designated by the government shall be responsible for attending to the duties related to the treatment of radioactive waste and the decommissioning of nuclear facilities. Based on the authorization of the government, the Public Agency for Radioactive Waste Management (PURAM) was established for the purpose to ensure the proper storage and disposal of radioactive waste and spent nuclear fuel elements. It is also responsible for activities related to the decommissioning of nuclear facilities, as well as for the operation of the Püspökszilágy Disposal Site (PDS) and the Interim Storage Facility for Spent Fuel (ISFS) located at the Paks nuclear power plant site. Its duties include preparation of the annual, intermediate and long-range plans for the Central Nuclear Financial Fund.

PURAM has undertaken geological investigations over a decade, focusing on a repository site in the south of the country. In mid 2005, the residents of Bataapati voted to approve construction of a repository for low and intermediate level wastes there and, in November of 2006, this was approved by the parliament. In December 2006, the government declared the Bataapati site an "investment of extraordinary significance", paving the way for accelerated licensing. [81] The preliminary approval of the Parliament confirms, that the construction of the repository for the final disposal of radioactive waste is serving the interest of the society, as the nuclear power plant is an existing facility, already in operation, thus the disposal of its radioactive waste has to be solved by the generation enjoying the advantages of the nuclear power plant. The licensing of the repository was facilitated by the Governmental Decree 257/2006. (XII.15) Korm. declaring that "the licensing procedure of the repository is of outstanding importance. In 2007, the repository received the license for building and water usage. After the public hearing of 29 March, 2007, the competent authority, the territorial Inspectorate for Environment, Nature and Water issued the environmental license, which – after an appeal – entered into force on 17 October 2007.

On the basis of the investment programme approved in May 2006 by the Minister disposing over the Central Nuclear Financial Fund the construction of the repository started. The competent authority, the South Transdanubian Regional Institute of the National Public Health and Medical Officer Service issued the construction license on 14 May 2008. In the first stage, by the fall of 2008 the most important surface parts of the National Radioactive Waste Repository were completed and the authority issued the license for their operation on 25 September 2008. Thus it became possible to accept a part of the radioactive waste from Paks nuclear power for predisposal storage. In the second stage, by 2011 all the surface facilities will be ready and below surface the first two disposal chambers for the disposal of wastes will be commissioned. The inauguration of the National Radioactive Waste Repository took place on 6 October 2008, and 80 drums with solid, low and intermediate level waste were transported in the repository from Paks nuclear power plant by December 2008.

The Public Opinion

It is important to stress that the public opinion in Hungary supports, in general, the use of nuclear energy for electricity generation, as we can infer from a vote that took place in the municipality of Bátaapáti, south of the Paks nuclear power plant in July 2005. According with this survey, the residents of the municipality voted overwhelmingly to approve the construction of the country's final repository for low and intermediate level radioactive waste in the area. The HAEA reported that with a voting turnout of 75%, residents voted 90.7% in favor of having the repository built at municipal plant in the south-central region of Hungary. Moreover, the results of the Eurobarometer on radioactive waste published in June 2005 show that Hungary is the country with the biggest share of citizens who support nuclear power (65%).

Looking Forward

In 1986, a preliminary decision was made by the Hungarian's government to expand the nuclear power programme by building two soviet PWRs units with a power capacity of 1 000 MWe each in Paks. Due to economic reason, a lower demand growth forecast and problems in providing the funds for such a large project, it was cancelled in 1989.

However, the possibility to build a second nuclear power plant in Hungary is again under serious consideration by the Hungarian's authorities. According with recent public information[77], the Hungary's Parliament paves the way to build new power reactors by approving, in principle, the construction of a new power reactor unit at the country's Paks nuclear power plant site. There were 330 votes in favor, six against and 10 abstentions. According to an explanation set out in the resolution approved by lawmakers, "preparations towards making the investment in the new unit will take at least five years to complete". The commissioning, planning and implementation of the construction project is expected to take a minimum of six years. Therefore, the overall period leading to construction will be at least 11 years. The explanation in the resolution stated that "some 73% of Hungarians support nuclear power generation in the country". Parliamentary approval was required before any discussions could begin on specific issues such as the reactor design. However, Hungary's prime minister Ferenc Gyurcsany said recently that, in his opinion, "about 2,000 MW of new nuclear generating capacity was needed at the Paks site". He suggested that "this could be a

---

[77] See NucNet on 31 March 2009 / News N° 33.

single 1 600 MW unit, or two 1 000 MW units". Experts will now review potential reactor designs and consider other aspects of construction, such as environmental and economic issues, before a final decision is taken to build a new nuclear power reactor. Due to the late closure of the four units in Paks[78], the Hungarian authorities have not yet adopted any decommissioning policy.

On a long-term basis, nuclear power and, to a limited extend domestic coal, are the only two energy alternatives for electricity supply in Hungary. Taking into account that there is no referendum or governmental or parliamentary decision to stop using nuclear energy for electricity generation in the future, and bearing in mind that the public opinion is, in general, in favor of continuing using nuclear energy for electricity generation, the future of the nuclear power programme in Hungary is assured, leaving the door open to the construction of new nuclear power reactors in the future, if necessary to satisfy any foresee increase in the energy demand.

THE NUCLEAR POWER PROGRAMME IN SWITZERLAND

Switzerland has a highly developed industrialized economy and one of the highest standards of living in the world. The main energy resources available are essentially hydropower. For this reason, the use of nuclear energy for electricity generation is an essential element of the energy balance of the country since 1969, when the first nuclear power reactor entered in operation.

The electricity consumption in Switzerland has been growing at about 2% per year since 1980. In 2004, electricity production was 65 billion kWh, matching demand. In 2006, the use of nuclear energy for electricity generation contributed 37.4% of Swiss total electricity generation (26.4 billion kWh). In 2007, the production of electricity using nuclear energy was 26.3 billion of kWh, representing 43% of the total electricity production in the country (around 61 TWh). Per capita electricity consumption in 2007 is 7 500 kWh per year, one of the highest in the world. In 2008, the nuclear share was 30.22%.

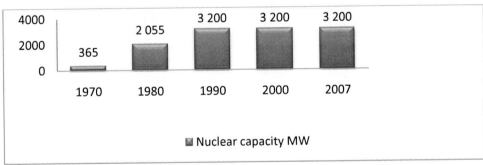

Source: CEA 8th Edition, 2008.

Figure 91. The evolution of the nuclear capacity in Switzerland during the period 1970-2007.

---

[78] The closure of the four units is foreseen during the period 2032-2037.

Switzerland has five nuclear power reactors in operation with a power capacity of 3 220 MWe. The net load factor of the nuclear power reactors in operation in the country in 2007 was, per type of reactor, the following: 92.7% for the BWRs and 95% in the case of the PWRs. The Swiss's government has propose to construct three new nuclear power reactors in the coming years with a total power capacity of 4 000 MW.

The evolution of the nuclear capacity in the period 1970-2007 in Switzerland is shown in Figure 91.

The Energy Policy and the Nuclear Law in Switzerland

Swiss's energy policy is focused on generating domestic electric power without combusting fossil fuels for already four decades. In 2007, around 57% of the electricity is generated in hydroelectric plants, which is possible due to the country's favorable topography; the remaining 43% are produced by the country's five nuclear power plants.

In 2003, a nuclear law was adopted by the Swiss's Parliament with the purpose to implement the country's energy policy. The Nuclear Energy Act of 2003, valid since first of February 2005, incorporates some important paragraphs for the future use of nuclear energy for the generation of electricity in the country. These paragraphs are the following: a) from mid 2006 to 2016 it is forbidden to reprocess spent nuclear fuel elements by a moratorium; b) all today's nuclear power plants got an unlimited license; c) to get a license for a new nuclear power plant a verification of documentation of feasibility of final waste disposal is needed; d) to get a license for a new nuclear power plant a facultative referendum is possible; and e) today's power plants have to verify the final waste disposal till 2015. (Homamnn and Streit, 2006)

One of the important components of the new energy policy adopted by the Swiss's government is the use of renewables and gas-fired power plants for the generation of electricity and an increase of energy efficiency, in addition to the use of nuclear energy and hydro for the same purpose. It is important to stress that, according to WANO sources, without new investment in the energy sector in the coming years a 20 billion kWh per year shortfall is predicted by 2020, representing a 25% of the future demand. This is due to phasing out of an electricity import arrangement from France, closure of the small Beznau and Muhleberg nuclear power reactors and the closure of a 355 MWe hydro plant, effectively removing 2 400 MWe. [95]

The Evolution of the Nuclear Power Sector in Switzerland

The evolution of the introduction of nuclear energy for the production of electricity in the country is briefly described in the following paragraphs. In 1957, the country's first research reactor, the 10 MW SAPHIR, started its operation and it ran until 1993. A second research reactor, DIORIT (30 MW) was designed and constructed indigenously. This second research reactor started operation in 1960 and it ran until 1977.

According with WANO sources, in 1960 the Swiss's government took over the research centre operating both reactors and, in 1988, this centre was transformed in the Paul Scherrer Institute - a flagship research centre in Switzerland. Construction of an experimental nuclear power reactor was commenced in 1962 at Lucens. This was a 30 MWt (7 MWe) heavy-water moderated gas-cooled unit located in an underground cavern. It started up in 1966 but experienced a core melt in 1969 and was written off since then.

At the same time, and according with the IAEA and CEA sources, the country's first nuclear power reactor built in Switzerland was Beznau 1, a Westinghouse PWR system with a net capacity of 365 MWe ordered by Nordostschweizerische Kraftwerke AG in 1965. The purpose of building this nuclear power reactor was to deal with the exhaustion of the hydroelectric potential, as well as to limit air pollution provoked by fossil fuel combustion for the generation of electricity. The construction work of Beznau 1 started in September 1965. The unit was connected to the electric grid in July 1969 and entered into commercial operation in September of the same year. In January 1968, a second unit with a net capacity of 365 MWe started its construction. Unit 2 was connected to the electric grid in October 1971 and entered into commercial operation in December of the same year. For the Mühleberg nuclear power plant, a General Electric BWR system with a net capacity of 355 MWe, was ordered in 1965 by BKW (Bernische Kraftwerke AG). Its construction started in March 1967. The unit was connected to the electric grid in July 1971 and entered into commercial operation in November 1972.

Following these three units, a consortium of utilities, Kernkraftwerk Gösgen (KKG), ordered a large PWR system from Siemens KWU for the Gösgen nuclear power plant with a net capacity of 970 MWe. The construction works started in December 1973. The nuclear power reactor was connected to the electric grid in January 1979 and entered into commercial operation in November of that year. Another utility consortium (KKL) ordered a similar-sized General Electric BWR system with a net capacity of 1 165 MWe for the Liebstadt nuclear power plant. The construction of the unit began in January 1974. It was connected to the electric grid in May 1984 and entered in commercial operation in December of the same year.

A further unit, with a net capacity of 950 MWe, was proposed to be built in Kaiseraugst near Basel, but this was abandoned following anti-nuclear opposition. The same occurred with the Graben nuclear power plant.

It is important to single out that all Swiss nuclear power plants have upgraded their power capacity. At the end of 2002, the nominal net powers were 365 MWe for Beznau, 355 MWe for Mühleberg, 970 MWe for Gösgen and 1 165 MWe for Leibstadt nuclear power plants. [130]

Table 42. Nuclear power reactors in operation in Switzerland

Reactors	Type	Net MWe	First power	Expected closure (approx)
Beznau 1	PWR	365	1969	2019
Beznau 2	PWR	365	1971	2021
Gösgen	PWR	970	1979	2029
Mühleberg	BWR	355	1971	2022
Liebstadt	BWR	1 165	1984	2034
Total (5)		3 220		

Source: CEA, 8th Edition, 2008.

Source: (Homamnn and Streit, 2006)

Figure 92. Location of the nuclear power plants in Switzerland.

In the other hand, it is important to note the following. Initially, the costs of the closing of the Swiss nuclear power plant were included in the costs of each reactor, but a different system that extend the obligations to a five year period after the definitive stop of the reactors was adopted by the government. Later on the costs of the closure and the cost of the waste management were added.

According to a study carried out in 2006 and presented recently by the Swiss Nuclear Forum, the total costs associated to the closure of the five Swiss nuclear power reactors currently in operation would be around US$1,269 million (2009), included the costs of the five years period after the stop of the operation of all units, the transfer of the spent nuclear fuel to the pools located in the nuclear power plant site, the supervision of the cooling of the reactor, the cost of the radiological protection and of the personnel in charge of the closure of the nuclear power plant, among others. On the other hand, it is important to know that the costs of the long-term closing should be paid annually by the electric companies to a national fund established in 1984. The costs of the waste management will be covered by another fund established in 2001. The study estimates in US$11,710 million (2009) the total costs of closing of the nuclear power reactors and waste management.

The president of the Swiss Nuclear Forum has said that the electric companies hope to operate their nuclear power reactors during 40 years. Nevertheless, the operator of the second older reactor at Mühleberg has requested a unlimited operation license when their current license ends up in 2012.

The CO_2 Law and the Nuclear Energy Act

The CO_2 Law is the central pillar of Swiss climate policy. The law entered into force on 1 May, 2000. The main objective of this law is "to reduce, by 2010, the emission of CO_2 to the atmosphere by 10% below the 1990 level, with the purpose to allow Switzerland to fulfill its international obligations under the International Climate Convention". To achieve this goals is indispensable to continue using nuclear energy for electricity generation in the country during the coming years.

After a two-year consultation phase, a new Nuclear Energy Act was adopted by the Swiss's Parliament in March 2003. The new Act started to be implemented by the Federal Council at the beginning of 2005. In order to fulfill its international obligations with the International Climate Convention, the Swiss government has made nuclear power a key element in its future energy plans. The new Nuclear Energy Act includes the following main points:

1) "the nuclear energy option is left open. In other words the government left open the possibility to build new nuclear power plants, and the possibility of carrying out a referendum on the construction of new nuclear installations;

2) intensified say of the site cantons and neighbor cantons as well as of the neighbor states in the preparation of a general license decision for new nuclear installations;

3) the introduction of a 10 year moratorium on the export of nuclear fuel for reprocessing covering the period 2006-2016;

4) the abstention from a legal time limit on the operation period of nuclear power plants.

5) provisions on the decommissioning of nuclear installations;

6) a concept of monitored long-term geological disposal of high-level radioactive waste (combines elements of final disposal and reversibility);

7) a stronger role of the government in managing disposal issues;

8) a funding system for decommissioning and waste management costs, including solidarity of operators of nuclear power plants;

9) the coordination of licensing procedures (pooling of all Federal and Cantonal licenses in one single license);

10) the general possibility to appeal against license decisions".

According with Swiss authorities, "the new law focus on energy efficiency, the increase use of renewable sources and large gas-fired power plants as well as nuclear power". The energy minister described energy-saving measures[79] "as the key to securing Switzerland's long-term supply of energy. Energy efficiency saves money, improves the efficiency of the economy and act as an incentive for investing in new technologies". Under the new strategy, the government also places considerable emphasis on the use of renewable energy for electricity generation, such as hydropower - Switzerland's most important domestic source of electricity.

With the new Nuclear Energy Act two major obstacles have to be overcome. The possibility of final waste disposal inside Switzerland has to be verified and during licensing process a facultative referendum is possible and in all probability. To be able to get a license for a new nuclear power plant one need the support from Swiss public. Therefore, the nuclear lobby has to do a lot of public relation work in the next years. (Homamnn and Streit, 2006)

---

[79] Swissinfo, dated 21/2/2007.

Nuclear Safety Authority and the Licensing Process

Nuclear safety is one of the major concerns of the Swiss's government. For this reason, a very strict licensing process is in place for the construction of a nuclear facility in the country. In this process the population of the site in which the nuclear facility will be built, has a important role to play. In the following paragraphs a brief description of the licensing process in force in Switzerland is included.

The construction and operation of nuclear facilities and any changes in their purpose, nature or size, require a general license prior to the granting of technical licenses. The general license determines the site and the main features of the project. It is essentially a political decision. The application for a general license must be accompanied by: a) a concept for the decommissioning of the installation or for the monitoring and the closure of the deep-geological depository; b) the demonstration of feasibility of disposal of the radioactive waste produced in the new nuclear installation; and c) the demonstration of the suitability of the site for deep-geological depositories.

After received the application for the construction of a new nuclear power plant or nuclear facility, the Federal Council transmits the application for consultation to the cantons, federal authorities and neighbor countries concerned. It also arranges for various expert reports to be prepared, mainly by the Swiss Federal Nuclear Safety Inspectorate. The application, the statements and experts' reports are made available to the public along with any supporting documents. Anyone may then submit written objections to the Federal Chancellery concerning the granting of the general license. The site canton, neighboring cantons and countries enjoy extended participation rights, as they must be involved in the general license granting procedure. Their concerns need to be considered as far as they do not unproportionally restrict the project. Finally, after having examined the application, the opinions given during the consultations, the experts' reports and any objection made, the Federal Council reaches a positive or negative decision; the granting of a general license must also be approved by the Federal Assembly. A referendum can be held against the approval by the Federal Assembly; 50 000 voters can demand a public vote on the project. If the Swiss electorate ratifies the project, the application for a construction license may be submitted. [96]

It is important to stress the technical character of the application for the construction, operation, modification or decommissioning of a nuclear installation, as well as licenses for geological investigations with regard to the construction of a deep-geological depository. The reason is that the main requirements are related to nuclear safety. The new provision is that all other procedures for non-nuclear licenses necessary for the realization of the project, will be integrated in the same procedure (e.g. cantonal licenses concerning construction and land use planning, protection of workers, etc.). Thus, there will be only one single license granted by the Federal Department of Environment, Transport, Energy and Communication. The expropriation procedure will also be partly integrated in this procedure. The application for a license for constructing, operating or modifying a nuclear installation must be particularly accompanied by a technical report (safety analysis report). All further documentation must be submitted according to the respective non-nuclear laws. [96]

One of the specific characteristics of the licensing process in Switzerland is that the all documents necessary for the license of a new nuclear installations is published for public consultation. All parties involved in the licensing process, according to the administrative procedure in force, have the possibility to appeal any decision adopted by the competent authorities. Also the canton where the installation is to be located will be consulted. If the canton rejects the application and if the Federal Department will nevertheless grant the license, the canton can appeal against this decision. The Federal Department decides on the application and on the appeals. There is a possibility to appeal against this decision at the Appeals Commission of the Federal Department for Environment, Transport, Energy and Communication (Federal Administrative Court). The Appeals Commission's decision can be appealed at the Federal Court. It is important to stress that there are no other countries in the world with a similar licensing process in which the population living in the future site of a nuclear installations have an important role to play in the process of approving a license for the construction and operation of a nuclear installation.

Nuclear Fuel Cycle and Waste Management

According to Swiss law, radioactive waste generated in Switzerland due to the operation of its nuclear power plants, or by other facilities related with the use of nuclear energy for peaceful purposes, has to be disposed off domestically, although exceptions may be granted by the government. All radioactive waste has to undergo geological disposal. The generators of radioactive waste, namely, the operators of the nuclear power plants and the federal state for the radioactive waste from medicine, industry and research, are responsible for the management of the nuclear waste, including disposal. No disposal facility is yet in operation in the country, thus all radioactive waste is kept in storage facilities. Each nuclear power plant has sufficient storage capacity for its own nuclear waste. Radioactive waste from medicine, industry and research is stored at the Federal Storage Facility operated by the research institute PSI.

It is important to single out that there is no national policy regarding reprocessing or direct disposal of spent nuclear fuel associated with the operation of the nuclear power plants. For this reason, utilities have been sending it for reprocessing it abroad so as to utilize the separated plutonium in Mixed Oxide (MOX) fuel. Reprocessing is undertaken by Areva, at La Hague in France, and by BNFL, at Sellafield in the UK, under contract to individual power plant operators. Switzerland remains responsible for the separated high-level wastes which are returned. About 1 000 tons of used fuel has been so far sent abroad for reprocessing, but the 2005 Nuclear Energy Act halted this for a ten years period from mid 2006. Used fuel is now retained at the reactors site or sent to Zwilag ZZL for interim above-ground storage, being managed as high-level waste. [95]

In 1972, the National Cooperative for Disposal of Radioactive Wastes (NAGRA) was established, involving power plant operators and the federal government. NAGRA submitted a feasibility report to the Swiss government in 2002 related with the disposal of spent nuclear fuel. The report showed that "used fuel elements, separated high-level waste and long-lived intermediate-level waste could be safely disposed of in Switzerland". In June 2006, the Federal Council concluded that "the legally required demonstration of disposal feasibility for all these had been successfully provided". Meanwhile, the 2005 Nuclear Energy Act required "the waste management and disposal programme to proceed and be reviewed by the federal authorities". Identification of site options for disposal is proceeding under this Act and the Spatial Planning Act with regional participation. Target date for repository operation is 2020.

It is important to stress that due to the opposition of the operation of the low and intermediate-level waste repository facility at Wellenberg, the activities in the mentioned site was blocked by a two cantonal referendums carried out in 1995 and in 2002.

At the end of 2006, the volume of packaged low and intermediate-level waste was 6 830 cubic meters. Added to this are the high-level waste and used fuel stored at the power plants and at Zwilag ZZL. In December 2006, there were eight containers with separated high-level waste from reprocessing and 17 containers with used fuel stored at Zwilag. Total costs of radioactive waste management are estimated at CHF11.9 billion. Nuclear plant owners have paid CHF8.2 billion towards final waste management and now pay into a national waste disposal fund created in 2000, which held CHF2.76 billion at the end of 2005. [95]

Regarding decommissioning policy, the Swiss's government adopted a decision to establish a Decommissioning Fund to cover all decommissioning activities that should be carried out on all nuclear power reactors currently in operation at the end of their lifetime. The fund was established in 1984 and power plant operators pay annual contributions to this fund. At end of 2005, it held over CHF1.25 billion, with projected requirement being CHF1.9 billion. The Decommissioning Fund and the Waste Management programme are funded under the Nuclear Energy Act by a fee of about CHF1 cent/kWh on nuclear power production. The resources available, at the end of 2006, reached CHF4.3 billion.

The Public Opinion

The nuclear controversy began in Switzerland in 1969 with the first signs of local opposition to the construction of a nuclear power plant at Kaiseraugst, near Basel. For 20 years, the Kaiseraugst project was the center of the nuclear controversy. The controversy involved site permit, local referenda, legal battles, site occupation by opponents in 1975, parliamentary vote in favor of construction in 1985 and, finally, parliamentary decision in 1989 to end the project definitively.

The Chernobyl nuclear accident had drastically affected the political climate in Switzerland with respect the use of nuclear energy for electricity generation. As a consequence of this nuclear accident and due to a strong public opposition the proposal to construct a nuclear power plant in Graben was also cancelled.

The controversy over the construction of new nuclear power plants led to several anti-nuclear initiatives at the federal level. These initiatives were the following:

a) an attempt to prohibit the operation of all nuclear power plants, as well as the prohibition to construct new ones was rejected by 51.2% of the voters in a referendum carried out in February 1979;

b) an attempt to prohibit the construction of new nuclear power plants, leaving untouched all nuclear power plants currently in operation was also rejected;

c) two initiatives differing only in the treatment to be applied to Leibstadt, then under construction, was rejected by 55% of the voters in a referendum carried out in September 1984;

d) a proposal to phase-out all nuclear power plants was rejected by 52.9% of the voters in a referendum carried out in September 1990;

e) a proposal to 10 year moratorium was accepted by 54.6% of the voters in a referendum carried out in September 1990;

f) in 1999, two new initiatives were presented one to ban the construction of new nuclear power plants until 2010 and the second to close all nuclear power plants after a 30 year live-span were rejected in two referendums carried out in May 2003 by 58.4% and by 66.3 % of the voters respectively.

Looking Forward

After the rejection of all initiatives to abandon the use of nuclear energy for electricity generation in the country in 2006, the Swiss authorities started looking for partners to build a large nuclear power plant using proven technology and probably at an existing nuclear plant site, preferable Beznau, were other two nuclear power reactors are currently operating.

In 2007, a joint venture was formed for the construction of two identical nuclear power reactors with a power capacity of 1 600 MWe each at Beznau and Mühleberg sites. It is important to note that in the case of the Mühleberg nuclear power plant personalities of the economic and industrial world of the cantons of Berne, Friburgo, Swears and Neuchâtel created, in May 2009, the "Forum Pro Mühleberg" to support the continuation of the operation of the nuclear power plant in Mühleberg, because the electricity produced by the plant is indispensable for the region. The closure of the Mühleberg nuclear power plant will stops the development of the whole region bearing in mind the growing increase of the electricity consumption foresee in the coming years and the population's favorable opinion toward the continuation of the operation of the nuclear power plant. The population in Mühleberg supports the construction of a new nuclear power plant in this site in the future.

Early in 2008, an announcement was made for a license to build a new nuclear power reactor at Gösgen with the purpose to satisfy, at least partially, a foresee shortage of generation of electricity by 2020. In opinion of some experts, this foresees shortage can only be partly met by the use of nuclear energy. The foresee gap should be met by the use of conventional technologies and imports.

To increase the contribution of nuclear energy to the production of electricity in the country and to improve the safety of the nuclear power reactors to be in operation in the future, the Swiss government announced, early in 2007, that the existing five nuclear power reactors will be replaced in due course with new units. The costs of decommissioning the existing nuclear power plants are estimated at CHF1.5 billion.

Summing up can be stated that the use of nuclear energy for the production of electricity will be part of any future energy plan to be adopted by the Swiss's government, not only to satisfy any foresee increase in the electricity demand in the future, but to fulfill the commitments assumed by the Swiss's government regarding the reduction of the emission of CO_2 to the atmosphere within the Kyoto's Protocol.

THE NUCLEAR POWER PROGRAMME IN THE CZECH REPUBLIC

Electricity consumption in the Czech Republic has been growing since 1994. In 2005, a total of 82 TWh was generated, more than half from coal, which is still the main energy source in final energy consumption in the country. In 2007, this production was almost the same (81TWh). Petroleum products due to road transport increase about 19%. The share of natural gas is almost the same in recent years, thanks to its use for households and heat production as required by current environmental legislation. In 2002, nuclear power provided 24.5% of the total electricity production in the country. In 2005, the production was 23.3 TWh, which represent 31% of the net total of electricity produced by the Czech Republic. In 2006, this share was 31.5% almost similar to 2005 but, in 2007, went down to 30.3% (20.1 TWh). In 2008, nuclear share was 32.45% a little bit higher than in 2007. In 2005, the country exported 16 TWh mostly to Germany and Austria. Electricity per capita consumption in the Czech Republic is about 5 300 kWh per year.

The Czech Republic has six nuclear power reactors in operation with a net capacity of 3 472 MW. The net load factor of the nuclear power reactors in operation in the country in 2007 was 78.7%. The Czech government has in plan the construction of two new nuclear power reactors in the near future with a total power capacity of 3 400 MWe.

In 2001, the structure of total electricity generation in the country was the following: fossil fuel (mostly coal) power plants provided 76.9% of total electricity generation, Dukovany nuclear power plant 19.8% and hydro power plants 3.3%. In 2006, the structure was the following: fossil fuels power plants provided 72,6% of total electricity generation; nuclear power plants provided 20,4% and hydro only 0,7%. Two research reactors are operated by the Rez Nuclear Research Institute (one is a 10 MW unit) and another by the Czech Technical University in Prague. The main one was originally a Russian design but has been extensively rebuilt; the other was designed locally. [79]

The Nuclear Energy Policy in the Czech Republic

The current energy policy of the Czech Republic with a horizon of 15 to 20 years was approved by the Czech government in January 2000. Compliance with the energy policy adopted is evaluated by the Ministry of Industry and Trade within intervals not longer than two years; the Ministry informs the government on the evaluation results and proposes eventual modifications. The key strategic targets of the energy policy include the determination of the basic conception of long-term development of the energy sector and determination of the essential legislative and economic environment, which would encourage electricity generators and distributors to prefer environment-friendly behavior.

The energy policy adopted by the Czech government is a basic document indicating the targets in the area of energy management according to the needs of economic and social development, including environmental protection. The long-term strategic targets of energy policy include a gradual reduction of the volumes of energy and raw materials needed by the Czech economy to meet the level of advanced industrial countries. The new sub-objectives up to 2020 on the demand side are: a) to remove price subsidies and distortions; b) to create competitive markets for electricity and gas; c) to achieve freedom of choice for consumers;

and d) to ensure energy efficiency enhancement. At present, the government energy policy is focusing on harmonizing the Czech energy sector standards with those in the EU. According with IAEA information, the main changes and priorities of the current energy policy in the Czech Republic are: a) increasing energy efficiency of energy use; b) protection of environment; c) respect of the principles of sustainable development; d) reliable and safe insurance of energy; e) economic competitiveness; f) rational introduction of new energy resources and identification of the role of domestic resources; g) finalization of energy sector restructuring and privatization; and h) energy policy based on identical goals as the energy policy of the EU. [22]

On the consumption side, the long-term strategic targets of the energy policy include a gradual reduction of the volumes of energy and raw materials needed by the economy to the level of advanced industrialized countries. This target should be achieved, in particular, by a support to new production technologies with minimum need for energy and raw materials and with maximum utilization of the energy and raw materials through national work. In the tertiary sphere, the need for energy should be reduced, mainly, through support to programmes leading to energy savings and to greater utilization of alternative energy and raw material sources in supplying the population with energy.

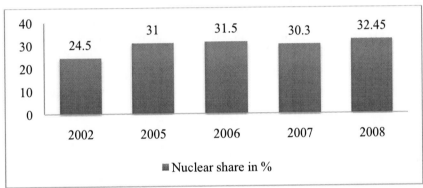

Source: CEA, 8th Edition, 2008.

Figure 93. Evolution of the nuclear share during the period 2002-2007.

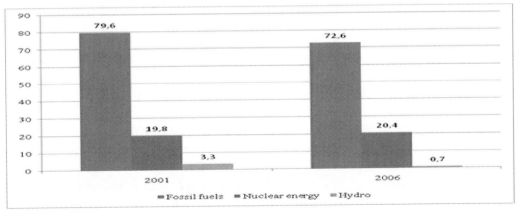

Source: IEA, IAEA.

Figure 94. Structure of the total generation of electricity in the country.

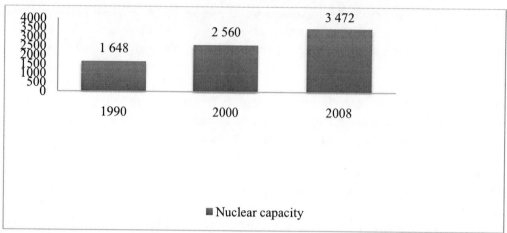

Source: CEA, 8th Edition, 2008.

Figure 95. The evolution of the nuclear capacity of the country during the period 1990-2008.

The main open issues of the energy sector, including the proposed solutions, are: a) adjustment of the prices of energy commodities and services and the tariff structure relating thereto (future development of the prices of electricity, natural gas and centralized heating); b) procedure of the privatization of the state ownership interests in the key energy companies while maintaining a reasonable state influence; c) laying down the rules for the development of the internal electricity and gas market (on the basis of EU Directives); d) creating a well-functioning, non-discriminatory, transparent and motivating system of support to possible energy savings, use of renewable resources and electricity and heat cogeneration units; e) nuclear power. [22]

Environmental protection in the energy area has been mostly focused on the removal of environmental damage, especially the damage caused by extremely high emissions of pollutants discharged into the air. Successive implementation, in the coal-fired power plants, of de-sculpturing and de-nitrifying projects, as well as installation of equipment to separate dust. It is important to stress that due to specific environmental measures adopted by the Czech government, today the country power-generation meet the limits defined in the Clean Air Act (Act 309/1991). Within the context of the efforts to alleviate the changes in the global climate, the Czech Republic should reduce the emissions of greenhouse gases by 8% before 2008-2012, compared with the level of the year 1990.

The Evolution of the Nuclear Power Sector in the Czech Republic

In 1958, due to the lack of oil reserves, the former Czechoslovakian's government started building its first nuclear power plant, a gas-cooled heavy water reactor at Bohunice (now in Slovakia).

It is important to note that the Czech company CEZ has communicated that the power of Unit 3 of the Dukovany nuclear power plant have been increase up to 500 MW after the modernization of the original design of this reactor. Unit 3 is the first of the four nuclear power reactors operating in the Dukovany nuclear power plant to reach that power. When the modernization process is extended to the other three units, the plant will reach a power of almost 2 000 MW. All the Czech nuclear power reactors had increased their original power of 440 MW to 456 MW between 2005 and 2008 throughout the substitution of the low pressure turbines for other more efficient. A new modernization process will includes new fuel to be used in the operation of the plant, improvement of the instrumentation and control systems, the substitution of the high pressure turbines and some modifications of the exit transformers. All the improvements will be carried out in Dukovany from 2013 and will allow a further increase in the production of electricity of the nuclear power plant up to 16 TWh.

The construction of the Temelin nuclear power plant with two nuclear power reactors with a net capacity of 912 MWe stared in February 1982. However, due to the collapse of the former Soviet Union, the division of the former Czechoslovakia, the introduction of several changes in the initial design of the nuclear power reactors, among other elements, slows down the progress of the work at Temelin nuclear power plant. In addition of what has been said before, some experts believe that the delay in the construction of the Temelin nuclear power plant was due also to the difficulties in installing a new digital instrumentation and control system, which called for complete recabling of the plant. Due to these changes, the cost escalated to nearly US$3 billion. Westinghouse was awarded contracts for the new instrumentation and control systems, the supply of nuclear fuel and associated components, the diagnostic and monitoring system and the radiation system. (Kovan, 2005)

In the other hand, an extensive modernization[80] of the Dukovany nuclear power plant started in 1995 and its completion is scheduled in 2010. The cost of the modernization programme is about US$425 million. Besides improving its competitiveness and meeting EU safety standards, the aim of the upgrade is to extend the Dukovany nuclear power plant operating license from 30 to 40 years. In 2002, finalized the introduction of almost 40 modifications of equipment increasing the nuclear safety of the Dukovany nuclear power plant, following a group of recommendations made by the IAEA, with the purpose to increase the safety of the plant to international standards. The investment made was 850 million CZK. One of the most important actions carried out was the renewal of supervision and control system equipment. The works started in 2002 at Unit 3 and are going to continue in sequence at the Units 1, 2 and 4 until 2010.

In 1982, construction work started in the Temelin nuclear power plant, with two WWER-1000 V-320 units, designed by Soviet organizations and Energoproject and built by VSB with engineering by Skoda-Praha. Construction was delayed and when it resumed in the mid 1990s, Westinghouse instrument and control systems were incorporated. The reactors started up in 2000 and 2002, with the upgrading having been financed by Czech Power Company (CEZ) with a loan from the World Bank[81]. A further two units were originally envisaged on the site, but are yet to be constructed.

It is important to single out the Czech's industry is capable to produce almost all of the main components of WWER design. The original design of both nuclear power reactors operating in the Czech Republic was from the Soviet Union nuclear industry but during construction of the Temelin nuclear power plant it was substantially changed by the Czech industry. Also at the Dukovany nuclear power plant a substantial improvement has been done by the Czech industry.

---

[80] Upgrading Programme of Nuclear Power Plant Dukovany Equipment.

[81] The Temelin nuclear power plant safety enhancement started, with the IAEA assistance, during the construction period. The combination of Eastern and Western technology was successfully completed.

Source: Photograph courtesy of Wikipedia (User:Japo).

Figure 96. Temelin nuclear power plant.

Table 43. Operating nuclear power reactors in the Czech Republic

Reactors	Model V=PWR	Construction date	Connection to the electric grid	Entered into commercial operation
Dukovany 1	V-213	January 1979	February 1985	May 1985
Dukovany 2	V-213	January 1979	January 1986	March 1986
Dukovany 3	V-213	March 1979	November 1986	December 1986
Dukovany 4	V-213	March 1979	June 1987	July 1987
Temelin 1	V-320	February 1987	December 2000	June 2002
Temelin 2	V-320	February 1987	December 2002	April 2003
Total (6)				

Source: WANO, 2009 and CEA, 8th edition, 2008.

Nuclear Safety Authority and the Licensing Process

According with the IAEA sources, the State Office for Nuclear Safety (SÚJB), is the Czech Republic's national regulatory authority in nuclear safety and radiation protection field. It was established as of 1st January 1993 by the Act No. 21/1992 Coll[82]. The SÚJB is carrying out state supervision and licensing activities. Main subjects of state supervision of nuclear safety, in accordance with SÚJB legal competence given by Act No. 18/1997 Coll. on the Peaceful Utilization of Nuclear Energy and Ionizing Radiation (Atomic Act)[83], is

---

[82] It is a follow-up organization of the former Czechoslovak Atomic Energy Commission.

[83] This Act was amended in 2002 with the purpose to harmonize national legislation on nuclear matters with the EU nuclear legislation.

observance of conditions established by the Atomic Act for activities related to nuclear energy utilization. That involves especially:

j)   activities performed in nuclear facilities with nuclear power reactor in operation, facilities for production, processing, storage and disposal of nuclear materials, radioactive waste repository and radioactive waste storage facilities;

k)   design, site location, construction and commissioning;

l)   operation, reconstructions and decommissioning;

m)   design, production, verification and systems modifications of nuclear facilities and their components;

n)   design, production, verification and reparation of packages for nuclear material management;

o)   radioactive material transportation;

p)   physical protection of nuclear facilities;

q)   training for selected personnel;

r)   research and development of activities related to nuclear energy utilization;

s)   technical safety of specified components.

SÚJB assess nuclear safety of facility performance. It includes also main component life verification, core and fuel status, maintenance and modification performance. Regulatory decisions of the SÚJB (except of fines) cannot be changed by any other governmental authority. The Chairman acts as the Nuclear Safety Inspector General. He appoints the SÚJB nuclear safety and radiation protection inspectors. The inspectors' authorities, to perform their function, are stipulated in the provisions of Act No. 18/1997 Coll.

Nuclear Fuel Cycle and Waste Management

The uranium for the fuel of the Dukovany nuclear power plant is supplied now by a domestic producer. However, conversion and enrichment services, together with fuel fabrication for the Dukovany nuclear power plant are provided by Russia. For the Temelin nuclear power plant, the fuel for the initial loading was already supplied and contracts for a few reloading are signed. The uranium for the initial fuel load was supplied by Russia, conversion and enrichment were provided by Russia and the United Kingdom and fuel fabrication took place in the USA. For the fuel reloads of the Temelin nuclear power plant, Czech uranium is ensured. Conversion will be done in France and Canada and enrichment and fuel fabrication in the USA (Westinghouse). The fuel for the Czech research reactors, including the uranium, comes from the Russian Federation. [22]

The Czech government has adopted a policy for the storage of spent nuclear fuel by which its originators should be responsible for its storage. In the case of nuclear spent nuclear fuel produced by the nuclear power plants, the spent nuclear fuel is storage in the site. According with IAEA information[84], at the Dukovany nuclear power plant an interim dry cask-type (CASTOR) spent nuclear fuel storage facility with capacity of 600 tons of uranium was put in operation in 1995. Currently, another interim storage facility for spent nuclear fuel

---

[84] See reference 22.

from the Dukovany nuclear power plant has been prepared. High-level waste and spent nuclear fuel classed as waste are unsuitable for disposal in existing repositories. The construction of a deep geological repository is proposed in the "Concept of Radioactive Waste and Spent Nuclear Fuel Management in the Czech Republic", prepared by the Radioactive Waste Repository Authority (RAWRA) in cooperation with a number of other organizations. The concept was completed by the Ministry of Industry and Trade for government discussion. Based on a preliminary timetable, approval for the final disposal facility site is expected in 2015, and the construction work of the repository will start approximately in 2030. The commissioning of the repository is scheduled for roughly 2065.

The issue of reprocessing spent nuclear fuel remains open, because there is no state policy on this important issue. The decision whether spent nuclear fuel is to be reprocessed or storage is, in principle, left to its owner. At present, the owner of the nuclear power plants does not consider the reprocessing of the spent nuclear fuel as economical. The preparation of a final repository of radioactive waste is the responsibility of the state, but the procedure regarding the reprocessing issue must be coordinated in the long-term between the state and the owner of the nuclear power plants. A decision to reprocess or directly dispose spent nuclear fuel (after its conditioning) as a waste is suspended for the time of its storage in the interim storage facility, envisaged for 40 to 50 years. A shallow land repository of radioactive waste is operated by the owner of the Dukovany nuclear power plant. It is designed to accommodate all future low and intermediate radioactive waste from both the Dukovany and Temelin nuclear power plants. A repository for low and intermediate radioactive waste is located in abandoned mine "Richard" near Litomerice on the north of the Czech Republic. [22]

Regarding decommissioning activities it is important to note that, according to the Atomic Act (Act No. 18/1997 Coll. amended with Act No. 13/2002 Coll. and Act No. 310/2002 Coll), the owner of the nuclear power plant should, under the state control, "prepare both financial and technical means for decommissioning of its nuclear facilities and it should provide payments to the Czech National Bank (the State Bank of the Czech Republic) to accumulate means necessary for a preparation and construction of a spent nuclear fuel repository". At present, the costs of future decommissioning of a nuclear power plant and disposal (deposit) of spent nuclear fuel are not directly reflected in the electricity prices and the nuclear power plant utility creates provisions based on its consideration, from the net profit to its reserve fund, under the state control.

The Public Opinion

After 2005, there is a majority of the Czech population that supports the use of nuclear energy for the production of electricity in the country in the future. However, the people that support the construction of new nuclear power reactors in the country do not reach 50%. In February 2007, a Czech public opinion agency (STEM) organized a poll among 1,222 respondents with an age over 18. The result of the poll show that after a temporary decrease in the support of the use of nuclear energy for the production of electricity during the period 2003-2005, the percentage of people who back a continuous development of the nuclear power sector in the Czech Republic has reached 60%. According with the results of the poll, roughly one-sixth of the adult Czech population supports the use of nuclear energy for the production of electricity. These supports vary slightly depending on the respondents' age and education (its increase slightly with higher education and with age). Between men and women the situation is the following: 65% of men support further development of the nuclear power sector in the country in the coming years, while 52% of women do so. However, the numbers of respondents, who are in favor of the construction of new nuclear power reactors for the electricity production is only 46%.

Another problem that the Czech's government has to face is the position of the public opinion not only within the country, but in some neighboring countries as well, is the conclusion of the construction of the Temelin nuclear power plant. During the 1990s, the Temelin project has aroused concern primarily because it will be an untried combination of old Soviet technology and new Western technology and because of its proximity to other countries[85]. Situated only 100 km from the Austrian border, nuclear fallout stemming from the Temelin nuclear power reactors would have devastating transboundary effects in Austria and, for sure, in other neighboring countries as well. Austria has attempted several time to persuade the Czech Republic to stop the operation of the Temelin nuclear power plant without success. However, the signature of the Melk Protocol by the Austrian and Czech highest authorities regarding the use of nuclear energy for electricity generation in the Czech Republic, reduced the tension between these two countries.

The 1986 Chernobyl nuclear accident, its far-reaching consequences and the fact that the reactors lie on the edge of an active seismic zone; have fueled the growing opposition to the operation of the Temelin nuclear power plant. The opposition to this project argues that the norms used during an important phase of the construction of the Temelin nuclear power plant were not the same as in the West and that energy saving measures that could be adopted by the Czech's government would be safer and cheaper than completing Temelin nuclear power plant. A study conducted by a Belgian consulting firm found that the Temelin nuclear power plant, which will produce 2 000 MWe of electricity, is the cheapest method for producing electricity now available in the country, but that energy efficiency could save 3 500 MWe of electricity, this means 1 500 MWe more that the electricity that will be produced by the Temelin nuclear power plant.

It is important to note that the decision to build the Temelin nuclear power plant was made despite a significant decline in the demand for power in recent years. In addition, a World Bank report indicated that between 1995 and 2010 there is a possibility that no increase in the demand for electricity occurred throughout Europe. Nonetheless, the nuclear industry expects that the demand of electricity grow annually by 2% in the coming years.

Looking Forward

The future of the development of the energy power sector in the Czech Republic is based on the following premises:

---

[85] It is important to note the following: The combination of Eastern and Western technology was successfully completed and verified by the commissioning process. The IAEA missions have so far confirmed that most of the safety issues have been resolved and works on the few remaining issues are in an advanced stage are not precluding safe operation of the Temelin nuclear power plant.

1) continued operation of the Dukovany nuclear power plant without limitation over the whole time horizon;

2) all the existing and newly built nuclear power plants and conventional power and heating plants are, or will be, equipped with facilities for the protection of environment, as required by the laws in force. No new large hydro power plants are planned to be built because the potential of water energy is already utilized at a high rate;

3) greater usage of renewable energy sources will be stimulated by the State Programme of Energy Savings and Usage of Renewable Sources. The spectrum of such sources includes both the traditional ones - mainly the small hydro power plants - and a wide range of their sources (biomass, wind energy, heat pumps and also geothermal energy and solar energy). Energy saving programmes will be strongly supported;

4) the general electricity generation system will rely on nuclear energy, on the exhaustion of the remaining reserves of coal, on the use of gas in cogenerating units, on the current level of electricity generation in hydropower plants and on support to more intensive usage of renewable resources;

5) the limited domestic availability of coal will not enable the existing coal-fired power plants to continue operating once their de-sulphuring units past their useful life. In the period of 2008-2020, it will be possible to retrofit only part of the existing capacity of the traditional coal-fired power plants, extending their useful life by another 15 years or so (until 2030 to 2035);

6) the scenario does not reckon with the possibility of releasing part of the coal reserves to which the environmental limits apply, hence, no new large power generating units which would use domestically extracted coal are planned to be built.

It follows from what has been said in the previous paragraphs that any new power plants which will be built after 2015 (2020) will have to use energy primary sources other than national available coal. These may include, in case of public acceptance, several nuclear power units as a stabilizing element of the national electricity system. [22]

The 2004 State Energy Policy envisages building two or more large nuclear power reactors, probably at Temelin, eventually to replace Dukovany reactors after 2020. Plans announced in June 2006 envisage one 1 500 MWe unit at Temelin after 2020, and a second to follow. However, in light of the countries large excess of base load electricity generation capacity, new nuclear power plant in additional to the current number of nuclear power reactors in operation is very unlikely to be built before 2020.

It is important to stress, that the Czech nuclear industry was involved in the production of different components for its nuclear power plants already built, including vessel and control rod drive mechanism and is also capable to produce almost all of the main components of WWER reactors design. The Czech industry is also the main nuclear supplier to other Eastern European countries[86].

Lastly, it is important to stress that the authorities of the Bohemian's region in which the Temelin nuclear power plant is located have reached an agreement with the state company CEZ by which the prohibition adopted in 2004 to build new nuclear power plants was

---

[86] Twenty reactor vessels were made by Skoda for countries in the Eastern European region in the last decades.

cancelled. In compensation, CEZ will give to the region US$200 million for infrastructure works up to 2018.

THE NUCLEAR POWER PROGRAMME IN GERMANY

Germany imported 61% of its primary energy supply in 2003, including oil and gas, which accounts for nearly 59% of its energy consumption. Germany has 17 nuclear power reactors in operation in 2009, with a total net capacity installed of 20 339 MW and supply 26% of the total electricity produced in the country[87]. The electricity consumption per capita, in 2004, was 6 224 kWh.

Out of the total nuclear power reactors, six are BWR and eleven PWR systems. All units were built by Siemens-KWU. One PWR has not-operated since 1988 because of a licensing dispute. The net load factor in 2007, per type of reactor, was the following: a) BWRs: 74.3%; and b) PWRs: 74.5%.

Table 44. Information on the nuclear power reactors in operation in Germany

Plant	Type	MWe (net)	Year start	Provisionally scheduled shut-down
Biblis-A	PWR	1 167	1975	2008
Neckarwestheim-1	PWR	785	1976	2009
Brunsbüttel	BWR	771	1977	2009
Biblis-B	PWR	1 240	1977	2009 or 2011
Unterweser	PWR	1 345	1979	2012
Isar-1	BWR	878	1979	2011
Phillipsburg-1	BWR	890	1980	2012
Grafenrheinfeld	PWR	1 275	1982	2014
Krummel	BWR	1 346	1884	2016
Gundremmingen-B	BWR	1 284	1984	2016
Gundremmingen-C	BWR	1 288	1985	2016
Gröhnde	PWR	1 360	1985	2017
Phillipsburg-2	PWR	1 392	1985	2018
Brokdorf	PWR	1 410	1986	2019
Isar-2	PWR	1 400	1988	2020
Emsland	PWR	1 329	1988	2021
Neckarwestheim-2	PWR	1 310	1989	2022
Total (17)		20 470[1]		

Note: According with WANO source in 2009 the net capacity was 20 339 MWe.
Source: CEA, 8th Edition, 2008.

---

[87] According with Nuclear Energy Data, in 2007 this share was 27% and the production was 133.2 TWh. The total production of electricity in the country in 2007 was 488 TWh.

After the reunification of Germany in 1990, all the Soviet designed nuclear power reactors operating in the Eastern part of the country were shut down for safety reasons and are being decommissioned. In general, 19 experimental and commercial nuclear power reactors have been shut down and decommissioned until 2008. Five of these are WWER-440 units at Greifswald, closed in 1989-1990 following reunification; five are BWRs, two are HTGRs (one is the large and relatively modern Muelheim-Kaerlich PWR shut down since 1988, and the second is the THTR-300 closed in April 1988), one is Stade PWR closed in November 2003, one is Obrigheim PWR closed in May 2005, one is the Rheinsberg PWR closed in June 1990, a prototype HWGCR Niederaichbach closed in July 1974, one is KNK II type of FBR reactor closed in August 1991, one is the MZFR type of PHWR reactor closed in May 1984 and one is a prototype WWER. Eleven involved full demolition and site clearance.

A summary of the information regarding the nuclear power reactors in operation in Germany is shown in Table 44.

The Energy Policy in Germany

Since the 1970s, a central focus of the West Germany government energy policy has been to shift electricity production away from imported oil and gas towards coal and nuclear power, taking into account the important brown coal reserves (seven in the top ten countries with the highest brown coal reserves in the world) that Germany has. As result of the application of this policy, the share of oil and gas in electricity production was reduced from the peak of 30% in 1975 to 11% in 2003, reducing Germany dependency of the Russian gas[88], while coal provides 55% of the country electricity in 2004.

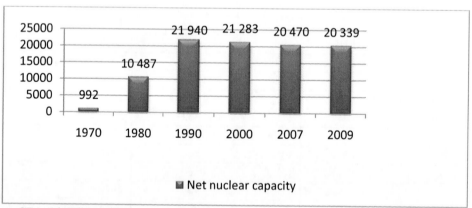

Source: CEA, 8th Edition, 2008.

Figure 97. The evolution of the nuclear capacity in the period 1970-2009.

During the same period 1975-2003, the share of nuclear has grown from 9% to 28%. In 2006, the share of nuclear power was 20.6%, providing almost one third of the total electricity produced by Germany, which was 158.7 TWh, a little bit lower than in 2003. In 2007, the

---

[88] Germany is the country in the world with the highest dependency of gas from Russia, reaching almost 44% of the national consumption of this type of energy source.

nuclear share increased again to 26% producing 140.9 TWh. In 2008, the nuclear share was 28.29%.

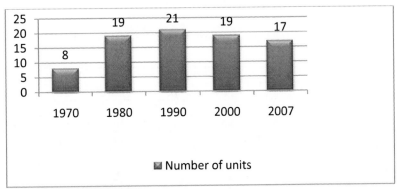

Source: CEA, 8th Edition, 2008.

Figure 98. The evolution in the number of nuclear power reactors in operation in Germany in the period 1979-2007.

Source: International Nuclear Safety Center at Argonne National Laboratory, USA.

Figure 99. Location of the German nuclear power plants.

To reduce the dependency from coal and nuclear energy for the production of electricity, the German government adopted the following energy policy: Since the 1990s, all federal governments promoted the utilization of renewable energies. The utilities are required by law to buy electricity generated by independent producers using renewables. In comparison with the generation price of nuclear or coal power plants, rather high minimum payments to the small producers are guaranteed. In 2003, the share of electricity production from renewables comes up to 8% - hydropower with 4.2%, wind-power with 3.2% - and it is intended to double this share by the year 2010. Nevertheless, in the medium-term large scale electricity production will continue to come from Germany's coal and nuclear power plants. [35]

The location of the German nuclear power plants is shown in Figure 99.

Table 45. Gross electricity production during the period 2004-2006 (TWh)

Energy Source	2006*		2005*		2004	
Nuclear	167.4	(26.3 %)	163.0	(26.3 %)	167.1	(27.1 %)
Lignite	152.0	(23.9 %)	154.1	(24.8 %)	158.0	(25.6 %)
Hard Coal	136.0	(21.4 %)	134.1	(21.6 %)	140.8	(22.8 %)
Gas	73.5	(11.6 %)	71.0	(11.4 %)	61.4	(10.0 %)
Oil	10.5	(1.7 %)	11.6	(1.9 %)	10.3	(1.7 %)
Water Power	27.9	(4.4 %)	27.3	(4.4 %)	27.8	(4.5 %)
Wind Power	30.5	(4.8 %)	27.2	(4.4 %)	25.5	(4.1 %)
Others (incl. Solar Power etc.)	38.0	(6.0 %)	32.0	(5.2 %)	25.3	(4.1 %)
Total	635.8		620.3		616.2	

Note: The final share of nuclear energy in 2006 was 31.4%. In 2007, according with Nuclear Energy Data, this share was around 27%. In 2008, the nuclear share was 28.29%.

Table 46. Nuclear power in Germany

	Number of nuclear units connected to the electric grid	Nuclear electricity generation (net TWh)	Nuclear percentage of total electricity supply
Germany	17	140.9	26
OECD Europe	150	929.3	27
OECD Total	346	2 278.1	23

* According with WANO, 2007.
Source: Nuclear Energy Data, 2007.

The Evolution of the Nuclear Power Sector in Germany

The evolution of the nuclear power sector in Germany was the following. After World War II, allied regulations prohibited any activity in the field of nuclear research and industrial development in the two parts in which Germany was divided.

After West Germany had officially renounced to produce, possess or use nuclear weapons, research and development of nuclear energy for peaceful purposes started with the support of Western countries. By that time, some countries were already–working in the development of nuclear technology for peaceful and military purpose alike, including the USA, Canada, the former URSS, Sweden, the UK, among others. To close the gap, an agreement was reached between the scientific, economic and political sectors in the country with the purpose to promote an extensive international cooperation in all possible areas of the peaceful use of nuclear energy. The German Atomic Programme included the construction of a series of prototype of nuclear power reactors for electricity generation, the concepts for a closed nuclear fuel cycle, and the disposal of radioactive waste in deep geological formations.

In 1955, the federal government established an atomic ministry (Bundesministerium für Atomfragen). Germany became a founding member of Euratom and of the NEA. Agreements for cooperation with France, the UK and the USA were concluded. With the assistance of US manufacturers, Germany started developing commercial nuclear power plants (Siemens/Westinghouse for PWR, AEG/General Electric for BWR). [35]

In 1957, the first nuclear research reactor entered in operation at Munich Technical University. One year later, a 16 MWe experimental nuclear power reactor was ordered from GE/AEG and reached criticality in 1960. In 1961, the first German designed nuclear research reactor was started up at Karlsruhe Nuclear Research Center. In the same year, the 15 MWe Pebble-bed High-Temperature nuclear power reactor in Kahl becomes the first German nuclear power reactor to feed electricity into the electric grid. Between 1958 and 1976, nineteen nuclear power reactors with a net capacity between 15 MWe and 670 MWe were ordered. Seventeen other nuclear power reactors were ordered between 1970 and 1982, with a net capacity between 806 MWe and 1 480 MWe.

The German nuclear industry received the first two orders for the construction of a nuclear power reactors in foreign countries from the Netherlands (Borssele) and from Argentina (Atucha) in the 1960s. The construction of the Borssele nuclear power reactor stared in July 1969. In the case of Atucha 1, the construction started in June 1986.

In January 1970, the construction of the world's then largest reactor, Biblis A, with a net capacity of 1 167 MWe, started in Germany. The reactor was connected to the electric grid in August of 1974 and entered into commercial operation in February 1975. Between 1970 and 1975, on the average, three nuclear power reactors were ordered annually with the purpose to increase the participation of nuclear energy in the generation of electricity in the country.

In 1969, Siemens and AEG founded Kraftwerk Union (KWU) by merging their respective nuclear activities. The domestic development of KWU nuclear power plants with PWRs started. On the basis of several years of operational experience, finally a standardized 1 300 MWe PWR ("Konvoi") was introduced in the market, mainly to speed up the licensing process. However, after some "pre-Konvoi" units, the construction of only three Konvoi units was actually realized (Isar 2, Neckarwestheim 2 and Emsland). The Konvoi units were ordered in 1982 and commissioned in 1988-1989, the last nuclear power plant project in Germany. [35]

In East Germany, its nuclear power programme started in 1955 with the assistance of the former Soviet Union. Research activities in nuclear physics began in 1956 in the Central Institute for Nuclear Physics at Rossendorf. In 1957, a research reactor supplied by the former Soviet Union started operation. The construction of the first East German-Russian reactor with a net capacity of 70 MWe started in January 1960 at Rheinsberg. This nuclear power reactor was connected to the electric grid in May 1966 and entered into commercial operation in October the same year. Between 1970 and 1976, the Greifswald nuclear power plant units 1-5 were constructed. The first unit was connected to the electric grid in December 1973 and entered into commercial operation in July 1974. The second unit was connected to the electric grid in December 1974 and entered into commercial operation in April 1975. The third unit was connected to the electric grid in October 1977 and entered into commercial operation in May 1978. The fourth unit was connected to the electric grid in September 1979 and entered into commercial operation in November of the same year. The fifth unit was connected to the electric grid in April 1989 and entered into commercial operation in November of the same year.

The first four nuclear power reactors at Greifswald were Russian WWER-440/W-230 type but Unit 5 was a WWER-440/W-213 type. Following the German unification, comprehensive safety assessments of the Soviet nuclear power reactors operating in East Germany were carried out. These analyses showed safety deficiencies in comparison with the current West German nuclear safety requirements. Due to technical and economic reasons, in particular uncertainties in the licensing process and also decreasing electricity consumption, it was decided not to upgrade these plants. They were prepared for decommissioning. All nuclear power reactors at Greifswald were shut down between March 1973 and April 1989. Also, work on the nuclear power plants under construction (Units 6, 7 and 8 at Greifswald with WWER-440/W-213 reactors and two WWER-1000 reactors near Stendal) was abandoned.

Two prototypes of advanced reactor design were developed in Germany: 1) the Pebble-bed High-Temperature Reactor[89]; and 2)the Fast Breeder Reactor (Schneller Natriumgekühlter Reaktor, SNR 300).

Due to economical and political reasons, the first prototype of Germany advanced nuclear power reactor design, after a successful commissioning and operation for some years, was shut down in April 1988. The second advanced nuclear power reactor was completed but never commissioned.

The German nuclear industry does not only participate in the development of the German nuclear power reactors design, but also with the French nuclear industry, played an active role in the development of the so-called "European Power Reactor (EPR)" design. For several years, German utilities together with Siemens/KWU and in close cooperation with its French counterparts (EdF and Framatome), had been developing an advanced PWR-EPR. The reactor design is evolutionary and shows enhanced safety features. The design includes provisions to control core meltdown accidents. German utilities also supported the development of an advanced BWR system (SWR 1000) by Siemens/KWU with additional passive safety features. [35]

The Phase-out Policy

In September 1998, the German government decided to phase out the use of nuclear energy for electricity production in the country. The most immediate effect of this policy was to terminate research and development activities on both the High-Temperature Gas-Cooled Reactor and the Fast Breeder Reactor after some 30 years of promising work. The second effect of this decision is a possible short supply of energy due to the fact that the current 17 nuclear power reactors in operation in the country are contributing with 140.9 TW/h to the electric grid and this amount cannot be supply through any other type of energy currently use in the country for electricity generation. The Deutsche Bank warned that "Germany will miss its carbon dioxide emission targets by a wide margin, face higher electricity prices, suffer more blackouts and dramatically increase its dependence on gas imports from Russia as a result of its nuclear phase-out policy, if it is followed through". Germany is committed with the Kyoto protocol to a 21% reduction of greenhouse gas emissions by 2010.

---

[89] The so-called Thorium-Hochtemperaturreaktor, THTR 300).

At the same time, in May 2007, the International Energy Agency (IAE) warned that "Germany's decision to phase out nuclear power would limit its full potential to reduce carbon emissions without a doubt[90]". The IEA urged the German government "to reconsider the phase-out policy in the light of adverse consequences". According with Table 44, four nuclear power reactors should be phased out in 2008-2009, totaling 3 963 MWe; four other units before 2015, totaling 4 388 MWe; seven additional units until 2020, totaling 9 354 MWe; and two units before 2022, totaling 2 634 MWe. The third effect was the lost of the exclusivity of the German supplier of nuclear power reactors with the start of the 21st century and the lost of the leading supplier of nuclear technology that country had in the past.

The German companies EnBW and RWE have confirmed that "they will maintain in operation the Neckarwestheim 1 and Biblis A reactors respectively beyond 2009, when a general election in the country could change the nuclear phase-out policy adopted by the federal government". Neckarwestheim 1 have a power of 785 MWe, while Biblis A a power of 1 167 MWe. In both units maintenance and modernization works were carried out in order to prolong their operation life after 2009.

All German nuclear operators have appealed to the justice to revoke the current government's plan that forces them to stop all nuclear power plant operations before 2021. Before adopting a final position regarding the future use of nuclear energy for electricity generation in the country, the following question need to be asks:

1) does Germany want to and can the country afford to take safe and highly efficient nuclear power reactors off stream without affecting the supply of electricity to the society?

2) is the phase out of nuclear power reactors a future-oriented strategy for the medium-term, or the decision has been taken only by ideological position?

3) is the decision to phase out nuclear power reactors under the present situation is ecologically justifiable?

4) is such a strategy also a convincing model and an example for other countries that are using nuclear energy for the production of electricity as well?

5) should Germany, a leader country in the field of nuclear technology and nuclear safety culture, refrain from the use of nuclear power?

6) who defines whether the exit from nuclear power is meaningful and successful?

7) what is the outlook for future energy supplies in Europe and worldwide?

According with the German Atomic Energy Forum, in 2006, the German nuclear power reactors were once again leading facilities by international standards in terms of efficiency, availability and reliability. The ten most productive nuclear power reactors worldwide included seven German nuclear power reactors, among them the Unit 2 in Neckarwestheim and Unit 2 in Philippsburg. [141]

Changing the nuclear phase out policy is the only decision that the government should adopt in the future in the nuclear field? The answer is no. One of the elements that should be considered urgently, if Germany decides to change its policy of phasing out nuclear energy

---

[90] According with IAEA sources, nuclear energy currently avoids the emission of about 170 million tons per year of carbon dioxide, compared with 260 Mt per year being emitted by other German power plants.

from the mix energy balance, is the dramatic situation the country is facing with respect nuclear education and specialized workforce development. A 2004 analysis of the situation of nuclear education and workforce development in the country was carried out by competent national authorities showed that the situation continues to erode rapidly, due to several reasons, in particular the implementation of the phase out policy. Employment is expected to decline in the nuclear sector, including the one dedicated to reactor building and maintenance by about 10% to 6,250 jobs in 2010. This situation is expected to stay without change until 2016.

The number of German academic institutions teaching nuclear related matters declined from 22 in 2000 to 10 in 2005 and is expected to decline further until five in 2010. While 46 students obtained their diploma in nuclear matters in 1993, they were zero in 1998. In fact, between the end of 1997 and the end of 2002, only two students successfully finished their nuclear studies. It is clear that Germany will face a dramatic shortage of trained staff whether in industry, utilities, research or public safety and radiation protection authorities in the coming years and this situation could have a negative effect in any future plans for the development of the nuclear industry in the country.

In October 2009, the new German government took the decision to prolong the life of its nuclear power reactors in operation, if the necessary safety measures are adopted. The government declared that the decision adopted will have only effect for those nuclear power plants that fulfill the demanding German and international safety standards. The new German government's resolution moves away from the decision adopted by the government, in 2000, establishing that the last nuclear power plants would be taken out of service between 2020 and 2022. The new government also establishes that the future abandonment of the nuclear power should be negotiated as soon as possible with the nuclear industry because the prolonged operation of the nuclear power plants should be in parallel to the establishment of quotas that the nuclear industry should pay to promote the use of other energy alternatives. The government has also agreed to continue with the studies of the Gorleben mine salt of Gorleben as possible final deposit of radioactive waste.

Nuclear Safety Authority and the Licensing Process

With respect to nuclear safety and waste management, the Atomic Energy Act, promulgated on December 23, 1959 right after the Federal Republic of Germany had officially renounced any acquisition, development or use of nuclear weapons, is the core of national nuclear energy regulations in Germany. The main purpose of the Act is "to protect life, health and property against any hazards caused by the use of nuclear energy for the production of electricity and the detrimental effects of ionizing radiation in the population and in the environment". In addition, the Act should provide compensation for any damage and injuries incurred by the use of this type of energy. The amendment of the Atomic Energy Act of 2002, introduce the phase out policy adopted by the government regarding the use of nuclear energy for commercial electricity production.

The Atomic Energy Act is supplemented by the Precautionary Radiation Protection Act, which was adopted by the German government as consequence of the Chernobyl nuclear accident. According to the Atomic Energy Act, it is obligatory to request a license to construct, operate, modify and decommissioning any nuclear installations in the country. A license is required for the construction, operation or any other holding of a stationary installation for the production, treatment, processing or fission of nuclear fuel or reprocessing of irradiated fuel. A license is also required for essentially modifying such installation or its operation and for decommissioning. The applicant can only be granted a license if he meets the individual requirements spelled out in Article 7 of the Atomic Energy Act as license prerequisites: a) trustworthiness and qualification of the responsible personnel; b) necessary knowledge of the otherwise engaged personnel regarding safe operation of the installation; c) necessary precautions against damage in the light of the state of the art in science and technology; d) necessary financial security with respect to legal liability for paying damage compensation; e) protection against disruptive actions or other interference by third parties; f) consideration of public interests with respect to environmental impact.

Concerning the safety of nuclear power plants, the Federal Environmental Ministry (BMU) has the competence in all matters related with the safety of all nuclear installations, including the safety of nuclear power plants. However, the execution of federal laws lies within the responsibility of the Federal States (Länder). For this reason, the licensing of nuclear installations is carried out by the Länder, where different ministries are responsible for licensing of construction, operation, essential modification and decommissioning of nuclear power plants. For technical matters in the licensing procedure and the supervision of nuclear facilities, the regulatory authorities of the Länder are supported by independent technical support organizations, in general the nuclear departments of the Technical Inspection Agencies (TÜV). [35]

Taking into account the federal character of the German's government, it is important to adopt all necessary measures to preserve the uniformity in the application of the licensing process in the whole territory. For this reason, the government designated the BMU to supervise the licensing and supervisory activities of the Länder authorities (so-called "Federal Executive Administration"). Supervision by BMU includes the right to issue binding directives. In performing its federal supervision, the BMU is supported by the Federal Office for Radiation Protection (BfS) in all matters concerning nuclear safety and radiation protection. The BfS is responsible - inter alia - for the construction and operation of nuclear waste repositories, subcontracting for this task with the Deutsche Gesellschaft zum Bau und Betrieb von Endlagern für Abfallstoffe mbH (DBE). As in licensing, the prime objective of the regulatory supervision of nuclear installations is to protect the general public and workers in these installations against the hazards connected with the operation of the installation. Officials of the supervisory authorities as well as the authorized experts working on behalf of the supervisory authority have access to the nuclear installation at all times and are authorized to perform the necessary examinations and to request any pertinent information. Nuclear installations are subject to continuous regulatory supervision. However, the Länder perform this supervisory procedure on behalf of the federal government. [35]

What happens in case of non-compliance with the requirements set up by the licensing permit? In this case the competent supervisory authority of the Länder is authorized by Article 19 of the Atomic Energy Act to issue orders stating: a) that protective measures must be applied and, if so, which ones; b) that radioactive materials must be stored at a place prescribed by the authority; and c) that the handling of radioactive materials, the construction and operation of nuclear installations must be interrupted temporarily or - in case of a revocation of the license - permanently suspended.

It is important to be single out that due to the high safety standards already applied to all nuclear installations in the country it make highly improbable that serious damage to the population and the environment would be caused by the operation of the nuclear power plants. Nevertheless, even in the case of high level of safety are in place in all nuclear installations, a nuclear accident cannot be excluded and provisions should be adopted by the owner of the installations and by the federal government to face this situation. Bearing in mind the potential degree of such a damage, it has always been an essential licensing prerequisite in Germany that sufficient financial security is provided for covering possible claims for damage compensation. The licensees are required to take out liability insurance policies for a maximum financial sum that is specified in the individual nuclear licensing procedure. The federal government and the Länder issuing the license jointly carry an additional indemnity which may be claimed by the damaged party. The maximum required financial security from liability insurances is limited to €2,500 million.

Nuclear Fuel Cycle and Waste Management

A small mine at Ellweiler operated in West Germany during the period 1960 to 1989. Now, all uranium used to ensure the operation of all German nuclear power plants, which is around 3 800 tons uranium per year, is imported from Canada, Australia and Russia, among other countries. According with WANO information, annual demand for enrichment is provided by Urenco's Gronau plant. Most of the balance is provided by Russia. Fuel fabrication is undertaken by Siemens, mostly at Lingen in Germany.

It is important to note that Germany has all the necessary facilities to close the nuclear fuel cycle. However, in 2009, only a few of them are in operation. At Gronau, the enrichment plant of URENCO expanded from a capacity of originally 400 SWU per year to 1 400 SWU per year within the last years and it is intended to increase the capacity further to 4 500 SWU per year. At Lingen, the fuel fabrication facility ANF is in operation. In 2002, the increase of the throughput capacity up to 500 tons of uranium per year was licensed. Three central interim storage facilities for spent nuclear fuel are in operation. These are: a) the Transport Flask Store Ahaus (TBLA) for irradiated fuel; b) the Transport Flask Store Gorleben (TBLG) for both, irradiated fuel and vitrified reprocessing products; and c) the interim Storage Facility Zwischenlager Nord (ZLN) exclusively for spent nuclear fuel from decommissioning of the nuclear power plants in Greifswald and Rheinsberg. The Waste Conditioning Facility (PKA) at the Gorleben site is now completed. According to the new German energy policy and to the respective amendment of the Atomic Energy Act, the waste management of nuclear power plants includes: a) the transport of spent nuclear fuel for reprocessing only until June 30, 2005 at the latest and the utilization of recovered nuclear fuel; b) from July 1, 2005, the use of the local interim storage facilities for spent fuel until a final repository will be commissioned; c) interim storage of spent nuclear fuel at central (external) interim storage facilities and, as soon as possible, at local interim storage facilities; and d) conditioning and interim storage of radioactive waste from operation and decommissioning of the nuclear power plants until a final repository be commissioned.

Based on the new energy policy, additional local interim storage facilities for spent nuclear fuel should be built on the nuclear power plant sites in order to managing the spent nuclear fuel produced by these plants. License applications have been introduced for 12 different sites. Meanwhile, one storage facility is in operation at the Gorleben site. Concerning the final repository of the spent nuclear fuel, the BMU has in mind that a future facility for all types of radioactive waste will be available around 2030. A working group on the site selection for a possible repository, set up by BMU, has produced a report on a comprehensive and suitable site selection procedure. [35]

Decommissioning Policy

The individual power utilities or their subsidiaries are the licensees of the nuclear power plant and, for this reason, responsible for the decommissioning of the plant after the expiration of its operational lifetime. To carry out the decommissioning and dismantling activities in the current nuclear power reactors in operation, the licensees should build up financial reserves to be prepared for the follow-up costs connected with the operation of a nuclear power plant such as the decommissioning and dismantling of the installations, and the treatment and disposal of radioactive material, including the spent nuclear fuel. So far, reserves amounting to €35,000 million have been set aside, of which about 45% are earmarked for decommissioning and dismantling and about 55% for waste management. It is expected that the decommissioning the currently operating nuclear power reactors will produce some 115 000 cubic meters of nuclear wastes.

The responsibility for the disposal of radioactive waste lies with the BfS, which is the legally responsible authority. All other radioactive waste management facilities, i.e. spent nuclear fuel interim storage, are within the responsibility of the waste producers. The Länder have to construct and operate regional state collecting facilities for the interim storage of radioactive waste originating, in particular, from radioactive applications in medicine, industry or universities. The protection objective of disposal of radioactive waste in a repository is laid down in the Atomic Energy Act and the Radiation Protection Ordinance. The Federal Mining Act regulates the aspects concerning the operation of a disposal mine. In addition, environmental legislation must be taken into account, in particular an environmental impact assessment has to be performed.

The Public Opinion

It is important to single out that German support for nuclear energy was very strong in the 1970s, following the oil price shock of 1974, due to their perception of the vulnerability of the country regarding energy supplies. The opposition to the use of nuclear energy for electricity production was not strong until early 1980s, but after the Chernobyl nuclear accident, the opposition to the use of this type of energy for the generation of electricity increase significantly. Due to this accident, the political consensus on the use of nuclear energy for electricity generation was lost. Since 1986, the quite successful German nuclear power programme faced a steadily increasing opposition. Many people were against the use of nuclear energy for electricity generation in the country. Violent demonstrations and occupation of potential sites took place, mainly at Brokdorf, Wyhl and Wackersdorf. Concerned citizens raised objections in administrative courts. Consequently, construction and licensing of nuclear power plants were considerably delayed due to ongoing litigation. Today, in Germany the construction of new nuclear power reactors for electricity generation is forbidden by law. [35]

However, German public sentiment in the last few years has swung strongly in support of nuclear energy again. A poll carried out late in 1997 showed that some 81% of Germans wanted existing nuclear power reactors to continue operating, the highest level for many years and well up from the 1991 figure of 64%. The vast majority of Germans expected nuclear energy to be widely used for the electricity generation in the foreseeable future. The poll also showed a sharp drop in sympathy for militant protests against transport of radioactive waste. A major poll carried out after the October 1998 election confirmed German public support for nuclear energy. Overall 77% supported the continued use of nuclear energy, while only 13% favored the immediate closure of nuclear power plants. [80]

When the government announced, in 2000, that the country was phasing out its nuclear power plants, it seemed that decades of anti-nuclear activism in Germany could be laid to rest. Indeed, protests against nuclear power virtually disappeared from the calendars of political activists. However, a series of revelations about leaks at a nuclear waste dump, combined with a fresh political debate about the nuclear phase-out policy has led to a revival of the movement. In December 2008, thousands of people turned out to disrupt the transport of radioactive nuclear waste from France to a dump in Germany. They were the biggest and most violent anti-nuclear protests in Germany since 2001, with activists setting fire to barricades and chaining themselves to train tracks. The protests delayed for hours the transport of the disputed nuclear waste to the long-term storage facility in Gorleben, located just south-east of Hamburg. This was a strong sign of the renaissance of the anti-nuclear movement, Jochen Stay, spokesman for one of the antinuclear groups, told reporters. The organization campaigns for the speedy phase-out of the country's remaining nuclear power plants. The opposition Greens and the far-Left Party had called on their members to join future protests. German papers see the protests as evidence that the old conflict over nuclear energy is far from over. The center-left Süddeutsche Zeitung writes: "In Gorleben it was much more than the repeat of a ritual. Mobilized by the Greens, many more people demonstrated this time. The non-violent blockades were carefully prepared and attracted many noticeably young demonstrators. In this way a movement that many believed was dead could be revived".

Many young citizens now take for granted the nuclear power phase-out that was pushed through by the SPD and the Greens. Anyone who now uses the climate crisis to put it in question, and presents nuclear power as environmentally-friendly energy, will not have any easy time of it -- rather they will provoke the renaissance of an old conflict. The conservative Die Welt wrote: "For a time, it seemed that rising prices of oil and gas and the great dependence on energy exports from Russia allowed for a heretical idea: rethinking the nuclear phase-out. It would be good, if this debate were to continue. However, the final storage question remains vital, and even those in favor of nuclear power have to concede this. In a heavily populated country like Germany -one that may be technically accomplished but also one which is notoriously fretful - this issue needs to be reexamined". The left-leaning Die Tageszeitung wrote: "The unresolved question of nuclear waste disposal is once again at the forefront of people's minds".

In 2005, according with special Eurobarometer, around 59% of the German people was against the use of nuclear energy for the generation of electricity and 38% was in favor. In January 2007, the percentage of German people that was against the use of nuclear energy for the generation of electricity dropped to 47%, around 31% support that the nuclear power plants in operation continuing the supply of electricity and 17% was in favor to cancel the nuclear power phase-out policy. However, it is important to note that, in 2007, around 60% of the German population did not favor of a phase out without alternatives. In the Eastern part of the country this percentage is almost 67%.

After three decades of opposition, the number of Germans who support nuclear energy is creeping upwards. According to surveys carried out by the Bielefeld-based Emnid Institute for both the atomic lobby Deutsches Atomforum and the anti-nuclear organization Greenpeace Germany, 48% of Germans are in favor of extending the remaining running times of the country's nuclear power plants, compared to 40% two years ago.

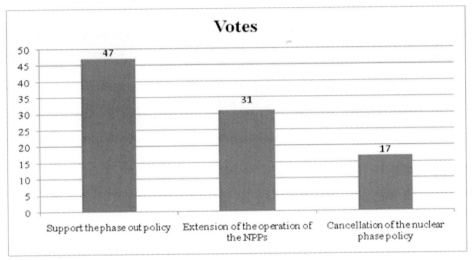

Source: Falter, 2007.

Figure 100. Position of the German people regarding the use of nuclear energy for the generation of electricity.

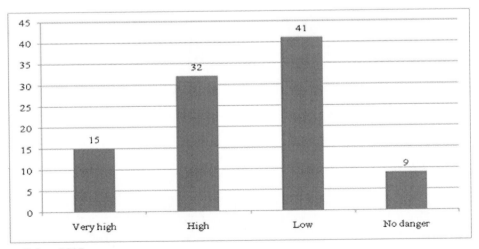

Source: Falter, 2007.

Figure 101. Position of the German population regarding the level of threat presented by nuclear power.

In April 2009, a poll carried out by the Forsa Institute showed that 57% of all Germans consider atomic energy "dangerous or very dangerous". Of those aged 18 to 29, only 49% are worried about the safety of nuclear power. Fears of a Chernobyl repeat have long dominated the nuclear debate in Germany, but Kemfert says "the generation that has no memories of that infamous accident sees things differently. Young people right now are pragmatic", she says, "and they are more worried about climate change than anything else".

What are the causes for this opposition? It would be plausible to speculate that it is anxiety, but if we look at the distribution of fear in the German population we find out, that fear of nuclear power risks is far less pronounced than one might think. Figure 100 shows that only about 47% of the population consider the danger presented by nuclear power to be "high or very high", whereas 50% thinks it is "low or that there is absolutely no danger at all". (Falter, 2007)

It is important to stress that the German nuclear industry was very concern with the adoption and future application of the nuclear phase out policy. Their concern was based on the fact that the application of this policy will have serious consequences regarding the supply of energy in the future and will not reduced the dependency of the country to energy sources outside the country. For this reason, in November 1998, Germany's electric utilities issued a joint statement pointing out that "achievement of greenhouse goals would not be possible without the use of nuclear energy for the production of electricity". A few days later, the Federation of German Industries declared that the "politically undisturbed operation of existing nuclear plants was a prerequisite for its cooperation in reaching greenhouse gas emission targets". A poll carried out early in 2007 found that 61% of Germans opposed the government's plans to phase out nuclear power by 2020, while 34% favored a phase out.

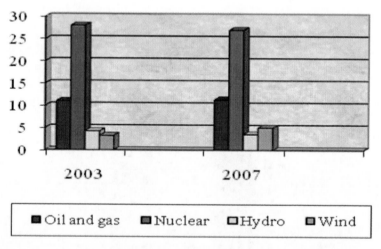

Source: IAEA.

Figure 102. Evolution of the percentage of different energy sources.

Looking Forward

Since the 1970s, a central focus of the German energy policy has been to shift electricity production away from imported oil and gas towards domestic available coal and nuclear power. In the past, the federal government encouraged the utilities to increasingly use domestic hard coal for electricity generation, this rose up to 45 million tons of hard coal per year in 1995. Subsidies were paid by the government and amounted to €5,000 million per annum in 1994, but were reduced continuously to €2,700 million in 2005. Lignite production in Germany is not subsidized, nevertheless, it accounts for nearly one third of the electricity production. To comply with environmental regulations since the mid 1980s, German utilities implemented state of the art technologies to reduce CO_2 emissions and of other contaminate gases to the atmosphere from electricity generation through coal. Also they invested in underground transmission networks.

Since the 1990s, the German government promoted the utilization of renewable energies for electricity production with the purpose to reduce CO_2 emissions. The utilities are required by law to buy electricity generated by independent producers using renewables energy. According with the IAEA, in 2003, the share of electricity production from renewables energy comes up to 8%, hydropower with 4.2%, wind-power with 3.2% and it is intended to double this share by the year 2010. Nevertheless, in the medium-term large scale electricity production will continue to come from Germany's coal and nuclear power plants.

A compromise agreement was worked out in mid 2000 and signed it in 2001 to limit the operational lives of nuclear power reactors to an average of 32 years, but deferring any immediate closures. Two key elements of the agreement reached were a government commitment to respect the rights of utilities to operate existing nuclear power reactors, and a guarantee that this operation and related nuclear waste disposal will be protected from any politically-motivated interference. Other elements of the agreement were the following : a) a government commitment not to introduce any one-sided economic or taxation measures; b) a recognition by the government of the high safety standards of German nuclear reactors and a guarantee not to erode those standards; c) the resumption of spent nuclear fuel transports for reprocessing in France and UK for five years or until contracts expire; and d) maintenance of two waste repository projects, one at Konrad and another at Gorleben.

The compromise agreement adopted although prohibit the construction of new nuclear power reactors for the time being, do not close the door to the future construction of new units. The agreement reached between government officials and representative of the German nuclear private industry was a pragmatic compromise, which in one hand limits the political interference of the government in the energy sector, while in the other, provides a basis and plenty of time for the formulation of a new national energy policy. This new energy policy should indicate very clear which will be the role to be played by nuclear energy in the energy balance of the country in the future, taking into consideration the current and future oil price, the situation of oil reserves, the increase in the price of gas and carbon and the negative impact in the environment of the CO_2 emissions due to the use of fossil fuel for electricity generation, among other elements. Germany is committed to a 21% reduction of greenhouse gas emissions by 2010 and this goal can only be achieved if nuclear energy continues to play an important role in the energy balance of the country in the coming years.

THE NUCLEAR POWER PROGRAMME IN FINLAND

Almost all activities that a modern society has to carry out to provide a high standard of living are based on a secure supply of electricity. Industries, school and universities, laboratories, research centers, hospitals, government offices, restaurants, etc, works consuming electricity. Finland's energy challenges are not very different from the ones that many other countries are facing today. One of these challenges is the following: Energy consumption is on the rise and this means the possibility to increases the CO_2 emission to the atmosphere at the time when the government has committed to reduce the country's greenhouse gas emissions due to the Kyoto Protocol. As a consequences of this commitment, the Finland's government has to close several coal-fired power plants and substituting their output with clean energy from alternative sources.

How to achieve this goal in a country without important energy sources? It is important to note that Finland has no natural oil or gas reserves to exploit, does not produce crude oil and, for this reason, should import crude oil to be processed by its refineries or import the end product. Finland does not produce any coal either. Russia is the most important supplier of crude oil, coal and gas to Finland. However, the country imports crude oil from Denmark and UK and hard coal from Poland and the USA.

According with 2001 statistics, in Finland the total primary energy consumption per capita was about 65% higher than the EU average and about 39% higher than the OECD average. According with the IAEA, this is mainly due to the weather, which demands space heating for most of the time, and the structure of the industry, which is energy intensive processing industry (wood, especially paper, heavy metal and chemical). A third factor is relatively high transportation requirements per capita caused by the low population density.

In Finland, the per capita electricity consumption is around 16 600 kWh per year. While some of it comes from hydro, much of it is either imported or generated from imported fuels. The primary indigenous energy resources in Finland are hydro power, wood, wood waste, pulping liqueurs and peat. Unexploited hydropower reserves are estimated to correspond to an annual production of the order of 7.0-8.5 TW/h. However, most of the unharnessed river areas are either nature reserves or frontier rivers or tiny waterfalls. The annual output of hydropower in Finland is approximately 3 000 MW which equals 13 TWh of electricity generation. The significance of hydropower has increased especially in the quick peaking power operation. Indigenous fuels and hydropower covered 29.3% of the energy demand in 2003. Finland imports all of its oil, natural gas, coal and uranium used for the operation of its nuclear power reactors.

Finland has four nuclear power reactors in operation in Loviisa (WWER type) and in Olkieluoto (BWR type) nuclear power plants, with a net capacity of 2 696 MWe in 2009. The net load factor of the nuclear power reactors in operation in the country, in 2007, by type of reactor was the following: BWRs: 95.5%; and PWRs: 95%. Finland's four nuclear power reactors are among the worlds most efficient, with average capacity factors over the 1990s of 94%. They have an expected operating lifetime of 50-60 years though were originally licensed was for only 30 years[91]. Finland is now constructing a new nuclear power reactor

---

[91] A 20-30 years license extension was recommended by the Radiation and Nuclear Safety Authority (STUK) and granted in mid 2007.

with a capacity of 1 600 MW and has plans to construct another unit with a capacity of 1 000 MW in the coming years[92].

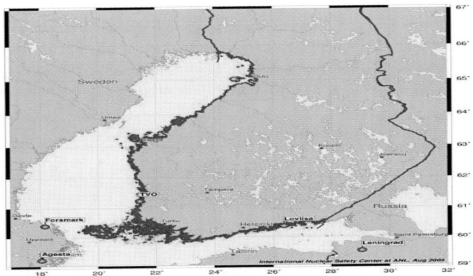

Source: International Nuclear Safety Center at Argonne National Laboratory, USA.

Figure 103. Location of the nuclear power plants in Finland.

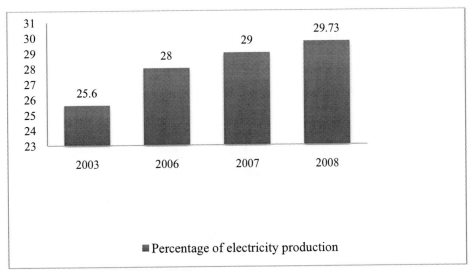

Source: IAEA and CEA, 8th Edition, 2008.

Figure 104. The evolution of the percentage of electricity production in Finland.

---

[92] According with TVO's officials, EIA presented a proposal to construct a new nuclear power reactor in Olkiluoto site to the government in February 2008. The new unit could be either a 1 000-1 800 MWe, PWR or a BWR unit. Fortum's plans for the construction of a 1 000-1 800 MWe unit at Loviisa site were submitted in June 2007. This unit may be a BWR, in which case the technology choice is between GE-Hitachi's ESBWR and Toshiba's Advanced BWR based on ABB's design already operating at Olkiluoto.

In 2003, nuclear energy produced around 25.6% of the total electricity supply in Finland. In 2006, nuclear energy generated 22 billion kWh, representing 28% of the total generating electricity. In 2007, the nuclear production of electricity stays almost the same as 2006, reaching 29% of the total production (22 billion of KWh). In 2008, the nuclear share was 29.73%.

The evolution of the percentage of the electricity production in Finland from 2003 to 2007 is shown in Figure 104. The percentage in the production of electricity increase during the period from 25.6% in 2003 to 29.73 % in 2008, this means an increase of more than 13%.

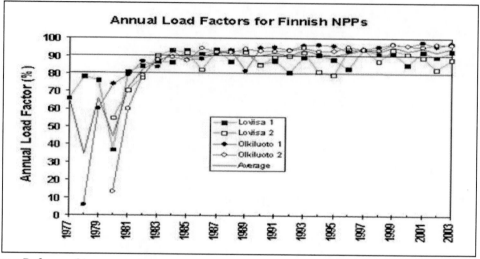

Source: Reference 31.

Figure 105. Annual load factors for the nuclear power reactors of Loviisa and Olkiluoto nuclear power plants.

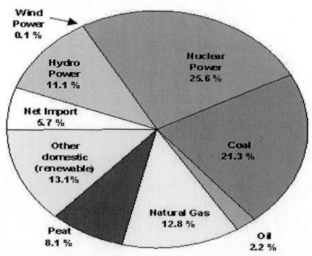

Source: Reference 31.

Figure 106. Electricity supply in 2003 by energy sources (total 85.2 TWh). In 2008, the nuclear energy share was 29.73%.

The nuclear power reactors in Loviisa were supplied by V/0 Atomenergoexport from the former Soviet Union, as well as the nuclear steam supply system and twin turbine generators. Imatran Voima Oy, IVO (predecessor of Fortum Power and Heat Oy) acted as its own architect engineer and coordinated the design and supply of equipment from several countries. This included the integration of West German instrumentation and, under Westinghouse license, an ice condenser containment system.

From Figure 106 it is easy to confirm that in Finland nuclear energy is the single energy sources that produce more electricity in comparison with other available energy sources.

The Energy Policy in Finland

The main objectives of the Finnish energy policy are the following: a) security of supply; b) effective energy markets and economy; c) environmental acceptability; and d) safety.

In Finland, energy supply decisions on energy systems take place at a fairly decentralized level with the exception of nuclear power. A substantial proportion of energy is imported and produced by private enterprises. The energy companies with majority ownership by state are also run on a purely commercial basis. [31]

In 1994, Finland ratified the Framework Convention on Climatic Change. According to the Kyoto Protocol, Finland's commitment is to return the emissions to the 1990 level. Meeting the emission limits, especially those of carbon dioxide, would be a challenging task without the expanded use of nuclear energy and renewable energy sources for electricity generation. Other actions that have been taken in this regards are the use of wood-based fuels, natural gas, a substitute for coal, as well as upgrading the current nuclear power reactors in operation in the country. Energy conservation has also played an important role within the actions already taken.

According with IAEA sources, the National Climate Strategy lays down guidelines as to how Finland will meet the targets for greenhouse gas reduction agreed upon in the Kyoto Protocol. Preparation work was started in 1999, and the Strategy was approved by the government in March 2001. Later the same year the Finnish Parliament endorsed it in a statement.

In the other hand, it is foresee, according with recent studies carried out by the Ministry of Trade and Industry, that the consumption of electricity is expected to evolve from the 2000 level of 79.2 TWh to 94.2 TWh in 2010 and further to 103.3 TWh in 2020, with an average annual economic growth of at least 2.5% for the next 30 years, with the purpose to ensure a high standard of living for the Finnish population (see Figure 107).

Within the above time frame, it is expected that the growth of primary energy use in the country to satisfy its energy needs slow down and even cease. However, new electricity production capacity is needed to meet the expected increased in power demand in the coming years and to compensate the decommissioning of older energy plants. Finland has to take a decisions about which forms of energy production it is going to support in the coming years. Increasing the use of nuclear energy for electricity production is currently under scrutiny, along with a debate about the viability of using different renewable energy sources, despite of the fact that recent studies have shown that the use of nuclear energy for electricity generation was the cheapest energy option for Finland.

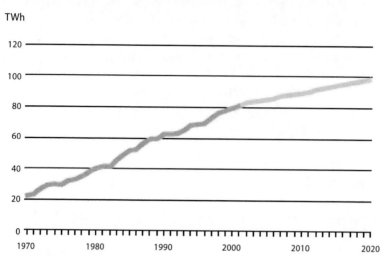

Source: Reference 69.

Figure 107. Expected evolution of the electricity consumption during the period 1970 to 2020.

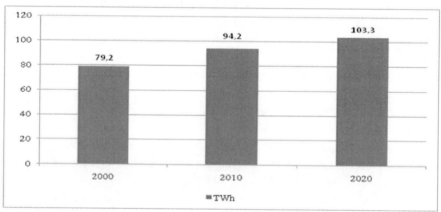

Source: Ministry of Trade and Industry data (Reference 69).

Figure 108. Expected evolution of the consumption of electricity from 2000 to 2020.

The government's role in nuclear power projects is to grant licenses and monitor the safety of the reactor's operation. According with the government's position regarding the use of nuclear energy for the production of electricity, a high level of safety is a precondition for the construction and operation of a nuclear power plant in Finland. "When it comes to the use of nuclear energy, safety always comes first," emphasizes Riku Huttunen, Deputy Director General, Ministry of Employment and the Economy.

But not only safety considerations are taken into account but also environmental impact, raw materials used and future self-sufficiency. All of these elements are key factors that could affect the choice of the energy source to be used for the production of electricity in the future.

It is important to note that Finland currently ratio of imported electricity is amongst the highest in Europe and this make the country very vulnerable to any political and economy changes in the situation of the countries where the energy is imported.

The Evolution of the Nuclear Power Sector in Finland

The historical evolution of the nuclear sector in Finland is briefly described in the following paragraphs. The units of the Loviisa nuclear power plant [93], located on the southern coast of the country and with a net capacity of 976 MWe, were ordered in 1969 (Unit 1) and 1971 (Unit 2). The construction of the first unit started in May 1975, was connected to the electric grid in February 1977 and entered into commercial operation in May of the same year. The construction of the second unit started in August 1978, was connected to the electric grid in November 1980 and entered into commercial operation in January 1981.

Table 47. Finland's nuclear power reactors

	Type	MWe net	Commercial operation	Licensed to
Loviisa 1	WWER-440	498	1977	2027
Loviisa 2	WWER-440	498	1981	2030
Olkiluoto 1	BWR	840	1979	2039
Olkiluoto 2	BWR	860	1982	2042
Total (4)		2 696		

Source: IAEA PRIS, CEA, 8th Edition, 2008.

Source: Photograph courtesy of Wikipedia (Annual report of Finnish radiation safety authority on nuclear safety control 2007).

Figure 109. Lovitisa nuclear power plant.

The units of the Olkiluoto nuclear power plant[94], located on the Western coast of the country and with a net capacity of 1 700 MWe, were ordered in 1972 (Unit 1) and 1974 (Unit 2). The construction of the first unit started in February 1974, was connected to the electric grid in September 1978 and entered into commercial operation in October 1979. The construction of the second unit started in August 1975, was connected to the electric grid in February 1980 and entered into commercial operation in July 1982. It is important to single

---

[93] Owned by Fortum Power and Heat Oy (Fortum).
[94] Owned by Teollisuuden Voima Oy (TVO).

out that the Olkiluoto Units 1 and 2 started up at 690 MWe gross (658 MWe net). They now produce 840 MWe and 860 MWe respectively and their lifetime has been extended to 50-60 years, subject to safety evaluation every decade. The government approved to progressively uprate the Olkiluoto nuclear power reactors further to 1 000 MWe each, starting with replacement low-pressure turbines in 2011. Loviisa units have been uprated 9.7%, from 465 MWe (gross) in 1977-1980 to 510 MWe (gross). The net capacity of the Loviisa nuclear power reactors is 498 MWe each.

A summary information regarding the nuclear power reactors operating in Finland is shown in Table 47.

It is important to stress that Finland is part of the deregulated Nordic system which faces shortages, especially in any dry years which curtail hydroelectric generation. With growing demand and the need to ensure reliable economic supply over the long term, various studies were carried out which showed that the use of nuclear energy for electricity production was the cheapest option for Finland.

Nuclear Safety Authority and the Licensing Process

The Nuclear Energy Act (990/1987) and the Nuclear Energy Decree (161/1988) gave the Finnish's Parliament the final say on permitting the construction of new major nuclear installations, including final disposal facilities for nuclear waste in the country. These two Acts also defines the licensing procedure and the conditions for the use of nuclear energy, including waste management, as well as the responsibilities and powers of the authorities. Each producer of nuclear waste in Finland is responsible for safe handling, management and disposal of waste and for meeting the costs of these operations. The funds required for future nuclear waste management must be raised gradually during the plant's operating time.

At the same time, the Finnish government has issued several general regulations related with the safety of nuclear power in Finland. These regulations are the following: 1) Decision of the Council of State on the General Regulations for the Safety of Nuclear Power Plants (395/1991); 2) Decision of the Council of State on the Physical Protection of Nuclear Power Plants (396/1991); 3) Decision of the Council of State for Emergency Response Arrangements at Nuclear Power Plants (397/1991); 4) Decision of the Council of State on the Regulations for the Safety of a Disposal Facility for Reactor Waste (398/1991); and 5) Decision of the Council of State on the Safety of Disposal of Spent Nuclear Fuel (478/1999).

In addition to the above regulations, it is important to take into account, during the licensing procedures, two special regulations. One is the Radiation Act (592/1991) and the Act on Environmental Impact Assessment (468/1994). The purpose of the first Act is to prevent and limit the detrimental effects of radiation on human health, while the second stipulates that environmental impact assessment is compulsory for nuclear facilities. Others laws in the general legislation also affect the nuclear sector.

Summing up can be stated that on nuclear matters the Ministry of Trade and Industry is responsible for the overall supervision of the use of nuclear energy in the country. The decision making-process for the construction of a nuclear facility is the following:

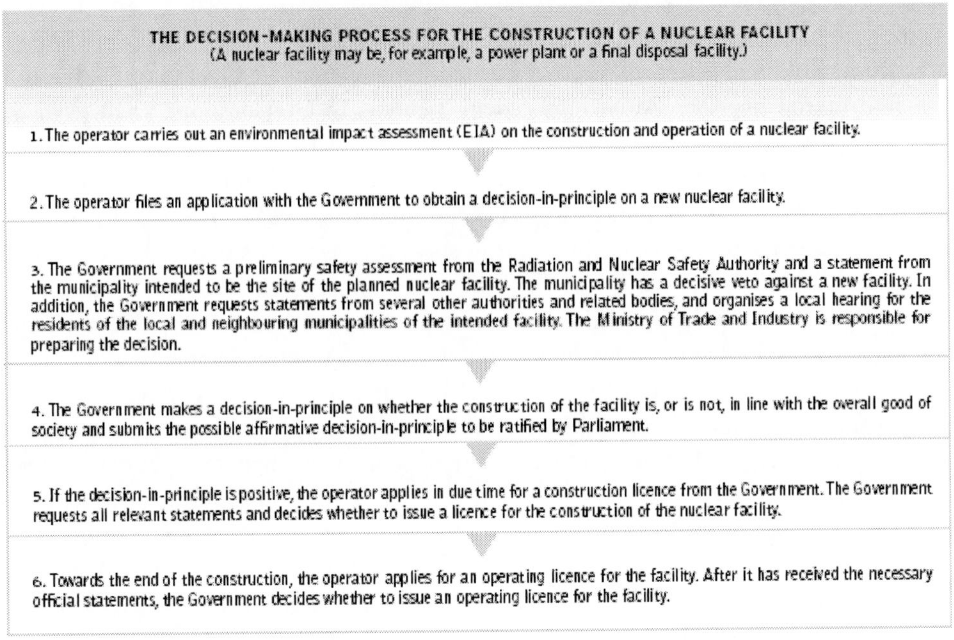

THE DECISION-MAKING PROCESS FOR THE CONSTRUCTION OF A NUCLEAR FACILITY
(A nuclear facility may be, for example, a power plant or a final disposal facility.)

1. The operator carries out an environmental impact assessment (EIA) on the construction and operation of a nuclear facility.

2. The operator files an application with the Government to obtain a decision-in-principle on a new nuclear facility.

3. The Government requests a preliminary safety assessment from the Radiation and Nuclear Safety Authority and a statement from the municipality intended to be the site of the planned nuclear facility. The municipality has a decisive veto against a new facility. In addition, the Government requests statements from several other authorities and related bodies, and organises a local hearing for the residents of the local and neighbouring municipalities of the intended facility. The Ministry of Trade and Industry is responsible for preparing the decision.

4. The Government makes a decision-in-principle on whether the construction of the facility is, or is not, in line with the overall good of society and submits the possible affirmative decision-in-principle to be ratified by Parliament.

5. If the decision-in-principle is positive, the operator applies in due time for a construction licence from the Government. The Government requests all relevant statements and decides whether to issue a licence for the construction of the nuclear facility.

6. Towards the end of the construction, the operator applies for an operating licence for the facility. After it has received the necessary official statements, the Government decides whether to issue an operating licence for the facility.

Source: Reference 69.

Figure 110. Decision making-process for the construction of a nuclear facility in Finland.

Nuclear Fuel Cycle and Waste Management

Along with additional construction in the nuclear sector, nuclear power companies are working on the final disposal of nuclear waste, because nuclear waste management preparations, funding and safe implementation are the responsibility of nuclear power plant operators.

The Finnish's law on waste management stipulates that radioactive waste produced in the country must be handled and disposed of domestically. With low- and medium-level radioactive waste, final disposal has progressed to the implementation phase. (Remes, 2008)

The operators of the nuclear power plants in Olkilouto and Loviisa have selected the Olkiluoto's site as the final repository for the spent fuel from their own nuclear power plants. According to TVO's Anneli Nikula, Senior Vice President, Corporate Communications and CSR, "we are among the frontrunners in the world in final disposal research. And the financial foundation for final disposal has been secured, because the financing needed for final disposal is collected in advance in the price of the electricity produced by the nuclear power plant".

In the field of nuclear fuel supply, Finland adopted the requirements of Euratom after its entry into the EU. Until the late 1990's, the supply of the nuclear fuel for the Loviisa nuclear power plant came from a Russian supplier. After the 1990s, the operator of the Loviisa nuclear power plant started to find an optional fuel supplier from Western sources besides the Russian fuel supplier. Together with the Hungarian Paks utility, test fuel assemblies have been bought from the British Nuclear Fuel Ltd (BNFL). First lead assemblies from BNFL were loaded in Loviisa in 1998 and from 2001 onwards part of the fuel for Loviisa nuclear power plant has been delivered by the BNFL. Uranium for Olkiluoto Units 1 and 2 comes (or has come) from Canada, Australia, Niger, China and Russia. Most of the enrichment has taken place in Russia, the rest in Western Europe. Fuel elements delivered to Olkiluoto have been manufactured by ABB Atom in Sweden, Siemens in Germany and Genusa in Spain. [31]

Finland' policy on storage of spent nuclear fuel promotes the storage of the spent nuclear fuel in pools at the reactor buildings for a limited number of years. After this period, spent nuclear fuel elements are transferred to interim spent nuclear fuel storage also at the nuclear power plant sites. Finland's four nuclear power reactors produce about 70 tons of spent nuclear fuel per year. According with the Ministry of Trade and Industry, the total amount of spent nuclear fuel to be produced by the four nuclear power reactors in operation in Finland in 40 years of operation will be around 2,700 tons. For the packing of this amount of fuel, over 1,500 containers for final disposal are required and about 15 kilometers of underground tunnels will be excavated for the containers. The network of tunnels to be built at a depth of several hundreds of meters will cover an area of a few dozens of hectares. If the operating life of the plants is increased to 60 years, the total amount of spent nuclear fuel for final disposal will be 4 000 tons.

A repository for medium and low-level wastes has been in use since 1992 at the Olkiluoto site and since 1998 a similar facility has been used for low-level waste disposal at the Loviisa site.

Regarding the final disposal of spent nuclear fuel and radioactive waste, Finland's policy on this important issue is implemented by Ministry of Trade and Industry. In order to find an acceptable solution to the problem of the final disposal of spent nuclear fuel, Finland authorities have approved a plan for the construction of a permanent geologic facility in a selected site. After spending a decade identifying possible locations, Finland selected four sites as possible locations for the geologic facility to be constructed. After working with the local community of the selected sites, the government gave permission to move forward with the project by the following year. The decision enjoyed broad support in Finland's Parliament, which voted 159 to 3 in favor of the plan. A construction license application should be submitted by 2016. It is expected that the commercial operations of the permanent geologic facility with a capacity to storage of 4 000 tons of spent nuclear fuel be ready by 2020.

The main elements of the Finland's policy on final disposal of spent nuclear fuel are the following: Power companies in Finland should pay annual contributions to the State Nuclear Waste Management Fund, which is a segregated fund operating under the auspices of the Ministry of Trade and Industry for the management of the spent nuclear fuel produced by the nuclear power plants operating in the country. The contribution covers the future treatment, storage and final disposal of spent nuclear fuel and radioactive waste, as well as decommissioning of the nuclear power plants. It is important to note that the power companies should take care by themselves of other actions related with the spent nuclear fuel management, including the management of medium and low-level operational wastes, the decommissioning of the nuclear power plants as well as the management of the thereby arising wastes. These companies are independently responsible for funding all nuclear waste management activities despite their cooperation on spent nuclear fuel disposal. To ensure that the financial liability is covered, the utilities must each year present cost estimates for the future management of nuclear wastes, including decommissioning of nuclear power plants. According with the IAEA, the latest estimate (2003) for the total fund target, including costs from management of existing waste quantities and from decommissioning of nuclear power plants, arises to €1,338 million with no discounting. The utilities are obliged to set aside the required amount of money each year to the State Nuclear Waste Management Fund. In rough terms, the cost for nuclear waste management, including plant decommissioning is €2.1/MWh (with no discounting), representing about 10% of the total power production cost.

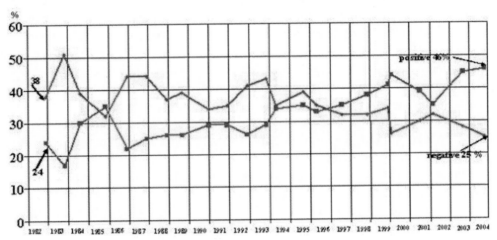

Source: TNS Gallup Oy; Finnish Energy Industries Federation.

Figure 111. Attitudes in Finland towards the use of nuclear power. The development of the acceptance of nuclear power 1982-2004.

The Public Opinion

Regarding public opinion towards the use of nuclear energy for electricity generation, the situation in Finland is briefly described in the following paragraphs. Since 1983, an independent university group has conducted public opinion surveys on energy alternatives. Figure 111 indicates how the attitudes towards nuclear power have evolved through the period 1982 and 2004. It can be seen from the mentioned figure that the public attitude was rather favorable towards nuclear power before the Chernobyl nuclear accident. The survey taken immediately after the accident showed a drastic change in opinions. At that time, only 14% were in favor of increasing nuclear capacity. The confidence lost in 1986 quickly returned by 1988 and the trend has been slowly improving since that. [31]

In 1999, another opinion poll was carried out in the country. One of the question asked was to collect the opinion of the inhabitants living in those municipalities were the candidate host communities for a spent fuel repository in Finland were located. The results showed that a clear majority of the people in Loviisa and Eurajoki agreed that a spent nuclear fuel repository could be sited in their home community, provided that studies can show that the encapsulation and disposal facility is safe.

The latest poll carried out in 2004 showed that 46% favored expanded use of nuclear energy for electricity generation, while 25% were opposed. The biggest uncertainty seems to concern about nuclear waste (see Figure 111).

Looking Forward

Why nuclear energy is attractive to Finland? The answer is very simple: Because the use of nuclear energy for electricity production is not only environmentally friendly, safe, affordable and largely domestically produced, but there is no other sufficient energy source available in the country, with the exception of hydro for the generation of electricity. In addition, Finland is struggling to reconcile a desire to reduce CO_2 emissions to diminish the negative impact of the use of fossil fuels for the production of electricity in the climate, while ensuring not only economic competitiveness but the highest possible standards of living for its population.

Under the new EU energy plan, Finland will be forced to reduce greenhouse gas emissions by 20%, increase renewable energy by 20% and increase efficiency by 20% by 2020. It is important to note that to meet such demands will inevitably cause an increase in energy prices. However, affordable energy is critical to sustaining economic competitiveness in economies with high labor costs, expensive environmental mandates, and other regulatory expenditures like the Finish's economy that depend on energy-intensive activities. For this reason, access to vast quantities of affordable energy is a top national priority for Finland.

Taking into account that nuclear energy is the single energy source with the highest level of electricity production in the country, there should be no doubt that this type of energy can provide electricity as affordable price and with zero CO_2 emissions. Based on these elements, the Finland's parliament in May 2002 voted 107-92 to approve building a fifth nuclear power reactor, with the goal to enter into commercial operation about 2009-2010. The fifth unit will be located at Olkiluoto nuclear power plant site. The new unit, selected on the basis of its operating cost, is the Framatome ANP's 1600 MWe (EPR[95]). The Finnish buyers, TVO, have chosen not to publish a detailed breakdown of the construction cost, but the order is described as 'turnkey' and company officials stated the cost was about €3 billion.

Assuming an output of 1 600 MW, this represents a cost of about £1,250/kW. The Olkiluoto order is widely seen as a special case and it has been suggested that the Framatome has offered price that might not be sustainable to ensure their new technology is demonstrated while the buyer, TVO, is not a normal electric utility. TVO is a company owned by large Finnish industry that supplies electricity to its owners on a not-for-profit basis. The plant will have a guaranteed market and will not, therefore, have to compete in the Nordic electricity market, although if the cost of power is high compared to the market price, the owners will lose money. The real cost of capital for the plant is only 5% per annum. (Thomas, 2005)

However, recent official declarations made by representative of the German-French Areva-Siemens company in charge of the construction of the EPR reactor in Olkiluoto, indicated that the construction of the unit will be delayed up to 2012, due to difficulties in the construction of the civil work. This delay will add around €2,000 million to the original construction costs.

When the Finnish Parliament approved the construction of the fifth unit the public opinion was 67% against of enlarging the number of nuclear power reactors operating in the country. Two were the main approaches that prevailed among the parliamentarian when the request was presented. One is the small cost for kilowatt and the second is to minimize the dependence of the country from external energy sources, particularly from Russia. The

---

[95] This is a third generation of PWR and the biggest reactor ever built in the world.

construction works of the fifth nuclear power reactor began in May of 2005 under the direction of French engineers. The enter into commercial operation of this new nuclear power reactor was expected to be in 2009 or 2010. Due to the enormous engineering work associated with the construction of this type of reactor, the first Generation III PWR system to be built in the world, the conclusion of the construction work suffered a serious delay. According to the last estimates, the nuclear power reactor is expected to entry in operation in June of 2012, that is to say, with three years of delay and its final cost can overcome the €5.000 million. According to TVO sources, the construction plans prepared by Areva were from the beginning not very realistic and the new generation of French's nuclear engineers didn't have sufficient practical experience in the construction of this new type of reactor. It is important to single out that Olkiluoto Unit 3 is also the first nuclear power reactor under construction in Western Europe after the Chernobyl nuclear accident in 1986[96].

One of the problems that is causing the delay in the construction of the fifth unit is the detection, by the Nuclear Security Agency of Finland, of around 1,500 problems related with the safety of the reactor, which are still waiting for the indispensable corrections before the last phase of the construction could start. The delay is so significant that Finland will not be able to fulfill its commitments to reduce the CO_2 emission to the atmosphere established by the Kyoto Protocol.

Source: Reference 69.

Figure 112. Possible site of the third nuclear power reactor in Olkiluoto.

For all of the elements mentioned in previous paragraphs, the electrical company TVO has demanded an indemnification of €2,400 million by damages to the nuclear group French Areva and the German partnership Siemens by a delay of three years in the construction of the EPR system in Finland. Due to this delay, Finland has to import energy until the new unit entry in operation. On the other hand, Areva-Siemens has demanded the Finnish company an

---

[96] France is also constructing a similar nuclear power reactor in the Framanvilles nuclear power plant (Unit 3) located in Normandy.

indemnification of €1,000 million because they consider that the delay in the construction is due to delays in the delivery of documentation by the Finnish company.

Despite of the problems that the government is facing with the construction of the fifth Finnish's nuclear power reactor, three applications for the construction of additional units were presented to the consideration of the Ministry of Economy, which is the government office in charge of the energy policy of the country. According to the Minister of Economy, Mauri Pekkarinen, the government will decide on these applications in the spring of 2010 and said the following: "I don't believe that the three applications are going to be accepted in one legislature", leaving the open door for the acceptance of perhaps two of the applications presented.

It is important to stress that the last public opinion survey indicated that 57% of the participants are in favor and 35% against the construction of new nuclear power plants in the country. The powerful Union Confederation, SAK, also supports the construction of new nuclear power reactors. "The construction of a new nuclear power plant, means work for our members. It is also a type of energy cleaner than fossil fuels and, therefore, can help in the fight against the climatic change", the secretary for the productive area of the SAK, Janne Metsämäki, expressed.

According with the latest Finnish's energy projections about 7 000-7 500 MWe of additional capacity will be needed in the country by 2030.

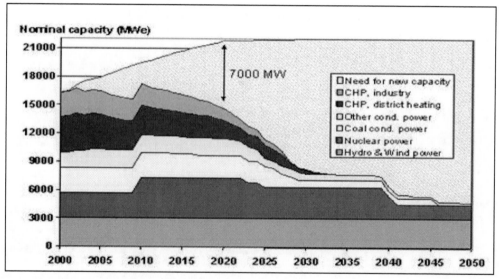

Source: Energy Visions 2030 for Finland.

Figure 113. The forecasted need for new power production capacity to meet the increased demand and to replace decommissioned power plants.

Despite of the intention of the nuclear industry to built at least three nuclear power reactors in the coming years, the Minister of Economic Affairs believes that, "at most, there is a need for one more reactor in addition to the reactor currently being built". The government wants the nation's focus to shift to conserving electricity and developing renewable energy sources, which currently provide one-fourth of Finland's total energy consumption and account for more than a fourth of its power generation[97].

The National Climate Strategy lays down guidelines as to how Finland will meet the targets for greenhouse gas reduction agreed upon in the Kyoto Protocol. Preparation work was started in 1999, and the Strategy was approved by the Government in March 2001 and the Finnish Parliament endorsed it in a statement.

According to the burden sharing between the EU countries, Finland's commitment is to return the emissions to the 1990 level. Meeting the emission limits especially those of CO_2 would be a challenging task to be accomplished without the expanded use of nuclear power and renewable energy sources.

The main actions aimed to restrict the CO_2 emissions growth have been increasing the use of wood-based fuels, natural gas, a substitute for coal, as well as upgrading the nuclear power plants. Energy conservation has also played a role. [31]

There are no domestic suppliers of whole nuclear power plants or nuclear steam supply systems in Finland. However, the Fortum Nuclear Services Oy provides technical support to the Loviisa nuclear power plant owner Fortum Power and Heat Oy and subsystem design services to Finnish and foreign customers.

During the construction of the existing nuclear power plants the collaboration with foreign vendors provided Finnish companies with the necessary experience in supplying certain mechanical equipments to be use in these plants such as simulators, fuel handling, storage equipment, radiation dosimeters and monitoring equipment. The design, deliveries and installation works in the connection of the modernization projects were carried out by domestic and foreign companies.

THE NUCLEAR POWER PROGRAMME IN SPAIN

The main indigenous energy sources in Spain are coal and hydro. Electricity consumption in the country has been increasing steadily in recent years. In 2003, electricity production in Spain grew by 6.5%, reaching a level of generation of 262 249 million kWh. The percentage of production by energy sources was the following: hydro (16.5%); fossil fuels (41.5%); nuclear power (23.6%); cogeneration (12.2%); wind (4.4%); solar (0.001%); and biomass (1.4%). In 2005, power production was around 295 billion kWh, 20% of this from nuclear power, 27% from coal and 27% from gas. The consumption per capita, in 2006, was about 5,500 kWh per year. In 2007, the nuclear production was 56.4 billion kWh, representing 17.4% of the total electricity production of the country (around 302,870 million kWh). In 2008, the nuclear share was 18.27%.

Spain has eight nuclear power reactors in operation with a capacity of 7 450 MWe. The net load factor of the nuclear power reactors in operation in the country in 2007 was the following: BWRs: 70.5%; PWRs: 83.4%.

Spain is essentially an island separate from the EU electric grid. There are only two countries that have electricity connection with Spain. These countries are France and

---

[97] See Remes, 2008.

Portugal. About 2% of power is imported from France and a similar amount exported to Portugal.

The Energy Policy in Spain

The energy policy adopted by Spain goes towards the complete liberalization of the energy markets. This goal was achieved in 1997. The main target was, at that time, the following: 1) to decrease the energy prices; 2) to ensure the energy supply and the quality of the supply; 3) to improve the energy efficiency; 4) to reduce the consumption of energy; and 5) to protect the environment.

The legal framework of the Spanish's energy policy is based in two laws. The first one is the Law 54/1997 on Electric Sector and the second the Law 34/1998 on Hydrocarbon Sector. In the field of electricity generation, the principle of freedom has been established in order to set up and to operate on a free market basis, although prior government authorization is required for the construction of new generating power plants. The authorization can only be denied on environmental grounds.

Law 34/1998 gave to the government the responsibility to establish the rules within the sector should work. This law considers the different stages in the hydrocarbon industry for liquid and gas. Law 34/1998 creates the Energy National Commission with the following mandate: a) to regulate the energy systems; b) to maintain the free competence among them and the transparency of the performance; and c) to benefice all the organizations working in the system and the consumers.

According with the IAEA, in September 2002 the Spanish government approved the document 2Gas and Electric Plan, Development of Transport Grids 2002-2011". On it is foreseen "to increase the electric consumption in an average of 3.75% per year, increasing the use of gas and maintaining the use of nuclear capacity". In accordance with the liberalization scheme, "the plan is not compulsory only contains indicative basis". Law 54/1997 on the Electric Sector admits "the right to the free installation in the field of generation". In relation with the construction of new nuclear power reactors, at this moment, there is no application to build a new unit, and the government is reluctant to the use of nuclear energy for the production of electricity in the country in the future.

The Evolution of the Nuclear Power Sector in Spain

Spain started the introduction of nuclear energy in different sectors of its economy in the early 1950s. At that time, the main Spanish's organization responsible for the introduction and development of the nuclear energy sector was the Nuclear Energy Board, an organ subordinate to the Ministry of Industry and Energy. The Nuclear Energy Board was in charge of the implementation of different activities associated with the introduction of nuclear energy in the country such as personnel training, raw materials procurement, basic scientific research and technology development. In 1964, the Law 25/1964 on Nuclear Energy was approved with the purpose to regulate the activities associated with the introduction and development of the nuclear power sector.

In June 1964, construction started on the first of three nuclear power reactors - Jose Cabrera, Zorita, a small PWR with a net capacity of 141 MWe. This nuclear power reactor was connected to the electric grid in July 1968, entered into commercial operation in August 1989 and was shut down in April 2006. Two years later, in May 1966, construction of Santa Maria de Garona, a medium-sized BWR with a net capacity of 446 MWe started. This nuclear power reactor, was connected to the electric grid in March 1971 and entered into commercial operation in May 1971. In June 1968, the construction of Vandellos 1[98], a medium-sized gas-cooled reactor similar to UK's Magnox units (MOX) with a net capacity of 480 MWe started. The reactor was connected to the electric grid in May 1972, entered into commercial operation in August 1972 and was shut down in July 1990. This first generation of Spanish nuclear power reactors - all turnkey projects - gave practical experience with three different designs, and led to a focus largely on PWR types of reactors in the 1970s.

In 1972, Enusa (now Enusa Industrias Avanzadas SA), a state-owned company, was set up to take over all of the nuclear front-end activities.

In the early 1970s, construction was started on a second group of Spanish's nuclear power reactors. Seven nuclear reactors were initially constructed, but only five of them were completed.

Table 48. Status of the Spanish nuclear power reactors

Station	Net capacity	Constru-ction date	Criticality date	Connection to the electric grid date	Commercial date	Shut down date
Almaraz-1	944	02-Jul-73	05-Apr-81	01-May-81	01-Sep-83	
Almaraz-2	956	02-Jul-73	19-Sept-83	08-Oct-83	01-Jul-84	
Asco-1	995	16-May-74	16-Jun-83	13-Aug-83	10-Dec-84	
Asco-2	995	07-Mar-75	11-Sept-85	23-Oct-85	31-Mar-86	
Cofrentes	1 064	09-Sept-75	23-Aug-84	14-Oct-84	11-Mar-85	
Jose Cabrera (Zorita)	141	24-Jun-64	30-Jun-68	14-Jul-68	13-Aug-69	31-Apr-06
Santa Maria de Garona	446	02-May-66	05-Nov-70	02-Mar-71	11-May-71	
Trillo-1	1 003	17-Aug-79	14-May-88	23-May-88	06-Aug-88	
Vandellos-2	1 045	29-Dec-80	14-Nov-87	12-Dec-87	08-Mar-88	
Vandellos-1	480	21-Jun-68	11-Feb-72	06-May-72	01-Aug-72	31-Jul-90
Total 10 (8-Operational)	7 448					

Sources: IAEA PRIS and CEA, 8th Edition, 2008.

---

[98] This is the only nuclear power reactor to have so far closed down in Spain.

In the early 1980s, construction of a third group of Spanish's nuclear power reactors started. Five nuclear power reactors were initially constructed but following a 1983 moratorium adopted by the Spanish's government, only one of them were completed, Vandellos 2. In 1994, the moratorium was definitively implemented and all units under construction were abandoned.

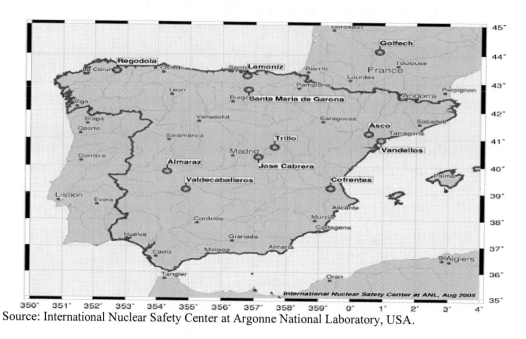

Source: International Nuclear Safety Center at Argonne National Laboratory, USA.

Figure 114. The location of the nuclear power plants in Spain[99].

During 1990s, Vandellos 1, a 480 MWe gas-graphite reactor, was shut down after 18 years of operation, due to a turbine fire which made the plant uneconomic to repair. In 2003, Enresa concluded phase 2 of the reactor decommissioning and dismantling project, which allows much of the site to be released. After 30 years, when activity levels have diminished by 95%, the remainder of the plant will be removed. The cost of the 63 month project was €93 million. In April 2006, the 142 MWe Jose Cabrera (Zorita) plant was closed after 38 years of operation. Dismantling the plant will be undertaken from 2009 by Enresa with a total cost estimated at €135 million. [93]

Spain is notable for power plant uprates. It has a programme to add 810 MWe (11%) to its nuclear capacity through upgrading its eight nuclear power reactors by up to 13%. For instance, the Almaraz nuclear power plant is being boosted by more than 5% at a cost of US$50 million. Some 519 MWe of the overall increase is already in place. Cofrentes was uprated 2% in 1988, another 2.2% in 1998, 5.6% in 2002 and 1.9% in 2003, taking it to 112% of its original capacity. Tentative plans will take it to 120% later in the decade. [93]

According with NEA and IAEA country file information, the main Spanish Nuclear Steam Supply System (NSSS) manufacturer is ENSA, which designs, produces and inspects nuclear power plants primary circuit equipment and components. This company is state owned through the Sociedad Estatal de Participaciones Industriales (SEPI), which controls

---

[99] In Figure 114, two nuclear power reactors were closed: Vandellos 1 and José Cabrera.

100% of its shares. ENSA is also constructing double purpose casks (called ENSA-DPT), to store and/or transport up to 21 PWR/KWU spent nuclear fuel assemblies. ENSA has provided primary circuit equipment, such as steam generators and reactor vessels, as well as components of the second and third generation Spanish nuclear power reactors. The company has export equipment to several countries, namely Argentina, Belgium, Germany, India, UK and USA, among others.

Nuclear Safety Authority and the Licensing Process

The licensing procedures in force in Spain are somehow similar to the ones in force in other EU countries with important nuclear power programme. According with the IAEA, the nuclear installation licensing procedure in force in Spain is regulated by the Law 25/1964 on Nuclear Energy and by Royal Decree 1836/1999 Nuclear and Radioactive Facilities Regulation, approved on December 1999. To license a nuclear installation, the following successive authorizations are needed: 1) Sitting authorization: It is a formal acknowledge of the purpose and the location submitted; 2) Construction authorization: It permits to start up the construction of the installation; and 3) Operation authorization: It permits to introduce nuclear fuel in the plant and to operate the plant. [124]

With respect to plant dismantling and plant modification, an authorization is also required prior to begin of the activity. These authorizations and permissions are granted by the Ministry of Economy, under previous and perceptive advice referring to nuclear safety and radiological protection issued by the CSN (Nuclear Safety Council).

According with IAEA sources, before granting a construction authorization, the project must be submitted for an environmental appraisal to the Ministry of Environment. As a consequence of this analysis, the project can be yield to certain requirements. To obtain theses authorizations and permissions, the documents determined in the current regulations must be submitted to the licensing authorities and the suitable tests, analyses and validations must be performed. Nuclear installations require authorizations granted by other administrative bodies, belonging to local administrations, according the rules of these bodies. Before granting the sitting authorization, a 30 days period is established for public hearings. During this period anyone can present allegations.

Finally, the Council of Minister will discuss, in December 2009, whether reform the current Law on Nuclear Safety with the purpose to introduce an environmental impact assessment to support future request of life extension of nuclear power reactors with more than 40 years of operation.

Nuclear Fuel Cycle and Waste Management

The provisions related with the nuclear fuel cycle were initially contained in the Royal Decree of 1979. Now the activities related with the nuclear fuel cycle are contained in the Royal Decree 1464/1999, which covers the front-end of the nuclear fuel cycle, while other decrees governs the activities associated with the back-end.

With regards the uranium mining, conversion and enrichment the situation in the country is the following: According with IAEA sources, due to the uranium lower market price, the

ENUSA mining activities in Saelices el Chico (Salamanca) stopped at the end of 2000. The uranium concentrates from ENUSA comes from COMINAK and from several foreign companies. COMINAK is a company from Niger, owned by several foreign companies including ENUSA, which holds 10% of its shares. There are no uranium conversion and enrichment plants in Spain. ENUSA owns 11% of Eurodif, European consortium with enrichment factories in France. ENUSA has signed several contracts with companies abroad for uranium conversion and enrichment activities.

Spain produces, through the Juzbado plant, nuclear fuel elements for most of the PWRs and BWRs systems operating in the country, as well as for some nuclear power reactors operating in Sweden, Germany, France, Finland and Belgium. According with the EC information, ENRESA[100], the state owned company, was set up in 1984 and is in charge of the radioactive waste management activities and the dismantling of nuclear installations. Its duties are as follows:

a) radioactive waste treatment and conditioning;
b) searching for locations, design, construction and operation of interim and final storage centers for high, medium and low level radioactive wastes;
c) management of the different operations related to the decommissioning of nuclear and radioactive installations;
d) to establish systems for collecting, transferring and transporting radioactive wastes;
e) to give support to civil protection services in case of nuclear emergencies;
f) final and safe conditioning of wastes derived from the mining and milling processes;
g) assuring of the long-term management of every radioactive waste storage facility;
h) to carry out the appropriate technical and economic studies, considering the deferred costs and to outline the proper economic policy.

With regards the management of nuclear waste it is important to single out the following: ENRESA draws up a proposal of General Plan for Radioactive Wastes, as established in the Royal Decree by which the company was constituted, and submits it to the Ministry of Industry and Energy for approval by the government. On July 31, 1999, the Spanish government approved the 5th General Plan for Radioactive Wastes. The basic hypotheses of this Plan, for the purpose of drawing up and performing the corresponding economic calculations, are: a) plant life for currently in operation nuclear power reactors: 40 years; b) 7 000 hours per year at 100% capacity; and c) discount rate 2.5%. [124]

The Plan approved foresee the management of 193 500 m³ of low and intermediate level of nuclear waste and around 15 000 m³ of high level spent nuclear fuel, including others kinds of high level radioactive wastes produced in the nuclear facilities operating in the country. The strategy settled for the management of the spent nuclear fuel considers a temporary storage facility, which has been constructed at Trillo nuclear power plant and will be in operation between 2002 and 2010. For this reason, the government will not adopt any decision with respect the final disposal of the spent nuclear fuel prior to 2010. Until that year, it will be necessary to conjugate two lines of research work regarding the final disposal of the spent nuclear fuel that are carried out by the Spanish competent authority. One of this line is

---

[100] Empresa Nacional de Residuos Radiactivos S.A.

to look towards a deep geological repository for the final disposal of the spent nuclear fuel. The second line of research is towards partitioning and transmutation of the spent nuclear fuel.

ENRESA has a medium and low-level radioactive waste storage installation located in El Cabril, Córdoba. It main tasks, which presently are carried out, are as follows: a) decommissioning of Vandellós Unit 1 to reach level 2; b) different activities concerning the spent fuel and high radioactive waste management and storage; c) management radioactive lighting rods; d) research and development activities; and e) international relation tasks.

In the nuclear fuel cycle front-end, the production activities of the Quercus plant were definitely finalized at the end of 2002. From the end of 2000, when uranium mining activities in the same site were stopped. [124]

Finally, it is important to single out the following. The Council of Ministers will discuss at the end of 2009 or beginning of 2010 a package of reforms regarding the final disposal of nuclear waste. The Executive will consider a report from the Ministry of industry on the need to build a single radioactive waste facility. Peoples who want to host this facility will have one month to discuss this issue and the decision to be adopted shall be approved a full municipal board. The Executive hopes to have chosen a site three months after the deadline for nominations to be presented. Until now, each nuclear power plants storage their nuclear waste in pools in the site of the nuclear power plant that produce the waste. In 2004, Congress unanimously called on the government to build a single facility for the final disposal of the nuclear waste. Although mayors deny, industry and public corporation of radioactive wastes (ENRESA) have maintained contacts with interested major of Yebra (Guadalajara), Tivissa and Ascó (in Tarragona) and Northern of Cuesta Urria (Burgos), areas where there are already a nuclear power plant in operation. In November 2008, Yebra City Council invited people to visit the Habog nuclear waste disposal facility in the Netherlands in order to see a nuclear waste disposal facility in operation The construction of a radioactive waste disposal facility will involve an investment of €800 million, a period of construction of five years and 500 new jobs.

The Council has plans to adopt a law that hardens the insurance conditions that must be subscribe by the operator of a nuclear power plant in case of a nuclear accident, The insurance prima will be increased from €700 to €1,200 million after becomes law.

The Public Opinion

In November 2007, the Forum of the Spanish Nuclear Industry carried out a survey in order to obtain information regarding the position of Spanish public opinion with respect to the use of nuclear energy for electricity generation. This was a continuation of the survey carried out in 2004 with the purpose to analyze changes in attitudes towards nuclear power. The main conclusion that may be drawn from the survey carried out in 2007 is as follows: "Spanish public opinion that is in favor of using the nuclear energy for electricity production has increased, but the number of people who are against has not decreased. In other words, some people who were or claimed to be undecided have moved to the ranks of those voting in favor, which now represent 23% of the population".

It is important to stress that most Spaniards support the use of renewable energy to address energy concerns, but generating one MWh of electricity using nuclear energy costs about a quarter of the price of using renewable sources. Additionally, Spain is required under the Kyoto Protocol to cut its current level of CO_2 emissions and this is something very difficult to achieve without the use of nuclear energy for electricity production. Spain is exceeding its 2008 target emissions by 50%.

The relative low support for the use of nuclear energy for electricity production is the result of a fear of a nuclear accident and contamination from nuclear waste, according to Carlos Bravo, director of Greenpeace's anti-nuclear campaign in Spain. This year's Eurobarometer survey on energy found 75% of Europeans aren't well informed about nuclear waste. In the case of Spain, the survey shows that it is one of the least-informed countries on this relevant issue, with 15% of people saying they are adequately informed on the subject. The lack of information is one of the reason why most Spanish people are afraid of the use of nuclear energy for electricity production.

It is important to note that, in 1976, the first elected government considered a plan for the construction of 49 nuclear power reactors to tackle soaring power demand and reduce the country's reliance on imports of fossil fuels. But in 1983, a combination of falling oil prices, opposition to use of nuclear energy for electricity generation, among other elements, limited the country's nuclear portfolio to 10 nuclear power reactors. Since then, one plant was closed after a fire and another was shut down in April 2006, the José Cabrera plant in Almonacid de Zorita, about 115 kilometers east of Madrid.

According to a poll carried out amount 1,000 adult Spanish people by Sigma Dos and published in *El Mundo* on August 2008, almost half of adults in Spain (48.3%) are against to eliminate the moratorium on the construction of new nuclear energy plant introduced in 1983, while 39.7% support ending this ban. The number of respondents that has no opinion on this matter reached 12%. In addition, 75.1% of the people would not approve the construction of a nuclear power plant in their own community, while 22% will approve it. Around 2.9% of the respondents have no opinion on this matter.

The current Spanish government has allowed the existing nuclear power plants to operate until their licenses expire. Once this happens, the decision adopted was to decommission all nuclear power reactors without exception. However, there is a strong political debate on this matter within the Spanish's political parties and within the society. The nuclear industry is strongly in favor not only in the extension of the lifetime of the current nuclear power reactors operating in the country, but of constructing new nuclear power reactors in the coming years.

Looking Forward

The future of nuclear energy for electricity generation is very uncertain in Spain. Recently, the President of the Spanish's government declared to the press[101] that "his country would choose renewable energy sources rather than the "easy option" of nuclear energy, as long as the serious problems posed by radioactive waste disposal have not been solved". The prime minister's anti-nuclear position has not change since the political agreement he signed with the Green Party prior to the legislative elections of 2004, in which he promised to gradually abandon nuclear energy in Spain in favor of safer, cleaner and cheaper alternatives. Two months later, as prime minister, he announced that "the Santa María Garoña nuclear

---

[101] Nuclear Power Debate Heats Up by José Antonio Gurriarán, March 26, 2008, IPS.

power station in Burgos would be decommissioned by 2009, to be followed by the remaining seven plants", and said also that "he would not authorize the construction of any new nuclear reactor".

The Santa María de Garoña nuclear power plant came on-stream in May 1971, with an estimated useful life of 40 years. Its owners, the ENDESA and IBERDROLA power companies, now say that it can operate without risk for another 20 years and are asking the government not to shut it down in 2011. It is interesting to mention that the Santa María de Garoña nuclear power plant, with its 85% net load factor in 2007, has been identified as the most efficient BWR system in Europe and the seventh at the world level. With these parameters in mind the closure of this plant is a political wrong decision that has been taken without any economic consideration. To sustain their position the two firms argue that in addition to the above parameters, nuclear energy is cleaner than fossil fuels, and point out that it is advisable not to depend on other countries for electricity generation.

The Ministry of the Environment took the decision on 14 January to give the go-ahead to the Declaration of Environmental Impact (DEI) for the decommissioning of the Zorita nuclear power plant (Guadalajara). The ministry considers that the environmental impact of the decommissioning of the Zorita nuclear power plant is conventional and acceptable. Of the three possible alternatives, the ministry chose the immediate total dismantling, which released the place for other uses. The process will include the decontamination and demolition, as well as the removal of waste.

It is important to stress that the high prices of oil reached in 2007 and 2008 as well as the foresee increase in the demand of energy in the coming years, are changing the minds of people in Europe, including Spain, who turned against nuclear power after the Chernobyl nuclear accident. The industry sector in Spain, particularly the nuclear industry, are pushing the government to abandon the policy of shut down nuclear power reactors and are asking authorization to extend the lifetime of several of the units currently in operation.

Table 54. Annual electrical power production for 2008

Total power production (including nuclear)	Nuclear power production
390 322 GWhe	52 486 GWhe

Source: IAEA PRIS, 2009.

According with the last study carried out by Foro Nuclear, Spain needs to build at least 13 000 MWe worth of nuclear power plants to achieve a feasible power generation mix by 2030. Nuclear power has made a comeback in some European countries, as a possible and realistic way to cut emissions of greenhouse gas carbon dioxide and to reduce dependence on uncertain supplies of imported fuels like gas and coal. The report estimated that Spain's installed generating capacity would need to rise to some 125 000 MWe by 2030, which compares to about 90 000 MWe at present. The report's main aim was to estimate the mix of different types of electrical energy Spain will need bearing in mind certainty of supply and quality of service to customers. Of the projected total, nuclear power would have to provide some 20 000 MWe, or 16% of installed capacity. The total includes 13 000 MWe to be provided by new nuclear power reactors, plus life extensions for existing ones. The purpose is that nuclear power plants should cover about 24% of projected peak winter demand of 82 700 MW.

THE NUCLEAR POWER PROGRAMME IN THE UNITED KINGDOM (UK)

The electricity production in the UK is about 400 TWh gross annually and the capacity installed is around 81 GWe. Net imports are about 8 TWh. Annual consumption is 355 TWh, or about 5 700 kWh/person. In 2006, the 19 UK nuclear power reactors[102] generated 19% of the country electricity (69 billion kWh out from some 380 billion kWh net), compared with 36% from gas and 38% from coal. In 2007, the nuclear capacity was 11 035 MWe and produced 52.5 TWh of electricity. This production represents 15% of the total electricity produced in the country in that year by all nuclear power plants. In addition, about 3% of UK electricity demand is met by imports of nuclear power from France, so overall nuclear share in the UK consumption is about 18%.

UK had in the past extensive coal deposits around the Eastern and Western edges of the Pennines, in South Wales, in the Midlands (Birmingham area), and in the Scottish Central Lowland. Easily accessible coal seams are now, however, largely exhausted. Fortunately for the energy-hungry British economy, large deposits of petroleum and natural gas under the North Sea came into commercial production in 1975 and, at present, the UK is self-sufficient in petroleum. However, the level of production of oil in the North Sea is decreasing as well as the oil reserve located in this area. This situation, as well as the commitment assumed by the UK's government to reduce the CO_2 emission to the atmosphere, has forced the government to look to renewable energy sources and, recently, to the use of nuclear energy for the generation of electricity in order to satisfy the expected growing electricity demand foresee in the future.

According with the opinion of UK's experts, nuclear appears to be a cost-effective form of carbon reduction –considerably cheaper than renewables according to the government's consultation paper. While such calculations inevitably depend on the assumptions chosen, the figures suggest that nuclear can deliver carbon reductions at a cost of around €25 per ton, while for renewables the cost is around 10 times as much, €250 per ton. The policy conclusion seems obvious: To meet tight carbon targets at minimum cost, the government should give nuclear support of the same sort as it gives to renewables, but at a much lower level. (Keay, 2007)

The Energy Policy in the UK

Until the 1980s, the UK's government energy policy encourage the use of nuclear energy for electricity production providing an increasing proportion of UK's electricity, with reprocessing of used fuel to recover fissile materials and increase the utilization of uranium. In 1988, uncertainties were expressed about the cost associated with the use of nuclear energy

---

[102] According with previous decision adopted by the UK's government, it is expected that all but one of the current number of reactors will be retired by 2023. This decision has been changed and an expansion of the UK nuclear power programme has been recently adopted.

for electricity generation in a white paper prepared by the UK's government. One year later, when the electricity system was privatized and deregulation began, the government announced that it would keep all nuclear power generation in the public sector. Then, in 1995, the UK's government published another white paper in which a throurougly revision of the nuclear power programme was made. The paper confirmed the government's commitment to the use of nuclear energy for electricity generation, but stating that no public sector will support the construction of new nuclear power reactors in the future. The UK's government stated very clearly that "the construction of new nuclear power plants should be a responsibility of the private energy industry".

The aim of the UK's government's energy policy is, according with IAEA source, to ensure secure, diverse and sustainable supplies of energy in the forms that people and businesses want, and at competitive prices. The government believes that this aim will best be achieved by means of competitive energy markets working within a stable framework of law and regulation to protect health, safety and the environment. Government policies also aim to encourage consumers to meet their needs with less energy input, through improved energy efficiency. The key elements of the UK's energy policy are:

1) to encourage competition among producers and choice for consumers and to establish a legal and regulatory framework to enable markets to work well;
2) to ensure that service is provided to customers in a commercial environment in which customers pay the full cost of the energy resources they consume;
3) to ensure that the discipline of the capital markets is applied to state owned industries by privatizing them where possible;
4) to monitor and improve the performance of the remaining state-owned industries, while minimizing distortion;
5) to have regard to the impact of the energy sector on the environment, including adopting policies and taking measures to meet international commitments;
6) to promote energy efficiency and renewable sources of energy;
7) to safeguard health and safety.

In 2006, the UK's government, bearing in mind the situation of the energy sector in the country, carried out a deep revision of its energy policy, with the purpose to put replacement of the country's nuclear power reactors firmly back on the national agenda. Two main reasons were considered: First due to energy security and second the need to limit carbon emissions due to the commitment adopted by the UK's government within the Kyoto Protocol. In November 2006, in a statement made by the Prime Minister to the parliament expressed his view that "in common with other countries around the world, particularly some countries within the EU, we need to put nuclear back on the agenda and at least replace the nuclear energy we will lose (from closing old reactors). Without it we will not be able to meet any of our objectives on climate change, or our objectives on energy security". [98]

However, it is important to single out that the current UK's energy policy regarding the construction of new nuclear power reactors is very clear: Any new unit would be financed and built by the private sector - with internalized waste and decommissioning costs. To achieve this, the government proposes to address potential barriers to new nuclear build, including design certification by Health and Safety Executive, to have new procedures, and streamlining planning permission for all large-scale energy infrastructure. In May 2007, the UK Planning Review considered the set of proposals for streamlining approval for major infrastructure projects, including the construction of new nuclear power reactors contained in the white paper prepared by the UK's government. It detached policy decisions from planning approvals and highlighted both the energy security challenge and the need to minimize carbon emissions in building 25-30 GWe of new capacity in the next two decades. The energy white paper stressed "that security of supply was now a major challenge and that rising fossil fuel prices coupled with costs on carbon emissions had changed the economic picture for clean electricity generation. It proposed stronger international and UK constraints on carbon emissions, more efforts on energy conservation and greater support for renewables, rising to £2 billion per year. Also it gave clear support for investment by the private sector in nuclear power capacity, so that nuclear power could play a significant role in UK's energy future. Excluding it from the 30-35 GWe of new generating capacity would be incurred high costs and major energy supply risks".

Lastly, it is important to note that in pursuit of the energy policy adopted by the UK's government, all former state-owned energy sector was privatized. The only part of the energy generating sector remaining in public ownership is the UK's older Magnox nuclear power reactors. Due to the privatization of the energy sector, the UK's government has no direct operational control over any part of the energy sector, which comprises private companies operating on the basis of their own commercial criteria and judgment.

The Evolution of the Nuclear Power Sector in the UK

British scientists were preeminent in the development of nuclear energy through 1940s. In 1953, the government approved construction of the first nuclear power reactor at Calder Hall. In 1955, a white paper announced the first purely commercial nuclear power programme, building up to 2 000 MWe of Magnox capacity and investigating the future use of fast breeder reactors. The first eight Magnox nuclear power reactors were small prototypes and initially dual-purpose, combining power generation with plutonium production for military purposes. However, the latter function was soon taken over by other facilities at Windscale and these Magnox reactors were reconfigured to provide power only[103].

It is important to single out that Britain's nuclear power plants are markedly different to that of any other country in the world. There are three nuclear reactor designs. The first is the Magnox design; the second is the Advanced Gas-cooled Reactors (AGR); and the third is the PWRs design. The reputation of the Magnox nuclear power reactors is very poor on the basis of their operating record. All are restricted to significantly less than their design rating and, in terms of their lifetime load factors, all fall in the bottom quarter of the world's plants. (Thomas, 2005)

In addition, the thermal efficiency of the Magnox nuclear power reactors is very poor. The first group of Magnox reactors had a thermal efficiency of 22%. The second group of this

---

[103] Magnox reactors use natural uranium metal fuel, have a graphite moderator and are cooled with carbon dioxide. Magnox fuel is so called because of its magnesium alloy cladding, and the chemical reactivity of this means that the fuel cannot be stored indefinitely but must be reprocessed.

type of reactors built in the UK rose to 28%[104]. The Magnox units were originally licensed for 30 years, but in some cases the license given was extended to 50 years.

The second type of reactors built in the UK was the AGR[105]. The AGR design was development based on a prototype which started up at Windscale in 1962. In 1964, after two years of operation, the AGR was adopted as the UK standard and 14 nuclear power reactors of this type were built at seven sites between 1965 and 1980. They are also graphite moderated and carbon dioxide cooled, but use enriched oxide fuel, which is burned up to low levels (relative to LWR fuel). The AGR reactors have a thermal efficiency of around 40%, due to very high coolant temperatures, well over 600°C (double the figure of most nuclear power reactors).

The seven AGR reactors built were of five separate AGR designs and they have a uniquely poor record both in construction over-runs and operating performance. The worst unit, Dungeness B1, took 24 years from the date in which the construction started until the date for it entered into commercial operation. It can only operate at about 90% of its design rating and its lifetime average load factor to the end of 2005 was 34.1%. Dungeness B2 has an average lifetime load factor of 40%. In general, the operating performance of the AGR reactors is very poor. Eight out of 14 AGR reactors fall in the bottom quarter of the table of the operating performance of the world's performance reactors, and only one is in the top half (Heysham B)[106]. Taking into account that each pair of AGR reactors was a unique design there was little standardization and operational problems were significant.

Summing up can be stated that most of the British nuclear power reactors cannot safely operate at their full design rating due to problems of corrosion and poor design. Like the Magnox units, the AGR reactors were designed and built by private industrial nuclear power consortia.

The information regarding the lifetime performance, construction, connection to the electric grid, enter into commercial operation and shut down of the nuclear power reactors in operation in UK is included in Tables 49-52.

Based on the data included in previous paragraphs and tables, can be concluded that the selection of the Magnox and AGR nuclear technology for the development of the nuclear power sector in the UK was not a good decision from the economical and technological point of view and could be used for those oppose to the use of nuclear energy for the production of electricity to demonstrate the ineffectiveness of the use of the nuclear technology for this purpose.

---

[104] Twenty six Magnox reactors were built in the UK. Two Magnox reactors were sold to Japan and Italy and similar units were built in France.

[105] The AGR design was based on the experience of the Magnox design, but included several significant improvements. High Temperature Gas-cooled Reactors offer a number of advantages over Light Water-cooled Reactors, for instance low operator radiation dose, on-load refueling capability and long operator response times to faults. However, the AGR design is undeniably more complex, more expensive and requires unusual materials for construction.

[106] 18th place.

Table 49. Lifetime performance of the UK nuclear power reactors

	Lifetime load factor % until 2005	Load factor in 2007
Dungeness A1	59.2	62.4
Dungeness A2	61.7	60.2
Oldbury 1	57.8	-
Oldbury 2	61.6	32.5
Sizewell A1	57.4	-
Sizewell A2	54.6	-
Wylfa 1	59.5	36.6
Wylfa 2	57.4	74.5
Dungeness B1	34.1	62.4
Dungeness B2	40.0	60.2
Hartlepool 1	56.8	63.2
Hartlepool 2	61.5	68.9
Heysham A1	58.1	48.8
Heysham A2	59.7	59.2
Heysham B1	74.0	85.7
Heysham B2	72.6	75.5
Hinkley Point B 1	68.7	44.9
Hinkley Point B 2	65.4	43.9
Hunterston B1	67.7	31.9
Hunterston B2	66.1	40.7
Torness 1	71.1	63.9
Torness 2	70.3	88.2
Sizewell B	83.5	98.5

Source: Nuclear Engineering International, June 2005 and CEA, 8th Edition, 2008.

In 1978, taking into account the poor outcome of the two nuclear reactors designs in operation in the UK, the government decided to build a PWR reactor in Sizewell. This nuclear power reactor was constructed using a Westinghouse design as reference. The construction started in July 1988. The unit was connected to the electric grid in February 1995 and entered into commercial operation in September 1995.

It is important to single out that the operating performance of the PWR Sizewell reactor is much better than any previous nuclear power reactor operating in the UK, with an average lifetime load factor of 83.5%. This is still some way below the performance of comparable nuclear power reactors in Germany and the USA, where the average is now 90% or more. The world ranking in terms of lifetime load factor is 49th, the best position between all UK nuclear power reactors[107].

---

[107] See Table 49.

Table 50. Information regarding the construction, connection to the electric grid and enter into commercial operation of the nuclear power reactors in operation in UK

Nuclear power reactor	Reactor type	Design size	Construction start	First power	Commercial operation
Oldbury 1	Magnox	300	May 1962	August 1967	December 1967
Oldbury 2	Magnox	300	May 1962	December 1967	September 1968
Wylfa 1	Magnox	590	September 1963	November 1969	November 1971
Wylfa 2	Magnox	590	September 1963	September 1970	January 1972
Dungeness B1	AGR	607	October 1965	June 1965	April 1985
Dungeness B2	AGR	607	October 1965	September 1965	April 1989
Hartlepool 1	AGR	625	October 1968	June 1983	April 1989
Hartlepool 2	AGR	625	October 1968	September 1984	April 1989
Heysham A1	AGR	611	December 1970	April 1983	April 1989
Heysham A2	AGR	611	December 1970	June 1984	April 1989
Heysham B1	AGR	615	August 1980	June 1988	April 1989
Heysham B2	AGR	615	August 1980	November 1988	April 1989
Hinkley Point B 1	AGR	625	September 1967	September 1976	October 1978
Hinkley Point B 2	AGR	625	September 1967	February 1976	September 1976
Hunterston B1	AGR	624	November 1967	January 1976	February 1976
Hunterston B2	AGR	624	November 1967	March 1977	March 1977
Torness 1	AGR	645	August 1980	March 1988	May 1988
Torness 2	AGR	645	August 1980	December 1988	February 1969
Sizewell B	PWR	1 118	July 1988	January 1995	September 1995

Source: Reference 53 and CEA 8th Edition, 2008.

Table 51. Expected shut down date for the nuclear power reactors operating in the UK

Reactors	Type	Expected shutdown
Oldbury 1 & 2	Magnox	Dec 2008
Wylfa 1 & 2	Magnox	Dec 2010
Dungeness B 1 & 2	AGR	2018
Hartlepool A 1 & 2	AGR	2014
Heysham 1 & 2	AGR	2014
Heysham 3 & 4	AGR	2023
Hinkley Point B 1 & 2	AGR	2016
Hunterston B 1 & 2	AGR	2016
Torness 1 & 2	AGR	2023
Sizewell B	PWR	2035
Total (19)		

Sources. IAEA PRIS and CEA, 8th edition, 2008.

Table 52. Decommissioned nuclear power reactors in the UK during the period 1977-2006

Reactors	Type	MWe each	Shut down
Berkeley 1 & 2	Magnox	138	1988-89
Bradwell 1 & 2	Magnox	123	2002
Calder Hall 1-4	Magnox	50	2003
Chapelcross 1-4	Magnox	50	2004
Dungeness A 1 & 2	Magnox	225	2006
Hinkley Pt 1 & 2	Magnox	235	2000
Hunterston A 1 & 2	Magnox	150	1989-90
Sizewell A 1 & 2	Magnox	210	2006
Trawsfynydd A & B	Magnox	195	1991
Windscale	AGR	32	1981
Dounreay FR	FBR	14	1977
PFR Dounreay	FBR	234	1994
Winfrith	SGHWR	92	1990
Total (26)		3 324	

Source: CEA, 8th Edition, 2008.

Nuclear Safety Authority and the Licensing Process

The licensing process in force in the UK since June 2006 has two stage. The first stage would focus on safety matters and will take three years to be implemented; the second would focus on the site and operator and will take less than a year to be implemented. Considering third generation of nuclear power reactors, a generic design authorization for each type would be followed by site-specific licenses. In February 2007, a judgment was brought against the government, notably the Department of Trade and Industry, that the public consultation process had been flawed. A further extended period of consultation took place and the government's support for new nuclear build was confirmed in January 2008. [98]

Source: International Nuclear safety Center at Argonne National Laboratory, USA.

Figure 115. Map with the location of the nuclear power plants in the UK.

The UK's government gave high priority to all aspects related with the safety of the nuclear power reactors in operation in the country after the Chernobyl nuclear accident. The main objective to be achieved is to ensure the safe operation of all nuclear power reactors to avoid a nuclear accident. The safety of UK nuclear installations and the protection of employees and the public from the potential hazards caused by them, is governed principally by provisions in the Nuclear Installations Act 1965, the Health and Safety at Work Act 1974, the Ionizing Radiation Regulations 1999 made under it and the Radioactive Substances Act 1993. No site may be used for the construction or operation of a commercial nuclear installation unless appropriate approval or planning permission has been given and a nuclear site license is granted by the Health and Safety Executive (HSE). The Nuclear Installations Inspectorate (NII) is that part of the HSE with delegated responsibility for administering the licensing function. [139]

The NII will request the operator of the propose nuclear power plant all necessary information to confirm that the future operator of the plant has the capacity to meet all the stringent safety requirements established by NII from the design of the nuclear power reactor through to decommissioning; in other words, the operator has to demonstrate to NII that the safety of the nuclear power plant operation will be properly controlled at all stages of the lifecycle of the plant.

However, according with international accepted principles, ultimate responsibility for the safety of a nuclear installation, including the safety of a nuclear power plant rest on the operating company. This principle applies independently whether the operating company is in the public or private sector. How to ensure the safety of the nuclear installations operation?

According with IAEA information, the NII carefully monitors the performance of nuclear installations against exacting standards and conditions. Should there be any doubt about a licensee's continued ability to meet its obligations, the Inspectorate has extensive powers. It can, for example, include additional license conditions at any time, direct the cessation of plant operation and ultimately direct that it be shut down altogether.

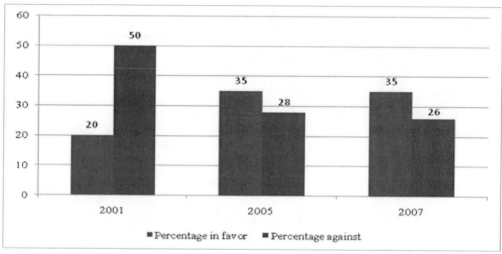

Source: IAEA Bulletin 50-1, September 2008.

Figure 116. Evolution of the support of nuclear power in the UK.

It is important to stress that an operating company may surrender a license, or it may be revoked by the NII, but in both cases still retains responsibility for all safety matters related with the nuclear installations until either a new license is issued or the HSE is satisfied with the measures adopted by the operating company.

Discharges to the environment from licensed sites of radioactive material in gaseous or liquid form are strictly controlled by two different agencies within the UK. One of these agencies is the Environment Agency in England and Wales and the second the Scottish Environment Protection Agency in Scotland. Disposal of intermediate and high level waste is a matter of ongoing policy development. There is close liaison between NII, the Environment Agency and the Scottish Environment Protection Agency under the terms of Memoranda of Understanding, which set out the lead roles of the organizations and requirements for liaison and consultation among them. As far as security regulation is concerned, security at sites operated by certain designated civil nuclear operators is regulated under a system of Ministerial Directions issued under the Atomic Energy Authority Act 1954 and the Nuclear Installations Act 1965. Nuclear power plants and associated laboratories are regulated separately under the Nuclear Generating Stations (Security) Regulations 1996.

Nuclear Fuel Cycle and Waste Management

UK as a nuclear weapon state has the control of all stages of the nuclear fuel cycle. Apart from raw uranium mining, the UK has an independent nuclear fuel cycle capability. The full range of the nuclear fuel cycle services - from fuel enrichment and manufacture through to spent nuclear fuel reprocessing, transport, waste management and decommissioning - are provided to the UK and international markets by British Nuclear Fuels Ltd (BNFL), which is wholly owned by the government. [139]

In 1995, the UK's government carried out a review of the nuclear power sector in which the government confirmed that BNFL would continue to offer the full range of nuclear fuel cycle services to national and foreign companies and restated the government's continuing support for the company in developing its overseas markets. This position of the UK's government reflect the evolution of the situation that exist with respect the use of nuclear energy for the electricity production within the EU and in other part of the world. Many countries with important nuclear power programmes are now thinking to expand these programmes in the light of the current climate change, the reduction of oil reserves, the increase in the price in fossil fuels, among others.

Regarding the management of spent nuclear fuel it is important to single out the following: Most low level waste is disposed of at either BNFL's Drigg surface disposal facility or at the disposal facilities at UKAEA's Dounreay site. Long-lived low level waste is stored and will be disposed of in Nirex' proposed facility. Intermediate level waste is currently stored, mainly at the centers of production, and will be disposed of in Nirex' proposed facility. High level wastes are currently stored, either raw or in vitrified form, mainly by BNFL at its Sellafield site, for a minimum of 50 years to cool. No decisions on final disposal have yet been taken and these will form part of a forthcoming government review, but the government is undertaking a research project to study this issue. Nuclear sites are licensed by the NII. Disposals of radioactive wastes may only be made under authorizations granted by the Environment Agency (or in Scotland by the Scottish Environment Protection Agency) but under operational agreements between them and the NII, the latter oversees waste operations on licensed sites. [139]

The Public Opinion

The UK's public opinion evolution on the question of maintaining the share of electricity provided by nuclear power and on the construction of new nuclear power reactors is the following: a) July 2001: According with a pool carried out by IPSOS MORI covering around 2,000 participants, around 20% support the use of nuclear energy for electricity generation and the construction of new nuclear power reactors, while 50% were against; b) November 2006: 35% support the use of nuclear energy for electricity generation and the construction of new nuclear power reactors; 28% were against; and c) end of 2007: 35% support the use of nuclear energy for electricity generation and the construction of new nuclear power reactors; 26% was against.

It is important to note that the support for the construction of new nuclear power reactors grew from around 30% in December 2003 to 36% in November 2007, while the opposition decreased from around 40% in December 2003 to 27% in November 2007.

In the UK's Parliament there was a substantial increase in support for building replacement nuclear capacity among ruling Labor party members in 2006. Opposition declined accordingly. Overall 61% of MPs supported new nuclear build and 22% opposed, while 66% said "nuclear should be a major contributor to UK 's energy future and 81%

thought it will be". Of Labor MPs, 60% supported new nuclear build (up from 35% in 2005) and 80% of Conservative MPs did so. By July 2007, the support had strengthened among both Labor and Conservative MPs to 72% overall and the opposition rose to 23%, based on concerns about wastes, costs and competition with renewables. The reasons for MP support in 2006 were energy security (37%), realization that renewables will not fill the gap (18%), reduce dependence on fossil fuels (15%), a good safety record (15%) and the need for balanced energy policy (12%), with very little difference between the parties except on the last, where Conservatives were stronger. Implementing effective policies on nuclear wastes before proceeding with new nuclear build was seen as necessary by 78% of MPs, but only 28% believed that there is already a clear way forward on this. Regarding MP's trusted sources of information on nuclear energy, academics rated 83%, nuclear industry leaders 51%, environmentalists 44% and media 24%. [98]

Lastly, in a report of the Nuclear Industry Association (NIA) prepared on December 2009 revealed that British women remain stubbornly opposed to the construction of new nuclear power reactors. Twenty-five years after Greenham Common, the figures show the split in attitudes to the use of nuclear energy for the generation of electricity between the male and female remains acute in Britain, and the industry is sufficiently worried by the trend to have commissioned a detailed study of female attitudes by Ipsos Mori. The most recent figures available show that about 46% of British men are in favor of building new nuclear power reactors, compared with just 26% of women. Although public support for the construction of new nuclear power reactors is growing in Britain amid concerns about climate change, expect plenty more resources to be thrown at the problem over the next few years as Britain's nuclear industry attempts to reduce the opposition of women to the construction of new nuclear power reactor. EDF, the UK's biggest nuclear operator, has already launched a campaign clearly targeted at women, featuring images of the company's female employees with the purpose to increase the support for the construction of new nuclear power reactors between female.

Looking Forward

The UK's oil and gas industry has a finite shelf-life. When the natural reserves are used up, sometime in the next 30 years, the country will face a hole in its energy resource that cannot be filled by renewable energy sources alone. The government hopes that nuclear energy will fill the power vacuum that will be created by the collapse of the indigenous oil and gas industry, while helping the UK cut CO_2 emissions to the desired 1990 levels of 60% by 2050.

According with UK's government information, in 2006 a review of energy policy was undertaken, with the purpose to reevaluate the role of nuclear energy in the energy balance of the country in the coming years. There are many reasons why the review should be carrying out. One of these reasons is energy security and the need to limit carbon emissions to the atmosphere, in order to fulfill the commitments adopted by UK in the framework of the Kyoto Protocol. The UK's government has committed to a 60% cut in carbon dioxide emissions by 2050. "If the UK is to meet its carbon reduction targets, a nuclear replacement programme is essential", said David White, energy spokesperson at the Institution of Chemical Engineers. However, the government expressed very firmly that "any new nuclear power plant would be financed and built by the private sector with internalized waste and decommissioning costs". In this regards, it is important to note the following: Almost all nuclear power reactors built until today have been funded either by government own power utilities, or by a combination of government and the private industry, or by major utilities where nuclear power plants is one of their portfolio of power assets operated by the utility. In all cases, the construction of such nuclear power reactors has not been funded against the future cash-flows of the project. Perhaps the only exceptions are the nuclear power plant already built in Dunkerque and the nuclear power reactor under construction in Olkiluoto (Unit 3). For this reason, the application of the new energy policy by the UK's government means that the new generation of nuclear power reactors to be built in the country can only be funded by existing major energy utilities with large balance sheets, such as Eon, REW of Germany and EdF of France, which already are major players in the UK's power market, or by an unprecedented project financing mechanism.

Based on what has been said above can be concluded that perhaps the intention behind the UK's nuclear power policy is to allow non-UK utilities to develop the energy sector of the country with the purpose to increase the efficiency of the nuclear power sector. The construction cost of new nuclear power reactor is estimated in around £2 billion.

In January 2008, the UK's government adopted a decision supporting the construction of new nuclear power reactors in the country in the coming years. The reasons for this change in the energy policy can be summarized as follows: 1) nuclear power generates electricity on a large scale but produces minimal greenhouse gases; 2) nuclear fuel is available in relative abundance, and is largely mined in stable economies such as Canada and Australia, both countries members of the Commonwealth and, therefore, it is considered a more secure energy source than other fossil fuels; and 3) modern nuclear power reactors are considered to be economically viable.

In connection with this new government energy policy, the British Energy called for private equity partners to help fund the construction of new nuclear power reactors. Eon and EdF both expressed interest in building new nuclear power reactors in the UK in the coming years and, for practical reasons, this would initially be on existing sites controlled by BE or British Nuclear Group. For convenience and finance's sake, it is expected that Britain's new nuclear power reactors will be built on or close to existing nuclear sites. Because of the high demand for power in the south east and London, Bradwell in Essex will join Sizewell, Dungeness and Hinkley Point as the most likely sites in England, closely followed by Wylfa in Anglesey, Hartlepool and Heysham. The two Scottish sites at Hunterston and Torness would be obvious candidates but the pro-oil and gas Scottish National Party has threatened to block their development, citing nuclear waste and the need to move towards renewable energies as their main concerns. The delays in the construction of new nuclear power reactors could delay somehow the return of the investment and this situation could limit the possibilities of the construction of new units in the coming years. However, Areva NP has hopes that the construction period of its ERP will be, in the case of the UK, around three-and-a-half years. According with WANO source, Areva NP in conjunction with EdF, applied for Generic Design Assessment (GDA) of its 1 600 MWe EPR design. A total of six utilities will be up to speed and able to support any site license application for EPR units, though only EdF will share costs of the GDA with Areva. EdF has said that "it wants to build several EPR units in the UK". The EPR received French design approval in 2004. Areva is expecting to build its first ERP unit in the UK by 2017, if planning procedures were improved and the government decisions were made on waste. GE-Hitachi Nuclear Energy has also applied for

GDA of its ESBWR system, supported by Iberdrola, RWE power and BE. Atomic Energy of Canada Ltd also applied for GDA for its ACR-1000 design, but later withdrew it to concentrate on Canadian projects.

Separately, British Energy has said that "it would support all four GDA applications and that it was conducting its own review of reactor designs from the four vendors which had submitted documentation to the regulator. It controls many of the likely sites for new nuclear power plants and has said that all of its sites would be suitable for new build, even considering possible sea level rise due to climate change. It has made transmission connection agreements with National Grid for possible new nuclear power reactors at Sizewell, Dungeness, Bradwell and Hinkley. The first three are in the southeast of England, with Hinkley in the southwest. The agreements will facilitate appropriate electric grid connections for a range of possible reactor types to be in place from 2016 onwards. [98]

In the other hand, the new White Paper on Nuclear Power prepared by the UK government in the first part of 2008, "put nuclear energy at the heart of the UK's government's response to the need for secure, safe, affordable, low-carbon energy supplies". The Prime Minister framed the new policy in terms of "taking determined long-term action to reduce carbon emissions, using nuclear power as a tried and tested technology, which has provided the UK with secure supplies of safe, low-carbon electricity for half a century".

Some 30-35 GWe of new generating capacity will be required in the next two decades, most of it base-load, in order to satisfy the foresee increase in the demand of energy in the country in the coming years. The government itself will take active steps to open up the way for the construction of new nuclear power plants by addressing sitting assessment criteria and progress in the GDA of new reactors over the next three years (up to 2011).

Based on the experience in the construction nuclear power reactors all over the world, the following areas could need some kind of government support to encourage the participation of the private sector in the construction of new nuclear power reactors in the UK. These areas are the following:

1) *construction cost*. The construction cost of a new nuclear power reactor would be high and there would be a significant risk of cost over-runs. The government might therefore have to place a cap on the cost a private investor would have to pay;

2) *non-fuel operations and maintenance cost*. Similarly, this is largely under the control of the owner and they may be willing to bear this risk;

3) *nuclear fuel cost*. Purchasing fuel has not generally been seen as a risky activity. Uranium can easily be stockpiled and the risk of increasing fuel purchase cost can be dealt with. The cost of spent nuclear fuel disposal (assuming reprocessing is not chosen) is, however, much more contentious and nuclear owners might press for some form of cap on disposal cost similar to the US arrangements;

4) *decommissioning cost*. The cost of decommissioning is very hard to forecast, but the costs will arise far into the future. Contributions to a well-designed segregated decommissioning fund appear relatively manageable, although if experience with decommissioning and waste disposal does reveal that current estimates are significantly too low, or if returns on investment of the fund are lower than expected,

contributions might have to be increased significantly. Private developers might therefore seek some 'cap' on their contributions. (Thomas, 2005)

The government's cost-benefit analysis used a figure of £1,250/kW for overnight capital cost and estimated that a nuclear power reactor generating 1 600 MWe would cost around £2.8 billion.

Taking into the information included in previous paragraphs a pre-development and planning period of 5.5 years is envisaged, so that construction of new nuclear power reactors could commence not before 2013. The UK's government is expecting that several new nuclear power reactors could be in operation by 2020. The number of new power reactors proposed to be constructed in the UK in the future is six, with a capacity of 9 600 MWe. However, the construction of up to 10 new nuclear power reactors are also mentioned by some UK's government officials and representatives of the press and the nuclear industry in the UK and France.

THE NUCLEAR POWER PROGRAMME IN THE RUSSIAN FEDERATION

Russia is a large country abundant in energy resources of various kinds. The country has a large reserves of oil, coal and gas. The energy sector is a well-developed and important part of the national economy, producing about 10% of national Gross Domestic Product (GDP). The dependency of the country of fossil fuels for the generation of electricity and for the consumption of energy is very high: up to 95% of the country's energy consumption is met by fossil fuel.

Until April 2009, Russia has 31 nuclear power reactors in operation with a capacity of 21 743 MWe, producing 150.1 TWh of electricity which represents around 16% of the total electricity production of the country. Russia has nine nuclear power reactors under construction with a net capacity of 5 980 MWe, a plan for the construction in the near future of 11 nuclear power reactors with a capacity of 12 870 MWe and a proposal to construct 25 additional units with a net capacity of 22 280 MWe. To achieve this goal it is necessary to build at least 2 nuclear power reactors per year.

The participation of the nuclear power sector in the energy balance of the country in the coming years should be around 25%, this means an increase of 9% from the level reached in 2007 with the purpose to satisfy the foresee increase in the demand of energy. To implement this ambitious plan the government will spend €18.3 billion in the construction of new nuclear power reactors until 2015 and another €20 billion with the same purpose through 2030. Russia is constructing two nuclear power reactors in Bulgaria and additional nuclear power reactors in Iran, China and India.

Table 53. Annual electrical power production for 2008

Total power production (including nuclear)	Nuclear power production
902 000 GWhe	152 057.79 GWhe

Source: IAEA PRIS, 2009.

The Energy Policy and Strategy in Russia

Despite its rich oil, gas and coal potential, Russia was one of the first countries to use nuclear energy for electricity generation. However, Russia is now moving steadily forward with plans for much expanded role of nuclear energy, with the purpose to at least doubling output by 2020. In 1996, nuclear power accounted for 13.1% of all electricity production in Russia. In 1999, nuclear power plants generated 14.5%. In 2006, the share of nuclear energy in the production of electricity was 15.9%, an increase of 13,9% in the last years. In 2007, nuclear power plants generated 16% of the total electricity produced in the country. In 2008, the nuclear share was 16.86%.

To achieve the above mentioned goals the Russian's government adopt two documents called "Nuclear Power Strategy of Russia for the Period to the Year 2020" and "Strategy for Development of the Russian Nuclear Power Sector in the First Half of 21th Century". The purpose of these two documents is to develop the nuclear energy sector in the country in the coming years by stating priorities as well as for the radical revision of structural and technological policies that pertain to the nation's energy supply for the period 1995-2010. Its main goal is to achieve the European level of per capita energy consumption and ecological safety of the population. The energy policy adopted make emphasis in the solution of regional energy supply problems that are affecting several countries. The main objectives of the approved strategy are the following: a) assuring the safety of operating nuclear power reactors, including those constructed in accordance with old regulations and the safety enhancement of nuclear power reactors under construction; b) development of improved new generation of nuclear power reactors; c) feasibility studies on the advanced nuclear power reactor concepts; d) research and development work on closed nuclear fuel cycle; e) research and development efforts on decommissioning of nuclear power reactor; f) development of cost-effective and environmentally safe spent nuclear fuel and radioactive waste management technology; g) safe operation of the research reactors, critical assemblies and other nuclear facilities; and h) remodeling research centers, experimental facilities and industrial units which support the nuclear industry development programme.

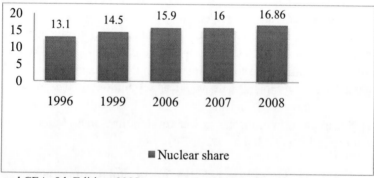

Source: IAEA and CEA, 8th Edition, 2008.

Figure 117. Evolution of nuclear share during the period 1996 to 2008.

The prognosis of energy sector development in the near future is based on: a) overcoming the national economic crisis and subsequent rise; b) new investment strategy; c) new price and taxation policies[108]; d) privatization and de-nationalization; and e) modernization of national laws and regulation in the energy sector.

The structural policy of the energy sector for the next 10-15 years aims:

1) enhancement of the efficiency of natural gas utilization and an increase its share of domestic consumption, especially in ecologically strained regions;
2) in-depth processing and comprehensive utilization of hydrocarbon raw materials;
3) enhancement of the coal quality, as well as the stabilization of coal production volumes;
4) reversal of the decline in, and moderate expansion of, oil production;
5) intensification of local and renewable energy resources development (hydro and wind power, peat, etc.);
6) priority in electricity generation development based on competitive and ecologically clean power plants;
7) safety and reliability enhancement of the first generations of nuclear power reactors and development of new advanced type of reactors.

According with the IAEA, the new technological energy policy is oriented toward: 1) radical enhancement of both the cost effectiveness and the energy efficiency of all stages of the extraction, conversion, distribution, and utilization of energy resources; 2) effective decentralization of the energy supply; 3) ecological and accident safety, as well as the reliability of the energy supply; and 4) development of qualitatively new technologies for the stable evolution of power industry: ecologically clean coal-fired power plants, safe nuclear power plants, efficient processes for the utilization of new sources of power, etc.

It is important to single out that regional energy policy adopted by the Russian government takes into account the main differences of energy supply conditions and structures of fuel resources that exist in various parts of the country.

The Evolution of the Nuclear Power Sector in Russia

A summary of the nuclear energy sector in Russia, based on the IAEA country information is included in the following paragraphs. In 1937, Russia started an active experimental studies on the structure of atomic nuclei. Production of pulse amount of neptunium and plutonium began in Leningrad Radium Institute in that year. Two years later, started the research into the feasibility of achieving a nuclear chain reaction. The largest cyclotron in Europe at that time was installed in the Leningrad Physical and Technical Institute. In 1940, the phenomenon of spontaneous nuclear fission in uranium-was discovered and a theoretical demonstration of the feasibility of energy release from a uranium nuclear fission chain reaction was carried out by a group of Soviets scientist. In 1943, a special physics laboratory, the No. 2 Laboratory in Moscow now known as the Russian Scientific Centre "Kurchatov Institute", was established. In 1945/1946, the production of metallic uranium was carried out in the country for the first time in Soviet's history. During this period

---

[108] Pricing and taxation constitute the core of the new Russia's Energy Policy.

the first experimental nuclear reactor[109] began its operation. In 1948, the first industrial nuclear power reactor started its operation after the controlled uranium fission chain reaction at the No. 2 Laboratory was achieved.

In 1954, Russia's first nuclear power reactor and the first in the world to produce electricity, the 5 MWe Obninsk nuclear power reactor, was connected to the electric grid and, in June of the same year, started the commercial production of electricity. Russia's first four commercial-scale nuclear power reactors started up its construction during the period 1957 - 1964. These were the Novovoronezh 1, that was shut down in 1988, Beloyarsky1 and 2, that were shut down in 1983 and 1990 respectively and Novovoronezh 2 that was shut down in 1990. In 1967, the first of today's production models were commissioned. These were water-moderated, water-cooled vessel-type (WWER) reactor at Novo-Voronezh. The construction of the first commercial boiling water-cooled graphite moderated reactor with nuclear superheating of the steam at Beloyarsk started in 1969. The reactor was connected to the electric grid in April 1980 and entered into commercial operation in November 1981.

In November 1974, the first commercial water-cooled graphite-moderated channel-type (RBMK) system at Leningrad started its operation. In 1974, commissioning of the world's first prototype-scale fast breeder reactor (BN-350) in Aktau for electricity generation and desalinated water production was given. In December 1976, the completion of the first nuclear central heating and power plant at Bilibino in the far northeastern part of Russia was achieved.

During the period 1984-1986, the commissioning of the Zaporozhie and Balakovo nuclear power plants with WWER-1000 units with full compliance to the new safety regulation was approved.

In Table 54 a summary of the information regarding the construction initial date, connection to the electric grid and entered into commercial operation of all nuclear power reactors currently in operation in Russia is included.

Summing up can be stated that Russia's nuclear power plants, with 31 operating nuclear power reactors totaling 21,743 MWe, consist of:

a) two first generation WWER-440/230 or similar PWRs;
b) two second generation WWER-440/213 PWRs;
c) six third generation WWER-1000 (V-320) PWRs with a full containment structure;
d) five WWER of different types (V-187, V-179 and V-338);
e) eleven RBMK light water graphite reactors. A further RMBK unit in Kursk is under construction;
f) four small graphite-moderated BWRs in eastern Siberia, constructed in the 1970s for cogeneration (EGP-6 model);
g) one BN-600 fast-breeder reactor.

In 1986, occurred the most serious accident in a nuclear power plant in operation in the world. The accident occurred at Unit 4 of Chernobyl nuclear power plant in Ukraine. This accident stop de implementation of the nuclear energy programme in the former USSR,

---

[109] This nuclear reactor was a high-purity graphite nuclear reactor.

reducing to only two nuclear power reactors connected to the electric grid between 1986 and 1993[110].

Table 54. Nuclear power reactors in operation in Russia

Reactor	Type V (PWR)	MWe net each	Commercial operation	Scheduled close
Balakovo 1-2	V-320	950	5/86, 1/88	2015, 2017
Balakovo 3-4	V-320	950	4/89, 12/93	2018, 2023
Beloyarsk 3	BN600 FBR	560	11/81	2010
Bilibino 1-4	LWGR EGP-6	11	4/74, 2/75, 2/76, 1/77	2009, 09, 11, 12
Kalinin 1-2	V-338	950	6/85, 3/87	2014, 2016
Kalinin 3	V-320	950	11/05	2034
Kola 1-2	V-230	411	12/73, 2/75	2018, 2019
Kola 3-4	V-213	411	12/82, 12/84	2011, 2014
Kursk 1-2	RBMK	925	10/77, 8/79	2021, 2024
Kursk 3-4	RBMK	925	3/84, 2/86	2013, 2015
Leningrad 1-2	RBMK	925	11/74, 2/76	2018, 2020
Leningrad 3-4	RBMK	925	6/80, 8/81	2009, 2011, +20 year
Novovoronezh 3-4	V-179	385	6/72, 3/73	2016, 2017
Novovoronezh 5	V-187	950	2/81	2010
Smolensk 1-3	RBMK	925	9/83, 7/85–10/90	2013, 2020
Volgodonsk 1	V-320	950	12/01	2030
Total: 31		21 743		

Source: IAEA and CEA, 8th Edition, 2008.

In 1999, Unit 2 of the Kursk nuclear power plant with monitoring of all fuel channels and with their partial substitution according to the check results was approved. In 2001, Rostov 1 (now known as Volgodonsk 1), the first of the delayed units started its operation, joining 21 GWe already on the electric grid. This greatly boosted morale in the Russian nuclear industry. It was followed by the conclusions of the construction of Kalinin 3 in 2004. In 2006, the government's decided to reinitiate its nuclear power programme with the purpose to add 2-3 GWe per year until 2030, as well as exporting nuclear power reactors to meet world demand for some 300 GWe of new nuclear capacity.

The location of the Russian's nuclear power plants is shown in Figure 118.

In the framework of the implementation of its nuclear power programme for industry development in the Russian Federation for 1998-2005 and for the period until 2010 authorized by the Decree of the Russian Federation Government No. 815, July 21, 1998, it is expected to extend the operation of the nuclear power reactors beyond the designed lifetime period of 30 years due to the implementation of a set of measures that will ensure their safety during continued operation. [113]

---

[110] Unit 3 in Smolensk nuclear power plant and Unit 4 in Balakovo nuclear power plant.

Table 55. Status of the Russian nuclear power reactors

Station	Type	Net capacity MWe	Status	Construction date	Criticality date	Grid date	Commercial date
Balakovo-1	WWER	950	Operational	01-Dec-80	12-Dec-85	28-Dec-85	23-May-86
Balakovo-2	WWER	950	Operational	01-Aug-81	02-Oct-87	08-Oct-87	18-Jan-88
Balakovo-3	WWER	950	Operational	01-Nov-82	16-Dec-88	25-Dec-88	08-Apr-89
Balakovo-4	WWER	950	Operational	01-Apr-84	March-93	May-93	22-Dec-93
Beloyarsky-3	FBR	560	Operational	01-Jan-69	26-Feb-80	08-Apr-80	01-Nov-81
Bilibino -1	LWGR	11	Operational	01-Jan-70	11-Dec-73	12-Jan-74	01-Apr-74
Bilibino-2	LWGR	11	Operational	01-Jan-70	07-Dec-73	30-Dec-74	01-Feb-75
Bilibino -3	LWGR	11	Operational	01-Jan-70	06-Dec-74	22-Dec-75	01-Feb-76
Bilibino -4	LWGR	11	Operational	01-Jan-70	12-Dec-75	27-Dec-76	01-Jan-77
Kalinin-1	WWER	950	Operational	01-Feb-77	10-Apr-84	09-May-84	12-Jun-85
Kalinin-2	WWER	950	Operational	01-Feb-82	25-Nov-86	03-Dec-86	03-Mar-87
Kalinin-3	WWER	950	Operational	01-Oct-85	Nov-2004	Dec-2004	Nov-2005
Kola-1	WWER	411	Operational	01-May-69	26-Jun-73	Dec-73	28-Dec-73
Kola-2	WWER	411	Operational	01-May-69	30-Nov-74	09-Dec-74	21-Feb-75
Kola-3	WWER	411	Operational	01-Apr-77	07-Feb-81	24-March-81	03-Dec-82
Kola-4	WWER	411	Operational	01-Aug-76	07-Oct-84	11-Oct-84	06-Dec-84
Kursk-1	LWGR	925	Operational	01-Jun-72	25-Oct-76	19-Dec-76	12-Oct-77
Kursk-2	LWGR	925	Operational	01-Jan-73	16-Dec-78	28-Jan-79	17-Aug-79
Kursk-3	LWGR	925	Operational	01-Apr-78	09-Aug-83	17-Oct-83	30-Mar-84
Kursk-4	LWGR	925	Operational	01-May-81	31-Oct-85	02-Dec-85	05-Feb-86
Leningrad-1	LWGR	925	Operational	01-March-70	12-Sep-73	21-Dec-73	01-Nov-74
Leningrad-2	LWGR	925	Operational	01-Jun-70	06-May-75	11-Jul-75	11-Feb-76
Leningrad-3	LWGR	925	Operational	01-Dec-73	17-Sep-79	07-Dec-79	29-Jun-80

Station	Type	Net capacity MWe	Status	Construction date	Criticality date	Grid date	Commercial date
Leningrad-4	LWGR	925	Operational	01-Feb-75	29-Dec-80	09-Feb-81	29-Aug-81
Novovoronezh3	WWER	385	Operational	01-Jul-67	22-Dec-71	27-Dec-71	29-Jun-72
Novovoronezh-4	WWER	385	Operational	01-Jul-67	25-Dec-72	28-Dec-72	24-Mar-73
Novovoronezh-5	WWER	950	Operational	01-March-74	30-Apr-80	31-May-80	20-Feb-81
Smolensk-1	LWGR	925	Operational	01-Oct-75	10-Sep-82	09-Dec-82	30-Sep-83
Smolensk-2	LWGR	925	Operational	01-Jun-76	09-Apr-85	31-May-85	02-Jul-85
Smolensk-3	LWGR	925	Operational	01-May-84	01-Dec-89	17-Jan-90	30-Jan-90
Rostov-1	WWER	950	Operational	01-Sep-81	17-Feb-01		30-Mar-01
Total	31	21 743					

Source: CEA, 8th Edition 2008 and IAEA PRIS.

In addition to the above activities carried out in 5 units in different nuclear power plants during the period 2004-2006, the following six units were also upgraded and their lifetime extended: in 2004, Kola 2 and Bilibino 2; in 2005, Leningrad 2 and Bilibino 3 and, in 2006, Kursk 1 and Bilibino 4 and Kola 1 and 2; for 10 years, the RBMK units in Leningrad 1 and Kurks 2. The related licenses were obtained from Gosatomnadzor of Russia for extended operation of the above power units.

Source: Reference 113.

Figure 118. The location of most of the Russian's nuclear power plants.

Russia's Nuclear Technology Development

Russia is one of the world's exporters of nuclear technology. Until 2000, Russia exported several nuclear power reactors to different countries within the Eastern European bloc. The main nuclear power reactors technology exported to countries outside Russia was the WWER-440 and WWER-1000 type of reactor. The RMBK reactor design was built, mainly, to produce electricity in the former Soviet Union. The main Soviet nuclear power reactor design being deployed is the V-320 version of the WWER-1000, from OKB Gidropress (Experimental Design Bureau Gidropress), with 950-1000 MWe net output. Advanced versions of this type of reactor with Western instrument and control systems have been built at Tianwan in China and at Kudankulam in India as AES-91 and AES-92 models respectively. An AES-91 model was bid for Finland in 2002. The AES-92 model was bid for Sanmen and Yangjiang in China in 2005 and was accepted for Belene in Bulgaria in 2006[111]. [89]

In 2005, Rosatom (the Federal Atomic Energy Agency) promoted the design for WWER-1500 PWRs. However, this plan was overtaken by development of the AES-2006 model incorporating a third-generation standardized WWER-1200 design with a capacity of 1 170 MWe. This is an evolutionary development model of the well-proven WWER-1000 in the AES-92 design, with longer life of 50 years instead of 30, greater power and greater efficiency. [89] The efficiency in the AES-92 model is expected to increase from 31.6% to 36.56%. The lead first units will be built at Novovoronezh 2 and Leningrad 2. Novovoronezh 2 should start commercial operation in 2012-13 and Leningrad 2 in 2013-14.

A new design of nuclear power reactor, the AES-2006, will consist of two nuclear power reactors expected to run for 50 years with a capacity factor expected to be around 90%. Capital cost could be around US$1,200/kW, but in the case of the first contract to build this type of reactor, capital cost is more like to be US$2,100/kW. The construction time is expected to be 54 months. The AES-2006 model have enhanced safety, including those related to earthquakes and aircraft impact with some passive safety features, double containment and core damage frequency of 1x10-7.

The main characteristics of the above mentioned soviet nuclear power reactor designs in comparison with Western design reactors, are summarizes in the following paragraphs:

1) Western design reactors employ the design principle of safety in depth, relying on a series of physical barriers, including a massive reinforced concrete structure called the containment, to prevent the release of radioactive material to the environment. With the exception of the WWER-1000 design, Soviet design reactors do not have such a containment structure;

2) Soviet design reactors are essentially variations on two basic designs: The WWER, or PWRs type, and the RBMK, the graphite moderated channel reactor;

3) three generations of Soviet design of WWER reactors, upgraded over time, are operating in Eastern Europe and the former Soviet Union. The first generation, the WWER-440 Model V230, operates at four nuclear power plant sites in three

---

[111] Major components of the two designs are the same, except for slightly taller pressure vessel in AES-91, but cooling and safety systems differ. The AES-92 has greater passive safety features but the AES-91 has extra seismic protection.

countries: Russia, Bulgaria and the Slovak Republic. The second generation, the WWER-440 Model V213, operates at five nuclear power plant sites in the same number of countries: Russia, Ukraine, Hungary, the Czech and the Slovak Republics. The third generation, the WWER-1000, operates at eight plant sites in three countries: Russia, Ukraine and Bulgaria.

4) at the time of the collapse of the Soviet Union, two advanced versions of the WWER-1000 were under development. Russia has continued the development of an upgraded WWER-1000, and has developed a new design for a 640 MW reactor with enhanced safety features;

5) three generations of RBMK reactors are operating in the former Soviet Union: 11 units in Russia, one in Ukraine and two in Lithuania (one unit was shut down in December 2004 and the second is expected to be shut down in 2009). Despite improvements to the RBMK design since the Chernobyl nuclear accident, concerns remain about these reactors, especially the first-generation ones. [126]

The main Russian PWRs in order of development are shown in Table 56.

The main differences between the all Soviet and Western nuclear power reactor designs can be summarizes as follows: Soviet design nuclear power reactors differ of the designs produced by other countries in many respects, including plant instrumentation and controls, safety systems and fire protection systems. While Soviet design reactors, like other nuclear power reactors, employ the design principle known in the West as "defense in depth," only one Soviet reactor design includes a containment structure as part of that principle. In the unlikely event that safety systems fail, reactors designed on the "defense in depth" principle rely on a series of physical barriers to prevent the release of radioactive material to the environment. At US plants: a) the first barrier is the nuclear fuel itself, which is in the form of solid ceramic pellets. Most of the radioactive by-products of the fission process remain bound inside the fuel pellets; b) these pellets are then sealed in rods, made of special steel, about 12 feet long and half an inch in diameter; c) the fuel rods are inside a large steel pressure vessel, which has walls about eight inches thick; d) at most nuclear power plants, this vessel is enclosed in a large, leak-tight shell of steel plate; all this is contained inside a massive steel and/or concrete structure, called the containment, with walls several feet thick.

Table 56. Type of Russian nuclear power reactors

Generic reactor type	Reactor model	Power plant
WWER-440	V-230	
"	V-213	
WWER-1000	V-320	
"	V-392	AES-91
"	V-392	AES-92
WWER-1200		AES-2006
WWER-1500	V-448?	

Source: WANO data base.

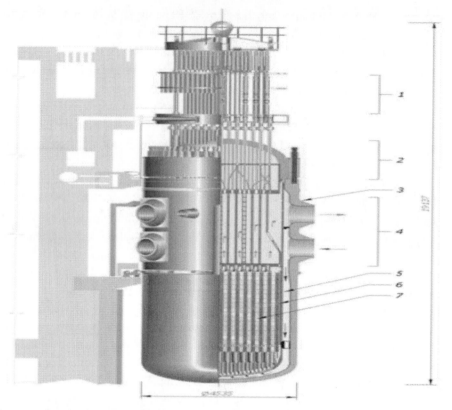

Note: 1 - control rods; 2 - reactor cover; 3 - reactor chassis; 4 - inlet and outlet nozzles; 5 - reactor
vessel; 6 - active reactor zone; 7 - fuel rods.
Source: Wikimedia Commons (Panther).

Figure 119. The WWER design.

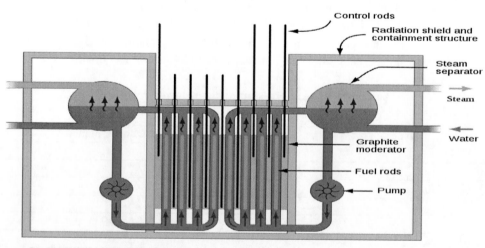

Source: Courtesy Wikipedia (Fireice).

Figure 120. The RMBK design.

Most Soviet design reactors employ similar features, but only the WWER- 1000[112] design has a containment structure like that of most nuclear power reactors elsewhere in the world. Without this protection, radioactive material could escape to the environment in the event of a serious accident. [126]

It is important to stress, regarding the WWER soviet nuclear power reactor design, the following: According with the IAEA source, although it shares a basic engineering concept with its counterparts, the Soviet PWR design is very different and does not meet Western safety standards. However, second- and third-generation of this design—the WWER-440 Model V213 and WWER-1000—are widely viewed as having a design safety basis sufficiently comparable to that used in the West to justify short-term and long-term safety and performance upgrades on both safety and economic grounds. The future Russia nuclear power technology is focus on the following four elements: a) serial construction of AES-2006 model; b) the development of the fast breeder BN-800 model; c) the serial construction of small and medium nuclear power reactors KLT-40 and WWER-300 models (100-300 MWe); and d) the construction of the HTGR model.

Following progress on these types of nuclear power reactors, the physically larger WWER-1500 design may be completed[113]. In the other hand, the BN-800 fast neutron reactor model being built by OKBM at Beloyarsk is designed to supersede the BN-600 unit, which is now operating in this site and will utilize MOX fuel with both reactor-grade and weapons plutonium. Further BN-800 units were planned to be built and a BN-1800 model has being designed for operation from 2020. [89]

Other development in the field of nuclear technology are, according with WANO source, the following: The VK-300 BWR design is being developed by the Research and Development Institute of Power Engineering (NIKIET) for both power (250 MWe) and desalination (150 MWe plus 1 675 GJ/hr). It has evolved from the VK-50 BWR model at Dimitrovgrad, but uses standard components wherever possible, e.g. the reactor vessel of the WWER-1000. A feasibility study on building four cogeneration VK-300 units at Archangelsk was favorable, delivering 250 MWe power and 31.5 TJ/yr heat. During the 1970-1980s, OKBM undertook substantial research on HTGRs. In the 1990s, it took a lead role in the international GT-MHR (Gas Turbine-Modular Helium Reactor) project based on a General Atomics (USA) design. Preliminary design was completed in 2001 and the prototype is to be constructed at Seversk (Tomsk-7, Siberian Chemical Combine) by 2010, with construction of the first four module nuclear power plant (4x285 MWe) by 2015. Initially, it will be used to burn pure ex-weapons plutonium, and replace production reactors which still supply electricity there. But in the longer-term perspective HTGRs are seen as important for burning actinides and later for hydrogen production.

---

[112] See Figure 119.

[113] Others nuclear power reactors with advanced safety features which were under development by Russia is the WWER-640 model developed by Gidropress jointly with Areva NP. The construction of the first nuclear power reactor of this type was stop due to lack of funds and is not included in recent plans for the construction of new nuclear power reactors in the country. The development of the MKER-800 model, an advanced RMBK design with improved safety system and containment, was stop and no further plans has been adopted for the construction of this type of reactors in the country in the future. The Brest Lead Cooled Fast Reactor model is another new type of nuclear power reactor that has been developed by Russia. the construction of the first unit has been proposed for the Beloyarsk nuclear power plant.

Since 2001, Russia has been participating very actively in the implementation of one of the most important programme promote by the IAEA in the field of nuclear power development, the IAEA project on Innovative Nuclear Reactors and Fuel Cycles (INPRO). In 2006, Russia joined the Generation-IV International Forum and the country is also a member of the NEA's Multinational Design Evaluation Programme, which is increasingly important in rationalizing reactor design criteria. In March 2008 AtomEnergoProm signed a general framework agreement with Japan's Toshiba Corporation under which they will explore collaboration in the civil nuclear power business.

Several feasibility studies were carried out with the purpose to identify specific areas of cooperation, such as design and engineering for the construction of new nuclear power reactors, manufacturing and maintenance of large equipment and front-end civilian nuclear fuel cycle business, among others. This cooperation could lead to the establishment of a strategic partnership between Japan and Russia in nuclear power.

Finally, it is important to single out the agreement signed, in October 2009, between Russia and China for the development of a fast breeder reactor. Russia and China have signed an agreement to begin the preliminary works on the design for two commercial fast breeder reactors of 800 MWe that will be installed in the coast of China. Last year a bi-national commission had propitiated the construction of a fast breeder reactor in China of similar characteristics to the BN 800 model that Russia is building in Beloyark. It is expected that this reactor should enter into commercial operation in 2012.

Improving Reactor Performance

It is important to note that Russia is also putting great emphasis in the improvement of the operation of present designs of nuclear power reactors with better fuels and greater efficiency in their use, closing much of the gap between Western and Russian reactors performance. One of the emphasis made is in the field of fuel developments, particularly the use of burnable poisons, gadolinium and erbium, as well as structural changes to the fuel assemblies.

According with WANO information, with uranium-gadolinium fuel and structural changes WWER-1000 fuel has been pushed out to four-year endurance and WWER-440 fuel even longer. For WWER-1000, five years is envisaged by 2010, with enrichment levels increasing nearly by one third (from 3.77% to 4.87%), average burn-up going up by 40% and operating costs dropping by 5%. With a 3 x 18 month operating cycle, burn-up would be lower but load factor could increase to 87%. Comparable improvements were envisaged for the later model of the WWER-440. For the RBMK design the most important development has been the introduction of uranium-erbium fuel at all units currently in operation, though structural changes have helped. As enrichment and erbium content are increased (e.g. from 2.6% to 2.8% and 0.6% respectively) increased burn-up is possible. Leningrad has implemented the higher-enriched fuel and Kursk is starting to do so. For the BN-600 reactor, improved fuel means up to 560 days between refueling. Beyond these initiatives, the basic requirements for fuel have been set as: a) fuel operational lifetime extended to 6 years; b) improved burn-up of 70 GWd/tU; and c) improved fuel reliability. In addition, many nuclear power plants will need to be used in load-following mode, and fuel which performs well under variable load conditions will be required. [89]

One of the main innovations introduced in the RMBK design[114] is the use of recycled uranium from WWER units in operation in several nuclear power plants in Russia. All RBMK models now use recycled uranium from WWER models and some has also been used experimentally at Kalinin 2 and Kola 2 WWER models. It is intended to extend the result of this experience to other WWERs reactors.

Nuclear Safety Authority and the Licensing Process

Nuclear safety and the management of nuclear waste were not a primary concern of the first type of Soviet reactor designs. According with WANO source, an early indication of the technological carelessness in the construction of nuclear power reactors and the management of nuclear waste was the substantial pollution followed by a major accident at Mayak Chemical Combine (then known as Chelyabinsk-40) near Kyshtym in 1957. The failure of the cooling system for a tank storing many tons of dissolved nuclear waste resulted in a non-nuclear explosion having a force estimated at about 75 tons of TNT. Up to 1951 the Mayak plant had dumped its wastes into the Techa river, whose waters ultimately flow into the Ob River and Arctic Ocean. Then they were disposed of into Lake Karachay until at least 1953, when a storage facility for high-level wastes was built - the source of the 1957 accident. Finally, a 1967 dust storm picked up a lot of radioactive material from the dry bed of Lake Karachay and deposited it on to the surrounding province. The outcome of these three events made some 26,000 square kilometers the most radioactively-polluted area on Earth by some estimates, comparable with Chernobyl. [89]

However, it is important to note that after the Chernobyl nuclear accident there was a significant change of culture in the Russian civil nuclear establishment, at least at the plant level. This nuclear accident put the whole nuclear industry into a long standby and Russia was strongly criticized for the way in which the country deals with the accident and for the failure in the design of this type of nuclear power reactor. This change in the culture of the safety of nuclear power plants was even more evident in the Eastern Europe countries that were using Soviet nuclear power reactor designs. By the early 1990s, after the collapse of the Soviet Union, a number of Western assistance programmes were in place, which addressed safety issues and helped to alter fundamentally the way things were done in these countries, including Russia itself. Design and operating deficiencies were tackled and a safety culture started to emerge. Several modifications were made to overcome deficiencies in the 12 RBMK reactors still operating in Russia and Lithuania, with the aim to remove the danger of a positive void coefficient response. Several automated inspection equipment has also been installed in these reactors. The other class of Soviet type of nuclear power reactors, which has been the focus of international attention for safety upgrades, is the WWER-440/230 model. These were designed before formal safety standards were issued in the Soviet Union and they lack many basic safety features that are common in other type of reactors used in several countries. Some of these nuclear power reactors are still operating in Bulgaria, Russia, among others.

Without any doubt can be stated the following: later Soviet design nuclear power reactors are very much safer and the most recent ones have Western control systems or the equivalent,

---

[114] See Figure 120.

along with containment structures. It is important to single out the relevant role played by the IAEA and WANO in contributing to huge gains in safety and reliability of Soviet era nuclear power plants. WANO comes into existence as a result of Chernobyl nuclear accident. In the first two years of WANO's existence, 1989-91, operating staff from every nuclear power plant in the former Soviet Union visited plants in the West on technical exchange and Western personnel visited every former Soviet Union nuclear power plants. A great deal of ongoing plant-to-plant cooperation and, subsequently, a voluntary peer review programme, grew out of these exchanges. In March 2007, Russia signed a cooperation agreement with the OECD's NEA, bringing it much more into the mainstream of world nuclear industry development. Russia has been participating for some years in the NEA's work on reactor safety and nuclear regulation and is hosting an NEA project on reactor vessel melt-through. [89]

The Russian Federal Supervision of Nuclear and Radiological Safety (Gosatomnadzor) is the nuclear regulatory authority of the Russian Federation with the headquarters in Moscow and seven regional offices throughout the country. The following regulations determine the procedure for nuclear power plant licensing: a) regulations on the order of special permission issued by Gosatomnadzor of Russia for examination of design and other materials and documents, substantiating safety of nuclear and radiologically dangerous installations and works (RD-03-12-94); b) regulations on arranging and carrying out examination of design and other materials and documents, substantiating safety of nuclear and radiologically dangerous installations and works (RD-03-13-94); and c) regulations on the order of issuing of special temporary permissions for designing nuclear and radiologically dangerous installations and works (RD-03-14-94).

The stages of obtaining the temporary permission (license) for nuclear power plant unit operation can be represented in brief as follows: a) license demand (submission of application documents); b) decision on the demand control; c) analysis of substantiating materials of demand; d) inspection at the nuclear power plant; e) conclusion on substantiating materials examination; f) conclusion on nuclear power plant inspection; g) general conclusion on obtaining temporary permission (license); and h) license (temporary permission). [113]

The main laws controlling nuclear power in Russia is the "Law About Utilization of Atomic Energy" and the "Law About State Policy in the Field of Radioactive Waste Management". Technical regulations created by Gosatomnadzor of Russia, which are in force today, are the legal framework for nuclear energy utilization. These regulations and rules address the aspects of safety assurance during site selection, designing, construction, operation and decommissioning of nuclear installations. All regulating documents developed by Gosatomnadzor have been compiled into a list of main scientific and technical documents, used by Gosatomnadzor for safety regulation and supervision during production and utilization of atomic energy, handling of nuclear materials, radioactive substances and articles on their base (P-01-01-03, Gosatomnadzor of Russia, 2003). Some aspects of nuclear related activity are regulated by decrees of the President or government of the Russian Federation. These decrees are the following:

Decrees of the President: a) Control of Export of Nuclear Materials, Equipment and Technologies of 27 March 1992; b) Utilities with Nuclear Power Plants of 7 September 1992; and c) Privatization of Enterprises under the Authority of the Ministry for Atomic Energy, and their Management in a Market Economy of 15 April 1993, etc.

Decrees of the government: a) Approval of Documents, Regulating Export of Equipment and Materials and of Corresponding Technology Used for Nuclear Purposes of 29 May 1992;

and b) Measures of Protection of the Population Living Adjacent to Nuclear Power Installations of 15 October 1992.

Nuclear Fuel Cycle and Waste Management

The Russian Federation, as a nuclear weapon state, has full control of all stages of the nuclear fuel cycle. The first civil uranium enrichment plant in the Russian Federation started operation in 1964 at Ekaterenburg. Three more plants came into operation later at Tomsk, Angarsk and Krasnoyarsk. At present, Minatom operates all four plants. The excess capacities that exist in these installations are offered to foreign utilities on a commercial basis.

Nuclear fuel fabrication is carried out at two plants: Electrostal and Novosibirsk. Electrostal produces fuel elements, assemblies, powder and pellets for WWER 440, BN 600, RBMK and PWR models. The Novosibirsk plant manufactures fuel elements and assemblies for WWER 1000 models. In the production of fuel assemblies for RBMK and WWER 1000 models, a quantity of fuel pellets is supplied from the Ust Kamenogorsk plant (Kazahstan). However, new lines for powder and pellet production at the Novosibirsk plant started operation in 2000-2002.

The reprocessing option is the one followed for dealing with spent nuclear fuel, with the exception of that originating from RBMKs, the spent nuclear fuel of which should be disposed of. Minatom operates the RT-1 Plant in Chelyabinsk for reprocessing fuel from WWER 440 models. The construction of a second reprocessing plant (RT-2) at Krasnoyarsk, which has a first line design postponed indefinitely[115]. Reprocessed uranium is used for RBMK fuel production. Plutonium obtained at RT-1 is temporarily stored on-site in dioxide form. Minatom operates several fuel storage facilities at RT-1 and RT-2 sites, and at several nuclear power plants. [113]

In the case of the reprocessing of the spent nuclear fuel, the policy adopted by the Russian government is to utilize recycled uranium and possible plutonium in MOX fuel, in order to close the nuclear fuel cycle as much as possible. However, no much progress has been achieved in the implementation of this policy and the spent nuclear fuel from RMBK and WWER-1000 models are stored mostly at nuclear power plant sites and not reprocessed. In the case of the spent nuclear fuel from WWER -440, the BN-600 and from naval nuclear power reactors, the spent nuclear fuel is reprocessed at the Mayak Chemical Combine plant, most common known as Chelyabinsk-65 at Ozersk near Kyshtym in the Urals. The government has plans to spend around US$5 billion on decommissioning activities and waste management in the future.

---

[115] The plant foresee to be dismantled in the future perhaps could be form part of the new Global Nuclear Infrastructure Initiative promoted by the Russian government in the framework of the multilateralization of the nuclear fuel cycle promoted by the IAEA and other countries. The purpose of this initiative is to reduce, as much as possible, the proliferation risk due to an increase in the construction of new nuclear power reactors in several countries around the world.

The Public Opinion

After the Chernobyl nuclear accident and until 2006, the majority of Russians believe that the use of nuclear energy for electricity production is a very high risk activity that may bring irreversible harm to the ecology and people's health. In a poll carried out in 2006 in Russia by the Public Opinion Foundation[116] covering 2,100 people in 45 regions, including Moscow, 13% of respondents spoke about fear and the feeling of being threatened and 9% specified their associations immediately mentioned the Chernobyl nuclear disaster, in the sense that such a catastrophe may happen again.

The problem of nuclear waste and the ecological harm arising from it was one of the concern mentioned by 3% of answers. Only 2% of respondents' answers contained a positive attitude toward nuclear power. A noticeable share of respondents feel calm about atomic energy, mentioning relatively neutral associations with today's nuclear power plants (9% of answers) or simply giving descriptions of the nuclear sector (6% of answers). (Vovk, 2006)

With respect to the opposition in Russia of developing nuclear power in the future, 16% of the respondents mentioned the bad ecological situation that could be produced by the use of nuclear energy for electricity production; 11% mentioned the general danger for people's health and lives; 11% mentioned the high probability of disasters and catastrophes; and 6% as the escape of radioactivity and the high daily doses of radiation that could be associated with the use of this type of energy for electricity production.

With respect to the positive aspects for Russia of developing nuclear power in the future, 17% of the respondents spoke about the possibility of producing more electric power; 3% in saving natural resources; 7% mentioned the low cost of atomic energy; and 7% about the earnings from power revenues and the general stimulating effect on the country's economy. Finally, 6% of the people spoke simply about the heating and lighting of houses and flats. Around 6% of the people see atomic energy development in our country as all-sufficient evidence of the high level of Russian science and technology. Russians have conflicting opinions on the balance between the usefulness and harm of nuclear energy development; 39% think it is useful rather than harmful for Russia, while 25% hold the opposite view. (Vovk, 2006)

According with the result of the mentioned poll, the majority of Russians are inclined to think that when the nuclear power is used properly and the equipment works in good condition, nuclear power-producing devices are relatively safe. Moreover, 29% of those polled share the opinion that when nuclear power plants work without problems, they are absolutely harmless to the environment. However, 51% of those surveyed think that nuclear power plants bring harm for ecology anyway, approximately half think that this type of plants are no more harmful to the ecology (23%) – or even less so (6%) – than thermal or hydroelectric power plants. Slightly more than one-quarter of respondents (27%), even non-operational nuclear power plants bring more harm to the ecology than steam power plants or hydroelectric power plants. The overwhelming majority of Russians (72%) would still respond badly to a decision to build a nuclear power plant in their area. Only10% of those surveyed would react positively to such a decision, with 9% expressing indifference. Around 83% of those surveyed, the aftereffect of the Chernobyl nuclear disaster is still causing harm to the ecology and the health of people living near this territory (only 7% believe that the

---

[116] See (Vovk, 2006).

consequences of that catastrophe have been already been eliminated). At the same time, only 12% of those surveyed think that in Russia the probability of such a large-scale catastrophe at an atomic power plant has decreased over the 20 years since the Chernobyl nuclear disaster, while 25% consider it to be the same as it was twenty years ago, and 24% think it has actually increased. (Vovk, 2006)

Large-scale public opinion polls conducted by different agencies between 2006 and 2008 show that the reputation of the nuclear industry, which seemed to have been desperately damaged by the Chernobyl nuclear accident, has been rapidly improving in the last two years. This situation is largely explained by society's support for the nation's development as a great energy-producing power, which was launched by President Vladimir Putin at the turn of the century. The overwhelming majority of Russians believe that the priority development of the power industry meets national interests. For this reason, almost half of them do not see any alternative to nuclear power as a source of energy when oil and gas reserves run out. Recent public opinion polls show that 27% of Russians are in favor of the construction of new nuclear power reactors, 42% favor the continued use of the 31 currently operating nuclear power reactors, and just 19% are completely against the use of nuclear energy for electricity generation. Though most of the population favors some type of nuclear power, 70% of Russians do not want a nuclear facility in their neighborhood.

Public concerns are most likely due to the technology being used for these new nuclear power plants. Though research is underway on new technologies such as fast breeder reactors, plans are to construct the new facilities with pre-Chernobyl Soviet designs and adding extra safety systems (none will be the RBMK design used at Chernobyl). People may have fewer concerns than they did 21 years ago at the time of the Chernobyl nuclear accident, but the reality of an accident still exists, and they desperately wish to keep themselves out of harms. This nuclear renaissance, encouraged by the global energy crisis, has played a serious role in the development of the nuclear power industry in Russia.

A survey carried out by Russia's Levada Center, in April 2008, has found that 72% of Russians were in favor of at least preserving the country nuclear power capacity and 41% thought that nuclear power was the only alternative to oil and gas as they depleted. Only two years ago, this figure was three times smaller. Hydro power and coal seen as other possible alternatives but only scoring 18% and 10% respectively. Summing up can be stated that the majority of the Russian's population support the use of nuclear power reactors currently in operation in the country in the future. However, the construction of new nuclear power reactors in the country is not supported by the majority of the Russian population, according with the result of the above mentioned polls.

Looking Forward

In 2000, the government of the Russian Federation approved the directives to regulate the development of the nuclear power industry[117]. According with these directives, "the electrical power generation using nuclear energy, at the best estimate, should be around 200 billion kWh in 2010, and around 230 billion kWh in 2020". The measures to be implemented with

---

[117] Nuclear Power Strategy of Russia for the Period to the Year of 2020 and Strategy for Development of the Russian Nuclear Power Sector in the First Half of 21th century.

the purpose to reach these targets are the following: a) lifetime extension up to 10-15 years for the operating first generation of nuclear power reactors; b) upgrading of the operating nuclear power reactors; c) increase of the availability factor; and d) putting new nuclear power reactors into operation.

For the next 5 years, the activities planned for the lifetime extension of several nuclear power reactors are the following: a) for 15 years: the WWER-440 first generation units (Novovoronezh-3 and 4 and Kola-1 and 2); and b) for 10 years: the RBMK units (Leningrad-1 and 2, and Kursk-1 and 2), and all four units of the Bilibino nuclear heat-and-power plant. [113]

The upgrading of the nuclear power reactors, including the upgrading of the turbine generators, is one of the measures adopted by the Russian government to improve the operational performance of the nuclear power reactors currently in operation in the country and to increase its efficiency. The increase of the availability factor to 85%, which corresponds to the world level, is to provide the additional power generation at the operated units up to 25 billion kWh per year. In parallel, with the shortened plant outage duration related with the scheduled maintenance activities, the availability factor could be increased due to the longer unit operation between the outages. The industry programme for effective fuel use for the period of 2002 to 2005 and up to 2010, provides for gradual transition of the WWER-1000 model to the 18-month fuel cycle. As the first stage of such a transition, the operation of the fuel has been started for the period of 350 effective days. In compliance with the effectiveness criteria for capital investments, the priorities of investments have been defined to support commissioning of the new units: a) before the year 2011: final construction of Kalinin 3, Kursk 5, Volgodonsk 2, Balakovo 5 and 6 and Beloyarsk 4 to the total capacity of 5.8 GW; b) before the year 2020: Final construction of Kalinin 4, Kursk 6, Smolensk 4, Bashkir nuclear power plant, Smolensk nuclear power plant, Novovoronezh nuclear power plant, Severo-Zapad nuclear power plant and BREST (nuclear technology of the fourth generation) to the total capacity of 13 GW; and c) before 2020: nuclear heat-and-power plants in Arkhangelsk, Voronezh, Saratov and Dmitrovograd to the total capacity of 5 GW (heat).

Upon implementation of this programme the electrical capacity of the nuclear power plants in 2010 will reach 28 GWe and in 2020 between 37 GWe to 41.4 GWe.

Another of the measures adopted by the Russian's government to increase the participation of nuclear power in the energy balance of the country is the increase in the performance of the current nuclear power reactors in operation. In this moment nuclear electricity output is rising strongly due simply to better performance of the nuclear power plants, with capacity factors leaping from 56% to 76% in the period 1998-2003. In gross terms, output is projected to grow from about 140-150 billion kWh in 2005 to 166 billion kWh in 2010 and 239 billion kWh in 2016 (18.6% of total), or more soberly to 230 billion kWh in 2020. Nuclear generating capacity is planned to grow more than 50% from 23 GWe gross (21.7 GWe net) in 2007 to 35 GWe in 2016, and at least double to 51 GWe by 2020. [89]

To implement this ambitious plan, Russia is planning to use third generation medium and large scale nuclear power units of improved safety, including advanced WWER 1000 (for domestic market and export), WWER 1500 (replacement of the first generation units and capacity growths), BN-800 (for plutonium utilization and solving of environmental problem) and BREST.

The Russian electric company Energoatom has adopted a complete programme to increase of the power of its nuclear power reactors and will enlarge their lifetime in 15 or 25 years. The extension of the lifetime of the current nuclear power reactors will allow, besides obtaining more energy, to win time to carry out the ambitious plan for the construction of new nuclear power reactors that has elaborated by Rosatom. According to Energoatom, and after the approval of the regulator, the oldest and smaller power reactors, the WWER-440 that had a life of 30 year old design, will be extended 15 years more and in the case of the WWER-1000 model, 25 years. On the other hand, the RBMK-1000 model will operate another 15 years.

According to the financial director of Energoatom, Yevgeny Konkov, "this company signed last year (2007) its first contract to fixed price to prolong the lifetime of Unit 4 of the Leningrad nuclear power plant, a RBMK-1000 model, for about US$500 million". The 2007 Energoatom economic report points out that "the extension for 15 years the lifetime of all the reactors is economically justified and it allows to have 2 112 TWh between 2007 and 2050". From the nuclear power reactors in operation, eight already have extended licenses. (Leningrado 1, Kursk 1, Novovoronezh 3 and 4, Kola 1 and 2 and Bilibino 1 and 2) producing additional 111 TWh between 2000 and 2007. The total investments programme foreseen for the period 2009-2015 is about US$45,600 million, 10% of which will be allocated for the modernization of the current nuclear power reactors in operation and the rest for the construction of new units. The current modernization works are concentrated on those units that would reach 30 years of their design life in 2013 (five RBMK-1000 model: Kursk 2 and 3, Leningrado 3 and 4, and Smolensk 3; a WWER-440 model: Kola 3; a WWER-100 model: Novovoronezh-5 and the fast breeder reactor of Beloyarsk.

Summing up can be stated that the aim of the Russia's energy policy is to reduce the use of natural gas for electricity and to double the contribution of nuclear power by 2020. The growth is to come from lifetime extension of the first generation units, by upgrading several of the current nuclear power reactors in operation, by increasing the availability factor to 85% average (and hopefully more), together with the construction of new nuclear power reactors.

According with press information, President Vladimir Putin has approved the construction of around 25 major new nuclear power reactors, which will almost double the share of nuclear power in Russia's electrical grid. The plan foresees an increase of 25% or more in the share of nuclear energy for electricity production for 2030. The plan envisions that overall domestic demand for electricity will increase 4% per year, and to satisfy this increase in the demand of electricity, is necessary to construct new nuclear power reactors in Tver, Nizhny Novgorod, Chelyabinsk and Yaroslavl or Kostroma regions. According with the plan approved, Russia will need to invest over US$282 billion in the construction of new nuclear power reactors by 2015, and another US$204 billion during the period 2016-2020. The government would earmark some US$24 billion for building new nuclear power reactors through 2015, and that Rosenergoatom, the state-controlled agency in charge of the nation's nuclear power plants, would provide another US$26 billion through 2030, as nuclear power generation becomes increasingly profitable.

One of the new nuclear power reactors to be built will be in Kaliningrado on the Baltic Sea. The Kaliningrado region is isolated from the rest of the Russian territory. The new nuclear power plant will have two nuclear power reactors with 2 300 MWe of capacity. The first unit will be built in the period 2010-2016 and the second in 2012-2018. The investments for the construction of the new nuclear power plant will be open to Russian and foreigners investors. At the moment, the energy supply of Kaliningrado depends exclusively on the imports of gas and of electric power.

It is important to note that, according to the country's main atomic agency spokesman, "Russia hopes to export as many as 60 nuclear power reactors in the next two decades. The plan approved includes the first build floating nuclear power reactor, set to begin operations in the White Sea by 2010. The Russian navy constructer of Baltiski Savod had begun officially last May in San Petersburgo the construction of the first floating nuclear power reactor. The company manufacturer Sevmach announced that the project had begun in April of 2007 in Severodvinsk in the White Sea, but its lack of capacity made that the contracting company Energoatom decided to replace it for Baltiski Savod. The ship called "Academic Lomonosov" is a lighter without autonomous propulsion and will be connected to the electric grid near the point in which the ship will be positioned. It will have two reactors KLT-40S[118] with a power of 35 MW each. They will have low-enriched fuel. The reactor core is normally cooled by forced circulation, but the OKBM design relies on convection for emergency cooling. Fuel is uranium aluminum silicide with enrichment levels of up to 20%. The assembly will be carried out in Viliutchinsk, south of the peninsula of Kamchatka. The cost of the project will be about €230 million and the lifetime of the floating nuclear power plant is considered to be 40 years.

In the ceremony of laying the foundation for the first floating nuclear power plant in the northern city of Severodvinsk held on 15 of April 2007[119], Federal Nuclear Agency director Sergei Kiriyenko said the following: "Today we are signing an agreement on the construction of six energy units of floating nuclear power reactors. The demand for them exists not only in Russia but also in the Asia and Pacific region where they can be used for water desalination". The first floating nuclear power plant will have a total capacity of 70 MWe and about 300 MW of thermal power. It will be nearly 150m long, 30m across and have 21,500 tons displacement. The estimate construction cost is US$336 million. Floating nuclear power plants can operate without fuel reload for 12-15 years and have enhanced radiation protection. Floating nuclear power plants are expected to be widely used in regions that experience a shortage of energy and also in the implementation of projects requiring standalone and uninterrupted energy supply in the absence of a development power system.

Lastly, it is important to single out that Russian and US companies were continuing joint research on a next generation of nuclear power reactors that would produce hydrogen as a byproduct. "An experimental reactor must be ready by 2015, and then work will start on an industrial reactor", Federal Nuclear Agency director Kiriyenko said. He also said that "the new reactor would cost some US$2 billion to build".

---

[118] The ship will uses modified OKBM KLT-40S nuclear power reactor derived from naval propulsion reactors used on the country's existing fleet of icebreakers in service for the last 50 years.

[119] The US produced a floating nuclear power plant in late 1960s early 1970s, with the purpose to supply 10 MW and desalinated water to the Panama Canal Zone.

THE NUCLEAR POWER PROGRAMME IN ROMANIA

Romania is a country with different energy sources such as crude oil with an old exploitation traditions, as well as natural gas, coal, especially cocking pit coal, lignite and brown coal.

Source: Green Cross Russia:

Figure 121. A picture of the floating nuclear power plant now under construction in Russia.

In the ceremony of laying the foundation for the first floating nuclear power plant in the The economic context in which the government is working is characterized by deregulation and competition, an increased demand of energy and more clear requirements for a clean and safe environment. The government is supporting a transition from fossil energy resources, expendable by definition, to new reliable sources and to a sustainable development, in order to prevent a future energy crisis and to reduce the impact in the climate due to the use of fossil fuels for the production of electricity.

There are not clear estimations referring to climate change. The consequences of global warming in Romania include changes in the severity and frequency of extreme weather events (temperatures in winter and summer, storms, flooding, etc.). The impact of the Kyoto Protocol in the current energy policy of the country is not significance due to a decrease by almost 50% of emissions of carbon dioxide during the 1990s. It is to be pointed out that there are efforts to mitigate emissions and to suitably modernize the industry but the emissions decay is mainly as a result of the severe economic recession by the end of 20th century. [111]

It is important to note that Romania was one of the first European countries that used electricity. According with IAEA sources, the Romanian electricity history goes back to 1862 when electric lighting was for the first time used in Bucharest. An electric power plant fitted with steam boilers and brush dynamos supplying direct current through a 2 kV line (underground cable) was commissioned in the downtown. The first European city endowed with electric street lighting was Timisoara, an exquisite town located in the western side of the country. This dates back to 1884.

Forty years later, in 1924, the Romanian's government approved what is considering the country's first energy law. This energy law specified explicitly that "the installations for production, transmission and distribution of energy were state property". The first energy law was amended and extended in 1930 and 1934, but not fundamentally altered. In 1938, the law for organizing the communal exploitation was enacted. After periods of amazing developments early in the 20th century, in-between the two world wars, the electricity sector knew a moderate development (an installed capacity of 501 MW and a production of 1.13 TWh in 1938). [111] Electricity consumption in Romania has been growing since 1999. The foresee growth of electricity consumption in the country in the future is about 4% per year for the period 2000-2020. Three scenarios have been considered by the IAEA with average annual growth rates for electricity of 2%, 2.8% and 3.8% respectively.

Table 57. Annual electrical power production for 2008

Total power production (including nuclear)	Nuclear power production
58 922.4 GWhe	10 333.6 GWhe

Source: IAEA PRIS, 2009.

Romania has two nuclear power reactors in operation with a net capacity of 1 310 MWe at Cernavoda nuclear power plant site (see Figure 123). It is important to stress that the Cernavoda nuclear power plant was built on the basis of a technology transfer from Canada (AECL), Italy and the USA, with CANDU-6 system. In 2004, nuclear power generated 5 144 GWh of electricity, 10% of the country's total. The Cernavoda nuclear power plant, which is owned and operated by Societatea Nationala Nuclearelectrica (SNN)[120], has performed extremely well since startup in 1996, when it first unit entered into commercial operation. In 2005, around 59.4 billion kTWh was produced, with net exports of 3 TWh; nuclear energy provides 8.6% of the total electricity produced in the country at very low cost; only hydro (29% of supply) is cheaper. In same year, 5.1 TWh was supplied to the grid from Unit 1 of Cernavoda nuclear power plant, which reached a capacity factor of 90.1%. It also supplied 176 GJ of district heating. In 2006, the share of nuclear energy electricity production was 9%. In 2007, the nuclear electricity production was 10.3 TWh representing 13% of the total generation. The production of electricity using nuclear energy was 7.1 TWh. In 2008, the nuclear share was 17.54%. Its net capacity factor has averaged over 86% so far, reaching more than 90% (both units) in 2007. The net load factor of Unit 1 was 96.2% and of Unit 2 was 93.2% in 2007. Operating and maintenance costs are US$1.25 cents/kWh. The units also provide district heating to Cernavoda township.

Cernavoda 1 is the first of five units planned for the Cernavoda nuclear power plant in the late 1970s. The lack of financial resources and a drop in power demand after 1990 resulted in the suspension of construction work on Units 2–5. In 2000, the government made the completion of Cernavoda 2 a high priority. The unit entered into commercial operation in 2007. In addition of Units 1 and 2, there are two nuclear power units under construction at Cernavoda nuclear power plant[121]. The construction of two new nuclear power reactors are planned with a capacity of 1 310 MWe. At least one additional nuclear power reactor was proposed to be built in the future.

It is important to single out that to produce the equivalent amount of electricity produced yearly by Cernavoda Unit 1, a lignite coal power plant requires about 6 000 000 tons of

---

[120] SNN was established in 1998.

[121] The government is thinking not to conclude the construction of Unit 5 but to start the construction of a second nuclear power plant in another site.

lignite and produces 1 500 000 tons of ash, of which 20 000 tons is fly-ash, about 4.5 million tons of CO_2 and significant quantities of SO_2 and NOx. For this reason, the nuclear option represents a good opportunity for Romania to reduce polluting emissions and to fulfill its commitments with the Kyoto Protocol.

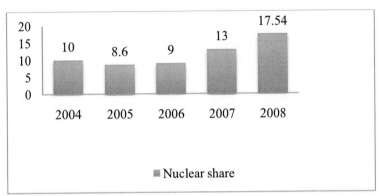

Source: IAEA PRIS and CEA, 8th Edition, 2008.

Figure 122. Evolution of the nuclear share in Romania.

Cernavoda 1 is the first of five units planned for the Cernavoda nuclear power plant in the late 1970s. The lack of financial resources and a drop in power demand after 1990 resulted in the suspension of construction work on Units 2–5. In 2000, the government made the completion of Cernavoda 2 a high priority. The unit entered into commercial operation in 2007. In addition of Units 1 and 2, there are two nuclear power units under construction at Cernavoda nuclear power plant[122]. The construction of two new nuclear power reactors are planned with a capacity of 1 310 MWe. At least one additional nuclear power reactor was proposed to be built in the future.

It is important to single out that to produce the equivalent amount of electricity produced yearly by Cernavoda Unit 1, a lignite coal power plant requires about 6 000 000 tons of lignite and produces 1 500 000 tons of ash, of which 20 000 tons is fly-ash, about 4.5 million tons of CO_2 and significant quantities of SO_2 and NOx. For this reason, the nuclear option represents a good opportunity for Romania to reduce polluting emissions and to fulfill its commitments with the Kyoto Protocol.

The Energy Policy in Romania

Since 2001, Romania's macroeconomic environment and business climate have improved but still the country is behind other European countries in many economic sectors. Romania is now in a transition period reducing the use of fossil fuel for the production of electricity and introducing new reliable energy sources in order to ensure a sustainable development and to prevent a energy crisis. The government is trying to cope with difficult economic circumstances the country is facing. The economic context is characterized by deregulation

---

[122] The government is thinking not to conclude the construction of Unit 5 but to start the construction of a second nuclear power plant in another site.

and competition, supported by the industry that is now under a full restructuring with concerning an increased demand of energy and more clear requirements for a clean and safe environment. [111]

In November 2002, the Romanian government approved "The National Strategy for the Development of the Nuclear Sector in Romania". The new strategy promotes "the increase of the nuclear share of electricity generation up to 20%-40% in the coming years, observing the requirements of sustainable development, price competitiveness and nuclear safety". The adopted strategy states that, "for the present conditions in Romania, taking into account the cost of the energy from the nuclear power plant versus the cost from the fossil fuel power plants, the investment should continue for the conclusion of Units 3,4 and 5 of the Cernavoda nuclear power plant during the period 2009-2020. Unit 3 was almost 15% completed when construction stopped, including the civil works for the reactor containment, turbine building and service buildings". Very little equipment or materials had been supplied.

It is important to single out that a large part of the electricity produced at these units will be available for export to other countries.

Source: Photograph courtesy of Wikimedia Commons (Photograph Bogdan Giuşcă).

Figure 123. Cernavoda nuclear power plant.

The Evolution of the Nuclear Power Sector in Romania

The development of the nuclear energy sector in Romania is, on the basis of available IAEA and WANO information, the following. In the late 1970s, a five unit nuclear power plant was planned at Cernavoda, on the Danube River. After considering carefully both Russian WWER-440 and Canadian CANDU technology, it was decided to adopt the later. [88]

Construction of the first unit started in July 1982. This unit was connected to the electric grid in July 1996 and entered into commercial operation in December the same year. Construction of the second unit started in July 1983. The second unit was connected to the electric grid in August 2007 and entered into commercial operation in October the same year. The Cernadova nuclear power plant was initially planned to have 5 nuclear power reactors.

However, in 1991, work on the remaining four nuclear power reactors was suspended in order to focus on the construction of Unit 1, responsibility for which was handed to an AECL-Ansaldo (Canadian-Italian) consortium.

In 2000, the government took a decision to complete Cernavoda 2 and allocated some €60 million for this purpose. During the period 2002-2003, additional €382.5 million package was allocated by the government to continue the construction work of Unit 2, including €218 million from Canada. In 2004, an Euratom loan of €223.5 million was approved by the EC for completion of Unit 2, including future upgrades. Total cost of the project to complete the construction of Unit 2 was estimate in €777 million. The conclusion of the construction of Unit 2 will increase the contribution of clean energy to electricity production in Romania, reducing CO_2 and other polluting emissions to the atmosphere, as well as the volume of solid waste resulting from burning coal. The new unit will reduce the dependence of Romania on external suppliers of primary resources, mainly natural gas and oil, geographically sited outside Europe, thus contributing to the increase of the security of energy supply of Romania and for ended of the EU. The conclusion of Unit 2 represents an economic and competitive source of base-load electricity compared with burning fossil fuels, consolidating the internal electricity market of the country.

In 2002, efforts got under way to resume work on Cernavoda Unit 3 and SNN commissioned a feasibility study from Ansaldo, AECL and KHNP (Rep. of Korea) in 2003. In August 2004, the government advertised for companies interested in completing Cernavoda Unit 3, a 700 MWe CANDU-6 unit, through a public-private partnership arrangement. This proved impractical and a feasibility study in March 2006 analyzed further options for both Units 3 and 4, including the possibility that SNN proceed with the conclusion of Unit 3. The companies involved in the construction of these two units are Arcelor Mittal's Romanian unit (MS), Czech Cez AS, Belgium's Electrabel, Italy's Enel SpA, Spain's Iberdola SA and Germany's RWE AG.

Table 58. Operating Romanian nuclear power reactors

Reactors	Model	Net MWe	Commercial operation
Cernavoda 1	CANDU- 6	655	1996
Cernavoda 2	CANDU- 6	655	2007
Total (2)		1 310	

Source: WANO, CEA, 8th Edition, 2008.

Nuclear Safety Authority and the Licensing Process

In Romania, there are several government institutions involved in the process of licensing of nuclear installations. These institutions are the following: a) the National Commission for Nuclear Activities Control (CNCAN; b) the Ministry of Water and Environmental Protection (MIR); c) the Ministry of Health; d) the Ministry of Interior; and e) the Ministry of Public Finance.

Each of these government institutions has an specific role to play within the nuclear power sector of the country.

In the field of nuclear safety, CNCAN "is the national competent authority in the nuclear field exercising the regulation, authorization and control powers provided under the Law 111/1996, on the Safe Deployment of Nuclear Activities". Since December 2000, CNCAN is an independent governmental body reporting only administratively to the Ministry of Waters and Environmental Protection. Actually, the president of CNCAN is a Secretary of State and the minister cannot interfere in CNCAN president's decisions. CNCAN is responsible for full surveillance and control in all issues relevant to nuclear safety regarding sitting, construction, commissioning, operation of nuclear power plants, research reactors and all nuclear facilities in Romania. In addition, CNCAN is in charge with full surveillance and control in all issues relevant to quality assurance, radiation safety, safeguards, export/import control, physical protection and emergency preparedness and monitoring the radioactivity of the environment. CNCAN is the national counterpart to the IAEA for nuclear safety, radiation safety, safeguards, physical protection, emergency preparedness, illicit trafficking events reporting. CNCAN plays the role of regulatory body integrator in the licensing process of nuclear installations. [111]

In today's world there are less and less possibility that an energy decision to be adopted by any particularly government can be taken without considering all environmental problems associated with it. In the case of Romania, the MIR is responsible for environmental protection legislation and regulations and for the licensing process from the environmental protection point of view. The MIR coordinates with the Pressure Vessel Authority (ISCIR), which is the responsible authority for licensing and control of pressure vessels, boilers and other pressure installations, including those from the nuclear field.

The decision to use one particular type of energy source will depend not only on economics, political, security of supply, environments impact, among others, but on the impact on the public health as well. In Romania, the Ministry of Health "is the responsible authority to organize the monitoring network of contamination with radioactive materials of food products over the whole food chain, inclusive drinking water, as well as other goods designated to be used by the population, according to the law. Also, the epidemiological surveillance system of the health condition of personnel professionally exposed, and of the hygiene conditions in units in which nuclear activities are deployed, are under its responsibility".

One of the major concern that the international community has in relation with the use of nuclear energy for the production of electricity is the physical protection of nuclear installations, in order to avoid the robbery of nuclear materials and terrorist activities. In Romania, the Ministry of Interior is responsible for the control of fire protection at nuclear installations and for the supervision of physical protection of nuclear installations and nuclear material. At the same time, the Ministry of Public Finance is the authority in charge of providing and controlling the financial support from governmental budgetary funds, sovereign guarantees, etc.

The situation regarding the main laws and regulations in force in the nuclear energy sector in Romania is the following: Romania has had laws in place governing the regulation of nuclear activities since 1974. They remained in force until 1996, when a new legislation was issued. In January 1998, important amendments to the Law 111/1996 have been approved. Under the umbrella of this new Nuclear Act, all related rules, practices and regulations in nuclear field were started to be assessed for compliance with applicable IAEA guides and standards. The licensing experience gained during construction, commissioning and initial operation of Unit 1 of the Cernavoda nuclear power plant was also carefully assessed and incorporated in the new legislative framework being now created in Romania. A comprehensive set of technical instructions, directives, regulations, procedures, industrial standards, nuclear design and safety guides, concerning the quality assurance and safe operation of nuclear facilities and nuclear power plants, cover activities such as project management, procurement, design, manufacturing, civil works, installation, commissioning and operation. All AECL design guides and safety design guides were endorsed by CNCAN. The IAEA Safety Series are also used as a basis for the CNCAN regulations. Most of the applicable industrial standards have been used during the licensing process of Unit 1 of the Cernavoda nuclear power plant. [111]

Nuclear Fuel Cycle and Waste Management

Romania is producing the nuclear fuel and the heavy water needed for the operation of the nuclear power plant at Cernadova. Unit 1 has been using 105 tons of natural uranium oxide fuel per year, which is fabricated by the SNN subsidiary Fabrica de Combustibil Nuclear (FCN) Pitesti fuel plant. The plant was qualified by AECL in 1994 and until today remains the only such plant producing CANDU fuel outside Canada.

At the end of 2003, it started making slightly heavier fuel bundles and, in preparation for Unit 2 commissioning, its capacity was doubled to 46 fuel bundles per day. Romania built a heavy water plant ROMAG situated at 7 km north-east of Drobeta Turnu Severin town, on the left side of the Danube River. ROMAG was projected to produce heavy water in two stages of development: 360 tons/year in the first stage, with four modules (90 tons/module), and 360 tons per year in the second stage, with another four modules, an unfinished investment. The factory produces heavy water of nuclear quality and it has the greatest capacity of Europe and the second in the world. The project was put into operation between 1980 and 1988. The first quantities of heavy water were produced on 17 July 1988.

Spent nuclear fuel is stored at the nuclear power plant site for up to ten years. It is then transferred to a dry storage facility for spent nuclear fuel based on the Macstor system designed by AECL. The first module was commissioned in 2003. Preliminary investigations are under way regarding a deep geological repository to be built in the country. A radioactive waste treatment facility operates at Pitesti. [88]

The Public Opinion

Support for nuclear power is stronger in Romania in comparison with the average within the EU, according with a poll carried out in 2005, a total of 55% of the Romanian people supports the use of nuclear energy for electricity generation compared with 37% average in the EU.

Looking Forward

In November 2002, the Romanian government approved a document containing the "National Strategy for the Development of the Nuclear Sector in Romania". According with this strategy, specific-measures should be adopted by the Romanian government to increase the nuclear share of electricity generation up to 20%-40% in the coming years, observing the requirements of sustainable development, price competitiveness and nuclear safety. It is important to note that early plans foresaw the construction of ten CANDU and three WWER-1000 units for Romania at different sites but, in March 2008, statement by the head of SNN reduced the initial plan to three more units by 2020. Two units of 1 310 MWe are already planned. One more unit with a capacity of 655 MWe was also proposed.

According with the statement made in May 2008 by the minister of economy and finance Varujan Vosganian, "the Romanian government assigned the necessary funds to carry out the feasibility study of the country's second nuclear power plant, and started the construction of Units 3 and 4 of the nuclear power plant in Cernavoda. The second nuclear power plant in Romania will have between two and four nuclear power reactors and the construction works are seen starting around 2020. According with a statement made by the director of Nuclearelectrica power plant, Teodor Chirica, "the construction of a second nuclear power plant would ensure 35% of Romania's energy needs. The new plant will probably be located in Transylvania, central Romania, near the Mures, Somes, or Olt Rivers. It will have a 1 310 MWe capacity, which is half the output of Romania's sole nuclear facility in Cernavoda". The new nuclear power plant is set to use different nuclear reactor technology. According with public information, the French company Areva will participate in the construction of the new nuclear power plant in Romania. "Romania needs a second plant as energy demand is permanently on the rise and fossil energy will be replaced", said Philippe Garderet, scientific director of Areva, who visited Bucharest to discuss the involvement of the company in the construction of the new nuclear power plant, including investment in uranium exploitation

In 2008, the Romanian's government approved the necessary funds to resume work on Cernavoda Units 3 and 4. Commissioning of Unit 3 is now expected to be in October 2014 at a cost of €1 billion and Unit 4 in mid 2015 at a cost of €1,2 billion[123]. The construction period is expected to be 64 months. The annual operating costs for Units 3 and 4 are estimated at €100 million and the cost of energy produced should be about €30-€35 per MWh, resulting in 9% to 11% profitability. Electricity cost is expected to be €2.8 to €3.25 cents/kWh.

When the new units start operation in 2015, around 35% of Romania's electricity demand could be supplied by nuclear power. The Romanian nuclear authorities abandoned the construction of Cernavoda Unit 5 and took a decision to build a new nuclear power plant at other site due to insufficient technical capacity at the plant site. In 2008, a statement made by the Romanian nuclear authorities said "that up to three more units should be constructed by 2020 at a new site. Romania's plan to build a second nuclear power plant is part of a strategy to reduce the country's dependency on Russian gas and concentrate more on nuclear power and renewable energy. Currently, around 40% of the gas in Romania used countrywide to produce electricity is provided by Russia. For the construction of the second nuclear power

---

[123] It is important to note that the Canadian nuclear regulator has announced it will no longer give construction permits for the Generation II CANDU technology in its own country, because it does not meet modern safety standards.

plant the Romanian authorities are searching the market to find the best suited nuclear technology which is now available. The EPR is a possibility to be considered for the new nuclear power plant and Areva is carrying out negotiations with Romanian's competent authorities on this subject.

THE NUCLEAR POWER PROGRAMME IN THE NETHERLANDS

The Netherlands is the world's eights biggest producer of natural gas. Other energy sources are oil, coal, nuclear energy and renewable energy.

The Netherlands has one nuclear power plant in operation in Borselle, with one PWR system with a net capacity of 485 MWe. The construction of the Borselle nuclear power plant started in July 1969. The Borselle nuclear power reactor produced, in 2002, around 4% of the total electricity generated in the country. In 2004, the Borselle nuclear power plant produced 3.6 TWh, which represented 3,8% of total power consumed in the Netherlands in that year. The electricity production of the Borselle nuclear power plant, in 2006, was about 3.5 TWh representing around 4% of the total electricity production and, in 2007, this production increased to 3.99 TWh, representing 4.1% of the total. In 2008, the nuclear share was 3.8%.

A second nuclear power plant in Dodewaard with a small BWR system with a net capacity of 55 MWe was shut down in March 1993. The construction of this nuclear plant began in May 1965. The decision to shut down of the Dodewaard nuclear power plant was adopted for the following reasons: Firstly, the SEP[124] felt that there was no longer any prospect of the Dutch government giving the go-ahead to the further development of nuclear energy in the Netherlands in the foreseeable future. Secondly, the Dodewaard nuclear power plant had been built, primarily, as a means of gaining experience with nuclear energy. It was never economic in the sense that revenues were higher than costs and this situation was likely to be exacerbated by the impending deregulation of the European electricity market. [133] The Dodewaard nuclear power reactor produced during its lifetime a total of 11 505.5 GWh.

It is important to note that the Borselle nuclear power plant was designed to operate with natural circulation and was fitted with an isolation condenser to remove excess heat. These properties later on became standard features of the new BWR design with passive safety characteristics. Originally planned to operate until 1 January 1995, its economic life was first extended to 1 January 1997, and later to 2004-2007 in order to justify a 450 million-guilder safety related back fitting programme carried out on the plant. In 2002, new election was held and, in 2003, the new conservative coalition government that took power decided to revert the previous decision to shut down the Borselle nuclear power plant moved the closure date back to 2013. In 2005, the phase-out decision was abandoned and, in June 2006, the Dutch government concluded a contract with the Borssele operators and shareholders to allow the operation of the reactor until 2034 on the following condition: it would be maintained to the highest safety standards and the stakeholders, Delta and Essent, agreed to invest €250 million towards sustainable energy projects. The government added another €250 million in the process avoiding the compensation claim they would have faced had they continued towards early shutdown. [85]

---

[124] N.V. Samenwerkende Slektricititeits Produktiebedrijven is the owner of the Dodewaard nuclear power plant.

The Energy Policy in the Netherlands

The new energy policy adopted by the Netherlands's government is built upon three pillars: a) liberalization; b) sustainability; and c) security of supply.

The aim of the liberalization energy policy is to break monopolies and to ensure that all customers are better served and to enhance economic efficiency. The liberalization is embedded in the energy policy adopted by the EU and will ultimately result in a single European energy market. [133]

Table 59. Annual electrical power production for 2008

Total power production (including nuclear)	Nuclear power production
103 560 GWhe	3 933 GWhe

Source: IAEA PRIS, 2009.

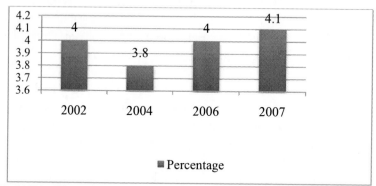

Source: WANO, IAEA, CEA, 8th Edition, 2008.

Figure 124. Evolution of the nuclear share in the Netherlands energy balance.

The Netherlands have ambitious targets on energy efficiency. The goal is to achieve, in 2020, an increase in energy efficiency at least 2% per annum and a 10% in the use of renewable energy within the energy balance of the country. The sustainability of this energy policy is determined significantly by climate change. The goals on energy efficiency and in the use of renewable energy for the production of electricity is to achieve 6% reduction of CO_2 emissions to the atmosphere in 2020 compared to 1990s level.

The issue of security of supply is of growing importance to the Netherlands government. The EU expects a growing dependency on import of energy of 70% by 2020. Oil prices show an increasing volatility and the recent supply developments in California raise the question whether this could happen in the Netherlands as well. Both liberalization of the internal market and raising the share of renewable energy in the long term can contribute to security of supply. [133]

During the elaboration of the new energy policy the future of the Borselle nuclear power plant was discussed. A summarized of the discussion about the future of the Borselle nuclear power plant is included in the following paragraphs.

During the autumn of 1994, when the new government took over, discussions took place in Parliament whether the lifetime of the Borssele nuclear power plant should be extended by three years from 2004 till 2007[125]. The previous government had issued licenses for the back fitting, which had already begun. This made it difficult for the present government to withdraw the permission. But, on 23 November 1994, Dutch Parliament passed a resolution calling on the government to "waive plans to permit lifetime extension" for the Borssele nuclear power plant. Meanwhile, the Minister of Economic Affairs had been negotiating with the electricity producers (SEP) for a politically and economically acceptable compromise. Finally, the Dutch government and the electricity producers agreed, in December 1994, that the Borssele nuclear plant is to be shut down at the end of 2003. In exchange, SEP will be compensated 70 million guilders allowing it to complete the safety-related back fitting programme. The updating and modification programme was completed by mid 1997 and consisted of the following features: 1) a comparison with modern safety regulations and practices, and the initiation of plant modifications where these were deemed useful or necessary, to enable the plant to comply with these regulations and practices insofar as was practical; this work covered design, operation and quality assurance; 2) the installation of hardware to help control or mitigate the effects of major accidents; such hardware, including a filtered containment vent and catalytic hydrogen recombiners; 3) a full-scope PSA, comprising levels 1, 2 and 3, with the purpose to identify plant vulnerabilities and to compare the plant risk with predefined quantitative risk objectives; and 4) a full-scope replica simulator for the training of plant staff. [133]

The Evolution of the Nuclear Power Sector in the Netherlands

The evolution of the nuclear power sector in the Netherlands is briefly described in the following paragraphs. In the 1930s, researchers at the Delft University of Technology became interested in the potential of nuclear energy and stockpiled natural uranium which was later kept hidden from occupying German forces. In the early 1950s, this uranium was a basis for collaboration with Norway's Institute for Energy Technology in the Halden heavy-water research reactor at Kjeller. In 1955, construction began on the Netherlands' own research reactor, the High Flux Reactor (HFR) at Petten. HFR was intended to help the country gain knowledge of nuclear technology and operations through materials research. The Ministry of Economic Affairs had a strategy to develop a national industry capable of designing, manufacturing and exporting nuclear power technology. The ultimate aim was that nuclear power would be introduced from about 1962 to gradually replace much fossil fuel electricity generation. [85]

The HFR research reactor of Petten was stop, in 2008, due to corrosion problems detected in some parts of the cooling primary system, but started again its operation by the middle of February of 2009, after a stop of more than six months. The HFR research reactor is one of the few nuclear research reactors in the world used for the supply of radioisotopes for medical treatment. The HFR research reactor provides 30% of the isotopes used in the world in certain medical treatment, especially for diagnosis purpose and for the treatment of several illnesses.

---

[125] According with the IAEA, the year 2004 was the original date for the reactor's decommissioning before Borssele's shareholders applied for a three-year life extension in order to justify a 450 million-guilder safety related back fitting programme.

In May 1965, construction started on the first nuclear power reactor in the Netherlands. The first nuclear power reactor was a 55 MWe BWR system at Dodewaard[126]. The reactor, intended as a test-bed for the national nuclear power industry The unit was connected to the electric grid in October 1968 and entered into commercial operation in January 1969. The main purpose of the Dodewaard nuclear power plant was to conduct nuclear experiments for commercial applications. It made the plant most suitable for verification experiments on a commercial scale. It was operated by Joint Nuclear Power Plant Netherlands Ltd (GKN) until March 1997, when it was shut-down for economic reasons.

The next nuclear power project was a commercial 452 MWe PWR system at Borssele, in the south west of the country. The construction of the nuclear power plant started in July 1969. The reactor was connected to the electric grid in July 1973 and entered into commercial operation in October of the same year. It was designed and built by Germany's Kraftwerk Union (Siemens) and operated by Electricity Generating Company for the Southern Netherlands (EPZ). In 2006, following an extension of its operating life to 2034, a turbine upgrade boosted the net capacity of the Borselle nuclear power reactor from 452 MWe to 485 MWe.

It is important to note that nuclear power has a small role to play in the Dutch electricity supply, with the Borssele reactor providing about 4% of total generation - 3.5 billion kWh of the 98.3 billion kWh total produced in 2006. In 2007, the nuclear production was 3.9 billion kWh. Another 19 TWh net of electricity is imported from other countries, and since some of that is nuclear-generated, official statistics put the nuclear share at 9-10%. [85]

Nuclear Safety Authority and the Licensing Process

In the Netherlands, basic legislation governing nuclear activities is contained in the Nuclear Energy Act of 1963, which has been amended on several occasions. Several governments' offices are involved in nuclear safety and in the licensing process of nuclear installations. Until 1999, these offices were the following: a) Minister of Economic Affairs; b) Minister for Public Housing, Spatial Planning and Environment; and c) Minister of Social Affairs and Employment.

During this period the Ministry of Economic Affairs played a coordinating role and also carried out "the main responsibility for the implementation of the Nuclear Energy Law, alongside the Ministry of Social Affairs and Employment and the Ministry of Housing, Spatial Planning and the Environment". According with IAEA information, as of 1 July 1999, agreement has been reached between the Minister of Economic Affairs and the Minister of Housing, Spatial Planning and Environment on the transfer of the coordinating role of the Minister of Economic Affairs. The latter will not have any longer the prime responsibility for the implementation and execution of the Nuclear Energy Act insofar as nuclear installations are concerned. Due to this agreement the Minister of Economic Affairs, retains the main responsibility for energy supply policy and share other responsibilities in nuclear matters with the Minister of Housing, Spatial Planning and Environment. According with the Royal Decree of 21 June 1999, "maintenance of the Nuclear Energy Act and all regulations based on it - with a few exceptions - have been transferred from the Ministry of Economic Affairs to

---

[126] The Dodewaard nuclear power reactor was the only one in the world which was cooled by natural circulation.

the Ministry of Housing, Spatial Planning and the Environment". One of the reasons to perform this transfer is the wish to comply with Article 8.2 of the Nuclear Safety Convention. This article demands "that appropriate steps are being undertaken to ensure that an effective separation between regulatory functions and other functions such as promotion and utilization of nuclear energy be maintained". By Royal Decree of 24 May 2000, the government decided to transfer the Nuclear Safety Department of the Ministry of Social Affairs and Employment to the Ministry of Housing, Spatial Planning and the Environment as well. This transfer has become effective as per 1 June 2000. The transfer of the Nuclear Safety Department to the Ministry of Housing, Spatial Planning and the Environment aims to utilize the limited human and financial resources available in the most efficient way.

Table 60. Dutch nuclear power reactors situation

Status of nuclear power plants	Type	Capacity MWe	First power	Status
Borssele	PWR	485	October 1973	Operational
Dodewaard	BWR	55	January 1969	Shut down

Source: WANO, 2009 and CEA, 8th Edition, 2008.

It is important to single out that the Nuclear Energy Act "requires a license issued by the regulatory authority for all activities involving radioactive materials, fissionable materials and ores when they exceed certain pre-set exemption levels". These activities include import, export, transport, preparation, use, storage, release and disposal of materials and construction and operation of facilities. In a revision of the Nuclear Energy Act, also decommissioning of nuclear installations will become an activity for which a license is required.

In addition to the Nuclear Energy Act, there are two other acts which have a bearing on the possibility of acquiring or modifying a license. One of these acts is the Environmental Protection Act, which stipulates "that an Environmental Impact Assessment be presented if an application for a license or, in certain cases, a modification of a license is made and the General Administrative Law Act, which sets out the procedures for obtaining a license including the specification of maximum terms for each decision step in the process and lays down the rights of the public to raise objections and appeals to a license".

Nuclear Fuel Cycle and Waste Management

Other legislations are in force in the country related with the use of nuclear energy. These legislations and government offices involved in its implementation are, in additional to what has been said before, the following: a) Ministry of Health, Welfare and Sport in charge of the protection of the patient undergoing medical examination and treatment; b) Ministry of the Interior, in charge of emergency response to large-scale accidents involving radioactive material.

All activities relative to the import, transport, use, storage, disposal and export of radioactive materials are subject to the provisions of the Nuclear Energy Act adopted in 1963 and revised in 1994. The Act is a framework act, which is enacted by separate decrees and ordinances, with the aim "to implement specific parts of the act". In the legislation a fundamental distinction has been made between activities related to the nuclear fuel cycle, which follow the most stringent regime, and activities related to the application of radioactive sources and electrical appliances emitting X-rays. [133]

In the field of waste management, there also specific regulations in force in the country with the aim "to regulate the management of all nuclear waste produced by the nuclear installations in operation". The regulations in force are the following: a) Nuclear Installations, Fissionable Material and Ores Decree adopted in1969 and revised in 1994; b) Radiation Protection Decree adopted in 1986 and revised in 1994; and c) Decree on the appointment of the Central Organization of Radioactive Waste (COVRA) as recognized waste management organization adopted in 1987. These regulations make emphasis on aspects associated with the protection of the population in general and the protection of the environment. The high-level waste produced in the country is stored in HABOG, a dry-storage facility operated by COVRA. Vitrified high-level waste returning from British Nuclear Fuels and Cogema (France) will be stored there, together with the spent nuclear fuel from the research reactors in Petten and Delft, for up to 100 years. COVRA also operates facilities for the treatment and storage of low and intermediate- level radioactive wastes at Borssele. In early 2002, the shares in COVRA were transferred from the utilities and research organization to the government. (Kovan, 2005)

The Netherlands had history of research activities into the gas centrifuge method of uranium enrichment. It is important to note that from the commercial point of view, uranium enrichment is the most important part of the fuel cycle for the Netherlands. Urenco Nederland BV has a license for a production capacity of 2 800 tons of separative work units per year and plans to increase it to 3,500 units per year. Similar uranium enrichment activities had been carried out in Germany and in the UK and, in 1970, an agreement (Almelo Treaty) was signed by the three countries on collaboration in the endeavor.

The result of the agreement was Urenco, a company jointly owned by the three governments. In 1979, site works on Urenco enrichment plants began at Almelo in the eastern Netherlands, and Capenhurst, UK. The plants began commercial production of enriched uranium in 1981 and 1982 respectively. In 1985, production began at Gronau in Western Germany and a Urenco plant based on the same technology is currently being built in New Mexico, USA. In addition, Urenco have the contract to supply centrifuges for Areva's new enrichment plant in Pierrelatte, France, that will replace the large existing diffusion enrichment plant. The production facility for Urenco's centrifuges is also located in Almelo in the Netherlands. Used nuclear fuel from Dodewaard was recycled at the UK's Thorp facility at Sellafield and that from Borssele at France's La Hague. Areva NC, operators of La Hague, hold a contract to recycle Borssele used fuel until 2015. [85]

The Public Opinion

The Three Mile Island and Chernobyl nuclear accidents provoked a strong reactions in the public opinions in most of the European countries, including the Netherlands. The Parliament adopted a decision to phase out the Borselle nuclear power plant taking into account the negative public opinion regarding the use of nuclear energy for electricity generation that exist in the country. However, the government voids this decision and took the decision to extend the operation license of the Borselle nuclear power plant until 2034 reflecting a change in the public opinion regarding the use of nuclear energy for electricity generation due to the negative impact in the climate produced by the use of fossil fuel for the same purpose.

Looking Forward

The exhaustion of some of the natural gas fields in the country, the current high price of oil, the increasing natural gas price, the change in the public opinion regarding the use of nuclear energy for electricity generation and the environmental impact of the use of nuclear power, marked shift in the position of some political parties in favor of building new nuclear power reactors in the country in the future.

In September 2006, the environment minister on behalf of the economics minister submitted to parliament a document entitled "Conditions for New Nuclear Power Plants". According with this document "the government wanted to move to a sustainable energy supply and that the abandonment of its earlier phase-out policy was part of a transition strategy adopted by the government". The construction of a new nuclear power plant[127] fitted into this transition model. In March 2008, the main advisory body of the Dutch government on national and international social and economic policy - the Social and Economic Council (SER) - said that "the government should consider expanding nuclear energy in two years when it is due to evaluate its climate policies".

Lastly, it is important to note that the company electric Dutch Delta has communicated to the Ministry of the Housing, Planning and Atmosphere the beginning of the necessary process for the construction of a new nuclear power reactor by means of the evaluation of the environmental impact. Delta hopes to obtain the license at the end of 2011, to request the construction permission in 2012 and to begin the construction in 2013, so that the reactor can be operative in 2018. The elected location is next to the current nuclear power plant of Borssele whose closing is foreseen for 2034. Although details have not been announced the cost will be between €5,000 and €7,000 million, including the provisions for the storage of residuals and the closing costs. Delta estimates that the cost is comparable to a thermal power plant using coal as fuel and smaller than a power plant using gas as fuel or a wind power plant and that it does not emits CO_2 to the atmosphere.

THE CURRENT SITUATION AND THE PERSPECTIVES OF THE NUCLEAR POWER SECTOR IN ITALY

Italy is poor in energy natural resources and, for this reason, depends heavily on imported energy supply[128]. Italy was among the first countries in the world to use nuclear technology for power generation purposes. In 2004, the total electricity consumption in Italy was 322 billion kWh. The per capita consumption of electricity in that year was 5 542 kWh. The local

---

[127] The new nuclear reactor to be built in the Netherlands should be a Generation III system with levels of safety being equivalent to those of Areva EPR system.

[128] In 2000, more than 80% of the Italian energy need was imported.

production of electricity was 300 billion kWh gross. The contribution of the different energy sources was the following: 42% from gas, 26% from oil, 15% from coal and 13% from hydro. Imports of 45.6 billion kWh (effectively some 16% of its needs) are required, mostly nuclear power from France. This is equivalent to output from about 6.5 GWe of capacity at 80%. [70]

From 1946 to 1965, the private industry played an important role in the introduction of a nuclear power programme in Italy. From 1966 to 1987, the government planned and supported the development of a nuclear power programme in the country with the purpose to reduce its energy dependency. However, after the Chernobyl nuclear accident, Italy not only halt the operation of all of its nuclear power reactors but started the decommissioning of all of them and approved a nuclear moratorium for the construction of new nuclear power reactors.

The Energy Policy in Italy

The last energy plan approved by the government in which nuclear energy was not included in the energy balance was in August 1998. It focused on a set of actions capable of yielding substantial results in terms of energy conservation, environmental protection, development of domestic energy sources, diversification of imported energy sources and their origins, and safeguarding the competitiveness of the production system. A five-years nuclear moratorium, following a popular referendum, which took place in 1987, officially expired in December 1993, nevertheless the government remains steadfast in excluding nuclear energy. [47]

Recently, the Italian government took the decision to reintroduce the use of nuclear energy for the production of electricity in order to reduce its dependency on energy imports and to satisfy the foresee electricity demand in the coming years without increasing the emission of CO_2 to the atmosphere. Diversification of energy sources is particularly challenging in this respect. Italy's energy mix is shifting from oil to more use of gas, with little probability of rapidly diversifying much further owing to the limited growth of renewable energy and local resistance to coal. [47] For this reason, the use of nuclear energy for electricity generation will be included in the energy mix of the country in the coming years. It is important to note that the dependency of the country on oil and gas from external supply sources, raises concerns about security of supply and the risk of high energy costs. Due to this situation, timely investment in energy production, transportation and interconnection, particular the introduction of the use of nuclear energy for electricity production, is essential to secure energy supply and more active competition.

The Evolution of the Nuclear Power Sector in Italy

Italy was a pioneer in the use of nuclear energy for the generation of electricity in the region. During the period 1963-1990, Italy built 4 nuclear power reactors. These reactors were the following: One BWR at Caorso nuclear power plant with a net capacity of 860 MWe. The construction of the nuclear power reactor started in January 1970, was connected to the electric grid in May 1978 and entered into commercial operation in December 1981. The nuclear power reactor was shut down in July 1990 after produce 29 031 GWh of electricity. One BWR at Garigliano nuclear power plant with a net capacity of 150 MWe. The nuclear power reactor started to be built in November 1959, was connected to the electric grid in January 1964 and entered into commercial operation in June 1964. The nuclear power reactor was shut down in March 1982 after produce 12 466 GWh of electricity. One GCR at Latina nuclear power plant with a net capacity of 153 MWe. The nuclear power reactor started to be built in November 1958, was connected to the electric grid in May 1963, and entered into commercial operation in January 1964. The nuclear power reactor was shut down in December 1987 after produce 26 654 GWh of electricity. One PWR at Enrico Fermi (Trino) nuclear power plant with a net capacity of 260 MWe. The nuclear power reactor started to be built in July 1961, was connected to the electric grid in October 1964 and entered into commercial operation in January 1965. The nuclear power reactor was shut down in July 1990 after produce 24 905 GWh of electricity.

Table 61. Nuclear power reactors shutdown in Italy

Reactors	Model	Net MWe	First power	Shut down
Latina	GCR	153	1964-87	December 1987
Garigliano	BWR	150	1964-82	March 1982
Trino Vercellese	PWR	260	1964-90	July 1990
Caorso	BWR	860	1978-90	July 1990
Total (4)		1 423		

Source: CEA, 8th Edition, 2008.

To support the introduction of a nuclear power programme in the country the Italian government adopted, in December 1962, the Law 1860 with the purpose to regulate the peaceful use of atomic energy in the country. This law assigned CNEN the role of the nuclear regulatory authority and foresaw the issuance of a subsequent law for radioactive protection of population and workers. In February 1964, the Italian government issued a complete set of Regulations (D.P.R. 230) to include details of the different aspects of nuclear safety and radiation protection. By this regulation, CNEN was confirmed as the official regulatory authority in nuclear matters. The safety criteria were adopted from countries exporting nuclear technology (mainly the UK and the USA). In 1962, after a long political struggle, the electric sector was nationalized and Enel was established as the sole utility. In 1964, the ownership of Latina nuclear power plant was transferred to Enel and, in 1966, as well as the Garigliano and Trino units, closing the first period of the Italian nuclear history. [47]

In December of 1966, Enel announced a huge nuclear programme forecasting 12 000 MW of nuclear power by 1980. A year later, in 1967, the Comitato Interministeriale per la Programmazione Economica - a committee in charge of coordinating the activities of ministries involved in the country's economic planning and of defining the nuclear programme of Enel - reorganized the nuclear sector and decides that Enel will maintain its position as the sole utility. ENI will be in charge of nuclear fuel and ANSALDO will be in charge of collaborating with foreign supplier(s) and will become the Italian nuclear components supplier. In 1967, an agreement was signed by CNEN and Enel for developing an Italian version of the Canadian CANDU. This reactor type, called CIRENE, was designed to use heavy water as moderator and boiling water as coolant. In 1972, ANSALDO got an order to build a 40 MWe prototype close to the Latina nuclear power plant, but the reactor never became operational due to technical problems and the lack of economic resources. Its construction was finalized in 1988. In 1974, following the oil crisis, the Ministry of Industry, Commerce and Crafts approved a National Energy Plan that foresaw the construction of 20 nuclear power reactors in order to reduce the contribution of oil on the Italian energy balance. The main effort during that period was to achieve a certain level of technological independence from the American licenser(s). Political indecision led the industry to spread technical and economic resources over four different reactor types, namely, the BWR of General Electric, the PWR of Westinghouse and Babcock, the CANDU of AECL, and the indigenous CIRENE. To attain the goals of the new energy plan, the Italian government joined the EURODIF consortium in 1973. [47]

In Italy, there are several research reactors operating, including AGN Constanza (since 1960), Uni of Pavia's LENA Triga II (since 1965), ENEA's Tapiro (since 1971), ENEA's Triga RC-1 (since 1960) and a subcritical assembly. Ansaldo Nucleare, in cooperation with Canada's AECL, participated in the construction of Cernavoda Unit 2 in Romania, and is also involved in international research and development activities related with new nuclear power reactor design systems. No facilities for the enrichment of uranium exist in Italy. However, there are several facilities capable to manufacture fuel elements but all of them were closed after the Chernobyl nuclear accident and the adoption of a moratorium on the use of nuclear energy for electricity generation in the country[129].

The Phase out Policy

All efforts to introduce and develop a nuclear power programme in Italy was stopped after the Chernobyl nuclear accident and as result of a referendum carried out in November 1987. Following the referendum, all operation of nuclear power reactors was largely stopped. In 1988, the government resolved to halt the construction of Montalto di Castro and Piemonte nuclear power plants, shut down the remaining nuclear power reactors and decommission them from 1990. Since that year, Italy remained largely inactive in the field of nuclear power.

Nuclear Fuel Cycle, Waste Management and the Licensing Procedures

With respect to waste management and disposal the situation in Italy is the following: According with IAEA information, the sources of radioactive waste in Italy include the nuclear power plants formerly operated by Enel, the fuel cycle plants operated by Fabbricazioni Nucleari S.p.A., ENEA research laboratories and experimental facilities and non-energy applications (e.g., biomedical and other uses). Criteria applicable to the classification, treatment and disposal of radioactive waste are set forth in ENEA/DISP's Technical Guide No. 26, issued in May 1988 and updated in 1997. These rules allow above ground disposal of treated low-level waste (Categories I and II) and prescribe suitable final disposal solutions (such as deep disposal) for high-level waste (Category III). As for categories I and II, solid low-level waste is to be super-compacted and cemented. Liquid low-

---

[129] In the middle of the 1990s, Enel decided to terminate nuclear fuel reprocessing activities in the country, on the basis of an economical and technical evaluation and to proceed with interim dry storage of the remaining spent fuel of light water reactors.

level waste is to be cemented in containers suitable for above ground storage. Until 2008, some 290 tons of spent nuclear fuel is in dry storage, and overall 5,500 m^3 of radioactive wastes are stored.

In November 2006, a bilateral French-Italian agreement cleared the way for SOGIN to sign a contract with Areva NC for reprocessing 235 tons of spent nuclear fuel now in storage. It is to be shipped to La Hague between 2007 and 2015 and the wastes are to be returned after 2020. Latina's Magnox spent nuclear fuel - about 1,400 tons in total - was shipped to UK for reprocessing at Sellafield. [94]

With respect to the licensing procedures in force in Italy can be stated the following: The National Agency for Environmental Protection and Technical Services (APAT) is the Agency which carries out technical and scientific activities of national interest related to the protection of the environment, the defense of the water and territory resources. It has operational and administrative autonomy under the directives and the control of the Ministry of Environment. For all nuclear activities, APAT acts as the technical body of the Ministry for Productive Activities (MAP). It is the responsible authority for the licensing process of nuclear installations and its responsibilities include: a) assessment of the safety analysis carried out by the operating organization; b) inspection of equipment and materials during the design, construction and operational phases for the systematic verification of facility operation safety; and c) enforcement action to remedy any failure to meet both the licensing conditions and any safety operation criteria.

The Technical Commission for Nuclear Safety and Health Protection from Ionizing Radiations, is an advisory body of APAT, which gives its technical advice on question of safety and health protection in relation to the main stages of the licensing procedure. It is composed of experts from ENEA, APAT, and various ministries. The Ministry of the Environment is the authority responsible for the decisions in the matter of environmental compatibility of nuclear projects, including decommissioning of nuclear power stations. [47]

Looking Forward

It is important to single out that, in 2004, a new energy law opened up the possibility of joint venture with foreign companies in relation to nuclear power plants and importing electricity from them. This resulted from a clear change in public opinion, especially among younger people favoring nuclear power for Italy. In 2004, Enel bought 66% of SE with its four WWER 440/V213 Bohunice and Mochovce reactors. Enel's subsequent investment plan approved in 2005 involves €1.88 billion investment to increase SE generating capacity, including €1,6 billion for completion of Mochovce Units 3 and 4 by 2011-12.

Later on in 2008, the Italian government adopted a decision to reintroduce a nuclear power programme to satisfy the foresee increase in the electricity demand in the future. Later in the year, in May 2008, the new Italian government confirmed that it will commence building new nuclear power reactors within five years, with the purpose to reduce the country's great dependence on oil, gas and imported power. The Italian's plan is to have 25% of its electricity generated from nuclear power by 2030. To reach this goal, no less than 8 to 10 large nuclear power reactors need to be built.

To support the implementation of such plan the government introduced a package of nuclear legislation, including measures to set up a national nuclear research and development entity, to expedite licensing of new reactors at existing non-operational nuclear power plant sites, and to facilitate licensing of new sites. The plans foresee the construction of eight to ten nuclear power reactors at one of three licensed sites: Garigliano, Latina, or Montalto di Castro. The first two sites had small early-model nuclear power reactors operating until 1982 and 1987 respectively. At Montalto di Castro two larger nuclear power reactors were almost complete when the country's November 1987 referendum halted construction of new units.

Lastly, in May 2009, the Italy's Senate approved measures aimed to bring back nuclear energy for electricity production. Under the new law on development and energy, the government will be given six months to prepare necessary legislation and select sites for new nuclear power plants. Apart from selecting sites, the government will have to define rules for nuclear waste storage, introduce streamlined procedure for new nuclear power plants' approval and set up an agency to supervise nuclear safety. The government would also has to adopt compensation measures for communities which may agree to host new nuclear power plants because local authorities have a final say on industrial projects' approval in Italy and public opinion has been hostile to nuclear energy since the Chernobyl nuclear accident. Opponents to the use of nuclear energy for electricity generation say densely populated Italy is not fit for the operation of nuclear power plants and has no funds for such costly projects. Its supporters say Italy needs to diversify energy supplies and reduce heavy dependence on fossil fuel imports as well as cut emissions of heat-trapping CO_2.

To support the reintroduction of a nuclear power programme in the country the Italian Minister's of Economic Development visited the USA, with the purpose to sign an agreement with its US counterpart on nuclear cooperation, steps which are indispensable to allow US companies to participate in the Italian nuclear power development. At the same time, the Italian company Ansaldo Nucleare and Toshiba-Westinghouse with base in the USA, decided to create an Italy-Japan consortium to work in the Italian market. This consortium, in turn, will joint venture with the Italian Enel and with the French EdF for the supplies of nuclear technology for the Italian nuclear power programme.

It is foreseen that four of the eight-ten reactors that Italy has in plan to build will be EPR type, which will be operated by Enel, Edf and probably also for other smaller partners. As for the remaining reactors, it is supposed that they will be of the Westinghouse AP1000 type, although it cannot discard the possibility that these units come from another design to be selected. The general manager of the Italian electricity company EDISON SpA declared that "its company is willing to invest, between 2015 and 2025, until US$5,700 million in the construction of new nuclear power reactors in Ital". The biggest Italian electric company, Enel, had already agreed with EdF to carry out a joint study for the construction and operation of four EPR systems in Italy once the government has made the pertinent decisions.

THE CURRENT SITUATION AND THE PERSPECTIVES OF THE NUCLEAR POWER SECTOR IN ALBANIA

Albania is one of the poorest countries in Europe, which still endures acute electricity shortages and almost daily blackouts, even in the capital. Albania has petroleum, natural gas, coal, and asphalt laying in the sedimentary rock formations of the country's southwestern regions. About 70% of Albania's territory is about 300 meters above sea level, twice the average elevation of Europe. These great heights, combined with normally abundant highland rainfall, facilitate the production of hydroelectric power along rivers.

With its significant petroleum and natural-gas reserves, coal deposits and hydroelectric-power capacity, Albania has the potential to produce enough energy for domestic consumption and export fuels and electric power to other countries. However, mismanagement led to production shortfalls in the early 1990s forcing the government to import both petroleum and electric power. Known petroleum reserves at existing Albanian drill sites totaled about 200 million tons but, in 1991, recoverable stocks amounted to only 25 million tons. The three refineries had a capacity of 2.5 million tons per year.

Albania's known natural-gas reserves have been estimated at 22,400 million m^3 and lie mainly in the Kuçovë and Patos areas. The country's wells pumped about 600,000 m^3 of natural gas annually during the late 1980s.

Albania's unprofitable coal mines produced about 2.1 million tons in 1987. The coal, mainly lignite with a low calorific value, was being mined mainly in central Albania near Valias, Manëz, and Krrabë; near Korçë at Mborje and Drenovë; in northern Tepelenë at Memaliaj; and in Alarup to the south of Lake Ohrid. Coal washeries were located at Valias and Memaliaj. Albania imported about 200 000 tons of coke per year from Poland for its metalworks. Albania used most of its coal to generate electric power.

About 80% of Albania's electric power came from a system of hydroelectric dams built after 1947 and driven by several rivers that normally carried abundant rainfall. Electric power output was estimated by Albanian officials at 3 984 000 MW hours in 1988. Outfitted with French-built turbines, Albania's largest power plant, the Koman hydroelectric plant on the Drin River, had a capacity of about 600 MW. The hydroelectric stations at Fierzë and Dejas, also on the Drin River, had capacities of 500 MW and 250 MW, respectively, and used Chinese-built turbines. Albania had no capacity to generate nuclear power, but in the early 1990s a research nuclear reactor was reportedly under construction with United Nations funds. With the purpose to reduce the participation of fossil fuels in the generation of electricity, the Albanian's government proposed, in 2007, the construction of a nuclear power plant for both domestic and export supply to Balkans and Italy. In 2009, Croatia supported the proposal and the two countries agreed to work together in carrying out a feasibility study related with the construction of a nuclear power plant at Lake Shkoder, close to the border with Montenegro. Bosnia and Montenegro were invited to participate in the project. The nuclear power reactor to be built will have a capacity of 1 500 MW and will have a cost of about US$5.3 billion.

It is important to note that in a referendum carried in Albania in 1987 led to a five-year moratorium on nuclear power and no government has since dared reopen the issue.

THE CURRENT SITUATION AND THE PERSPECTIVES OF THE NUCLEAR POWER SECTOR IN PORTUGAL

In 2007, Portugal's electricity production was about 47 billion kWh/yr (gross). The cost of electricity in the country is expected to remain close to the present value of approximately US$0.14 per kWh, but the future liberalization of the electricity market could have an effect on the cost of electricity.

The main sources of energy for the generation of electricity in Portugal are coal and gas (27%) and hydro (22%). According with public available information, there was a clear trend for an increase in the energy demand during the period 1990-2004 and it is expected that this trend will continue in the coming years.

Portugal is a country which a high dependence on the imports of energy supply, which account for almost 87% of the country's energy needs. The net imports of electricity from Spain are about 5 TWh per year. Most of the country's supply comes from fossil fuels – oil in particular. For this reason, there are four main options to satisfy the foresee energy demands in the coming years: a) the increase use of fossil fuel. In this case the high price of oil and gas should be taken into consideration. In the last five years the price of gas increased from US$3 to US$9 per MBtu; b) the increase in the use of hydropower; c) the increase in the use of renewable; and d) the introduction of a nuclear power programme. The introduction of carbon taxes of US$100- US$200 per ton would improve the competitiveness of nuclear power and makes the use of this type of energy for electricity generation more attractive.

Portugal's lack of nuclear capabilities sharply contrasts with that of the EU, particularly since the largest electricity contribution in the EU in 2003 came from nuclear power (21% of installed capacity and 33% of electricity production). An interest in developing these capabilities has been expressed in Portugal, but no action has been taken up to date to put into practice this interest. Portugal's future energy and environment policy scenario must be explored in the context of its membership in the EU. The EU has set as targets an increase in the renewable energy technologies' share of electricity production to 22.1% by 2010, coupled with the integration and liberalization of the European electricity market. Furthermore, it has established as common objectives sustainable development, competitiveness and security of supply as basis for a European Energy Policy framework. It is also important to mention that Spain and Portugal are working on merging their electricity markets to conform the Iberian Electricity Market (MIBEL). (Ponce de León Baridó, 2006)

In 2004, the government rejected a proposal to construct a nuclear power plant but this decision is now under reviewed due to a foresee increase in the price of natural gas, coupled with an expensive carbon tax. This situation will be conducive to making nuclear power a more attractive energy production option in Portugal in the future.

THE CURRENT SITUATION AND THE PERSPECTIVES OF THE NUCLEAR POWER SECTOR IN NORWAY

Norway's electricity is almost entirely hydro. In 2007, the total electricity generated in the country was 137.7 billion kWh gross, from 29.5 GWe of capacity, almost 100% produced by the hydroelectric power plants. During a year with normal level of precipitation the net energy import/export exchange between Denmark, Finland, Norway and Sweden is relative small, but during years with low precipitation and accordingly low hydropower generation, the net energy import/export exchange between these countries increases significantly. In 2004, energy net imports were 11.5 TWh and, in 2005, net exports were 12 TWh, mostly to and from Sweden. In 2006, imports and exports largely balanced. Norway's electricity per capita is about 25,000 kWh per year, one of the highest in the world.

Taking into account the almost total dependence of Norway on hydro for the generation of electricity, industrial Norwegian leaders are thinking to introduce a nuclear power programme to supplement hydro in the production of electricity in the coming years. They believe nuclear power based on thorium, which Norway has plenty of, will prevent future energy crises. In February 2008, the Norwegian's government received a report from the Thorium Report Committee indicating the possible use of thorium as a fuel in future nuclear power plants to be built in the country[130]. The reported entitled "Thorium as an Energy Source – Opportunities for Norway" expressed that "building thorium-fuelled power reactors was a possibility that need to be study deeply by the national competent authorities and the energy private sector". The country's Halden research reactor currently in operation in the country could be used to carry out all necessary test related with the use of thorium as a fuel. The Committee notes "that Norway has one of the major thorium resources in the world, a potential energy content which is about 100 times larger than all the oil extracted to date by Norway, including the remaining reserves".

A Norwegian expert claims that thorium produces 250 times more energy per unit of weight than uranium used in the present types of fission nuclear power reactors. In addition, the thorium lobby stresses that thorium fuel in contrast with uranium fuel does not produce any plutonium and that the spent thorium fuel would be much less radioactive than conventional nuclear waste. Also they claim that the half-lives of the radioactive waste products are in the range of hundreds of years instead of thousands of years in the case of conventional spent nuclear fuel. Another often used argument is that thorium reactors will not be based on moderated chain reactions like in fission nuclear power reactors, but on accelerator-driven systems (ADS). ADS could be the third stage of the three-staged thorium based fuel cycle. [137] Despite of the fact that a thorium reactor does not produce any weapons grade plutonium, it does produce weapons grade uranium-233 which is even a more effective fissile material than uranium-235.

The committee's report also indicated that "the country should strengthen its international collaboration in the peaceful use of nuclear energy, particularly for the production of electricity, and should develop its human resources in nuclear science and engineering so as to keep the thorium option open as complementary to the uranium option". After receiving the report, Norway's Minister of Petroleum and Energy, Åslaug Haga, said: "I register that the report neither provides grounds for a complete rejection of thorium as a fuel source for energy production, nor does it offer enough reason for embracing it as such. The government's viewpoint has not changed, meaning that there exist no plans to allow construction of nuclear power plants in Norway". Apparently, financial and technical uncertainties in developing a thorium fuel cycle infrastructure have made the Norwegian government very careful to make a clear decision. [137]

The Norwegian Radiation Protection Agency has licensed an underground repository inside a mountain for radioactive waste from the country's oil and gas industry. It will hold 6 000 tons of waste, and 400 tons has already been placed there. This underground repository could be used also for the disposal of the spent nuclear fuel coming from the nuclear power

---

[130] Thorium is a naturally occurring radioactive trace element found in most rocks and soils. It was discovered in 1828 by the Swedish chemist Jons Jakob Berzelius, who named it after Thor, the Norse god of thunder. Australia and India each have around one quarter of the world's reserves, while both Norway and the USA have 15%. (137)

plants that could be built in the country in the future, if the Norway's government take the decision to introduce a nuclear power programme. However, until today there are no indication that the Norway's government are considering the introduction of a nuclear power programme to satisfy its energy needs in the future.

THE CURRENT SITUATION AND THE PERSPECTIVES OF THE NUCLEAR POWER SECTOR IN POLAND

In 2007, Poland produced some 159 billion kWh (gross) from 32 GWe installed capacity. The main energy source for the production of electricity in the country is coal. In 2007, the production of electricity using coal as a fuel was 93% of the total electricity consumed by the country. The per capita consumption of electricity is 2 700 kWh per year, but with a very high CO_2 emissions. Poland is one of the largest coal producers in the world and has the largest reserves of coal in the EU estimate between 14 and 20 billion tons located in the Upper Silesian coal basin. Poland is a net electricity exporter – 11 billion kWh in 2006, mostly to Czech Republic and Slovakia. The country is also exporting electricity to the Scandinavian countries.

According with IAEA sources, about 33% of Poland's natural gas needs is being supplied by domestic sources, the remaining 67% is imported from Russia (via Belarus and Ukraine). Oil reserves are insignificant and are sufficient for only about 4% of current demand. The vast bulk of oil and liquid fuels is, therefore, imported either from Russia or from the North Sea and Arab countries. There are no uranium ore in Poland. Potential hydro power capacities are estimated at 14 TWh of yearly electricity production.

The forecast of Poland's electricity consumption is to grow by 90% to 2025, but the EU has placed stringent restrictions on CO_2 emissions, which could limit the use of coal as the energy source to satisfy the foresee demand of electricity in the country in the coming years.

The New Energy Policy in Poland

On January 4, 2005, the Polish's government adopted a new energy policy until 2025. The new energy policy is in line with the EU's Energy Policy. The main goals of the new Polish's Energy Policy are the following: 1) assuring the energy security of the country; 2) assuring growths of competitiveness and energy efficiency of the national economy; and 3) protecting the environment against negative impact of the energy sector.

The general goals of the Polish's Energy Policy are to be achieved, inter alia, by:

a) enhancing by legal and regulatory instruments a balanced structure of the primary energy supply with preferences to use domestic coal and lignite resources to the extent which is harmonized with the ecological requirements;

b) further implementation of competitive markets of electricity and gas in accordance with the EU's Energy Policy;

c) assuring a rational level of energy costs in the national economy through improving its energy efficiency, also in the energy sector;

d) legal enforcement of the role of local governments in preparing and implementing local plans for electricity, gas and district heat supply;

e) improving the present system of promotion of generation of electricity and heat by renewable sources and in cogeneration mode by the introduction "green certificate" and allowing for their trade;

f) balancing interests of energy consumers and suppliers in order to improve security and quality of energy supply;

g) using the regulatory, as well as the restructuring instruments, in companies being still owned by the State Treasury for building a market oriented structure of economically sound energy enterprises. [105]

For the first time since the change of Poland's national economy, the following statement on nuclear energy has appeared in the energy policy document of Poland: "Regarding the necessity of diversification of primary energy carriers and limitation of emission of greenhouse gases to the atmosphere, the introduction of nuclear energy into the Polish energy system is justifiable. However, the implementation of nuclear power will need a public acceptance. The forecasts show a necessity to get electricity generated by nuclear power plants already in the second decade of the century. Therefore, taking into consideration the necessary lead time for preparation and construction of the first nuclear power plant in Poland, it is an inevitable need for launching a public debate of that issue as soon as possible".

It is important to note that Poland started, in the 1980s, the construction of four 440 MWe Russian WWER-440 units at Zarnowiec in the north of the country, but these were cancelled in 1992 due mainly to the Chernobyl nuclear accident and a strong public opinion against the use of nuclear energy for the production of electricity. The components of the unfinished nuclear power plant were sold. However, the site remains a leading contender for at least 1 500 MWe of capacity.

Nuclear Safety Authority

The need to draft an atomic law and to prepare construction license for the Zarnowiec nuclear power plant force the establishment of a Nuclear Regulatory Task Force in the Central Laboratory for Radiological Protection (CLOR) in 1984. Another group of CLOR staff has been involved in licensing and inspections of radiation sources users in Poland The new group performed regulatory review of available safety documentation and prepared construction license. It was involved also in drafting Atomic Law Act that finally was adopted in April 1986. In 1987, a new Department for Nuclear Regulatory Tasks (DNRT) was established in CLOR, as well as a new department named "Principal Nuclear Safety Inspectorate". In 1992, State Inspectorate for Nuclear Safety and Radiological Protection (PIBJOR) was established by the President of NAEA, as organization separated from both, NAEA and CLOR. In 1996, PIBJOR with its entire staff was incorporated into NAEA and formed 2 departments for licensing and inspection, one dealing with activities and the other with facilities. In 2004, the Radiation Emergency Centre was established.

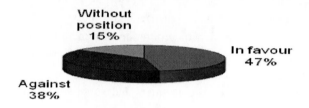

Source: National Atomic Energy Agency, 2006.

Figure 125. Results of the poll.

Following IAEA recommendations nuclear regulatory function is separated from functions related with promotion or management of facilities and activities subject to regulatory control, in particular in the area of research reactors, radioactive wastes and spent nuclear fuel. Regulatory functions are performed according to acts and regulations clearly defining the obligation of applicant and inspected entity, as well as the obligation of NAEA inspectors, regarding both licensing process and conducting inspections.

Before the accession of Poland to the EU in 2004, a full revision of Poland's laws and regulations in the nuclear and radiation safety area was completed. It provides for full harmonization of Poland's laws and regulations in this area with the EU's directives on nuclear matters. New Atomic Law Act of 29 November 2000, in force since January 2002 (amended in 2008), incorporates into law the IAEA Basic Safety Standards for Radiological Protection and provides also, for practical implementation of the requirements of the Convention on Nuclear Safety, Joint Convention, and others international instruments in force in the field of nuclear safety. The Act applies to nuclear power reactors as well as other nuclear and radiation facilities and activities involving ionizing radiation threats. The Act established that "the President of NAEA is the only national nuclear regulatory authority"; this means one institution will control all radiation sources or facilities from their sitting until the final disposal as waste (decommissioning and dismantling).

Operation without a license is prohibited by the law in force. NAEA President and Chief Inspector have tools for enforcement, including power to modify or revoke of any authorization previously given to operate a nuclear facility. Regulatory inspections and assessments will be carried out to ascertain compliance with regulations and authorization conditions approved by NAEA. Inspectors carrying out inspections activities have power to give on the spot orders to stop any activity, if safety is jeopardized. Monitoring of effects is performed to measure achieved safety level, radiation situation of the country, radiation doses (occupational and public) and environmental measurement results.

Nuclear Fuel Cycle and Waste Management

With respect nuclear fuel cycle facilities located in Poland it is important to note that, according with IAEA sources, the only facility of the back-end of nuclear fuel cycle is located in Rózan. However, this near-surface repository for short-lived radioactive waste that every year it received about 300 m^3 of solidified, compacted low and medium level radiation waste from medicine, industry, research and the research reactor in Swierk was closed, due to a release of tritium. At the moment, there is no permanent storage facility for high radioactive waste. The government will resume work started in 1994 to develop a new storage facility for low and medium level radioactive waste and plans to identify storage options for radioactive waste and spent fuel in deep geological formations.

The Public Opinion

A public opinion poll carried out in December 2006 by the National Atomic Energy Agency showed "that 60% supported the construction of nuclear power plants in the country to diminish the country's dependence on natural gas from Russia and to reduce CO_2 emissions to the atmosphere". However, another poll carried out in December 2008 with 1,002 respondents shows "that 47% of the respondents were in favor to build nuclear power plants to satisfy energy demand, 38% were against and 15% has no opinion (see Figure 125).

Looking Forward

In July 2006, Lithuania invited Poland to join with Estonia and Latvia in building a new large nuclear power plant in Lithuania with the purpose to replace the Ignalina nuclear power plant. Polish participation would justify a larger and more economical unit such as an EPR system. In February 2007, the three Baltic States and Poland agreed to build a new nuclear power plant at Ignalina, initially with 3 200 MWe. At least one unit of the project is expected to be operating by 2015. Total cost will be some €6 billion. Poland wish to has access to at least 1 200 MWe of the total capacity built to justify its participation in the project.

In May 2008, the government formed the Lithuanian Energy Organization (LEO) to build the new nuclear power plant and to be in charge of the transmission of electricity to Sweden and Poland. In July 2008, LEO with energy companies from Latvia, Estonia and Poland (Latvenergo, Eesti Energia and Polska Grupa Energetyczna) established the Visaginas project development company for the new 3 200 -3 400 MWe nuclear power plant. Though located close to the Ignalina nuclear power plant site, the new one will be called "Visaginas" due to the name of the nearby town of the same name. Lithuania wants, at least, 34% of the production of the new nuclear power plant (1 090 -1 160 MWe), Poland wants 1 200 MWe, while Latvia and Estonia want 400-600 MWe each. It is important to note that recently the Estonian and Poland Prime Ministers have urged Lithuania to accelerate work on its planned new nuclear power plant, which is due to draw investment from the three States plus Latvia. Poland's Prime Minister Donald Tusk urged Lithuania "to accelerate the speed for the preparation of the construction of the new Ignalina nuclear power plant and said his country was, for the first time, considering building its own nuclear power facilities regardless of the Ignalina project". In March 2009, Lithuanian President Valdas Adamkus said "that the work on the stalled project to construct a new nuclear power plant to replace Lithuania's Ignalina nuclear power plant could start this autumn". However, the new Lithuanian government is now questioning the construction of a new nuclear power plant in the country and had ordered a new study on this subject. The new nuclear power plant was initially due to be up and operated by 2015, but is now formally targeted for 2018 due in part to disagreement with

Estonia, Latvia and Poland over shares of power output. The experts suggest 2020 as a more realistic target for the opening of the new nuclear power plant.

In parallel to the possible construction of a new nuclear power plant in Lithuania, and taking into account the EU policy on CO_2 emissions, the Polish cabinet decided, in 2005, that for energy diversification and to reduce CO_2 and sulfur emissions to the atmosphere the country should consider seriously the introduction of a nuclear power programme, so that an initial nuclear power plant might be operating soon after 2020. In July 2006, the new Prime Minister reaffirmed the need to build nuclear power plants. A feasibility study carried in 2006 suggested "that not less than 11.5 GWe nuclear capacity need to be installed in the medium term in order to satisfy the foresee increase in electricity demand and to reduce the CO_2 emissions to the atmosphere", but this was considered by the Polish's government an unrealistic recommendation. For this reason, a total of 4.5 GWe nuclear capacities to be installed by 2030 were then targeted. In 2007, a draft energy policy proposes to add 10 MWe of indigenous nuclear capacity by 2030, with the purpose to provide around 10% of electricity. To reach this goal an increase of 7.5% should be achieve by 2022. In 2008, the Economy Minister said that "the first nuclear power plant was proposed by 2023 at the northern town of Zarnowiec". The government then brought the target date for first nuclear power to 2020.

In January 2009, state-owned Polska Grupa Energetyczna SA (PGE)[131], announced plans to build two nuclear power plants, each with a capacity of 3 000 MWe, one in a site located in the North part of the country and another in a site located in the East part. PGE estimates that the cost of the construction of these two nuclear power plants would be between €2,500 and €3,000/kW. The energy security strategy, approved by the Polish's government in January 2009, aims that the first of the two nuclear power plants to be built by PGE should be finished by 2020.

Lastly, in September 2009, a plan was adopted by the government for the construction of the first nuclear power plant. The construction works will begin in 2016. The plan establishes four basic stages: In the first stage, to be completed by December 2010, the government will define in detail the Polish programme for the introduction of a nuclear power programme in the country and will start consultation with competent government offices and with the population on this important issue. In the second stage, covering the period January 2011- December 2013, the construction of the new nuclear power plant will be negotiated and the contract signed. In the third stage, covering the period January 2014- December 2015, the technical designs of the nuclear power plant will be prepared and the necessary licenses will be negotiated. In the fourth stage, covering the period January 2016- December of 2020, the construction of the nuclear power plant will start. During all phase a wide campaigns on public information will be carried out in order to keep inform the population about the implementation of the programme and on the importance for the country the introduction of a nuclear power programme to satisfy the foresee increase in the demand of electricity and for the reduction of the CO_2 emission to the atmosphere in the coming years.

---

[131] State-owned Polska Grupa Energetyczna SA (PGE) is Poland's largest power group by generating capacity.

THE CURRENT SITUATION AND THE PERSPECTIVES OF THE NUCLEAR POWER SECTOR IN BELARUS

Belarus produces around 32 billion kWh per year from 7 GWe installed capacity. Most of the electricity produced came from gas-fired power plants. The per capita electricity consumption in the country is around 3 330 kWh per year. The country imports 90% of its gas from Russia, much of it for electricity generation. A single nuclear power plant would be expected to reduce gas imports by US$200-US$400 million per year.

Belarus started, in the 1990s, to construct near Minks a nuclear power plant using Russian WWER-1000 system but construction was abandoned in 1988 after the Chernobyl nuclear accident. The planning activities for the construction of another nuclear power plant near the northeastern city of Vitebsk were also halted for the same reason. Later in the same decade, Belarus considered the possibility to reintroduce a nuclear power programme in the country by procuring a CANDU system from Canada and initiated discussions with Russia, Canada, France and the USA on this subject. Due to lack of funds, the idea of constructing a nuclear power plant was abandoned. Other studies on the construction of a nuclear power plant using Russian technology, or as alternative, Belarus participation in a nuclear power plant to be built at Smolensk or Kursk in Russia, was carried out by the Belarus's government.

The intention of the Belarus's government to build a nuclear power plant was not abandoned despite of the failure of the first two attempts. In the middle of 2006, the government approved a plan for the construction of an initial 2 000 MWe PWR nuclear power plant in the Mogilev region in the Eastern Belarus and near the Russian border. This plant is expected to provide electricity at half the cost of that from Russian gas and to provide some 30% of the foresee demand of electricity by 2020. The construction cost of the nuclear power plant is about €4 billion[132].

Several international nuclear power reactor vendors were invited to present concrete offers to build the planned nuclear power plant in the coming years. In August 2008, the energy ministry announced that proposals had been received from Atomstroyexport (Russia), Westinghouse-Toshiba (USA), Areva (France) and Guangdong Nuclear Power Corporation (China) for the construction of the nuclear power plant. After studying all proposals mentioned above, Russia's Atomstroyexport has emerged as the most likely supplier for the two 1 000 MWe units of Generation III type of reactor. Operation of the first unit is envisaged for 2016 and the second for 2018. Two further units are proposed to be constructed with the purpose to entry into commercial operation by 2025. In June 2007, Russia's Eximbank offered a US$2 billion credit line to enable purchase of equipment from Russia's Power Machines Company as a major part of the overall cost.

In order to create the necessary conditions to introduce a nuclear power programme in the country the Belarus's government has been in contact with the IAEA in order to seek its support in selecting a site, strengthening the country's nuclear regulatory system, assisting in the drafting of nuclear energy legislation and training specialists for future nuclear-related jobs.

---

[132] According with January 2008 estimate.

In November 2007, a presidential decree defined the organizations responsible for preparing for the construction of the nuclear power plant and budgeted money for engineering and site selection. The candidate sites were Krasnopolyansk and Kukshinovsk, both in the Mogilev region near Russia and Ostrovetsk in the Grodno region near Lithuania. Site works are expected to begin in 2009 and the construction work possibly in 2010. A Directorate for the construction of the nuclear power plant will be established under the Ministry of Energy and a Nuclear and Radiation Safety Department will be set up as part of the Emergencies Ministry. This department will act as the state nuclear regulator and licensing authority. The state-run Belnipienergoprom enterprise has been designated as the general designer of the nuclear power plant and will be responsible for negotiating and signing contracts with the Russian counterpart will carry out feasibility studies and preparing tender documents.

In January 2008, the Belarus's government confirmed that the country will begin the construct of its first nuclear power plant, possibly, in 2010 with the purpose to provide some 30% of the electricity demand by 2020.

The Public Opinion

Despite of the decision already adopted by the Belarus's government regarding the construction of a nuclear power plant in the country in the coming years, it is important to single out that the public opinion in Belarus is still affected by the consequences of the Chernobyl nuclear accident. According to a poll carried out in 2005 by the government, 47% of the Belarusian people still do not support the use of nuclear energy for electricity generation, while only 28% approved the construction of nuclear power plants in the country in the future. However, the evolution of the public opinion regarding the use of nuclear energy for electricity production has change and today only small group of people are openly campaigning against the construction of a nuclear power plant in Belarus.

THE CURRENT SITUATION AND THE PERSPECTIVES OF THE NUCLEAR POWER SECTOR IN ESTONIA

Estonia's energy policy objectives until the year 2020 and 2018 are included in the following two documents: "The Energy Economy Development" and the "Electricity Economy Development Plans", respectively. Until 2009, Estonia has no nuclear power plant in operation.

The Electricity Economy Development Plan foresees the reconstruction of the energy production of electricity in Estonia within the next 10-15 years. The plan foresee that the combined production of heat and energy should be extended from the current 200 MW to 300 MW by the year 2014 and the renovation of two more blocks in the Narva Power Plant with the total capacity of up to 600 MW by end of 2015. The plan also contemplates the construction of a nuclear power plant in Estonia by the year 2023. According with plan approved, the energy supply should be more diversified in Estonia in the future and more different sources of energy should be used for the production of electricity as compared to the current situation. While 60% of energy is currently produced from oil shale, in 15 years the proportion of oil shale should remain below 30%. The proportions of other sources of energy should be increased as well as the use of nuclear energy for electricity generation.

The most important steps to be taken by the Estonia's government in the next few years in the energy sector are the development of legislation for the construction of a nuclear power plant, the development of terms and conditions for the further renovation of Narva Power Plant, the creation of new energy connections with other European countries, the development of an action plan on renewable energy and the national development plan of the heating sector.

The amount of state expenditures on the activities planned in the Development Plan of the energy sector will be approximately 32 billion kroons until 2020 and for the activities of the Development Plan of the electricity sector approximately 17.5 billion kroons until 2018. Together with the involvement of private capital and loan capital, the full implementation of the Development Plan for the Energy Sector in the course of the next 15 years will cost more than 100 billion kroons.

The Current Situation and the Perspectives of the Nuclear Power Sector in Ireland

Ireland has repeatedly rejected the construction of a nuclear power plant. A big victory was won here when the public opinion stopped the then Fianna Fail government going ahead with their plans to build not one, but four nuclear power reactors at Carnsore Point in Wexford in the late 1970s.

In April 2006, the state agency, Forfás, has warned "that Ireland will face a liquid fuel crisis in the next 10 to 15 years and may have to build a nuclear power plant in the future in order to satisfy its growing electricity needs". In a new report, the agency warned "that Ireland is now more heavily dependent on imported oil for their energy requirements than almost every other European country in a moment when the world is approaching a point termed "Peak Oil", where global oil production can no longer be increased". The Agency warns also "that a sudden and more imminent onset of "Peak Oil" could require the suspension of the Ireland's Kyoto environmental targets, because it would become more reliant on coal, gas and peat. The report stress "that Ireland is now using 50% more oil per person than in 1990 and that the country is the sixth most dependent on oil for electricity generation out of 25 EU countries". The report says "it will take up to 10 years to significantly reduce Ireland's dependence on imported oil and, for this reason, the government may have to consider the development of a small-scale nuclear power plant in the future and for greater use of renewable energy sources such as wind, wave, biomass and tidal energy".

Despite of all previous actions taken by the Irish's government to introduce a nuclear power programme, there are no current plans for the use of nuclear energy for the generation of electricity in the near future. It is important to single out that, at the same time that the government was considered the possibilities of introducing a nuclear power programme in the country, local anti-nuclear groups sprang up all over the country, forcing the government to put aside their idea to use nuclear energy for the generation of electricity. According to every single one of the polls done at the time there was a large majority against going ahead with the construction of a nuclear power plant at Carnsore. With major opposition all over the island and several thousand determined to physically stop any building work, the Carnsore plan was quietly dropped.

The Current Situation and the Perspectives of the Nuclear Power Sector in Turkey

There are no nuclear power plants in operation or under construction in Turkey until now. However, studies to build a nuclear power plant in the country started in 1965. Between 1967 and 1970, a feasibility study was undertaken by a foreign consultant company to build a 300-400 MWe nuclear power reactor with the purpose to enter into commercial in operation in 1977. Unfortunately, because of the problems relating to site selection and other issues, the project was not implemented.

In 1973, the Turkish Electricity Authority (TEK) decided to build an 80 MWe prototype nuclear power reactor. However, in 1974, the project was cancelled because it could delay the construction of a bigger nuclear power plant. Instead of this prototype reactor, TEK decided to build a 600 MWe nuclear power reactor in Southern Turkey. Site selection studies were made in 1974 and 1975 and the Gülnar-Akkuyu location was found suitable for the construction of the first nuclear power reactor in the country. In 1976, the Atomic Energy Commission granted a site license for Akkuyu. In 1977, a bid was prepared and Asea-Atom and Stal-Laval companies were awarded the contract as the best bidders. Contract negotiations continued until 1980. However, in September 1980, due to the Swedish government's decision to withdraw a loan guarantee, the project was cancelled. A third attempt to build a nuclear power plant was made in 1980. Three companies were awarded the contract to build four nuclear power reactors with one CANDU system to be supplied by AECL, one PWR system to be supplied by KWU in Akkuyu, and two BWR systems to be supplied by GE in Sinop. Due to Turkey's request to apply the Build Operate Transfer (BOT) model, KWU resigned from the bid. Although AECL accepted the BOT model, it insisted upon a governmental guarantee of the BOT credit. The Turkish government refused to give such a guarantee and as a consequence the project was cancelled.

In 1993, the High Council of Science and Technology identified nuclear electricity generation as the third highest priority project for the country. In view of this decision, the Turkish Electricity Generation and Transmission Company (TEAS) included the construction of a nuclear power reactor in its 1993 investment programme. In 1995, TEAS selected the Korean KAERI as the consultant for the preparation of the bid specifications. The bid process started in 1996. Three consortiums offered proposals in 1997: AECL, NPI and Westinghouse. After a series of delays, in July 2000, the government decided to postpone the project.

In the spring of 2006, Turkey's energy ministry called for bids for the construction of the country's first nuclear power reactor, to be built in Akkuyu near Mersin on the Mediterranean Sea. Plans for a second nuclear power reactor in Sinop, on the Black Sea, are already in the works. Turkish Atomic Energy Agency (TAEK) President Oktay Cakiroglu confirmed, on 12 of April 2006, that the country's first nuclear power reactor will be built in the Black Sea province of Sinop. The facility is expected to help meet the country's energy demand over the next 15 years. A 100 MW pilot nuclear power reactor will be constructed by 2009, followed by three nuclear power reactors with a total capacity of 5 000 MW.

It is important to note that Turkey has been debating the pros and cons of building a nuclear power plant for almost 30 years. Previous governments tried to get such a project going three different times, but ended up shelving their projects in the face of opposition from environmental groups. Environmentalists, as well as the opposition Republican Peoples' Party, have also objected to the Sinop project, citing its high price and security concerns. Despite of the opposition to the introduction of a nuclear power programme in the country, Energy and Natural Resources Minister Hilmi Guler says "authorities are determined to pres and going ahead with their plans".

According with the latest available public information, the governments of Russia and Turkey has signed an agreement for the construction of 14 nuclear power reactors in Turkey in the coming years.

Chapter 5

THE CURRENT SITUATION AND PERSPECTIVES IN THE USE OF NUCLEAR ENERGY FOR ELECTRICITY GENERATION IN ASIA AND THE PACIFIC

The arguments in favor of the nuclear energy option for electricity generation all over the world have been propped up by the concern that wake up the CO_2 emissions and the negative impact of such emissions in the world climate. The nuclear energy option also has been favored by increasing restlessness on power security and by the significant increase in the prices of fossil fuels in 2007 and 2008, particularly oil, among other factors.

The progress in the approval and construction of new nuclear power plants are moving slowly in most of the developed world, except in the Asia region, which is the region with the highest nuclear power reactors under construction in the world in 2009[133]. However, it will take some time before the nuclear industry can revert its descendent tendency in terms of participation in electricity generation in almost region. Nuclear power is an industry still in crisis in the West in 2009, due to the panic produced by the nuclear accidents at Three Mile Island and Chernobyl and because of the multi-billion-dollar cost overruns that have plagued nuclear energy projects throughout Europe and the USA. But if the nuclear energy almost died in the West in the past, it has being reinforce in Asia and reborn in other parts of the world. The booming fuel-hungry nations across the Asian region are ordering and building new nuclear power reactors at a rate not seen in decades. Why this boom in the construction of new nuclear power reactors in Asia? The answer is very simple: Asian leaders had said they have no choice but to use nuclear energy for the production of electricity due to the explosive rate at which the demand for energy is growing in several countries. According with the IAEA projections, power demands in Asia will triple by 2015.

In the last years it has been spoken much of the renaissance of the nuclear industry, although the generating capacity of the global number of reactors hardly increased in the last decade: less than 2% of the new total generating capacity. According with IAEA information, the world nuclear generation capacity increased little more than 21 GWe between 2000 and 2007, to add less than 372 GWe, or around 15% of the world-wide total electricity generation. In the specific case of Asia and the Pacific, the region produced, in 2007, a total net of 5 546

---

[133] There are 34 nuclear power reactors under construction in five countries in Asia in 2009 with a net capacity of 21 921 MWe.

GWh of electricity out of which 523 GWh came from nuclear energy representing 9.4% of the total electricity produced in the region by all energy sources. In Asia and the Pacific, they are 103 nuclear power reactors in operation in six countries, including Taiwan, with a net capacity of 72 539 MWe and 34 units under construction in five countries with a net capacity of 21 921 MWe. China and India alone have 25 nuclear power reactors under construction and have plans to construct 128 new nuclear power reactors in the future, in order to elevate the proportion of nuclear energy in the generation of electricity in both countries. In the specific case of China, the new nuclear power reactors planned and proposed are 103 with a net capacity of 91 060 MWe, for the time being, the biggest construction programme in the world.

There should be no doubt that 2006 and 2007 were years of increasing activities in the field of nuclear power in the Asia and the Pacific region, being the only region in the world where electricity generating capacity and, specifically, nuclear power capacity is growing significantly. Important plans for the expansion of current nuclear power programmes were announced in some countries in Asia and the Pacific, while others are considering plans for the introduction of nuclear power programmes for the first time in the coming years. In contrast, North America and most of Western Europe[134], where growth in electricity generating capacity and particularly nuclear power leveled out for many years, a number of countries in the Asia and the Pacific region, particularly in the East and South Asia sub-region, are planning and building new nuclear power reactors to meet their increasing demands for electricity. [5]

Countries that have announced plans for significant expansion in the use of nuclear energy for electricity generation in the Asia and the Pacific region are the following: China, India, Japan, Pakistan and the Republic of Korea. Other countries such as Indonesia, Vietnam, Bangladesh and the Democratic People Republic of Korea have plans for the introduction, for the first time, of nuclear energy for electricity generation in the coming years.

Table 62. Nuclear power in Asia and the Pacific

	Power reactors in operation	Power reactors under construction	Power reactors planned or proposed
China*	11 (6)	19 (2)	103
India	17	6	25
Indonesia	-	-	6
Japan	53	2	14
S. Korea	20	6	5
N. Korea	-	-	1
Pakistan	2	1	4
Thailand	-	-	6
Vietnam	-	-	10
** Total	103	34	174

Note: *Including six nuclear power reactors in operation and two under construction in Taiwan.
Source: IAEA and CEA, 8th 2008 Edition, 2008.

[134] Only Finland and France are now constructing new nuclear power plants in Western Europe.

It is important to single out that the Southeast Asian economies, themselves beneficiaries of an oil and gas export bonanza through the 1970s-1990s, also find themselves in an energy crunch as once ample reserves run down and the search is on for new and cleaner energy supplies. For this reason, regional leaders at the 13th ASEAN Summit meeting held in Singapore in November 2007 issued a statement promoting the use of nuclear energy for electricity generation alongside renewable and alternative energy sources in the future.

According with the IAEA, CEA and other public sources, 103 nuclear power reactors are operating in six countries of the region, including Taiwan, 25 units under construction in six countries and plans to build about another 174 units in the coming years (see Table 62).

In addition, there are about 56 nuclear research reactors in 13 countries of the region. The only two major Pacific Rim countries without any nuclear power and research reactor in operation are Singapore and New Zealand.

Until 2010, the projected new generating capacity in Asia and the Pacific is around 38 GWe per year. For the period 2010 to 2020, the projected new generating capacity is 56 GWe. Much of the growth foresees will be in China, Japan, India and the Republic of Korea. According with WANO sources, the nuclear share in the region in 2020 is expected to be at least 39 GWe and maybe more, if environmental constraints limit fossil fuel expansion.

According with the projection made by the International Energy Outlook 2007 (IEO 2007), nuclear power generation in the non-OECD countries is projected to increase by 4% per year from 2004 to 2030. The largest increase in installed nuclear generating capacity is expected in non-OECD Asia, where annual increases in nuclear capacity average 6.3% and account for 68% of the total projected increase in nuclear power capacity for the non-OECD region as a whole. Of the 58 GWe of additional installed nuclear generating capacity projected for non-OECD Asia between 2004 and 2030, 36 GWe is projected for China and 17 GWe for India[135].

Much of the growth in world economic activity between 2004 and 2030 is expected to occur among the nations of non-OECD Asia. China, non-OECD Asia's largest economy, is expected to continue playing a major role on both the supply and demand sides of the global economy. IEO 2007 projects an average annual growth rate of approximately 6.5% for China's economy over the period 2004 to 2030. The country's economic growth is expected to be the highest in the world. India is another Asian country with a rapidly emerging economy. Average annual GDP growth in India over the 2004 to 2030 projection period is 5.7%. In the rest of non-OECD Asia, economic activity has remained robust, with exports increasing in response to a rebound in global demand for high-technology products and stronger import demand from China. Over the medium term, national economic growth rates in the region are expected to be roughly constant over the period 2004 to 2015, before tapering off gradually to an average of 4.3% per year from 2015 to 2030 as labor force growth rates decline and economies mature.

It is expected that nuclear power plants continuing to have in Asia and the Pacific in the future a 'front-loaded' cost structure, i.e. they are relatively expensive to build but relatively inexpensive to operate. For new nuclear power plants constructions in the Asian and the Pacific region, however, the economic competitiveness of these type of plants will depends on the different energy alternatives available, on the overall electricity demand in the different countries and how fast it is growing, on the market structure and investment environment, on

---

[135] Other projections for these two countries are 40 GWe for China and 20 GWe for India.

environmental constraints and on investment risks due to possible political and regulatory delays or changes, among others. Thus, economic competitiveness is not the same in all countries and it depends on the specific situation of each of them. For example, in Japan and the Republic of Korea, the relatively high cost of using energy alternatives benefits nuclear power's competitiveness. In other countries, such as India and China, rapidly growing energy needs encourage the development of all energy options. (IAEA, GC (51)/INF/3, 2007)

There are different motivations which drove nuclear power developments in different periods in Asia and the Pacific. Three periods can be identified: a) the US Atoms for Peace initiative in the 1950's; b) the oil crisis of the 1970's; and c) the economic growth and need for energy in the 1990's after the end of the Cold War.

There are several forces that have limited the use of nuclear energy for electricity generation in Asia and the Pacific. These forces are the following: a) response to the Three Mile Island nuclear incident; b) the Indian nuclear explosion in 1974; c) restraints on international nuclear cooperation by supplier states; and d) the Chernobyl nuclear accident.

Looking ahead, there are six countervailing political movements that might impact nuclear power's future in Asia and the Pacific. First, is the concern for global warming which might promote the use of nuclear energy for electricity generation. Second, is the growing public distrust of nuclear safety which reduces support for the use of this type of energy for electricity generation. Third, are the disposal of high nuclear waste and the lack of a final disposal location for the disposal of this type of waste in the region. Fourth, is the high capital cost involved in the construction of a nuclear power plant, despite of the reduction in the construction time achieved by Japan. While nuclear power uses a safe and proven technology, and even if it gained complete public trust, electric utilities, especially in deregulated markets, will want to generate electricity with the technology that makes the most fiscal sense. In comparison to the 1970's, energy security is relatively less important than energy costs. In several countries nuclear power plants cannot compete with fossil fuel fired power plants, because modern fossil fuel technology has lowered costs in comparison to nuclear options. Fifth, the increase use of nuclear energy for electricity production in Asia and the Pacific will depend of the availability of new generation of nuclear power reactors. Sixth, is the so-called "proliferation risk" that is involved in the use of nuclear energy for electricity generation, particularly if the country decided the development of their own nuclear fuel cycle.

It is important to note that in Northeast Asia, where nuclear industry has reached a relatively high level of development, the most urgent problems are those related to the so-called "back-end" of the nuclear fuel cycle, especially low and high-level radioactive wastes and spent nuclear fuels. These problems inevitably cause anxiety among the public, often adversely affecting further development of nuclear power programs. This is particularly true in Japan, Korea and Taiwan, as already noted. (Kaneko, Suzuki, Choi, and Fei, 1998)

For countries in Southeast Asia where nuclear power development has just started or is about to start, the problems are not as urgent as in Northeast Asia, but would become serious in the near future. For instance, in some countries the problems are mostly related to the "front-end" of the nuclear fuel cycle, particularly the building of basic infrastructure, necessary to support the introduction of a nuclear power programme. For this group of countries, one of the most important issue is international technical assistance in nuclear safety, environmental management of radioactive wastes, physical protection of nuclear materials/plants, nuclear legislation, manpower training and public education. Some

Southeast Asian countries also require special financial assistance to build nuclear power plants.

East Asia, with its enormous appetite for electricity, is turning to nuclear energy for the production of electricity because it is a proven, available and, in many cases, economically competitive source of energy and because, at least for many East Asian countries, alternatives are not consistently, cheaply, or conveniently available. Coal, for example, is in short supply in Japan, North and South Korea, and virtually non-existent in Taiwan. China's supply, while abundant, is in Northwestern provinces far from the centers of population and industrial/commercial developments in the South and Southeastern coasts. Oil and natural gases are available primarily from import, therefore, expensive. Hydropower has limited potential for many of these East Asian countries with small geographical area. It is being developed in China, but it cannot meet the country's appetite for electricity. That makes nuclear energy one of the most accessible, practical and economic choices for large base load power plants, and East Asia is currently the only region in the world that has plans to rapidly expand nuclear power as a major energy source within the next century. (Kaneko, Suzuki, Choi, and Fei, 1998)

THE ASIAN NUCLEAR SAFETY NETWORK (ANSN)

The introduction of a nuclear power programme or its extension always raises a series of issues that transcend national borders. These issues are, among others, the following: 1) the safety of nuclear energy production; 2) the dangers associated with reprocessing (i.e., risk of diversion for military purposes); 3) the challenges of disposal of spent nuclear fuel and nuclear waste; and 4) the issues related to the physical protection of nuclear materials and facilities.

Therefore, it is not surprising that, as nuclear energy has developed in Asia and the Pacific, particularly in Northeast and Southeast Asia, there has been a parallel growth of multilateral cooperative initiatives, including from governments in the region. However, it is not clear that these efforts have yet addressed the full range of concerns encompassed by what the IAEA calls the "new realities" of the fuel cycle. (Kaneko, Suzuki, Choi, and Fei, 1998)

One of these initiatives is the establishment of the Asian Nuclear Safety Network (ANSN). The ANSN has been established in order to promote a safety culture among states operating nuclear power plants (See Figure 166). The objectives of the ANSN are "to pool and share existing and new technical knowledge and practical experience in Asia to improve the safety of nuclear installations". It aims at the sustainability of national and regional nuclear safety infrastructures and optimum use of existing and new nuclear safety information and is implemented in the frame of the IAEA's Extra Budgetary Programme (EBP) on the Safety of Nuclear Installations in the South East Asia, Pacific and Far East Countries. The participating countries in ANSN are the following: China, Indonesia, Japan, Republic of Korea, Malaysia, Philippines, Thailand and Viet Nam. Australia, France, Germany and USA provide in-kind and/or financial support for the implementation of the project. The IAEA plays a coordinating and facilitating role in the ANSN implementation.

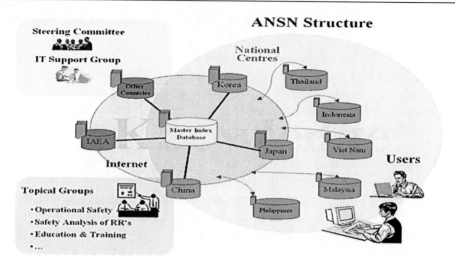

Source: IAEA.

Figure 126. The Asian Nuclear Safety Network.

There is a shared view among the countries participating in the ANSN that this network should be, in the future, a platform for addressing policy and technical safety issues for maintaining sustainable nuclear safety in the whole region. This is an important position adopted by the participating countries in the work of ANSN, due to the implementation of significant nuclear power programmes in several Asia and the Pacific countries.

According with the IAEA Safety Fundamental, "the most pressing challenge for the nuclear power industry in the region is ensuring that adequate safety infrastructure is in place to support the nuclear power plant design, construction, operation, maintenance and decommissioning of nuclear power reactors, as well the regulatory activities associated with all of these actions. As commercial nuclear power reactors continue to be built in the region, increasing demand and competition for limited resources will challenge most state's nuclear safety infrastructure as they undertake new activities or activities such as sitting that have not been performed for many years".

In the Asia and Pacific region, a group of countries is developing new nuclear power reactor designs and, for this reason, there is a good opportunity to ensure that appropriate security features are incorporated into their design, construction and operation. To do this successfully will require effective national strategies, supported by appropriate international guidance.

Another important challenge is the effective dissemination of the experience in the operation of nuclear power reactors. The effective dissemination of operating experience is an ongoing challenge. The commitments to report issues and lessons learned and to take action based on experience of others need to be enhanced to prevent events from recurring. Sharing of utility and regulatory actions in response to events, and also the sharing and replication of good practices, may assist in precluding the occurrence of significant issues. (SF-1 IAEA, 2006)

The work of ANSN could assist countries in the region in promoting and supporting effective dissemination of operating experience of several countries that are now operating nuclear power plants.

THE ASIAN NETWORK FOR EDUCATION IN NUCLEAR TECHNOLOGY (ANENT)

One of the major concerns of the IAEA and nuclear national authorities of countries thinking to expand or initiate a nuclear power programme, is not only the availability of qualified personnel in different areas associated with the use of nuclear energy for electricity generation, but the adequate preparation of the personnel involved in the implementation of such programmes. With much of the forecasted growth for nuclear power taking place in Asia and the Pacific region, the IAEA met and consulted with representatives from several Asian member states on establishing the Asian Network for Education in Nuclear Technology (ANENT), a regional network for higher education in nuclear technology. Created in 2003, ANENT establishes a central point for exchange of information and materials for education and training, a base for distance learning opportunities and a mechanism to support the exchange of students, educators and researchers in the nuclear field. ANENT also helps Asian nuclear education programmes by providing a mechanism for transfer of academic credits and regional recognition of degrees. (Sacchetti, 2008)

The ANENT will play an important role in supporting the expansion of nuclear power programmes planned in several countries of the region. To support the expansion of nuclear power programmes in several Asia and the Pacific countries, it is important to have a number of qualified and experience professional and technicians in nuclear matters, particularly in the operation of nuclear power reactors, nuclear safety, and maintenance of nuclear power reactors, among others.

TECHNOLOGICAL ADVANCED IN FISSION'S REACTORS IN ASIA AND THE PACIFIC

The increase use of nuclear energy for electricity production in Asia and the Pacific will depend of the availability of new generation of nuclear power reactors. The technological advances in new generation of fission reactors in the region are briefly described in the following paragraphs.

Japan

According with the IAEA[136], in Japan, ABWR systems benefit from standardization and construction in series. The first two ABWRs began commercial operation in 1996 and 1997, and two more began commercial operation in 2005 and 2006. Two ABWRs are under construction in Taiwan, China. A development programme was started in 1991 for ABWR-II with the goal of significantly reducing generation costs, partly through increased power and economies of scale. Commissioning of the first ABWR-II is foreseen for the late 2010s. In addition to the work carried out bay Japan in the development of the ABWR systems, the

---

[136] See document IAEA GC(51)/INF/3 of 2007.

basic design of a large advanced PWR has been completed for the Japan Atomic Power Company's (Tsuruga 3 and 4 units). A larger version of the ABWR system, the APWR+, is in the design stage.

At the same time, Japan is testing a 30 MWth HTTR system that began operation in 1998 and work continues on safety testing and coupling to a hydrogen production unit. A 300 MWe power reactor prototype is also under consideration. It is scheduled for commissioning by September 2010. It expects to use some 20 GWe of nuclear heat for hydrogen production by 2050, with the first commercial plant expected coming on line in 2025.

With respect the test of fast breeder reactors the situation in Japan is the following: In 2005 preparatory works on necessary modifications to the 280 MWe prototype fast breeders MONJU prior to its restart has been introduced. To develop advanced fuels, materials and technology for minor actinide burning and transmutation, the JOYO reactor, an experimental fast breeder reactor, will begin irradiation of oxide dispersion strengthened ferity steel of uranium plutonium. (IAEA GC(51)/INF/3, 2007)

Finally, it is important to single out the following: The most important Japanese nuclear companies are developing small and medium nuclear power reactors to be use in industrialized countries as well as in developing countries. The Toshiba Corporation is developing an extra-compact nuclear power reactor of 10 MWe and it has begun the necessary procedures for its approval in the USA. The new reactor, the Toshiba 4S, is designed to reduce to the minimum their supervision and maintenance necessities by means of automatic functions that assure the security in the occurrence of certain events. On the other hand, Mitsubishi Heavy Industries Ltd. has completed the design of a PWR system of about 300 MWe, while Hitachi Ltd plans to develop a BWR system of 400-600 MWe that could be used in relative small electrical grid and in developing countries.

Republic of Korea

According with the IAEA[137], the Republic of Korea benefits of standardization and construction of nuclear power reactors in series through the Korean Standard Nuclear Plant (KSNP) concept. Eight KSNPs are now in commercial operation. The accumulated experience is the basis for developing an improved KSNP, the Optimized Power Reactor (OPR), with the first units planned for commercial operation in 2010 and 2011. The Korean Next Generation Reactor, for which development began in 1992, is now named the Advanced Power Reactor 1400 (APR-1400) and will be bigger to benefit from economies of scale. The first APR-1400 is scheduled to begin operation in 2012.

In addition to the construction of new generation of nuclear power reactors in the country, the Korea Atomic Energy Research Institute has conducted research, technology development and design work on the 600 MWe KALIMER-600 advanced fast reactor concept. The conceptual design was completed in 2006. From 2007 the development of sodium cooled fast reactor (SFR) technology entered in a new phase within the framework of the Generation IV collaboration project.

---

[137] See document IAEA GC(51)/INF/3 of 2007.

India

According with the IAEA[138], in 2005 and 2006, India connected the first two units using its new 540 MWe HWR design at Tarapur. India is also designing an evolutionary 700 MWe HWR and is developing the Advanced Heavy Water Reactor (AHWR), a heavy water moderated boiling light water cooled vertical pressure tube type reactor, which has passive safety systems and optimized to use thorium as fuel. A Fast Breeder Test Reactor (FBTR) has been in operation since 1985 and a 500 MWe Prototype Fast Breeder Reactor (PFBR) is now under construction at Kalpakkam. It is scheduled for commissioning by September 2010.

China

According with the IAEA Nuclear Technology Review for 2007, in China, work continues on safety tests and design improvements for the 10 MWth High Temperature Gas Cooled Reactor (HTR-10), with pebble bed fuel, which started up in 2000. A commercial prototype HTR based on it is expected to start up in 2010. Plans are in place for the design and construction of a power reactor prototype (HTR-PM). At the same time, a 25 MWe sodium cooled pool type Chinese Experimental Fast Reactor is under construction, with first criticality foreseen for mid-2009 and grid connection in mid-2010. The next two stages of development will be a 600 MWe prototype fast breeder reactor, for which design work started in 2005 and a 1000-1500 MWe demonstration fast breeder reactor.

China is also participating, since 2006, in the activities carried out by the Generation IV International Forum (GIF), a group of countries which are developing the next generation of nuclear power reactors[139].

THE PUBLIC OPINION

The future expansion in the use of nuclear energy for electricity generation in Asia and the Pacific will depends on several factors. One of these factors is the public opinion. The results of opinion polls are notoriously dependent on the particular question asked. They can also be very volatile. A number of themes may, however, be detected. Whether a particular person or group of people tends to be pro- or anti-nuclear at a particular moment depends on a number of features, including: a) perceptions of the need for the technology; b) perceptions of risk – nuclear power tends to be less popular in the immediate aftermath of an accident, while people who are more familiar with the technology, perhaps through having lived near an operating nuclear power plant for some years, tend to be less worried than those who are not; c) social, political and psychological factors. Political parties within a single country can hold radically different views on nuclear technology; and d) people whose jobs depend on the local nuclear facility tend to be more pro-nuclear than those who do not. (Grimstom, 2002)

It should be stressed, though, that there is always a range of opinions in favor and against regarding the use of nuclear energy for electricity generation among people from the same

---

[138] See document IAEA GC (51)/INF/3 2007.
[139] For more information on the Generation IV of nuclear power reactors see Chapter I.

country and apparently very similar backgrounds. Many developing countries have anti-nuclear movements, even if they may be small and environmental pressure groups are increasingly establishing themselves in the developing world. A number of specific explanations have been suggested for the apparent special unease felt about nuclear power in many countries. They include:

a) links to the military, both real (the development of shared facilities) and perceptual;

b) secrecy, coupled sometimes with an apparent unwillingness to give straight answers in part, perhaps, because of links to military nuclear operations in some countries, and in part because of commercial issues;

c) the historical arrogance of many in the nuclear industry, dismissing opposition, however well-founded or sincerely held, as irrational;

d) the apparent vested interest of many nuclear advocates to be contrasted with the apparent altruism of opponents who, for example, are often not funded to take part in public inquiries;

e) the perceived potential for large and uncontainable accidents, and other environmental and health effects, notably those associated with radioactive waste;

f) the overselling of nuclear technology, especially in its early days and, in particular, with regard to its economics, leading to a degree of disillusionment and distrust;

g) a general disillusionment with science and technology, and with the "experts know best" attitude of mind that was more prevalent in the years immediately after the Second World War;

h) the wider decline of "deference" towards "authority", including, for example, politicians and regulatory bodies. (Grimstom, 2002)

It is important to single out that negative public opinion about the use of nuclear energy for the generation of electricity, whether justified or not, can be extremely expensive for investors in nuclear power programme, and can even act as an absolute barrier. For example, public opposition to the construction or operation of a nuclear power plant could increase the costs of nuclear generated electricity in a number of ways. There may be delays during the construction of the nuclear power reactor or in achieving an initial operating license, or interruptions in operation, among others. In India, for example, nearly 60 years after the introduction of nuclear energy for peaceful purpose, the nuclear industry and government nuclear authorities has failed to deliver what the pro-nuclear lobby had promised.

In the case of Japan, the only country in the world that was attacked with nuclear weapons, nuclear accidents[140], official misdeeds, the falsification of safety reports by Tokyo Electric Power Company, which led to the temporary closure of many of its nuclear power reactors and political missteps during the 1990s, have activated public opinion to question the wisdom of Japan's nuclear policy. Although the Japanese people are not ready to abandon nuclear power, they are also not willing to accept all the arguments in its favor. The Japanese

---

[140] The nation's first accident occurred in 1999, when radioactive material escaped a uranium enrichment plant in Tokaimura village, North-East of Tokyo, killing two employees. The latest accident was recorded in 2004 at Kansai Electric Power Co's Mihama nuclear power plant in Fukui Prefecture, North-West of Tokyo. In this accident, non-radioactive steam leaked and killed four workers.

government is been forced as never before to justify its policies. Already in several instances, it has been forced to back down from proposed nuclear sites and plans.

In Vietnam, there is a certain opposition to the use of nuclear energy for electricity generation, which somehow is understandable. Whenever people think about nuclear energy, they usually see it as very dangerous form of energy and the risks of nuclear energy are real. People are obsessed with the Chernobyl nuclear accident, which caused severe damage to millions of people and to the environment. Moreover, Vietnam is one of the few Asian nations blessed by large deposits of fossil fuel it has huge tracts of offshore oil, so why are the reasons to use nuclear energy for electricity generation? The answer could be that perhaps the Vietnamese leaders are eager not to be left behind in the nuclear power binge sweeping Asia, or they do not want to use the fossil fuel reserves now for electricity production increasing the negative impact in the environment by the emission of CO_2, if they can use other type of clean energy for the same purpose.

The Greenpeace International movement against nuclear power has also affected Vietnamese people. The organization worries about nuclear waste and the proliferation of nuclear weapons. In the most extreme cases, fears of public reaction can lead to a fully completed plant being refused an operating license, or for a government to take steps to prevent nuclear construction or to close down existing facilities before the end of their technical lifetimes. (Grimstom, 2002)

The Philippines is one example of a country in the Asia and the Pacific region with a nuclear power plant fully completed but to which operating license was not given due to strong opposition of public opinion and, for this reason, no electricity has ever been generated by the plant.

LOOKING FORWARD

According with the IAEA, CEA and other public information sources, there are 34 nuclear power reactors under construction in five countries in Asia and the Pacific and plans to build around 174 units in the coming years. It is expected that, in 2020, the nuclear share in the region be at least 39 GWe and maybe more, if environmental constraints limit fossil fuel expansion. The largest increase in installed nuclear generating capacity is expected in non-OECD Asia, where annual increases in nuclear capacity average 6.3% and account for 68% of the total projected increase in nuclear power capacity for the non-OECD region as a whole.

The 2010 projection of new generating capacity in Asia and the Pacific is 38 GWe per year. From 2010 to 2020, it is expected to be 56 GWe per year, but up to one third of this is for replacing shut down power plants. This is about 36% of the world's new capacity (current world capacity is about 3 500 GWe, of which 368 GWe is nuclear). Much of this growth will be in China, Japan, India and Korea.

Summing up can be stated that non-OECD Asia is expected to lead the world in the installation of new nuclear power capacity accounting for 51% of the projected net increment in nuclear capacity worldwide. China is projected to add 40 GWe of nuclear capacity by 2020 and India 20 GWe by 2050[141]. Strong growth in nuclear capacity in China and India will help

---

[141] Other sources estimate the electricity generation projections in China and India to be 36% and 17% respectively.

both countries improve fuel diversification in their power sectors, although thermal generation will continue to dominate in both countries. In China, the nuclear share of total electricity generation is projected to rise up to 5% in 2030, and in India it is projected to rise up to 8%. (IEO, 2009)

As can be easily see in Figure 127 India will be the country in the Asia and the Pacific with the highest nuclear share for the period 2004-2030, followed by China.

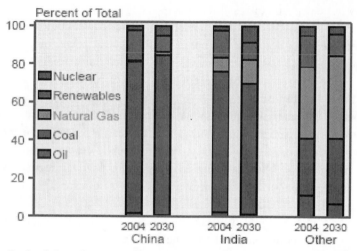

Sources: 2004: Derived from Energy Information Administration (EIA), International Energy Annual 2004 (May-July 2006), web site www.eia.doe.gov/iea. 2030: EIA, System for the Analysis of Global Energy Markets (2007).

Figure 127. Net electricity generation in non-OECD Asia by fuel for the period 2004 and 2030.

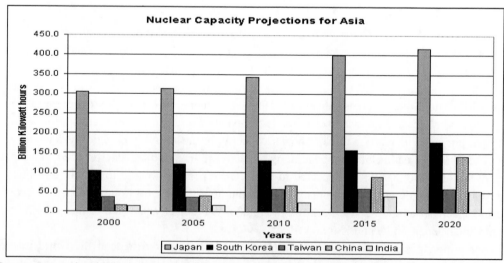

Source: International Energy Outlook, 2002, Energy Information Administration (IAE).

Figure 128. Nuclear capacity projections for Asia and the Pacific.

From Figure 128 can be stated that Japan and the Republic of Korea (South Korea) will be the two countries of the Asia and the Pacific region with the highest nuclear capacity during the period 2010-2020. However, it is foresee that China could reduce the gap in nuclear capacity with the Republic of Korea, if the ambitious plan for the construction of several nuclear power plants is definitively implemented.

THE NUCLEAR POWER PROGRAMME IN JAPAN

Energy security is a major public energy policy concern in Japan. Although Japanese energy consumption is among the highest in the world, the country lacks significant domestic energy resources, forcing the country to imports over 81% of primary energy requirements and 99% of fossil fuel requirements. This dependence exposes Japan's economy to disruptions in international energy markets and makes the country very vulnerable to energy crisis.

It is important to single out that Japan is an island nation that cannot exchange energy supplies with neighboring countries through transmission lines or pipelines, and that it is poor in energy resources and relies on foreign suppliers for most of its energy requirements. To this end, efforts are needed to diversify overseas suppliers of major energy resources and to create an adequate stockpiling base in order to provide for all contingencies, including supply cutoffs. Moreover, because, among economically and industrially advanced nations, Japan has the highest rate of primary energy dependence on oil, and an alarmingly high rate of reliance on Middle Eastern oil imports, an important strategy involves the use of alternative energy sources to the extent feasible. [58]

The second priority is to minimize the environmental load from energy consumption. To this end, steps have been taken to promote energy conservation to the fullest possible extent and to accelerate the introduction of energy sources that impose relatively small loads on the environment, such as nuclear energy and renewable energy sources. Particularly important are measures to reduce emissions of greenhouse gases such as carbon dioxide and methane. Japan, pledged to achieve a target of reducing greenhouse gas emissions by 6% on average from the 1990 level, for the five-year period from 2008 to 2012. In order to achieve this target, Japan is required to make manifold efforts to improve energy utilization efficiency, such as through the introduction of cogeneration systems, which are useful as dispersed energy supply sources, and to promote energy-saving technology by encouraging people to make changes in their lifestyles, which consume much energy. At the same time, a set of policy measures must be put in place to shift power sources from fossil fuels to nuclear power and renewable energy, which involve smaller amounts of carbon dioxide emissions and, as an interim step, to shift from oil to natural gas.

Prompted by energy security concerns, Japan has promoted energy efficiency, becoming one of the most energy efficient countries in the world. Japan has also diversified its primary fuel requirement away from oil. Oil consumption declined from 77% of Japan's total primary energy use in 1973 to about 52% in 2002. Moreover, oil consumption has been relatively stable in recent years, rising only 0.5 million barrels per day; this means from to 4.8 million barrels a day (b/d) in 1988 to 5.3 million b/d in 2002. This is a stark contrast to trends in neighboring China and the Republic of Korea, where oil consumption has more than doubled over the same period. (Medlock III and Hartley, 2004)

Wealthier nations like Japan, which lack their own natural energy resources and have already developed large nuclear industries, are more concerned about their potential vulnerability to world oil and gas markets and the security of energy supply. For this reason, diversity of fuels enhances Japanese economic security in times of disruption or crisis due to the following elements:

a) diversity of fuel sources increases flexibility to keep overall costs low during sudden or prolonged disruptions or demand spikes;

b) conversely, heavy reliance on one or two fuel types can raise the economic stakes of a major disruption;

c) electricity prices are, on average, lowest when there are no constraints to construction of new capacity. While it is true that greater fossil fuel use increases exposure in the aftermath of an oil crisis, eliminating the use of fossil fuels is not an efficient policy. Certain types of generation capacity are best suited to meet particular loads;

d) related to the previous point, encouraging the use of any one fuel beyond its efficient level will generally raise overall electricity costs to the national economy. Moreover, using subsidies to encourage artificially the use of certain fuels can raise overall electricity costs to the national economy. These costs should be weighed carefully against the value of the benefit to the public good of promoting cleaner or more secure fuels such as nuclear power. There is a level of nuclear capacity for Japan that is cost-minimizing. Movement toward a level of nuclear capacity that is either above that level or below it will raise the overall costs of electricity generation in Japan;

e) the most cost-effective fuel to replace nuclear power from an energy security point of view would be coal, disregarding environmental considerations. Since coal is imported from different sources than oil and gas, greater use of coal could also contribute generally to security benefits. On the other hand, while new clean coal technology avoids problems of SOX and NOX pollution, a potential issue is that the clean coal process does not eliminate CO_2 emissions. (Medlock III and Hartley, 2004)

It is important to single put that nuclear power can provide stable fuel costs on a day-to-day basis and protect overall national economic performance during times of disruption of energy supply and, for this reason, nuclear power generation offers a technical advantage over other energy sources in that nuclear fuel has a higher energy density and can be stockpiled easily. In addition, the fuel supply is stable because, unlike oil resources, uranium resources are dispersed in politically stable countries and regions. When practical technologies capable of using uranium more highly efficiently in fast breeder or other advanced reactors become available sometime in the future, nuclear power generation will be able to supply energy for an even longer period of time. It will then stand truly as a promising technical option for providing humankind with its much-needed energy. [58]

In 2009, Japan has 53 nuclear power reactors in operation with a net capacity of 45 199 MWe, generating 27.5% of the total electricity of the country. Japan has also two units under construction with a net capacity of 2 285 MWe; 13 units planned with a total capacity of 17

915 MWe and one unit propose with a capacity of 1 300 MWe. By 2015, nuclear contribution is expected to increase, especially if emission targets under the Kyoto Protocol are met. The goal is to have over 40% of the total electricity generation using nuclear energy before 2020. However, Japanese executives say "proposals have been scaled down and the WANO is not convinced that all will be built, taking into account the economic situation of the country". Japan has also 17 research reactors.

The nuclear power reactors most recently started up in Japan include third generation advanced reactors, with improved safety systems. The first of these units was connected to the electric grid in 1996. Japan is committed to reprocessing its spent nuclear fuel to recover uranium and plutonium for re-use in electricity production, both as MOX fuel in conventional reactors and in fast neutron reactors.

In 2003, a number of Japan's nuclear power reactors were shut down over several months for checking, following inspection irregularities. The last of these units restarted in 2005. In total, Japan has shut down completely three nuclear power reactors with a net capacity of 320 MWe in 1976, 1998 and 2003 respectively.

Supporters of nuclear power point out to its low operating costs and the historically stable costs for uranium fuel, especially when compared to oil or natural gas. They assert that the stable costs in particular demonstrate that nuclear power contributes to Japan's energy security, as it does not face the same commodity risk of other fuels for power generation. In addition, Japan has been able to source uranium imports from different and arguably more reliable foreign suppliers. (Medlock III and Hartley, 2004)

It is important to stress that the future expansion in the use of nuclear energy for electricity generation in Japan will depend on the public opinion, on the technological advanced in the design of new nuclear power reactors and in the absence of new nuclear accidents.

Table 63. Summary of the information regarding the use of nuclear energy for electricity generation in Japan

	Number of units connected to the electric grid	Nuclear electricity generation MWe	Nuclear percentage of total electricity supply (%)
Japan	53	45 199	27.5%

Source: WANO, 2009.

Table 64. Annual electrical power production for 2008

Total power production (including nuclear)	Nuclear power production
964 931.7 GWhe	240 518.5 GWhe

Source: IAEA PRIS, 2009.

The Energy Policy in Japan

Japan's energy policy has been driven by considerations of energy security and the need to minimize dependence on current energy imports. For this reason, the first priority in any Japanese energy supply policy is securing the steady supply of energy necessary to support the lives of the people. The new nuclear policy should be promoted with a primary focus on the following: a) nuclear safety; b) disaster prevention; c) public trust; d) coexistence with local sitting communities; e) adherence to the peaceful use of nuclear energy; and f)international understanding;

The main elements contained in the energy policy adopted by Japan regarding nuclear power are the following: 1) continue to have nuclear power as a major element of electricity production; 2) recycle uranium and plutonium from spent nuclear fuel, initially in low radioactive wastes, and have reprocessing domestically from 2005; 3) steadily develop fast breeder reactors in order to improve uranium utilization dramatically; and 4) promote nuclear energy to the public, emphasizing safety and non-proliferation. [104]

In March 2002, the Japanese government announced that it would rely heavily on nuclear energy in the future to achieve greenhouse gas emission reduction goals set by the Kyoto Protocol and committed by the government.

A ten year energy plan, submitted in July 2001 to the Minister of Economy Trade and Industry (METI), was endorsed by the Japanese cabinet. It called for an increase in nuclear power generation by about 30% (13 000 MWe), with the expectation that utilities would have nine to twelve new nuclear power reactors operating by 2011. The main elements of the ten year energy plan are, according with a report of the Japan Atomic Energy Commission, the following: a) based on the history of nuclear energy in the 20th century, the new plan should define problems to be solved and prescribe long-term strategies to be carried out to develop the diverse potential of nuclear energy. By returning to the starting point, the significance and role of nuclear energy should be reexamined in comparison with other energy options, taking into consideration changes in lifestyles, societal values, views of the international community and developments in science and technology; b) amid growing public concern and distrust of nuclear energy, a prerequisite for any nuclear energy policy is winning the understanding and confidence of the Japanese people, society and the international community. In order to do so, the new plan should not only provide specific guidelines to parties involved in nuclear energy, but should also send a clear message domestically and to the world. In this regards, it is important to single the following: With the "Monju" fast breeder reactor accident as the stimulus, the Atomic Energy Commission of Japan has been promoting public participation in the policy-making process. In order to ensure the support of the public to the expansion of the Japanese nuclear power programme, the government must make continued efforts to encourage citizens to take part in the policy-making process by holding public hearings on policy options, and take advantage of opportunities to demonstrate its accountability. These processes should be reviewed in a flexible manner to meet changes in the social situation. In order to continue to hear what citizens have to say and to reflect that in nuclear energy policy, a study should be made on creating a new forum for listening to the people, similar to that of the Round-Table Conference; and c) the new plan should determine the roles of the government and the private industry energy sector, with emphasis on concepts and policies that must be adhered to and pursued steadily, now and in the future.

In June 2002, a new energy policy law was adopted by the Japanese Parliament. The law set out the basic principles of energy security and stable energy supply, giving greater authority to the government in establishing the necessary energy infrastructure to ensure the economic growth of the country foresee for the coming years. It also promoted greater efficiency in energy consumption, a further move away from dependence on fossil fuels and energy market liberalization.

The strategy choice by Japan for the development of the nuclear power option was the use of the LWR as the near-term power reactor to be followed, as soon as possible, by the introduction and deployment of the FBR. The FBR, which uses much less uranium than an LWR of the same capacity, was a crucial part of the strategy because uranium was then believed to be a scarce resource. This strategy, based on the LWR producing the startup fuel for the FBR, implicitly included spent nuclear fuel reprocessing, plutonium recycle and disposal of separated wastes in geologic repositories, all of which have serious implication from the proliferation point of view. Nations with limited indigenous energy reserves as Japan, made particularly strong commitments to this strategy, independently of the proliferation risk involved in its implementation. However, it is important to note that this strategy was chosen in response to the believe that uranium was in critically short supply and that fossil fuel prices would soon rise sharply, that nuclear power would become the dominant energy source, and that the costs of the FBR and its fuel cycle would actually be less than that of the LWR. All of these assumptions have proven to be false and risky. Other nuclear technology developed so far has proven to be more effective, less costly and less risky from the proliferation point of view.

In November 2002, the Japanese government announced that it would introduce a tax on coal for the first time, in order to encourage the use of other type of energy sources with less negative impact in the environment. At the same time, METI would reduce its power-source development tax, including that applying to nuclear generation, by 15.7%. The taxes introduced by Japan were originally designed to improve Japan's energy supply mix and to address Kyoto goals by reducing carbon emissions.

In 2004, Japan's Atomic Industrial Forum released a report on the future prospects for nuclear power in the country. It brought together a number of considerations, including 60% reduction in carbon dioxide emissions and 20% population reduction but with constant GDP. Projected nuclear generating capacity in 2050 was 90 GWe. This means doubling both nuclear generating capacity and nuclear share to about 60% of total power produced. In addition, some 20 GW (thermal) of nuclear heat will be utilized for hydrogen production. Hydrogen is expected to supply 10% of consumed energy and 70% of this will come from nuclear plants. In July 2005, the Atomic Energy Commission reaffirmed the energy policy for nuclear power in Japan. The main goal is to achieve a 30-40% nuclear share in total electricity generation after 2030, including replacement of current nuclear power reactors with advanced LWRs. FBRs will be introduced commercially, but not until about 2050. In April 2007, the government selected Mitsubishi Heavy Industries (MHI) as the core company to develop a new generation of FBRs. This was backed by government ministries, the Japan Atomic Energy Agency (JAEA) and the Federation of Electric Power Companies of Japan. These are concerned to accelerate the development of a world-leading FBR by Japan. MHI has been actively engaged in FBR development since the 1960s as a significant part of its nuclear power business. [82]

In April 2006, the Institute of Energy Economics of Japan forecast for 2030 that while primary energy demand will decrease 10%, electricity use will increase and nuclear share will be around 40%, from 63 GWe of capacity. Ten new nuclear power reactors should come on line by 2030 and Tsuruga 1 would be retired. In May 2006, the ruling Liberal Democratic Party urged the government to accelerate development of FBRs, calling this "a basic national technology". It proposed increased budget, better coordination in moving from research and development to verification and implementation, plus international cooperation[142].

It is important to note that METI's 2008 electricity supply plan shows nuclear capacity growing to 61.5 GWe (23.4% of the total) by 2017, and the share of supply growing from 2007 depressed 262 TWh (25.4%) to 458.3 TWh (41.5%) in 2017.

The Evolution of the Nuclear Power Sector in Japan

In 1954, Japan budgeted 230 million yen for the introduction of nuclear energy in the country, marking the beginning of the programme related with the peaceful uses of nuclear energy in different sector of the economy, including the energy sector. One year later, the Japanese Parliament adopted the Basic Atomic Energy Law, called the "Basic Law" of 19 December 1955. The Basic Law states that "its objectives are to secure energy resources for the future and to promote the research, development and use of nuclear energy for peaceful purposes". It goes on to establish "a framework for the regulation of nuclear activities, specific aspects of which are to be dealt with in subsequent separate laws".

Article 2 of the Basic Law sets out "the three basic principles of democracy, independence and public disclosure governing the peaceful use of nuclear energy". Article 4 provides "for the establishment of the Atomic Energy Commission and the Nuclear Safety Commission in the Cabinet Office to ensure a democratic approach to nuclear energy administration".

In June of 1957, the Japanese's Parliament adopted the "Law for the Regulation of Nuclear Source Material, Nuclear Fuel and Reactors" (the "Regulation Law"). This law governs "the location, construction and operation of nuclear facilities in Japan". The purpose of the Regulation Law is "to ensure the peaceful use of nuclear source material, fuel and reactors, and to ensure public safety by preventing the hazards that arise from these materials and reactors. Specifically, it establishes "the licensing regime governing all stages of nuclear activities and sets out essential rules to observe in international conventions". The Regulation Law also provides "for control over nuclear material and equipment and restricts the transfer, importation and exportation of nuclear fuel material to those who are engaged in refining, manufacturing or reprocessing of such material and to operators of nuclear facilities". Concerning nuclear materials, the Regulation Law provides "that premises where nuclear fuel is present may be subject to inspection by the safeguards division at any time and operators of nuclear facilities are responsible for establishing security rules and procedures at their installations and in relation to specified nuclear materials contained therein. The operators must report any loss or theft of nuclear materials".

---

[142] Japan is already playing a leading role in the Generation IV initiative, with focus on sodium-cooled FBRs, though the 280 MWe Monju prototype FBR remains shut down.

The last law adopted by the Japanese's Parliament in 1957 was the "Law Concerning Prevention from Radiation Hazards due to Radioisotopes" (the "Prevention Law"). The aim of the Prevention Law is to "regulate the use, sale, disposal or any other handling of radioisotopes and radiation-emitting equipment in order to prevent radiation hazards and to secure public safety".

After the accident in Tokaimura, the Japanese's Parliament decided to revise the "Law for the Regulation of Nuclear Source Material, Nuclear Fuel and Reactors" (the "Revised Regulation Law"). The Revised Regulation Law was adopted on 13 December 1999. The reason for the adoption of a revised law was to take into account the need to drastically enhance nuclear safety regulations in the wake of a nuclear accident.

In addition, to the adoption of the Revised Regulation Law, a new special "Law on Emergency Preparedness for Nuclear Disaster" was adopted on 17 December 1999, based on lessons learned from current nuclear disaster prevention measures. The Revised Regulation Law entered into force on 1 July 2000, and the special Law on Emergency Preparedness for Nuclear Disaster took effect as a special law to modify and complement the "Basic Law for Countermeasures against Disaster" on 16 June 2000.

The first commercial nuclear power reactor in Japan, Tokai Power Station of the Japan Atomic Power Company Ltd., with a net capacity of 159 MWe, started to be built in March 1963, was connected to the electric grid in November 1965 and commenced operations in July 1966. The construction of nuclear power reactors reached its peak in the 1970s and 1980s, at a time when Japan's export-driven and energy-hungry industries were also expanding at their fastest. During this period, 37 nuclear power reactors became operational (one unit in the 1960s, 20 units in the 1970s and 16 units in the 1980s). By the end of the 1990s, Japan had 51 operational nuclear power reactors. At the beginning of the 1990s, nuclear power provided only 9% of Japan's energy needs but by the end of the decade nuclear energy provided 32% of the country's needs an increase of 23%.

In the 1970s, the first light water nuclear power reactors were built in cooperation with US companies. These reactors were bought from GE and Westinghouse companies with contractual work done by Japanese companies, who would later get a license themselves to build similar nuclear power reactor designs. Developments in nuclear power since that time has seen contributions from Japanese companies and research institutes on the same level as the other big users of nuclear power in several countries.

Japan's nuclear industry was not hit as hard by the effects of the Three Mile Island and the Chernobyl nuclear accidents. For this reason, construction of new nuclear power reactors continued to be strong through the 1980s, 1990s and up to the present day. Since 1966, when the construction of the first nuclear power reactors started to be built, Japan has pushed ahead actively with the introduction of nuclear power as a form of energy replacing oil, starting up new reactors at an average rate of 1.5 units a year. However, starting in the mid-1990s, there were several nuclear related accidents and cover-ups in Japan that eroded public perception of the nuclear industry, resulting in protests and resistance to the construction of new nuclear power reactors in the country. These accidents included the Tokaimura nuclear accident in 1999, Japan's worst nuclear radiation accident that took place at a uranium reprocessing facility in Tokai, Ibaraki, Ibaraki prefecture, Northeast of Tokyo, on 30 September, 1999; the Mihama steam explosion operated by Kansai Electric Power Company in the town of Mihama, Fukui, in the Fukui Prefecture, about 320 km West of Tokyo; and the cover-ups after an accidents at the MONJU, Japan's only fast breeder reactor, located in Tsuruga, Fukui

Prefecture. Other circumstances unfavorable to nuclear power include the advance of deregulation in the electric power industry, the installation of wide-area supply networks for electricity and natural gas, a trend toward energy conservation, the accelerated introduction of renewable energy sources and participation of the political parties advocating an anti-nuclear policy in the governments. [58]

It is important to note that after this group of accident, the safety culture in Japan's nuclear industry came under greater scrutiny and a number of nuclear power reactors ordered were canceled. Canceled orders include, in 1994, the Hohoku nuclear power reactor at Hohoku, Yamaguchi; in 1997, the Kushima nuclear power reactor; in 2000, the Ashihama nuclear power reactor at Ashihama; and, in 2003, the Maki nuclear power reactor at Maki, Niigata (Kambara) and the Suzu nuclear power plant at Suzu, Ishikawa[143].

These cancellations reflect to some degree the safety concerns that surfaced after the Monju cover-up and the Tokaimura nuclear accident and could be compared to the situation in the USA where there was a large number of nuclear power orders cancellations after the Three Miles Island and the Chernobyl nuclear accidents.

A summary of the information regarding the current and future of the use of nuclear energy for electricity generation in Japan is shown in tables and figures.

Table 65. Nuclear power reactors operating in Japan

Reactor	Type	Net capacity MWe	Commercial operation
Fukushima I-1	BWR	439	March 1971
Fukushima I-2	BWR	760	July 1974
Fukushima I-3	BWR	760	March 1976
Fukushima I-4	BWR	760	October 1978
Fukushima I-5	BWR	760	April 1978
Fukushima I-6	BWR	1 067	October 1979
Fukushima II-1	BWR	1 067	April 1982
Fukushima II-2	BWR	1 067	February 1984
Fukushima II-3	BWR	1 067	June 1985
Fukushima II-4	BWR	1 067	August 1987
Genkai-1	PWR	529	October 1975
Genkai-2	PWR	529	March 1981
Genkai-3	PWR	1 127	March 1994
Genkai-4	PWR	1 127	July 1997
Hamaoka-3	BWR	1 056	August 1987
Hamaoka-4	BWR	1 092	September 1993
Hamaoka-5	ABWR	1 325	January 2005
Higashidori-1 Tohoku	BWR	1 067	December 2005
Ikata-1	PWR	538	September 1977
Ikata-2	PWR	538	March 1982
Ikata-3	PWR	846	December 1994

---

[143] The Suzu nuclear power plant may continue operating sometime in the future, if economic factors turn more in its favor, though there has been no sign of this happening.

Reactor	Type	Net capacity MWe	Commercial operation
Kashiwazaki-Kariwa-1	BWR	1 067	September 1985
Kashiwazaki-Kariwa-2	BWR	1 067	September 1990
Kashiwazaki-Kariwa-3	BWR	1 067	August 1993
Kashiwazaki-Kariwa-4	BWR	1 067	August 1994
Kashiwazaki-Kariwa-5	BWR	1 067	April 1990
Kashiwazaki-Kariwa-6	ABWR	1 315	November 1996
Kashiwazaki-Kariwa-7	ABWR	1 315	July 1997
Mihama-1	PWR	320	November 1970
Mihama-2	PWR	470	July 1972
Mihama-3	PWR	780	December 1976
Ohi-1	PWR	1 120	March 1979
Ohi-2	PWR	1 120	December 1979
Ohi-3	PWR	1 127	December 1991
Ohi-4	PWR	1 127	February 1993
Onagawa-1	BWR	498	June 1984
Onagawa-2	BWR	796	July 1995
Onagawa-3	BWR	796	January 2002
Sendai-1	PWR	846	July 1984
Sendai-2	PWR	846	November 1985
Shika-1	BWR	505	July 1993
Shika-2	BWR	1 304	March 2006
Shimane-1	BWR	439	March 1974
Shimane-2	BWR	789	February 1989
Takahama-1	PWR	780	November 1974
Takahama-2	PWR	780	November 1975
Takahama-3	PWR	830	January 1985
Takahama-4	PWR	830	June 1985
Tokai-2	BWR	1 060	November 1978
Tomari-1	PWR	550	June 1989
Tomari-2	PWR	550	April 1991
Tsuruga-1	BWR	340	March 1970
Tsuruga-2	PWR	1 110	February 1987
Total: 53 units		45 199	

Note: Fukushima 1 = Fukushima Daiichi 1.
Source: CEA, 8th Edition, 2008.

Table 66. Japanese nuclear power reactors under construction

Reactor	Type	Gross Capacity MWe	Construction start	Operation
Tomari-3	PWR	912	Nov 2004	Dec 2009
Shimane 3	ABWR	1 373	Dec 2005	Dec 2011
Total (2)		2 285		

Source: CEA, 8th, Edition 2008 and IAEA PRIS, 2009.

Japan has shut down six nuclear power reactors. These reactors are the following:

Table 67. Nuclear power reactors shut down in Japan

Name	Type	Status	Net	Gross	Connected
Fugen atr	HWLWR	Permanent Shutdown	148	165	1978/07/29
Hamaoka-1	BWR	Permanent Shutdown	515	540	1974/08/13
Hamaoka-2	BWR	Permanent Shutdown	806	840	1978/05/04
JPDR	BWR	Permanent Shutdown	12	13	1963/10/26
Monju	FBR	Long-term Shutdown	246	280	1995/08/29
Tokai-1	GCR	Permanent Shutdown	137	166	1965/11/10

Source: IAEA PRIS, 2009.

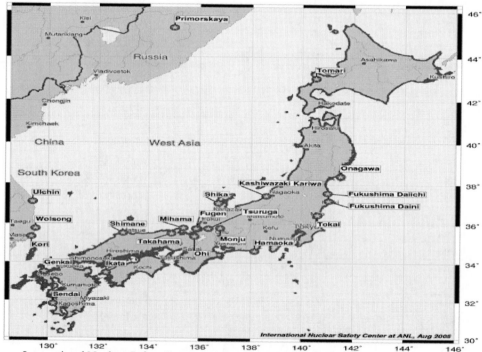

Source: International Nuclear Safety Center at Argonne National Laboratory, USA.

Figure 129. The location of the nuclear power plants in operation in Japan.

Table 69 contained relevant data related with the evolution of the net load factor of all nuclear power reactors operating in Japan during the period 2000-2007.

Looking back at how nuclear power generation has developed in Japan, can be easily seen that, during the 1970s, nuclear power reactors were generally operating at low availability due to a succession of problems. Based on investigations of the causes, increasing availability factor then improved gradually, thanks to the implementation of drastic measures, improvements and standardization efforts. In the late 1990s, the overall availability factor of nuclear power reactors remained above 80% for the PWR type and below 80% for the BWR type.

Table 69. Evolution of the net load factor (%)

Reactor type	2000	2001	2002	2007
BWR	79.0 (28)	78.6 (28)	71.9 (29)	53,3 (32)
PWR	83.3 (23)	84.3 (23)	87.3 (23)	81,4 (23)

Note: ( ) number of operating reactors.
Source: METI nuclear data (August 2003) and CEA, 8th Edition, 2008.

Table 68. Japanese nuclear power reactors planned

Reactor	Type	MWe gross (each)	Start construction	Start operation
Ohma 1	ABWR	1 383	5/2008	11/2014
Fukushima I - 7 and 8	ABWR	1 380	4/2010	10/2015 10/2016
Higashidori 1 and 2	ABWR	1 385	11/2009 and 2012	10/2017 2019
Tsuruga 3 and 4	APWR	1 538	10/2010	2016-17
Kaminoseki 1 and 2	ABWR	1 373	2010 and 2013	2015-16 2020-21
Hamaoka 6	ABWR?	1 490?	-	2018-22
Higashidori 2	ABWR	1 385	2014	2020
Namieodaka	BWR	825	2014	2020-21
Sendai 3	APWR	1 590	2013	2019-20
Total (13)		17 915		

Source: CEA: 8th Edition, 2008.

From Table 69 can be stated that the net load factor of the BWRs system are declining in the period considered moving from 79% in 2000 to 53,3% in 2007; this represent a reduction of 25,7% in seven years. By the contrary, the net load factor of the PWRs system stay over 80% during the same period.

Nuclear Technological Development

As an island country, it is not possible for Japan to exchange energy with neighboring countries through power transmission lines or pipelines. Japan is also energy-scarce, depending on foreign countries for about 80% of its energy resources. Based on these facts, the Japanese' government concludes that it is rational to continue making the fullest possible use of nuclear power generation as one of the mainstays of the nation's energy supply.

Nuclear power generation contributes to improved energy sufficiency and to the stability of the energy supply, in addition to play an important role in reducing Japan's carbon dioxide emissions as committed to the Kyoto Protocol.

The technological advanced achieved until now in the field of nuclear reactor design is an important factor for the successful continuation of the use of nuclear energy for electricity production in the country. In the 1970s, a prototype Advanced Thermal Reactor (ATR) was built at Fugen. This had heavy water moderator and light water cooling in pressure tubes and was designed for both uranium and plutonium fuel, but particularly to demonstrate the use of plutonium. It was the first thermal reactor in the world to use a full MOX core. The 148 MWe unit was operated by the Japan Nuclear Cycle Development Institute (JNC) and was finally shut down in March 2003. Construction of a 600 MWe demonstration ATR was planned at Ohma but, in 1995, a decision was adopted not to proceed with its construction.

Mitsubishi Heavy Industries (MHI) is now marketing its 1700 MWe APWR design in the USA and Europe, and lodged an application for US design certification in January 2008[144]. The US-APWR design has been selected by TXU (now Luminant) for Comanche Peak, Texas. The APWR system is also in the process of being licensed in Japan with a view to the first 1 538 MWe units being constructed at Tsuruga (Units 3 and 4) from 2009. Approval by Fukui prefecture was given in March 2004. It is simpler than present PWRs, combines active and passive cooling systems to greater effect, and has over 55 GW days per ton (GWd/t) burn-up. Design work continues and will be the basis for the next generation of Japanese PWRs. The APWR+ system is 1 750 MWe and has full-core MOX capability. In mid 2005, the Nuclear Energy Policy Planning Division of the Agency for Natural Resources and Energy instigated a two-year feasibility study on development of next generation of LWRs. The new designs, based on ABWR and APWR designs, are to lead to a 20% reduction in construction and generation costs and a 20% reduction in spent nuclear fuel quantity, with improved safety, thirteen-year construction and longer life. The project is expected to cost US$520 million over eight years to develop one BWR and one PWR designs, each of 1 700-1 800 MWe. [82]

In September 2007, it was announced that government with the support of the private industry and utilities would develop new LWR designs for deployment about 2020. These will have at least 5% enriched fuel and an 80 year operating life.

In Japan, ABWR units benefit from standardization and construction in series. The first two ABWRs began commercial operation in 1996 and 1997, and two more began commercial operation in 2005 and 2006. Two ABWRs are under construction in Taiwan, China. A development programme was started in 1991 for ABWR-II system with the goal of significantly reducing generation costs, partly through increased power and economies of scale. Apart from the Fugen experimental Advanced Thermal Reactor (ATR), this would be the first Japanese reactor built to run solely on MOX fuel incorporating recycled plutonium. [82]

In addition, the basic design of a large advanced PWR system has been completed for the Japan Atomic Power Company's Tsuruga 3 and 4 Units and a larger version, the APWR+ system, is in the design stage.

Two High Temperature Engineering Test Reactor (HTTR) are under consideration. The first one is a 30 MWth that began operation in 1998 and work continues on safety testing and coupling to a hydrogen production unit. The second one is a 300 MWe power reactor prototype. It is scheduled for commissioning by September 2010.

In 2005, began the preparatory related with the introduction of all necessary modifications to the 280 MWe prototype fast breeder Monju reactor prior to its restart. To develop advanced fuels and materials and technology for minor actinide burning and transmutation, the Joyo reactor, an experimental fast breeder reactor, will begin irradiation of oxide dispersion strengthened ferritic steel, of uranium–plutonium.

---

[144] MHI also participated in developing the Westinghouse AP1000 system, but now that Westinghouse has been sold to Toshiba, MHI will develop PWR technology independently.

Japan has a High Temperature Test Reactor, which has reached 950°C, high enough to enable thermo chemical production of hydrogen. It expects to use some 20 GWe of nuclear heat for hydrogen production by 2050, with the first commercial plant coming on line in 2025.

Nuclear Safety Authority and the Licensing Process

Nuclear safety is one of the major concern of the Japanese people. The situation regarding nuclear safety and the process for licensing nuclear installations is briefly described in the following paragraphs. The main government offices in charge of safety issues are: a) the Nuclear and Industrial Safety Agency (NISA) within the Ministry of Economy Trade and Industry; b) the Nuclear Safety Commission; c) the Atomic Energy Commission; and d) the Nuclear Safety Network and the Japan Nuclear Technology Institute. NISA is responsible for nuclear power regulation, licensing and safety. It conducts regular inspections of safety-related aspects of all nuclear power plants. The Nuclear Safety Commission is a more senior government body set up in 1978 under the Atomic Energy Basic Law and is responsible for formulating policy, alongside the Atomic Energy Commission. Both are part of the Cabinet Office.

Following the 1999 Tokai criticality accident, electric power companies along with enterprises involved with the nuclear industry established the Nuclear Safety Network (NSnet). The network's main activities are "to enhance the safety culture of the nuclear industry, conduct peer reviews and disseminate information about nuclear safety". In 2005, this was incorporated into the Japan Nuclear Technology Institute, based on the US Institute of Nuclear Power Operations. The NSnet division cooperates with INPO and WANO and arranges peer review activities. [82]

Because of the frequency and magnitude of earthquakes in Japan, particular attention is paid now to seismic issues in the sitting, design and construction of nuclear power reactors. In May 2007, revised seismic criteria were announced which increased the design basis criteria by a factor of about 1.5 and required utilities to undertake some reinforcement of older nuclear power reactors.

Nuclear operators should bear the primary and heavy responsibility for securing the safety of their nuclear installations. Nuclear operators should effectively secure safety on their own initiative through activities to maintain safety and their executives are expected to make every effort to see that a safety-first operation policy permeates throughout their organizations. In training researchers, technicians and engineers, steps should be taken to enhance educational programmes related to safety. Following the criticality accident at a uranium processing plant in 1999, the Nuclear Safety Network and certain other organizations were established by parties involved with nuclear energy. It is strongly hoped that efforts will be exerted to promote safety consciousness and to share relevant information and experience among all members of various industrial circles, in general, and to upgrade the ethics of the entire nuclear power industry in particular, through these networks and organizations. [58]

Nuclear Fuel Cycle and Waste Management

Japan has no indigenous uranium and its requirements should be met by purchasing uranium from different countries such as Australia, Canada, Kazakhstan, among others. For energy security reasons, and notwithstanding the low price of uranium for many years, Japanese policy since 1956 has been to maximize the utilization of imported uranium, extracting an extra 25-30% of energy from nuclear fuel by recycling the unburned uranium and plutonium as mixed-oxide fuel. According with the strategy adopted regarding the use of nuclear energy for the production of electricity, Japan has been progressively developing a complete domestic nuclear fuel cycle industry, based on imported uranium[145].

At present, spent nuclear fuel from nuclear power plants in Japan is cooled in storage pools at the plant sites for a certain period of time, and then reprocessed. So far, most spent nuclear fuel from Japanese nuclear power plants has been entrusted to overseas reprocessors but, in the future, such fuel will be reprocessed at a commercial reprocessing plant located in the village of Rokkasho, in Aomori Prefecture. Spent nuclear fuel beyond the capacity of that reprocessing plant will be stored at the power plants sites, or will be delivered to spent fuel intermediate storage operators, who will store it safely outside the power plant compounds, until it can be reprocessed. High-level radioactive waste remaining after useful substances such as plutonium and uranium are separated from spent nuclear fuel during the reprocessing will be vitrified into a stable form, stored for 30 to 50 years for cooling, and then disposed of in deep geological formations. The recovered plutonium will be used in existing LWRs in a process of MOX fuel utilization and in research and development on, for example, FBRs technology. [58]

JAEA operates a small uranium refining and conversion plant, as well as a small centrifuge enrichment demonstration plant, at Ningyo Toge, Okayama prefecture. While most enrichment services are still imported, Japan Nuclear Fuel Ltd (JNFL) operates a commercial enrichment plant at Rokkasho. This plant began operation in 1992 using indigenous technology. Now is testing a lead cascade of its new Shingata design, and expects to re-equip the plant with this new technology. At the same site; active testing of a vitrification plant commenced in November 2007, with separated high-level wastes being combined with borosilicate glass. The plant takes wastes from the adjacent reprocessing plant, after uranium and plutonium are recovered from spent nuclear fuel for recycle, leaving 3% of the spent nuclear fuel as high-level radioactive waste.

According with WANO sources, Japan has 6 400 tons of uranium recovered from reprocessing and stored in France and the UK, where the reprocessing of spent uranium was carried out until now. In 2007, it was agreed that Russia's Atomenergoprom would enrich this for the Japanese utilities who own it. At Tokaimura, in Ibaraki prefecture North of Tokyo, Mitsubishi Nuclear Fuel Co Ltd operates a major fuel fabrication facility, which started up in 1972. Further fuel fabrication plants are operated by Nuclear Fuel Industries (NFI) in Tokai and Kumatori, and JAEA has some experimental mixed oxide fuel facilities at Tokai for both the Fugen ATR and the FBR programmes, with capacity about 10 tons per year for each. Also at Tokai, JAEA has operated a 90 tons per year pilot reprocessing plant using Purex technology which has treated 1 116 tons of spent nuclear fuel between 1977 and its final

---

[145] Indeed, it was Japan's hunger for reliable energy supplies that, in part, drove its military expansion into Asia in the 1930s and 1940s and the involvement of Japan in the Second World War.

batch early in 2006. JNFL's Rokkashomura reprocessing plant is due to start operation in May 2008, following a 22 month test phase, plus some delay at the end of 13 years construction at a cost of US$ 20 billion. The new Rokkasho plant will treat 14 000 tons of spent nuclear fuel stockpiled there to end of 2005, plus 18 000 tons of spent nuclear fuel arising from 2006, over some 40 years. It will produce about four tons of fissile plutonium per year. JAEA operates spent nuclear fuel storage facilities there and is proposing a further one. It has also operated a pilot high-level waste vitrification plant at Tokai since 1995. [82]

It is important to note that the Federation of Electric Power Companies has said that nine member companies will use plutonium as MOX fuel in 16-18 nuclear power reactors from 2010 under the programme adopted by Japan. About six tons of fissile plutonium per year is expected to be loaded into nuclear power reactors operating in the country. Meanwhile, MOX fuel fabricated in Europe from some 40 tons separated reactor grade plutonium from Japanese spent nuclear fuel can be used for the same purpose in the selected nuclear power reactors.

It is important to single that out that Japan's plutonium stocks has been increasing in the last years reaching 41 tons of separated reactor grade plutonium (about 35% fissile) stored and awaiting use in MOX fuel in 2004. At the end of 2007, there was 26.4 tons of fissile plutonium held by Japanese utilities: 13.9 tons in France, 11.3 tons in UK and 1.2 tons held domestically. It is estimated that 5.5 to 6.5 tons of fissile plutonium will be used each year from about 2012. [82]

In April 2005, the Aomori prefecture approved the construction of a MOX plant at Rokkasho, adjacent to the reprocessing plant. An agreement was signed by the Governor of the prefecture, the Mayor of Rokkashomura and the head of JNFL. The Governor urged the Federation of Electric Power Companies "to step up their efforts towards realization of the MOX use programme". There should be no doubt that the approval of the MOX plant is a significant step forward in closing the fuel cycle in Japan and was strongly supported by the federal government, Atomic Energy Commission and utilities. JNFL has applied for a license to build and operate the 130 ton per year J-MOX plant. It is expected that the plant begin operation by 2012. It is estimate that the cost of the construction of the mentioned plant will be US$1.2 billion.

At the same time, the head of the branch of the French company Areva in Japan has informed, in October 2009, on the signature of a contract with Japan for the construction of a new enrichment plant for the transformation of enriched uranium in dioxide of uranium (UO_2). With this new facility, the capacity of Japan will be duplicated and will be in a position to satisfy the increase in the demand of UO_2 that is expected in coming years. The plant will be built as a joint-venture of Areva with Mitsubishi Corp. and Mitsubishi Heavy Industries, association in which the French company possesses 30% of the actions.

In 2005, Tepco and JAPC announced that a Recyclable Fuel Storage Centre would be established in Mutsu, operating from 2010 with 5 000 tons capacity. It will provide interim storage for up to 50 years before spent nuclear fuel is reprocessed.

Regarding nuclear waste and taking into account the concern of the people on how this waste is managed, it is important to stress the following: For a given amount of energy, nuclear power produces a smaller amount of waste than other forms of electricity generation, which, for example, does not require extensive space for storage or disposal. However, because that waste is radioactive, how ultimately dispose of it is a top-priority issue that must be resolved in order to press ahead with nuclear power development. When a nuclear power plant with a capacity of one million kW is operated for one year, several hundred 200 liter metal drums of low-level radioactive waste are generated. In addition, about 30 150 liter stainless steel containers of vitrified assemblies of high-level radioactive waste, the residue after useful substances such as plutonium and uranium are separated from spent nuclear fuel through its chemical reprocessing result. Other radioactive waste will result when the plant is, in time, decommissioned. Low-level radioactive waste is already disposed of by underground burial. High-level radioactive waste, because its activity level remains high for such a long period of time, must be maintained safely for a long period of time, so that its radioactivity will not have any significant effect on the environment where people live. For high-level radioactive waste, the leading concept in most countries is disposal several hundred meters deep, in stable geological formations. [58]

In May 2000, the Japanese Parliament approved the "Law on Final Disposal of Specified Radioactive Waste" (the "Final Disposal Law") which mandates "deep geological disposal of high-level waste (defined as only vitrified waste from reprocessing spent reactor fuel)". In line with this, the Nuclear Waste Management Organization (NUMO) was set up in October 2000 by the private sector to progress plans for disposal, including site selection, demonstration of technology there, licensing, construction, operation, monitored retrievable storage for 50 years and closure of the repository. Some 40 000 canisters of vitrified high-level waste are envisaged by 2020, needing disposal, all the arising from the Japanese nuclear power plants until then. Repository operation is expected from about 2035, and the US$28 billion cost of it will be met by funds accumulated at 0.2 yen/kWh from electricity utilities (and hence their customers) and paid to NUMO. This sum excludes any financial compensation paid by the government to local communities.

In mid 2007, a supplementary waste disposal bill was passed which says that "final disposal is the most important issue in steadily carrying out nuclear policy". It calls for the government "to take the initiative in helping the public nationally to understand the matter by promoting safety and regional development, in order to get the final disposal site chosen with certainty and without delay. It also calls for "improvement in disposal technology in cooperation with other countries, revising the safety regulations, as necessary, and making efforts to recover public trust by, for example, establishing a more effective inspection system to prevent the recurrence of data falsifications and cover-ups".

In 2004, METI estimated the costs of reprocessing spent nuclear fuel, recycling its fissile material and management of all wastes over 80 years from 2005. METI's Electricity Industry Committee undertook the study, focused on reprocessing and MOX fuel fabrication including the decommissioning of those facilities (but excluding decommissioning of nuclear power reactors). Total costs over 80 years amount to some 19 trillion yen, contributing almost one yen (US$ 0.9 cents) per kilowatt-hour at 3% discount rate. About one third of these costs would still be incurred in a once-through fuel cycle, along with increased high-level waste disposal costs and increased uranium fuel supply costs. Funding arrangements for high-level waste were changed in October 2005 under the new "Back-end Law", which set up the Radioactive Waste Management Funding and Research Centre (RWMC) as the independent funds management body. All reserves held by utilities will be transferred to it and companies will be refunded, as required, for reprocessing.

The Public Opinion

Prior to the 1990s, public opposition to the use of nuclear energy for electricity generation did exist but was localized, usually temporary, and often led by left-wing groups or local trade associations. The focus of this opposition was usually health-related and limited to citizens living near a particular nuclear facility. Why the general public has considerable apprehension about nuclear power? The answer is the following: Because radiation cannot be perceived with the five senses, because the effects of radiation on health are uncertain and because safety arrangements provided in nuclear facilities are unclear to the public. In addition, ordinary people are in no position to foresee the effects and dangers of accidents or problems at a nuclear power plant. Nevertheless, the Japanese people are not ready to abandon nuclear power, but they are also not willing to accept all the arguments in its favor. For this reason, the Japanese government is being been forced as never before to justify its policies. Already in several instances, it has been forced to back down from proposed sites and plans for the construction of new nuclear power reactors.

It is important to single out that opinion polls throughout the 1980s and 1990s showed that a majority of Japanese found Japan's nuclear power plants safe, or somewhat safe, and a large majority of those who supported development of nuclear power believed that this type of energy was a key energy to ensure the energy independence of the country. These views allowed officials to discount protests as short-term, selfish economic anxiety.

Public expressions against nuclear power stand out in two referendum carries out in Japan in 1996. The first referendum was held in the town of Maki, Niigata Prefecture. In the non-binding referendum, the citizens of Maki rejected the construction of a nuclear power plant in their township. The outcome of the referendum was not legally binding, but the town's Mayor did, as promised, bar the sale of municipal land to the electric company. Second, Masayasu Kitagawa, the Governor of Mie Prefecture, outright rejected the construction of the Ashihama reactor in his prefecture when he demanded cancellation of the project in a speech to the Mie Prefectural Assembly. On the same day as the speech, the president of Chubu Electric Company announced that the company intended to cancel plans to construct the reactor. These two events ended a 37-year dispute over the plant's sitting, which included the gathering of 810,000 signatures, or half of Mie's electorate, in a 1996 petition against the building of the nuclear power reactor. (Kotler and Hillman, 2000)

The opposition of the public opinion to the expansion of the country nuclear power programme has risen significantly since 1990s, following several accidents that occurred at different nuclear power plants, the falsification of safety reports at Tokyo Electric Power Company, which led to the temporary closure of many of its nuclear power reactors and several earthquakes that hit the country in the last several decades. The nuclear accidents have contributed greatly to shattering public confidence in government and corporate nuclear oversight. People feeling very uneasy about nuclear power went from 21% before the Tokaimura accident to 52% afterwards. In an October 1999 Japan Public Opinion Company survey, only 11% supported government plans to increase the use of nuclear energy for electricity generation in the country. Fifty-one percent said keep the present situation, and 33% said reduce or stop the use of nuclear energy for electricity generation. Given a choice, the public preferred non-nuclear options (solar/wind generation 62%, conservation 54.9%, compared to 20% for nuclear power). In other words, the public does not completely accept the government's arguments that nuclear power is safe, necessary for Japan's energy security and ecologically sound because it does not emit smoke.

In a poll conducted in November 1999 by *Mainichi Shimbun* found "that 53% of people have feelings of distrust regarding the Japanese government's nuclear energy policy, while 38% have feelings of trust for it". The results of the polls demonstrate the public's withdrawal of support from current nuclear policy. More than just voicing safety concerns, the polls suggest a public expressing displeasure at the violation of the trust it had in the government to insure the safe use of nuclear materials. In another survey conducted at the end of 1999 by the Tokaimura government planning section, the number of respondents supporting nuclear power after the Tokai accident fell to 32.2% from 52% before the accident. Those calling for the discontinuation of its usage, however, rose from 11.7% to 40% after the accident. In addition, the number of respondents who said that nuclear power facilities are unsafe after the Tokai accident was 78.2%. Only 32.1% of the same respondents said they felt the facilities were unsafe before the accident. (Kotler and Hillman, 2000)

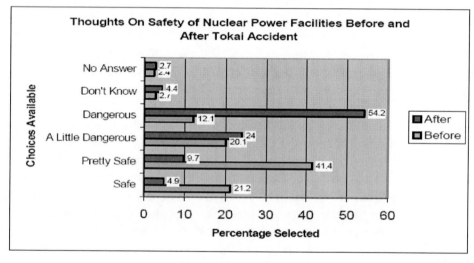

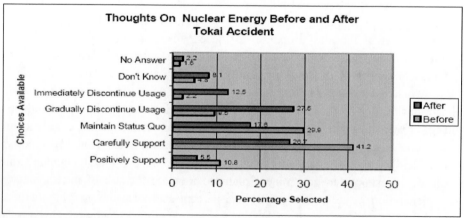

Source: (Kotler and Hillman, 2000)

Figure 130. Public opinion before and after Tokai nuclear accident.

The polls mentioned in previous paragraphs present important evidence that a change in public opinion regarding Japan's nuclear energy policy is occurring at both the national and local levels. The levels of opposition to the use of nuclear energy for electricity generation, all greater than the corresponding levels of support, show consistent opposition across Japanese society. Though not yet an overwhelming opposition, this consistent view on nuclear energy is hard for Japan's leaders to continue to ignore in the future.

It is important to note, during the analysis of the position of the Japanese's people towards the use of nuclear energy for electricity production, that Japan is the only country to have suffered a full-scale nuclear attack, and the only country in the world to have suffered massive casualties from radioactive fallout. However, this particularly situation of the country does not has impeded Japan to be one of the country with the highest number of nuclear power plants in operation in the world after the USA and France.

There are several issues regarding the use of nuclear energy for the production of electricity that need to be considered when the public position is analyzed. One of these issues is the acute public sensitivity to nuclear power bearing in mind the nuclear attacks on Hiroshima and Nagasaki. The second issue is the lack of indigenous energy resources that could be used in Japan for electricity generation. Aside from some small-scale geothermal power plants, the country has no other significant sources of energy such as oil and coal. This lack of indigenous energy resources pushes Japan to support the introduction of nuclear energy in different sectors of its economy, particularly for electricity generation after the USA proposal "Atoms for Peace". The third issue was safety. Nevertheless, it is important to note that during this period concern over nuclear safety was not the most important issue for the public opinion and the Japanese were accustomed to placing great faith in their engineers, who had learned to build skyscrapers, roads, bridges and sea walls that could withstand earthquakes. The construction of nuclear power plants also fitted into the Japanese model of spending heavily on infrastructure to boost development in the country. This was a time when Japan's government laid down the blueprint for the country's energy development, with little dissent from most of its citizens.

The nuclear accidents at Three Mile Island and Chernobyl, prompted more Japanese to question their own nuclear industry, but they remained a tiny and powerless minority. The real catalyst for the growth of the anti-nuclear movement in Japan was a series of accidents, safety lapses and cover-ups which have led to the end of the support of the public opinion to the Japanese's nuclear industry. Despite of the promise made by Japan's nuclear operators to improve safety procedures after the series of serious nuclear incidents occurred in several nuclear power plants, 12 power companies admitted thousands of irregularities in reporting past problems to the competent Japanese's authorities. As a result, residents across Japan have started resisting the construction of new nuclear power reactors and other nuclear facilities in their locations and, in some cases, have taken legal actions to suspend operation in existing nuclear power plants.

The fourth issue regarding the use of nuclear power is the resistance of the current nuclear power plants in operations and other nuclear facilities to earthquakes. Existing regulations require nuclear power plants to be able to withstand an earthquake of magnitude 6.5, although the government now wants to raise that to 6.9, taking into account the magnitude of the earthquakes that affected Japan in recent years. The Hamaoka nuclear power plant, on the coast south of Tokyo, is built directly on top of a major fault line. A shortage of suitable land, most of Japan is very mountainous, forces the nuclear power companies to build in places like Hamaoka. The nuclear power reactors there could well be the strongest anywhere in the world they sit in massively-reinforced concrete bunkers, supposedly able to withstand a quake up to 8.5 in magnitude, but the people living near the plants is afraid that the plant collapse in case of a strong earthquake. Hamaoka's operator says "this encompasses every conceivable tremor in Japan", but the earthquake that triggered the 2004 Indian Ocean tsunami was measured at more than 9.0.

Those that favor the use of nuclear energy for electricity generation argue that there have been remarkably few serious accidents around the world, considering the number of nuclear power reactors in service in more than 30 countries and the five decades or so they have now been operating. According with some expert's opinion, Japan's record suggests that future accidents are more likely to arise from human error than natural disasters.

Despite of what has been said in previous paragraphs, the urgent need to reduce carbon emissions in the world's second-largest economy will probably eclipse all these concerns, force Japan to continue relying on the use of nuclear energy for the generation of electricity in the foreseeable future, even in the case that the majority of Japanese people do not support the use of this type of energy for this purpose. The only thing that the Japanese people can do is to pray that the government will force the nuclear industry to apply a more entrenched culture of safety than it has shown in the past.

Looking Forward

The Japanese government has maintained strong support for nuclear power. After the Tokaimura nuclear accident, many reorganizations of the government funded research institutions occurred and stricter controls were enforced, but the size and scope of research in nuclear power topics has continued to expand. While the number of nuclear power reactors is expected to increase in the coming years, the focus of new developments will shift to the advanced fuel cycle and next generation of nuclear power reactors. For this reason, Japan is a major player in the GNEP, an international partnership to promote the use of nuclear power and close the nuclear fuel cycle in a way that reduces radioactive waste and the risk of nuclear proliferation, and has joined the ITER an international tokamak research/engineering proposal for an experimental project that could help to make the transition from today's studies of plasma physics to future electricity-producing fusion power plants project. Furthermore, a USA-Japan Joint Nuclear Energy Action Plan was adopted aimed at putting in place a framework for the joint research and development of nuclear energy technology, which is indicative of the commitment the Japanese government has to the development of new nuclear technologies for the generation of electricity in the most efficient manner

Expansion of nuclear power has been a cornerstone of Japanese energy policy over the past two decades and, without any doubt, it will continue to play the same role in the coming years. It is important to note that to maintain or increase the current level of nuclear power generation that is 30% to 40% of the total electricity generation from now on: 1) the existing nuclear power reactors should be used efficiently and, as long as possible, possibly 60 years of life time on the premise of ensuring safety; 2) efforts in constructing new nuclear power reactors under planning should be continued on the basis of understanding and consensus of the public, including local residents; 3) replacement of existing nuclear power reactors should start around 2030, using advanced model of the current LWRs with large output power and/or some medium-sized of this type of reactor; and 4) commercial FBRs should be adopted from around 2050.

Japan has plans to increase nuclear generation capacity by up to 30% through 2010. However, nuclear accidents, such as the major incident in 1999 at Tokaimura, have undermined public confidence in nuclear power. If the additional nuclear power reactors are not built, Japan faces an eventual shortfall of as much as 28 GW in the coming years, which will require turning to other energy sources to meet the deficit. This could translate into additional imports of up to 1.2 million barrels per day of oil or 186.7 billion cubic meters per day of liquefied natural gas, thus increasing Japan's exposure to potential supply disruptions. (Medlock III and Hartley, 2004)

The future development of the nuclear power sector in Japan is based on short, medium and long term objectives. The short term objectives are the following: 1) to enhance the operating efficiency of existing nuclear power reactors by improving plant availability and by pursuing power up-rating and life extension; 2) to promote the utilization of the plutonium recovered from the spent nuclear fuel and to construct interim spent nuclear fuel storage facilities in parallel with the completion of the commissioning test of the Rokkasho Reprocessing Plant.

With respect the medium term the main objective is "to prepare advanced nuclear power reactor designs with improved performance as candidates for the replacement of the retiring reactors, taking into consideration the prediction that significant number of nuclear power reactors in operation will start their retirement in 10-30 years".

The Japanese government is proposing, based on comprehensive consideration of public interests, the establishment of an appropriate environment for inducing the private sector to make long-term investments on the abovementioned subject, identify and characterize good elements of innovative technology platforms related to various improvements and fund for the development of such technology in a timely fashion. It also proposes that manufacturers should strengthen the business structure and achieve the scale and competitiveness to be able to compete in international market by developing unique and innovative technologies and significantly improving the efficiency of business execution by strengthened mutual cooperation.

The long-term objectives include the development of FBRs and its fuel cycle systems. The purpose of this long-term objective in the coming ten years or so is to explore and clarify these concepts with the aim to have this type of reactor commercially available in 2050s with enhanced safety, reliability and economic performance consistent with the requirement of neighbor friendliness; with sufficient security in terms of proliferation resistance and physical protection and by which the nation can enjoy the benefits of effective fuel utilization and more sophisticated nuclear waste management system, consistent with a national goal of pursuing sound material-recycle based society through reducing, reusing and recycling.

In the research and development policy area, the Japanese government is proposing "to pursue a set of research and development activities across several different time frames in parallel; including: a) exploratory research for innovative system concepts such as fusion, nuclear hydrogen, etc.; b) Technology development for commercialization of innovative system such as FBRs and its fuel cycle systems; and c) engineering development for commercialization of innovative system such as advanced LWRs, in addition to generic research and nuclear safety research and those aiming at the improvement of design and operation of existing plants.

In March 2008, Tokyo Electric Power Company announced that the start of operation of four new nuclear power reactors would be postponed by one year due to the incorporation of new earthquake resistance assessments. Units 7 and 8 of the Fukushima Daiichi nuclear power plant would now enter commercial operation on October 2014 and October 2015, respectively. Unit 1 of the Higashidori nuclear power plant is now scheduled to begin operating in December 2015, while Unit 2 will start up in 2018 at the earliest.

Japan plans to build 14 nuclear power reactors with the purpose to increase its nuclear generation capacity from 25% in 2008 to 40% in 2030 and to reduce its energy dependence of oil and of the contamination considered responsible for the global heating.

THE NUCLEAR POWER PROGRAMME IN THE REPUBLIC OF KOREA

The Republic of Korea has carried out a very ambitious nuclear power programme since the 1970s, in parallel with the nation's industrialization policy. The nation has maintained also a strong commitment to nuclear power development as an integral part of national energy policy, which aims at reducing external energy vulnerability and insuring against global fossil fuel shortages. Currently, Korea has one of the most dynamic nuclear power programmes in the world. At the beginning of the implementation of its nuclear power programme, power plants were constructed mostly through "turn-key" contracts, providing little opportunity for domestic industries to participate. Since then, however, domestic participation in the overall civil construction, management, design and equipment supply has continuously increased through the adoption of the "non turn-key" approach. [83]

It is important to single out that the power demand in the Republic of Korea has increased by more than 9% per year since 1990 and it is foresee that the demand of electricity wills growth 2.5% per year until 2020. Per capita electricity consumption in 2006 was 8 400 kWh, up from less than 1 000 kWh in 1980; this means an increase of more than 8 times. Over the last three decades, the Republic of Korea has enjoyed 8.6% average annual growth in GDP, which has caused corresponding growth in electricity demand from 36 billion kWh in 1978 to 388 billion kWh in 2006. [84]

The country has an electricity generation capacity of 62 GWe in 2005 and it is expected to grow to 88 GWe in 2017. Around 26.6 GWe (30%) of this growth will be nuclear, supplying 47% of foresee demand (214 TWh). In 2020, the foresee nuclear capacity will be around 27.3 GWe and it is expected to supply 226 billion kWh, 43.4% of the total production of electricity and, by 2035, the government expects that nuclear power will supply 60% of the foresee power demand.

In 2009, the country has 20 nuclear power reactors in operation with a net generation capacity of 17 716 MWe. In 2007, the nuclear power reactors produced 144.3 TWh of electricity. This production represents around 35.3% of the total electricity production in the country in that year. In 2008, the nuclear share was 35.62%. Korea has six nuclear power reactors under construction with a net capacity of 5 350 MWe, three nuclear power reactors are planned to be constructed in the future with a net capacity of 4 050 MWe and two proposed with a net capacity of 2 700 MWe. The country has two nuclear research reactors. The Republic of Korea imports more than 97% of its energy demand. A summary of the information regarding the nuclear power reactors operating in Korea is included in Table 70.

Table 70. Nuclear power reactors operating in the Republic of Korea

Reactor	Type	Net capacity MWe	Start construction	Connected to the electric grid	Commercial operation	Planned close
Kori 1	PWR	569	8/1972	6/1977	4/1978	2017
Kori 2	PWR	637	12/1977	4/1983	7/1983	
Wolsong 1	PHWR	578	10/1977	12/1982	4/1983	
Kori 3	PWR	964	10/1979	1/1985	9/1985	
Kori 4	PWR	966	4/1980	11/1985	4/1986	
Yonggwang 1	PWR	942	6/1981	3/1986	8/1986	
Yonggwang 2	PWR	936	12/1981	11/1986	6/1987	
Ulchin 1	PWR	940	1/1983	4/1988	9/1988	
Ulchin 2	PWR	937	7/1983	4/1989	9/1989	
Yonggwang 3	PWR (Syst 80)	987	12/1989	10/1994	3/1995	
Yonggwang 4	PWR(Syst 80)	987	5/1990	7/1995	1/1996	
Wolsong 2	PHWR	683	9/1992	4/1997	7/1997	
Wolsong 3	PHWR	681	3/1994	3/1998	7/1998	
Wolsong 4	PHWR	685	7/1994	5/1999	10/1999	
Ulchin 3	PWR (KSNP)	995	7/1993	1/1998	8/1998	
Ulchin 4	PWR (KSNP)	992	11/1993	12/1998	12/1999	
Yonggwang 5	PWR (KSNP)	990	6/1997	12/2001	5/2002	
Yonggwang 6	PWR (KSNP)	993	11/1997	9/2002	12/2002	
Ulchin 5	PWR (KSNP)	995	10/1999	12/2003	7/2004	
Ulchin 6	PWR (KSNP)	994	9/2000	1/205	6/2005	
Total: 20		17 451				

Source: CEA, 8th edition, 2008.

From Table 70 can be stated that during the 1970s only one nuclear power reactor entered into commercial operation, eight units during the 1980s, seven units during the 1990s and four units during the 2000s.

A summary of the information regarding the use of nuclear energy for electricity generation in Korea is shown in Tables 72-73.

Table 71. South Korean nuclear reactors under construction or on order

Reactor	Type	Net capacity MWe	Start construction	Commercial operation
Shin Kori 1	OPR-1000	960	June 2006	12/2010
Shin Kori 2	OPR-1000	960	June 2007	12/2011
Shin Wolsong 1	OPR-1000	960	November 2007	10/2011
Shin Wolsong 2	OPR-1000	960	June 2008	10/2012
Shin Kori 3	APR-1400	1 340	October 2008	9/2013
Shin Kori 4	APR-1400	1 340	October 2009	9/2014
Shin Ulchin 1	APR-1400	1 350	March 2011	12/2015
Shin Ulchin 2	APR-1400	1 350	March 2012	12/2016
Total 8		9 220		

Source: CEA 8th edition 2008 and WNA data January 2008.

Table 72. Nuclear power in Korea

	Number of nuclear units connected to the electric grid	Nuclear electricity generation MWe	Nuclear percentage of total electricity supply (%)
Korea	20	17 451	35

Source: Tables 76 and 79.

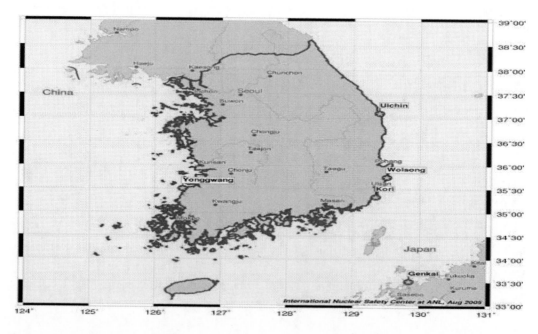

Source: International Nuclear Safety Center, Argonne Nuclear Laboratory, USA.

Figure 131. Nuclear power plants location in the Korean Peninsula.

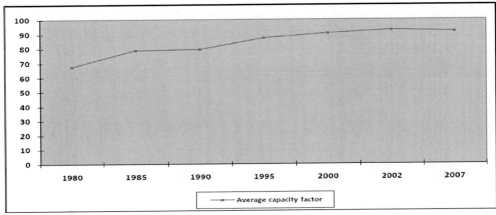

Source: IAEA.

Figure 132. Average capacity factor of the Korean nuclear power plants during the period 1980-2007.

From Figure 132 can be stated that the average capacity factor of the Korean nuclear power plants during the 1995-2007 was above 85%.

The Energy Policy in the Republic of Korea

In order to achieve the goal of the Atomic Energy Act, the Atomic Energy Commission adopted, in July 1994, the "Direction to Long-term Nuclear Energy Policy Towards the Year 2030". The Direction emphasizes "the safe and peaceful use of nuclear energy under a spirit of pursuing a better life in harmony with nature for the Korean people". It describes four primary objectives contributing to the economic, technological development and ultimately improvement of human welfare. These objectives are the following: 1) "to enhance the stability in energy supply by promoting nuclear energy as a major energy source of domestic electricity generation; 2) to achieve self-reliance in a nuclear power reactor and proliferation-resistant nuclear fuel cycle technology through comprehensive and systematic nuclear energy research and development; 3) to foster nuclear energy for electricity generation as a strategic export industry by securing international competitiveness through the advancement of nuclear technology, on the basis of active participation and initiatives of the civil sector; 4) to play a leading role in the improvement of human welfare and the advancement of science and technology by expanding the use of nuclear technology in agriculture, engineering, medicine and industry and by enacting basic nuclear technology research".

For the effective achievement of these four objectives, 10 basic directions of a long-term nuclear energy policy were established. These are the following:

Table 73. Annual electrical Power production for 2008

Total power production (including nuclear)	Nuclear power production
404 981 GWhe	144 254.5 GWhe

Source: IAEA PRIS, 2009.

a) "to continue expanding the development and utilization of nuclear energy in the future, unless an epoch-making alternative energy source becomes available in the foreseeable future;

b) to develop and utilize nuclear energy for peaceful purposes only and to consistently uphold this policy;

c) to further strengthen the efforts to improve nuclear safety, recognizing the fact that securing nuclear safety is a prerequisite to the development and utilization of nuclear energy;

d) to improve the economy and to strengthen the international competitiveness of domestic industries through the advancement of nuclear technology;

e) to increase the public's understanding of and support for nuclear energy while respecting the public's right to know under the ideals of democracy and openness;

f) to implement the nuclear energy policy in such a way as to promote a balanced development of the entire spectrum of both nuclear industries and technologies;

g) to promote creative research and development activities so that nuclear energy can play a leading role in demonstrating the possibilities of technological innovation and to challenge new areas of science and technology, as an integral part of the national science and technology policies;

h) to conduct nuclear research and development activities in collaboration with industries, universities and research institutes by rational division of the responsibilities between the governmental and the non-governmental sectors, in view of the specialization, complexity and the immensity of nuclear research;

i) to implement the nuclear energy policy on the basis of international understanding and co-operation in order to keep up with international harmonization;

j) to consistently implement the nuclear energy policy on the basis of long-term perspectives on the techno-economic and socio-political environment".

In order to achieve the objectives of the long-term nuclear energy policy, the government established a legal basis to formulate a "Comprehensive Nuclear Energy Promotion Plan (CNEPP)" every five years through the amendment to the Atomic Energy Act in January 1995. The CNEPP includes long-term nuclear policy objectives and basic directions, sector-by-sector objectives and a budget and investment plan.

The Atomic Energy Act stipulates that the Minister of Science and Technology and the heads of the concerned ministries shall formulate sector-by-sector implementation plans for those areas under their jurisdiction every five years in accordance with the CPPNE, and shall establish and implement annual action plans according to the sector-by-sector implementation plans. The first CNEPP was formulated in June 1997. In July 2001, the Korean government formulated the second CNEPP which included an implementation plan for the five years from 2002 to 2006, and a direction for nuclear energy policy towards the year of 2015. The 10 promotion areas of the CNEPP are as follows;

1) "nuclear electricity generation and reactor development;
2) the nuclear fuel cycle;
3) utilization of radiation and radioisotopes;

4) fostering and promotion of the nuclear industry;
5) enhancement of public understanding and site acquisition;
6) nuclear safety and radiation protection;
7) radioactive waste management;
8) basic and fundamental nuclear research and development;
9) training of nuclear manpower;
10) nuclear diplomacy and international co-operation".

In 2002, and with the purpose to accelerate Radiation Technology (RT) development, the government adopted the "Act on the Utilization of Radiation and Radioisotopes". This act foresees the creation of a Radiation and Radioisotopes Research and Development Centre under the KAERI that was established in 2005. The act also aims to secure RT research funding, to formulate related industries and to develop human resources.

The Evolution of the Nuclear Power Sector in Korea

One of the most important elements in the Korean nuclear power development programme is the increase participation of the national industry in its implementation.

The first three commercial units - Kori 1 and 2 and Wolsong-1, constructed in the 1970s were bought as turnkey projects. The next six nuclear power reactors, Kori 3 and 4, Yonggwang 1 and 2, Ulchin 1 and 2, the country's second generation of nuclear power reactors, involved local contractors and manufacturers in the construction of these reactors. At that stage the country had six PWR units in operation from Combustion Engineering in USA, two units from Framatome in France and one unit from AECL in Canada of radically different design.

In the mid 1980s, the Korean nuclear industry adopted a plan to standardize the design of nuclear power reactors in operation with the purpose to achieve much greater self-sufficiency in building them. In 1987, the industry entered a ten-year technology transfer programme with Combustion Engineering (now Westinghouse) to achieve technical self-reliance. This programme was extended in 1997. As result of this programme, three more CANDU-6 PHWR units were ordered from AECL in Canada, with the purpose to complete the Wolsong nuclear power plant. These three units were built with substantial local input and were commissioned during the period 1997-1999.

In 1987, the industry selected the CE System 80 steam supply system as the basis for the standardization of nuclear power reactors. Yonggwang 3 and 4 were the first units to use this new steam system with great success. A further step in standardization process was the Korean Standard Nuclear Plant (KSNP) concept, which from 1984 brought in some further CE System 80 features and incorporated many of the US Advanced Light Water Reactor design requirements. It is the type used for all further 1000 MWe units as well as the two nuclear power reactors under construction in the DPRK (North Korea). In the late 1990s, to meet evolving requirements, a programme to produce an improved KSNP, or KSNP+, was started. This involved design improvement of many components, improved safety and economic competitiveness and optimizing plant layout with streamlining of construction programmes to reduce capital cost. Shin-Kori 1 and 2 will be the first units of the KSNP+

programme to be constructed. These units are expected to be among the safest, most economical and advanced nuclear power reactors in the world. [84]

The Advanced Pressurized Reactor APR-1400 draws on CE System 80+ innovations, is more evolutionary rather than radical. The new reactor design has US Nuclear Regulatory Commission design certification as a third-generation reactor. The APR-1400 was originally known as the Korean Next-Generation Reactor when work started on the project in 1992. The basic design was completed in 1999. It offers enhanced safety with improve seismic design and has a 60-year design life. Cost is expected to be 10-20% less than the construction cost of the KSNP/OPR-1000 reactor. The first two APR-1400 units are licensed for construction and it is expected to entry into commercial operation in 2013 and 2014 respectively.

The KAERI has designed an integrated desalination plant based on the 330 MWt SMART reactor to produce 40 000 m³/day of water and 90 MWe at less than the cost of gas turbine. The first of these reactors is likely to be built on Madura island, Indonesia about 2010. KEPCO has signed an agreement with Indonesia's PLN power utility to conduct a feasibility study for Indonesia's first nuclear power plant. This plant will probably include one or more OPR-1000 units. The Indonesian government earlier confirmed in principle approval of four 1 000 MWe units on the Muria peninsula, 450 km east of Jakarta in central Java, with a view to commissioning in 2016. [84]

Nuclear Safety Authority and the Licensing Process

The main governments agencies involved in safety issues in the Republic of Korea are the following: a) the Atomic Energy Commission (AEC); b) the Nuclear Safety Commission (NSC); c) the Ministry of Science and Technology (MOST); d) the Korean Institute of Nuclear Safety (KINS); e) the Korea Atomic Energy Research Institute (KAERI); f) the National Nuclear Management and Control Agency (NNCA); g) the Ministry of Commerce, Industry and Energy (MOCIE).

The AEC is the highest decision-making body for nuclear energy policy and is chaired by the Prime Minister. It was set up under the Atomic Energy Act adopted in 1994.

The NSC, chaired by the Minister of Science and Technology, is responsible for nuclear safety regulation. It is independent of the AEC and was set up by amendment of the Atomic Energy Act in 1996. The regulatory framework is largely modeled on the NRC of the USA.

The MOST has overall responsibility for nuclear research and development, nuclear safety and nuclear safeguards. In September 1994, the MOST issued the "Nuclear Safety Policy Statement" containing five regulatory principles of nuclear safety. These principles are the following: 1) independence; 2) openness; 3) clarity; 4) efficiency; and 5) reliability.

The Nuclear Safety Policy Statement declares that "securing safety is a prerequisite to the development and utilization of nuclear energy, and that all workers engaged in nuclear activities must adhere to the principle of priority to safety". It emphasizes "the importance of developing the nuclear safety culture that the IAEA has referred to". It also prescribes that "the ultimate responsibility for nuclear safety rests with the operating organizations of nuclear installations, and is in no way diluted by the separate activities and responsibilities of designers, suppliers, constructors, or regulators". Finally, it prescribes that "the government shall fulfill its overall responsibility to protect the public and the environment from radiation hazards that might accompany the development and utilization of nuclear energy.

In 2000, the Atomic Energy Act was amended, to strengthen nuclear safety as follows: a) increase the number of NSC members in order to guarantee more specialists participate in the policy decision making process; b) introduction of the Periodic Safety Review (PSR) to ensure that the safety of operating nuclear power plants is maintained at current safety standards and practices; c) introduction of the Standard Design Certificate to streamline the licensing process for the construction of nuclear power plants with same design; and d) introduction of the ICRP Publication 60 on a step-by-step basis with full implementation starting in January 2003.

The KINS, an expert safety regulator, comes under MOST, as does the KAERI, responsible for research and development.

The NNCA is the responsible agency for nuclear material accounting and the international safeguards regime. Since 2006, actions were adopted to strengthen its independence.

MOCIE is responsible for energy policy, for the construction and operation of nuclear power plants, nuclear fuel supply and radioactive waste management.

According to the Atomic Energy Act (as amended), regulation and licensing procedures for nuclear power plants in Korea are divided into three stages: a) in the site selection stage; b) for the construction permit; and c) when the utility requests an operating license (the MOST must confirm that the built plant conforms to the submitted).

Nuclear Fuel Cycle and Waste Management

Korea's demand for uranium and nuclear fuel cycle service has continuously increased with the expansion of its nuclear power capacity. The demand is projected to account for more than 5% of the world's demand from the year 2000 onwards. Korea imports uranium concentrates from Australia, Canada, U.K, France, Russia, USA and South Africa. In 2002, Korea imported a total of 6 million pounds of uranium. KHNP, as the sole consumer of nuclear fuel in Korea, has basic guidelines to ensure the stable supply of nuclear fuel and to pursue economic efficiency at the same time by purchasing through international open bids. For uranium concentrates, KHNP has tried to maintain the optimal contract conditions, through both long-term contracts and spot-market purchases. Conversion and enrichment services come from the USA, UK, France, Canada and Russia via long-term contracts. Fuel fabrication services are fully localized to meet domestic needs. [83] KAERI has developed both PWR and CANDU fuel technology and with Korea Nuclear Fuel Company (KNFC) has supplied PWR fuel since 1990 and CANDU PHWR fuel (unenriched) since 1987. The capacity production of KNFC is 550 tons per year for PWR fuel and 700 tons per year for CANDU PHWR fuel.

In 2006, enrichment demand was 1.8 million SWU supplied from overseas. Tenex, Urenco and Usec have previously supplied this but, in mid 2007, KHNP signed a long-term contract (more than 10 years) for €1 billion with Areva NC for enrichment services at the new Georges Besse II plant in France. [84]

With respect of radioactive waste management, the situation in Korea is the following: KHNP is responsible for managing all its radioactive wastes. The Atomic Energy Act of 1988 established a 'polluter pays' principle under which KHNP is levied a fee based on power generated. A fee is also levied on KNFC. The fees are collected by MOST and paid into a national Nuclear Waste Management Fund. A revised waste programme was drawn up by the Nuclear Environment Technology Institute (NETEC) and approved by the AEC in 1998. Spent fuel is stored on the reactor site pending the construction of a centralized interim storage facility by 2016. The capacity of the centralized interim storage will be around 20,000 tons. Long-term, deep geological disposal is envisaged to be constructed before 2010. Low and intermediate-level wastes are also stored at each reactor site, the total being about 60 000 drums of 200 liters. Volume reduction (drying, compaction) is undertaken at each site. A 200 ha central disposal repository is envisaged for all this eventually with capacity for 800,000 drums. It will involve shallow geological disposal of conditioned wastes, with vitrification being used on intermediate-level wastes from about 2006 to increase public acceptability. [84]

It is important to single out that NETEC took over the task of finding repository sites after several abortive attempts by KAERI and MOST carried out during the period 1988-1996. In 2000, it called for local communities to volunteer to host a disposal facility. Seven did so, including Yonggwang county with 44% citizen support but, in 2001, all local governments vetoed the proposal. MOCIE then, in 2003, selected four sites for detailed consideration and preliminary environmental review with a view to negotiating acceptance with local governments from 2004.

The area selected for the low and intermediate nuclear waste facility will get US$290 million in community support according to "The Act for Promoting the Radioactive Waste Management Project and Financial Support for the Local Community" of 2000. The aim of this is to compensate for the psychological burden on residents, to reward a community participating in an important national project, and to facilitate amicable implementation of radioactive waste management. In November 2005, after votes in four provincial cities, Kyongju /Gyeonju on the East coast 370 km South East from Seoul was designated as the site. Almost 90% of its voters approved, compared with 67% to 84% in the other contender locations. In June 2006, the government announced that the Gyeongju repository would have a number of silos and caverns some 80 m below the surface, initially with capacity for 100,000 drums and costing US$730 million. Further 700,000 drum capacity would be built later, total cost amounting to US$1.15 billion.

The Public Opinion

There is no serious public controversy on the use of nuclear energy for electricity generation in Korea. For the contrary, the public accepts the use of nuclear energy for the generation of electricity as main driving force for satisfy the foresee increase in electricity demand in the country in the coming years. However, this no means that public opinion in Korea has no concern about the use of this type of energy not only for electricity generation but in other peaceful applications in different economic sectors as well. These concerns are related with two issues: radioactive wastes disposal and the safety of the nuclear power reactors in operation.

One of the most difficult problems that the nuclear industry is facing is sitting the nuclear facilities, particularly nuclear power plants and the depositary of nuclear waste, and obtaining the public consensus when the nuclear enterprise should be expanded. For this reason, in order to improve the public acceptance of nuclear energy for the production of electricity and to correct opinions of anti-nuclear movement, the Korean's authorities are given all necessary support to provide appropriate education to the public through regular national education system and are promoting transparent information to be provided by utility and the mass media.

As part of this purpose, an "Act for Supporting the Communities Surrounding Power Plants2 has been adopted in the 1990s by the Korean authorities in order to help improve the living standards of the population living around the nuclear facilities.

In general, it appears that the people have more or less good image about the use of nuclear energy for electricity generation. Support for nuclear power appears to be highest in the Republic of Korea, where a majority of respondents in a survey carried out in October 2005 by Global Public Opinion on Nuclear Issues and the IAEA, 52% say that it is safe and that interested countries should build new nuclear power reactors. Only one in ten South Koreans say that nuclear power is dangerous and two-third agree with expanding the use of nuclear power to meet the world's growing energy needs and to help combat climate change.

Looking Forward

According to "The Basic Plan of Long-Term Electricity Supply and Demand", which was finalized in August 2002, ten new nuclear power units will be constructed by 2015, including eight units that are currently under construction or planned. The capacity of the nuclear power reactors under construction and planned is 5 350 MWe and 4 050 MWe respectively. The number of nuclear power reactors proposed is two with a net capacity of 2 700 MWe. The share of nuclear power capacity and nuclear power generation will be increased to 34.6% and 46.1% respectively by 2015.

The "Direction to Long-Term Nuclear Energy Policy Towards the Year 2030"[146], emphasizes "the safe and peaceful use of nuclear energy under a spirit of pursuing a better life in harmony with nature. It describes four primary objectives contributing to the economic, technological development and ultimately improvement of human welfare and 10 basic directions of a long-term nuclear energy policy were established.

In the field of new nuclear power reactors development the situation in Korea is the following: KAERI has been developing the SMART (System-integrated Modular Advanced Reactor) - a 330 MWt PWR with integral steam generators and advanced passive safety features, designed for generating electricity up to 100 MWe and/or thermal applications such as seawater desalination. Design life of the SMART system is 60 years, with a three-year refueling cycle. KAERI has also submitted the VHTR design to the Generation IV International Forum, with the purpose to produce hydrogen from it. The VHTR design is envisaged as 300 MWt modules each producing 30,000 tons of hydrogen per year. KAERI expects that the VHTR engineering design be ready in 2014, construction is expected to start in 2016 and the enter into operation is expected to be in 2020.

In 2005, KAERI embarked upon a US$1 billion research, development and demonstration programme aiming to produce commercial hydrogen using nuclear heat around 2020. KAERI has close links on hydrogen with the Institute of Nuclear and New Energy Technology (INET) at Tsinghua University in China, based on China's HTR-10 reactor, and

---

[146] See reference 83 for additional information on this issue.

is forming other links with its counterpart in Japan. In 2005, it set up a South Korea-US Nuclear Hydrogen Joint Development Center involving General Atomics. It plans to develop the sulfur-iodine (SI) process for hydrogen production while also developing high-temperature reactors and the alloys enabling them to be used with heat exchangers for chemical plants. Prototype SI hydrogen production is expected about 2011, followed by a pilot plant in 2016, which will then be connected to a high-temperature reactor. [84]

Another type of nuclear power reactors under development is the KALIMER (Korea Advanced Liquid Metal Reactor), a 600 MWe Pool type Sodium-Cooled Fast Reactor designed since 1992 to operate at 510°C. A transmuter core is being designed and no breeding blanket is involved. Future deployment of KALIMER as a Generation IV type is envisaged.

Beyond fission, KSTAR (Korea Superconducting Tokamak Advanced Research) was launched in December 1995 and began operating in September 2007 at Daejeon. The US$330 million facility is the world's eight fusion device and will be a major contribution to world fusion research, contributing to the ITER project taking shape in France. [84]

Finally, it is important to note, in the framework of strengthening the nuclear power sector in Korea, that the Doosan Heavy Industries and Constructions (DHIC), the main heavy industry of Korea, has reached an agreement, in 2004, with Skoda Holding in the Czech Republic, for the purchase of 100% of their filial Skoda Power with the rights to use the technology of the turbines Skoda. This acquisition will be added to the capacities of Doosan in nuclear technologies for boilers, turbines and generators - the three main components for modern electric power plants - locating the company in a leader's position in the industry of global equipment for electric power plants. With the technology of turbines acquired by DHIC it will be able to improve and expand its competitiveness in the electric power plants business. The purchase of the filial Skoda Power reached US$730 millions.

THE NUCLEAR POWER PROGRAMME IN THE DEMOCRATIC PEOPLE REPUBLIC OF KOREA (NORTH KOREA, DPRK)

The DPRK developed dual-purpose gas-graphite reactors as its main activity in the field of the use of nuclear energy for electricity generation. The gas-graphite reactor started its operation in 1985. The small gas-cooled graphite-moderated (metal), named "Experimental Power Reactor", has 25 MW thermal capacity and is located in Yongbyon. It has all the features of a plutonium production reactor for weapons purposes and produced about 5 MWe.

The DPRK also made substantial progress in the construction of two larger nuclear power reactors designed on the same principles. One of the reactors is a prototype of about 200 MWt (50 MWe) at Yongbyon. The second is a full-scale reactor of about 800 MWt (200 MWe) at Taechon. The construction of both reactors was halted in 1994 when the US/DPRK Agreement Framework "froze" the entire gas-graphite reactor programme and associated fuel cycle. DPRK developed and produced many items for its gas-graphite reactor programme, including reactor components, nuclear-grade graphite, equipment for measuring personal dosimetry, environmental monitoring equipment and instrumentation and control equipment. (Albright, 2007)

In addition of the two units partially built but subject to political delays, the country has one research reactor. DPRK was close to commissioning one small power reactor, but international concern focused on attempts to develop illicit weapons capability caused this to be halted.

The USA and the Republic of Korea offered assistance in the construction of two nuclear power reactors at Kumho, which would not produce weapons-grade plutonium, if the gas-graphite reactor programme and associated fuel cycle are stop. The agreement for the supply of these two nuclear power reactors and for stopping the gas-graphite reactor programme and associated fuel cycle was signed late in 1995. Under the Agreed Framework, the DPRK was to receive two 1 000 MWe LWRs. They are KSNP) type developed by the Republic of Korea. However, the construction of the these two nuclear power reactors was suspended late in 2003, renewed in 2004 and 2005 but terminated the project in May 2006.

The most worrisome aspect of the DPRK nuclear power programme is the prospect of a diversion of spent nuclear fuel and its reprocessing to extract plutonium for nuclear weapons. To alleviate this concern, a dialog started within the Six Party Talks and among NGOs and the DPRK to ensure that verification and other measures, such as rapid removal of any spent nuclear fuel or multilateral control over the reactor, are adequate to minimize the chance of misuse of the reactor and its fuel. (Albright, 2007)

The Public Opinion

Is very difficult to know the real public opinion about the use of nuclear energy for electricity generation in the DPRK, due to the impossibility to carry out any independent survey on this subject and the full government control of all newspapers, journals, magazines and other media that circulate in the country.

Looking forward

The development of a nuclear power programme in the DPRK will depend of the full implementation of all agreements adopted in the framework of the Six Parties Talks, on the interest of the Republic of Korea and other nuclear suppliers countries to provide financial and technical support of this programme in the future, and on the position of the DPRK government regarding the continuation of the construction of its nuclear power programme. At this moment, there are no real possibilities to know exactly which will be the outcome of the nuclear power programme development in the DPRK.

THE NUCLEAR POWER PROGRAMME IN THE PEOPLE REPUBLIC OF CHINA (CHINA)

China depends, in almost 80% for the production of electricity, in the use of fossil fuels, particularly coal. This is due to the huge coal reserves that the country has, being third after the USA and Russia, with 13% of the world coal reserves[147]. Hydro power generates about 18% of the total electricity production. Rapid growth in energy demand has given rise to power shortages and the reliance on fossil fuels has led to much air pollution. The economic loss due to pollution is put at 3-7% of GDP. [77]

While coal is the main energy source in China, most reserves are in the North or Northwest of the country far from the main development areas of the country. This situation represents an enormous logistic problem and a factor that increase the cost of using coal for electricity generation.

In 2005, about 519 GWe was installed in China, generating 2 475 billion kWh. In 2006, some 102 GWe of generating capacity was added- a 20% growth, and a further 91 GWe was added in 2007. About three quarters of the power produced is used in the industry sector. At the end of 2007, there was reported to be 145 GWe of hydro capacity, 554 GWe fossil fuel, 9 GWe nuclear and 4 GWe wind, totalizing 712 GWe capacity available in the country.

It is important to single out that due of the heavy reliance on old coal-fired power plants electricity generation, China is the second-largest contributor to energy-related carbon dioxide emissions after the USA. According with IEA report of 2004, its share in global emissions - mainly from the energy power sector - would increase from 14% in 2002 to 19% in 2030, but this now looks conservative. To reduce the current level of the country's air pollution one of the available energy alternative is the use of nuclear energy for the generation of electricity in order to shut down the most contaminated old coal power plants.

China offers the greatest potential for new development in the use of nuclear energy for electricity generation, especially in the coastal areas remote from the coalfields and where the economy is developing rapidly. In April 2009, China has 11 nuclear power reactors in operation with a net capacity of 8 587 MWe. In 2007, they provided 65.3 billion kWh, representing 1.9% of the total electricity produced by the country. In 2008, the nuclear share was 2.15%. China has 19 nuclear power reactors under construction with a capacity of 18 920 MWe[148], 26 units planned to be constructed in the near future with a capacity of 27 660 MWe and 77 units proposed to be constructed in the future with a net capacity of 63 400 MWe. The country has 13 research reactors.

A summary of the information regarding the use of nuclear energy for electricity generation in China is shown in Tables 74-76.

It is important to stress that of the 11 nuclear power reactors currently in operation, three units use domestic nuclear technologies, two units are equipped with Russian nuclear technology, four units with French nuclear technologies and two units are Canadian designed. All nuclear power reactors employ second generation nuclear power technologies.

Table 74. Nuclear power situation in China in 2009 (excluding Taiwan)

	Number of units connected to the electric grid	Nuclear electricity generation. MWe	Nuclear percentage of total electricity supply (%)
China	11	8 587	2.15

Source: CEA, 8th Edition, 2008 and WANO.

---

[147] China coal reserves are 126,2 billion distributed in the following manner: Bituminous and Anthracite: 68,6 billion of tons, Sub- Bituminous: 37,1 billion of tons and Lignite Total: 20,5 billion of tons.

[148] Including two nuclear power reactors in Taiwan.

Table 75. Operating nuclear power reactors in China

Units	Province	Type	Net capacity MWe	Commercial operation
Guangdong 1	Guangdong	PWR	944	February 1994
Guangdong 2	Guangdong	PWR	944	May 1994
Qinshan-1	Zhejiang	PWR	288	April 1994
Qinshan-2 -1	Zhejiang	PWR	610	April 2002
Qinshan 2-2	Zhejiang	PWR	610	May 2004
Lingao-1	Guangdong	PWR	938	May 2002
Lingao- 2	Guangdong	PWR	938	January 2003
Qinshan- 3-1	Zhejiang	PHWR	650	December 2002
Qinshan- 3-2	Zhejiang	PHWR	650	July 2003
Tianwan-1	Jiangsu	PWR (WWER)	1 000	May 2007
Tianwan-2	Jiangsu	PWR (WWER)	1 000	August 2007
Total 11			8 572*	

Source: CEA, 8th edition, 2008.

*Note: According with the IAEA, in 2009 the total net capacity installed in China (excluding Taiwan) was 8 587 MWe.

Table 76. Annual electrical power production in China for 2008

Total power production (including nuclear)	Nuclear power production
3 043 800 GWhe	65 325 GWhe

Source: IAEA PRIS, 2009.

In the case of Taiwan, the information regarding the nuclear power reactors in operation in the country is shown in Table 77.

Table 77. Nuclear power reactors in operation in Taiwan

Units	Type	Net capacity MWe	Commercial operation
Chin Shan 1	BWR	604	December 1978
Chin Shan 2	BWR	604	July 1979
Kuosheng 1	BWR	985	December 1981
Kuosheng 2	BWR	948	March 1983
Maanshan 1	PWR	890	July 1984
Maanshan 2	PWR	890	May 1985
Total 6		4 921	

Source: CEA, 8th Edition, 2008.

It is important to single out that nuclear power development in China has been going on for more than 30 years, but its development has never been incorporated into the national electric power plan. Instead, power projects were arranged individually and built in a scattered fashion. An opportunity arose finally in 2003. With frequent occurrence of electricity shortages, the Chinese government decided to adjust the section on electric power in its 10th Five-Year Plan. This provided the "appropriate development of nuclear power an opportunity for "quantitative growth". The 16th National Congress stated that "China's GDP should be quadrupled by 2020, estimated on the basis of the economic development target of US$4 trillion, by then the nation will need generating installed capacity of around 800-900 million kW". China's current installed capacity stands at 350 million kW, so it needs newly added capacity of 450-550 million kW in the following years to reach this target.

The increase in the use of nuclear energy for electricity generation form part now of a longer-term plan to raise China's nuclear capacity six fold; this means to at least 50 GWe by 2020 and then a further three to four fold increase to 120-160 GWe by 2030. The US$30 billion development programme will require the construction of about two nuclear power reactors a year, similar in scale to the large French nuclear construction programme undertaken in the 1980s. However, it is important to note that the long-term nuclear power programme in China will constitute only 4% of the national electricity generation capacity by 2020.

The sites of the nuclear power plants in China, including Taiwan, are shown in Figure 133.

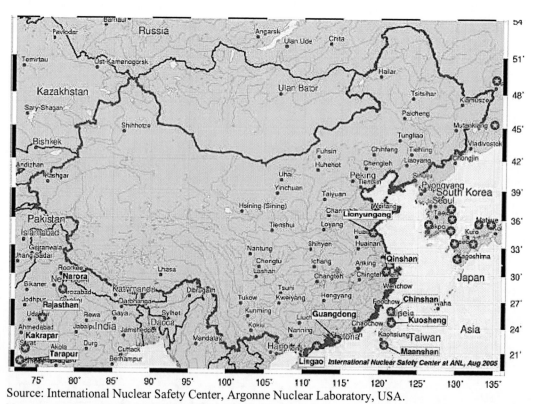

Source: International Nuclear Safety Center, Argonne Nuclear Laboratory, USA.

Note: According with the latest information there are 17 nuclear power reactors in operation in China (11) and Taiwan (6), 19 under construction, 26 planned and 77 proposed.

Figure 133. The location of the nuclear power reactors in operation, under construction and planned in China(including Taiwan).

The Evolution of the Nuclear Power Sector in China

According with WANO and IAEA sources, Taiwan started to build nuclear power plants in 1970s and China in the 1980s. During the 1970s, Taiwan started the construction of all of its six nuclear power reactors currently in operation in the country. China built three units in the 1980s, seven units during the 1990s and one unit in the 2000s. The first two units connected to the electric grid in Taiwan were Chin Shan 1 and 2 in November 1977 and December 1978 respectively. Both units entered into commercial operation in December 1978 and July 1979 respectively. The next two units, Kousheng 1 and 2 started to be built in November 1975 and in March 1976 respectively. Both units were connected to the electric grid in May 1981 and June 1986 and entered into commercial operation in December1981 and March 1983 respectively. The last two units built in Taiwan were Maashan 1 and 2 in August 1978 and February 1979 respectively. Both units were connected to the electric grid in May 1984 and in February 1982 and entered into commercial operation in July 1984 and May 1985, respectively.

In August 1987, the construction of the first of the two nuclear power reactors in Daya Bay, Guangdong 1, started. The construction of the second unit started in April 1988. Both reactors were supplied by Framatome with GEC-Alstom turbines with a net capacity of 944 MWe. EdF managed the construction of the Daya Bay reactors, with the participation of Chinese engineers. Unit 1 was connected to the electric grid in August 1993 and Unit 2 in February 1994. Commercial operation of Units 1 and 2 was in February and May 1994 respectively. The nuclear power plant produces about 13 billion kWh per year, with 70% transmitted to Hong Kong and 30% to Guangdong.

The construction of the Lingao 1 and 2 units started in May and November of 1997 respectively[149] with a net capacity of 938 MWe. These units were connected to the electric grid in February and December 2002 and entered in commercial operation in May 2002 and in January 2003 respectively. The two Lingao units use French technology supplied by Framatome ANP, but with 30% of Chinese participation. The type of reactor is CPR-1000. They are reported to have cost US$1,800 per kilowatt[150].

The construction of the Qinshan-1 nuclear power reactor started in March 1985, in Zhejiang province 100 km South West of Shanghai. The Qinshan 1 nuclear power reactor is China's first indigenously-designed and constructed nuclear power reactor with the pressure vessel supplied by Mitsubishi, Japan. Design of the nuclear power reactor was made by the Shanghai Nuclear Engineering Research and Design Institute (SNERDI). [77] Unit 1 with a net capacity of 288 MWe was connected to the electric grid in December 1991 and entered into commercial operation in April 1994. In October 2007, Qinshan 1 nuclear power reactor was shut down for a major upgrade. The upgrade involved the replacement of the entire instrument and control system along with the reactor pressure vessel head and control rod drives. Areva NP was in charge of supervising the upgrade work. The purpose of this upgrade is extending the lifetime of Qinshan 1 reactor beyond 30 years. The construction of Qinshan 2-1 unit, with a net capacity of 610 MWe, representing phase II of the Qinshan project,

---

[149] Lingao nuclear power reactors are virtually replicas of adjacent Daya Bay units constructed by Framatome in Guangdong province.

[150] It is important to single out that Daya Bay and Lingao nuclear power plants together comprise the "Daya Bay nuclear power base" under the common management of Daya Bay Nuclear Power Operations and Management Co (DNMC), part of China Guangdong Nuclear Power Group (CGNPC).

started in June 1996. The unit was connected to the electric grid in February 2002 and entered into commercial operation in April 2002. Unit 2-2 started up in April 1997, with a net capacity of 610 MWe, was connected to the electric grid in March 2004 and entered into with commercial operation in May of the same year. Both units are of CNP-600 type and are locally-designed and constructed two-loop reactors, scaled up from Qinshan1. The construction of Qinshan 3-1 unit, with a net capacity of 650 MWe, started in June 1998 and of Unit 3-2, with a net capacity of 650 MWe, in September the same year. The construction of these two units represent phase III of the Qinshan project. The first unit was connected to the electric grid in November 2002 and entered into commercial operation in December the same year. Unit 2 was connected to the electric grid in June 2003 and entered into commercial operation in July the same year. Both reactors are of CANDU-6 PHWR type, supply by AECL. The construction of Qinshan 3-1 and 3-2 units was a turnkey project.

In October 1999, started the construction of the Tianwan 1 nuclear power reactor at Lianyungang city in Jiangsu province. In October 2000, started the construction of the Tianwan 2 nuclear power reactor. Both are Russian AES-91 WWER type with a net capacity of 1 000 MWe, constructed under a cooperation agreement between China and Russia. The first unit was connected to the electric grid in May 2006 and entered into commercial operation in May 2007. Unit 2 was connected to the electric grid in May 2006 and entered into commercial operation in August the same year. The cost is reported to be US$3.2 billion, with China contributing US$1.8 billion of the total cost. The reactors incorporate Finnish safety features as well as Siemens Areva instrumentation and control systems. Design life is, for both reactors, 40 years.

The Energy Policy in China

China has set the following points as key elements of its nuclear energy policy: 1) PWRs will be the mainstream of the nuclear power programme but not the sole reactor type to be installed; 2) nuclear fuel assemblies should be fabricated and supplied from local sources; 3) domestic manufacturing of nuclear power plant and equipment will be maximized, with self-reliance in local design and project management; and 4) international cooperation will be encouraged on the base of the above mentioned points.

Nuclear Technology Development in China

The technology base for further development of nuclear power reactors remains officially undefined, though according to Nucleonics Week "the two year struggle between the established CNNC pushing for indigenous technology and the small but well-connected State Nuclear Power Technology Corp (SNPTC) favoring imported technology was won by SNPTC". In particular SNPTC proposes use of indigenized 1000+ MWe with advanced third generation technology, arising from Westinghouse AP1000 designs built at Sanmen and Haiyang.

In September 2006, the head of the China Atomic Energy Authority said that "he expected large numbers of third generation PWRs units derived from foreign technology to be built from about 2016, after experience is gained with the initial Sanmen and Haiyang

AP1000 units". A report suggests that the Westinghouse AP1000 will be boosted, perhaps to 1 400 MWe, under the Shanghai Nuclear Engineering Research and Design Institute (SNERDI) for large-scale deployment and possibly export, with Westinghouse agreement. The agreement with Westinghouse is for the company to transfer technology over the first four units so that SNPTC can build the following ones on its own. However, it would not be able to export the units unless the design was substantially modified by SNERDI. [77]

It is important to note the following regarding the participation of the Chinese nuclear industry in the implementation of the nuclear power programme approved by the government. CNNC had been working with Westinghouse and Framatome Areva NP at the SNERDI since the early 1990s to develop a Chinese standard three-loop PWR design, the CNP-1000 based on Qinshan units, with high burn-up and 18 month refueling cycle. CNNC has been keen to create its own brand of advanced second generation reactor such as this with intellectual property rights and wanted to build two initial CNP-1000 reactors at Fangjiashan near Shanghai in the 11th Economic Plan, though the design probably would not be ready. In early 2007, the CNP-1000 development was stalled indefinitely, and SNERDI resources transferred to the AP1000 program, though this will create a problem in relation to export plans for two CNP- 1000 units to Pakistan. Guangdong Nuclear Power's indigenous focus has been on the French-derived three-loop units such as at Lingao, without major modification, and now called CPR-1000 or "Improved Chinese PWR". However, Areva retains intellectual property rights for this. It seems that the CPR-1000 will be widely and relatively quickly deployed for domestic use under CGNPC leadership.

In September 2005, AECL signed a technology development agreement with CNNC which opened the possibility of it supplying further CANDU-6 units. The agreement foresees the possibility that Qinshan phase III two units be replicated for 25% less cost. Any replication would be on the basis of involving local engineering teams, but not on a turnkey basis, despite of the fact that the technology used in the construction of these units is now well understood by the Chinese counterpart and the decades-old CANDU-6 design would likely pose less problems for technology transfer than state of the art third generation designs from Westinghouse and Areva NP.

In February 2006, the State Council announced that the large advanced PWR and the small High Temperature Gas Cooled Reactor (HTR) are two high priority projects for the next 15 years. The former will depend on Sino-foreign cooperation, in order to master international advanced technology on nuclear power and develop a Chinese third generation large PWR. CNNC has confirmed this, while pointing longer-term to FBRs. The small HTR units will be 200 MWe reactors with pebble bed fuel, similar to that being marketed by South Africa, but plans have evolved to make them twin 100 MWe units driving a single steam turbine. [77]

Longer-term FBRs are seen as the main technology to be developed by China and included in its nuclear power programme to be implemented in the coming years. CNNC expects that the fast breeder technology to become predominant by mid of the current century.

Finally, it is important to single out the following: Despite of the nuclear energy policy already adopted by the government favoring PWR design within the implementation of its nuclear power programme for the coming years, China has expressed interest in the ABWR design development by GE. The US company is offering to the Chinese government its two BWR types - the ABWR which is operating in Japan and under construction there and elsewhere, and the newer ESBWR which features strongly in US plans for new capacity. GE Nuclear and its Japanese partners have been in discussion with CNNC and provincial governments the possibility to include theses type of reactors within the Chinese nuclear power programme.

Nuclear Safety Authority and the Licensing Process

The main Agency offices in charge of safety issues and involved in the licensing process in China are the following: 1) China Atomic Energy Agency (CAEA); 2) National Development and Reform Commission (NDRC); 3) State Nuclear Power Technology Corporation (SNPTC); 4) National Nuclear Safety Administration (NNSA); 5) China National Nuclear Corporation (CNNC); 6) China Power Investment Corporation (CPI, formed from the State Power Corporation and inheriting all its nuclear capacity); 7) China Guangdong Nuclear Power Holding Company (CGNPC); 8) China Hua Neng Group (CHNG); 9) China National Uranium Corporation (CNUC) and Sinosteel; 10)Beijing Institute of Nuclear Engineering (BINE); 11) China Nuclear Engineering and Construction Group (CNEC or CNECG); 12) China Nuclear Engineering Co; 13) Nuclear Power Institute of China (NPIC); 14) China Nuclear Power Engineering Corporation (CNPEC); 15) China Nuclear Energy Industry Corporation (CNEIC); 16) China Nuclear Power Design Co; 17) Areva Dongfang; 18) Shanghai Nuclear Energy Research & Design Institute (SNERDI); 19) China Institute of Atomic Energy; 20) China Institute for Radiation Protection; 21) Ministry of Science & Technology (MOST); and 22) Chinese Nuclear Society.

Under the State Council of Ministers, CAEA is responsible for planning and managing the peaceful use of nuclear energy and promoting international cooperation. Since it constitution in 1998 CAEA has been the key body planning and managing civil nuclear energy and reviewing and approving feasibility studies for new nuclear power plants. It is under the control of the Commission for Science, Technology and Industry for National Defense under the State Council.

NDRC is the economic planning agency established in 1998 responsible for new nuclear power plant project approval reporting to the Commission for Science, Technology and Industry for National Defense under the State Council.

In 2004, SNPTC was set up with the purpose to take charge of technology selection for new nuclear power reactors being bid from overseas. SNPTC is directly under China's State Council and closely connected with it.

In 1984, NNSA was set up under CAEA and now is the licensing and regulatory body which also maintains international agreements regarding safety issues associated to the operation of nuclear power reactors and other nuclear facilities. It now reports to the State Council directly

SEPA is responsible for radiological monitoring and radioactive waste management. A utility proposing the construction of a new nuclear power plant should submits feasibility studies to the CAEA, sitting proposals to the NNSA and environmental studies to SEPA.

In 1998, the CNNC was established. It is a self-supporting economic entity that combines military with civilian production, taking nuclear industry as the basis while developing nuclear power and promoting a diversified economy. CNNC controls most nuclear sector business, including research and development, engineering design, uranium exploration and mining, enrichment, fuel fabrication, reprocessing and waste disposal. CNNC designed and built Qinshan 1-3 unit and controls the full Qinshan power plant operation.

CPI is a major power generator and is the largest state-owned nuclear power investment and operating organization. At the end of 2004, it was reported to have assets of US$12.8 billion. It was at the forefront of discussions on new nuclear power reactors for the 11th five-year plan and, by mid 2005, had submitted nuclear power projects with the total capacity of 31 460 MW to the State Development and Planning Commission for approval.

CGNPC, comprising some 20 companies with an assets of RMB 60 billion, is playing a leading role for Daya Bay, Ling Ao, Yangjiang, Hongyanhe and Ningde nuclear power plants as well as further projects in the Guangdong province and outside it. CGNPC was established in 1994 and is 45% owned by the provincial government (via Guangdong Nuclear Power Co), 45% by CNNC and 10% by CPI. There is 25% Hong Kong equity in the Daya Bay plant. In Guangdong, CNPEC, as part of CGNPC was set up in 2004, with the purpose to plays the leading reactor engineering role. China Nuclear Power Design Co is another CGNPC subsidiary, responsible for feasibility studies and designs.

CHNG is one of China's major electricity generators, with about 50 GWe in operation. It is becoming involved with nuclear power, with two projects in Shandong province. It is an independent state-owned but incorporated business entity focused on power generation with the aims to have 80 GWe installed by 2010 and 120 GWe by 2020.

CNUC is responsible for CNNC's uranium exploration in the country. In December 2006, China Nuclear International Uranium Corporation was set up by CNNC to acquire uranium resources internationally. It is active in Niger and is investigating prospects elsewhere. Sinosteel is another State-owned entity with equity in an Australian uranium explorer and 60% joint venture with it in developing a mine, hoping to sell product to the Chinese nuclear industry.

In 1998, CNECCG, a major state entity was established. It is responsible for nuclear power plant construction, including those to be constructed in Pakistan. CNECG is closely linked with BINE, a CNNC subsidiary responsible for basic reactor design. It is based in the Haidian university precinct north of Beijing.

China Nuclear Engineering Co was set up by CNNC in 2006 to rationalize design work for new nuclear power reactors, as well as to assist in win overseas orders for the construction of new units. It is built on the technology basis of BINE and is also responsible for the construction, equipment procurement, trial testing and operational maintenance of nuclear power plants. Future project design will move from BINE to China Nuclear Engineering Co, allowing BINE to concentrate on technology planning. [77]

NPIC is based in Chengdu, Sichuan Province and is part of CNNC. It was set up in 1958 for nuclear reactor engineering research, design, testing and operation. Its research and development activities are concentrating in the CANDU design used at Qinshan nuclear power plant and, in particular, aspects of its fuel cycle.

SNERDI was part of CNNC and worked with BINE and NPIC in detailed design work for the AP1000 projects but now has been reassigned to SNPTC. It remains dedicated to AP1000 design work. It also worked closely with AECL on reactor engineering for the Qinshan CANDU units.

CNEIC is a CNNC subsidiary established in 1980 as a trading company authorized to carry out import and export trade of uranium products, nuclear fuel cycle and nuclear power and technology equipment. It acted as agent in establishing Qinshan and Tianwan nuclear power plants.

Areva Dongfang is a joint venture of Areva with DongFang Electric Corporation (DEC). It is a high-profile state-owned company specializing in power equipment manufacturing. It had supplied 110 GW of generating equipment over 20 years.

The China Institute of Atomic Energy is responsible for research and development on vitrification of high-level waste and the China Institute for Radiation Protection is responsible for research and development on decommissioning.

Finally, MOST is responsible for planning major nuclear energy research projects and the Chinese Nuclear Society, established in 2006, is responsible for nuclear science popularization and education.

Nuclear Fuel Cycle and Waste Management

China, as a nuclear weapon state, has full control of the whole nuclear fuel cycle. China's known uranium resources of 70 000 tons are theoretically sufficient to fill the requirements for the mainland nuclear power programme for the short-term. Production of some 840 tons per year, including that from heap leach operations at several mines in Xinjiang region, supplies about half of current needs. The balance is imported from Kazakhstan, Russia and Namibia. [78]

Recently, China Guangdong Nuclear Power Holdings Co., one of the country's two nuclear-energy firms, said "it will need more than 100 000 metric tons of uranium between 2009 and 2020 to feed its growing fleet of nuclear power plants, a huge jump that underscores the scope of China's nuclear-energy ambitions". Guangdong Nuclear Power's uranium needs will jump to 10 000 tons a year in 2020, from 2 000 tons this year, Zhou Zhenxing, chairman of the company's uranium-supply unit[151]. It is easy to see that domestic uranium output is by far not enough to meet China's needs now and in the future.

China has two conversion plants. One of these plants is operating at Lanzhou, of about 1 500 t per year, and another at Diwopu, also in Gansu province, of about 500 tons U per year.

A Russian centrifuge enrichment plant at Hanzhun, in the Shaanxi province, was set up under 1992, 1993 and 1996 agreements between Minatom/Tenex and CNEIC. The first modules at Hanzhun came into operation in 1997-2000. In November 2007, Tenex undertook to build a further processing capacity completing the 1990s agreements. Up to 2001, China was a major customer for Russian 6th generation centrifuges and further supplies of these are scheduled from 2008. The Lanzhou enrichment plant in Gansu province started in 1964 for military use and operated commercially from 1980 to 1997 using Soviet-era diffusion technology. A Russian centrifuge plant started operation there in 2001 and it is designed to replace the diffusion capacity. [77]

A contract signed with Urenco ensure the supplies from Europe of 30% of the enrichment for the Daya Bay nuclear power plant and Tenex has agreed to supply SWU as low-enriched uranium to China from 2010 to 2021.

---

[151] According with WANO information, the global uranium consumption is 65,000 tons a year. China's five operational uranium mines - two in Jiangxi province and one each in Shaanxi, Liaoning and Xinjiang - together produce about 840 tons a year.

CNNC's PWR fuel fabrication plant at Yibin, Sichuan province, supplies Qinshan 1 with 11 tons a year of fuel assemblies. A second production line was established in the same factory to supply 26 tons per year of fuel assemblies to the Daya Bay units. Over 2003-2006 enrichment for Lingao is being increased from 3.2% to 4.45%. WWER fuel fabrication for Tianwan is due to begin in 2009. The Yibin plant, operated by CNNC subsidiary China Jianzhong Nuclear Fuel Co Ltd, is expected to keep expanding its current capacity by 2020. [77]

For CANDU, after the initial core loading of Canadian fuel at the PHWRs, subsequent fuel assemblies will be supplied by Baotou fuel factory. The CANDU fuel production line project was launched in April 1999. In order to meet its goal of being self-sufficient in nuclear fuel supply, additional fuel production capacity will be required. However, the fuel for Taishan being supplied to CGNPC by Areva, comprising the two first cores and 17 reloads will be fabricated in France.

With respect to CANDU technology it is important to stress the Chinese strategy towards recycling uranium for use again in CANDU reactors. After supplying two CANDU 6 units for the Qinshan phase three development, AECL has struck a deal with the Qinshan Nuclear Power Company, China North Nuclear Fuel Corporation and the NPIC to develop the technology for the use of uranium recovered from spent nuclear fuel from other Chinese reactors for use at Qinshan 3 and 4 units. These two units are PHWRs designed to run on natural uranium fuel - that is, composed of about 0.7% uranium-235 and not enriched to increase this. They differ from the bulk of China's current and future nuclear power reactors, which is based on PWRs running on low-enriched uranium (about 5% U-235).

Once the fuel from the PWRs has been used for a certain time, its effective enrichment level drops beyond the point it is useful for generating electricity. However, the mix of uranium within the spent nuclear fuel could in theory still be used in a CANDU reactor. In this way, PHWR and PWR units could be operated in a symbiotic way which recycles uranium. This is what the companies hope to achieve. AECL president and CEO Hugh MacDiarmid said that "CANDU nuclear technology has the potential to make a major contribution to reducing China's dependence on imported nuclear fuel resources. However, it is not yet clear whether the Chinese project could be a success or whether intermediate steps of chemical and physical processing would be carried out.

When China started to develop nuclear power, a closed fuel cycle strategy was also formulated and declared at an IAEA conference in 1987. The spent nuclear fuel activities involve: at-reactor storage; away-from-reactor storage; and reprocessing. CNNC has drafted a state regulation on civil spent nuclear fuel treatment as the basis for a long-term government programme.

Based on expected installed capacity of 20 GWe by 2010 and 40 GWe by 2020, the annual spent nuclear fuel arising will amount to about 600 tons in 2010 and 1 000 tons in 2020, the cumulative arising increasing to about 3 800 tons and 12 300 tons, respectively. The two CANDU units, with lower burn-up, will discharge 176 tons of spent nuclear fuel annually.

Construction of a centralized spent nuclear fuel storage facility at Lanzhou Nuclear Fuel Complex in Gansu Province began in 1994. The initial stage of that project has a storage capacity of 550 tons and could be doubled in the future.

A pilot reprocessing plant using the Purex process with a capacity of 50 ton per year was opened in 2006. A large commercial reprocessing plant based on indigenous advanced technology is planned to follow and begin operation about 2020.

In November 2007, Areva and CNNC signed an agreement to assess the feasibility of setting up a reprocessing plant for spent fuel and a MOX fuel fabrication plant in China, representing an investment of €15 billion. High-level wastes will be vitrified, encapsulated and put into a geological repository some 500 meters deep. Site selection is focused on six candidate locations and will be completed by 2020. An underground research laboratory will then operate for 20 years and actual disposal is anticipated from 2050. Early in 2008, CCNC signed an agreement with AECL to undertake research on advanced fuel cycle technologies such as recycling recovered uranium from spent PWR fuel and Generation IV systems. Reconstituted PWR spent nuclear fuel with up to 1.6% fissile content is used directly as CANDU fuel. [77]

There is already industrial-scale disposal of low and intermediate-level wastes at two sites, in the Northwest and at Bailong in Guangxi autonomous region of South China.

The Public Opinion

China general support for nuclear energy as a means of reducing fossil fuel reliance, whether standalone or in combination with renewable reached 62%, one of the highest in the world, according with the result of a survey carried out in March 2009 by Global Management Consultancy Accenture. When asked if their country should start using or increase the use of nuclear power, either outright or if their concerns were addressed, respondents in China reached 91%. However, these results does not means that in specific cases the China public opinion could be against to the construction of a nuclear power plants in specific location. For example, a planned nuclear power plant in Southeast of Shandong Peninsula in the Yellow Sea, East China has generated controversy among local residents who allege the facility will located only a few kilometers away from a popular holiday resort, will damage the beauty of a coastal area and pose a potential environmental threat. Residents have voiced concern over the planned nuclear power plant since 2005, when the Shandong provincial government unveiled the plan. Local residents' concerns grew when they learned there are three nuclear power plants planned along the province's 120-kilometer coastline.

The State Environmental Protection Administration (SEPA) made a statement saying the project must be examined and approved by the administration before construction could begin. The national environmental watchdog confirmed it had not received any application from the local authority. Four Chinese companies - China Nuclear Engineering and Construction (Group) Corporation (CNECC), Shandong Luneng Development Group Co Ltd, Huadian Power International Corporation Ltd and Shandong International Trust and Investment Corporation - invested in the project, in which CNECC owns a 51% stake.

Looking Forward

China has the biggest plan for the development of the nuclear power sector in the world. The programme foresee the construction of 103 nuclear power reactors in the coming years with the purpose to increase the nuclear share in the future energy balance of the country. According with WANO information, the government had planned to increase nuclear generating capacity to 40 GWe by 2020. Despite of the ambitious nuclear power programme adopted by China, the goal of the government is to have not less than 4% of the total generation of electricity from nuclear energy in 2020.

In May 2007, the National Development and Reform Commission announced that its target for nuclear generation capacity in 2030 was 160 GWe. In March 2008, the newly-formed State Energy Bureau (SEB) said that the target for 2030 should be at least 5% of electricity from nuclear power, requiring at least 50 GWe to be in operation by that year.

More than 16 provinces, regions and municipalities have announced intentions to build nuclear power reactors in the 12th five year plan 2011-2015, most of which have preliminary project approval by the central government but are not necessarily scheduled for construction. Provinces will put together firm proposals with reactor vendors and submit them to the central government's National Development and Reform Commission (NDRC) for approval before 2010.

In July 2004, the State Council formally approved the construction of two additional units at Lingao nuclear power plant (Units 3 and 4). Construction started in December 2005 and the 1 080 MWe units, based on Lingao 1 and 2 reactors designs, are expected to be in operation in 2010 and 2011 respectively. Unit 3 will have 50% of Chinese participation and Unit 4 around 70%. Both units will be constructed under the project management of CNPEC.

The construction of Qinshan 2-3 and 2-4 units started in March 2006 and in January 2007, respectively. Chinese participation in the construction of these two 650 MWe CNP-600 reactors will be more than 70%. The construction work is expected to take 60 months.

In September 2004, the State Council approved plans for the construction of two nuclear power reactors at Sanmen, followed by the construction of six units at Yangjiang with a capacity of 1 000 or 1 500 MWe each. The Sanmen and Yanjiang nuclear power plants were subject to an open bidding process for Generation III reactor designs. Three bids were received for the construction of four Sanmen and Yangjiang nuclear power reactors. The bids were presented by Westinghouse (AP1000 design), Areva (EPR) and Atomstroyexport (V-392 version of WWER-1000). In December 2006, the Westinghouse AP1000 design was selected for the four units at Sanmen and Yangjiang, later changed to Haiyang in the more northerly Shandong province. This will give China a leading position with late Generation III reactor technology and provide the platform for China's further nuclear technology development. The first unit is expected to be operating at Sanmen in 2013.

The US Export-Import bank approved US$5 billion in loan guarantees for the Westinghouse bid. The US NRC gave approval for Westinghouse to export equipment and engineering services as well as the initial fuel load and one replacement for the four units. Bids for both nuclear power plants were received in Beijing on behalf of the two customers: CGNPC for Yangjiang, and CNNC for Sanmen in Zhejiang province.

At the end of February 2007, a framework agreement was signed between Westinghouse and SNPTC selecting Haiyang in Shandong province as the site of the second pair of AP1000 units.

In July 2007, Westinghouse, along with consortium partner Shaw, signed the contracts with SNPTC, Sanmen Nuclear Power Company, Shangdong Nuclear Power Company (a subsidiary of CPI) and China National Technical Import and Export Corporation (CNTIC). Specific terms about the cost of the sale and the construction work were not disclosed but the figure of US$5.3 billion for the deal was widely quoted. Sanmen site works commenced in February 2008 and full construction for Unit 1 started in 2009.

It is important to note the following. China, on April 2009, started the construction of its first third generation PWRs using AP 1000 technologies developed by Westinghouse. The reactors, located in Sanmen in Zhejiang Province, will be the first in the world using such technology. The Sanmen nuclear power plant will be built in three phases, with an investment of more than US$5.88 billion injected in the first phase.

The first phase project will include two units each with a generating capacity of 1.25 million kW.

The first generating unit will be put into operation in 2013 and the second in 2014. The plant will eventually have six such units. "It is the biggest energy cooperation project between China and the USA", said Zhang Guobao, vice-minister in charge of the National Development and Reform Commission and also head of the National Energy Administration. "It will contribute to the human kind's peaceful use of nuclear power", he said. Westinghouse became the winner after China signed a memo with the USA on the introduction and transfer of third generation nuclear power technologies in December 2006. The final agreement was reached between China's State Nuclear Power Technology Corporation and Westinghouse in July 2007, according to which China will buy four third generation PWRs from Westinghouse. The agreement also involves technology transfer to China. Two of the four PWRs will be installed in Sanmen in Zhejiang Province and two in Haiyang City in Shandong Province. Based on this agreement, China can replicate the experiences of the third generation nuclear power technologies and build more such reactors in the future. Speaking at Sunday's inauguration ceremony of the first-phase project of the Sanmen nuclear power plant, Chinese Vice Premier Li Keqiang urged "making more efforts to develop new energy to ensure the country's energy security and boost economic growth". He underscored "innovation as the key to nuclear power development, calling for enterprises to adopt advanced technology and enhance self innovation". He said also that "it was inevitable that China would need to improve energy structure and enhance energy conservation and emission cuts when resources and environment issues took their toll on economic development".

In February 2007, it was reported that EdF had entered a cooperation agreement with CGNPC to build and operate two EPR nuclear power reactors. This deal was not expected to involve the technology transfer which is central to the Westinghouse contracts, since the EPR has multiple redundant safety systems rather than passive safety systems and is seen to be more complex and expensive, hence of less long-term interest to China. However, negotiations with Areva and EdF dragged on and in August 2007 it was announced that the EPR project had been shuffled to Taishan so that CPR-1000 units could be built at Yangjiang as soon as possible. Yangjiang will be the second nuclear power plant of the Guangdong Nuclear Power Group, which has long preferred French technology. Development of all six units of the plant was approved in 2004, and site works are well under way. Now that the CPR-1000 has been confirmed as the type of reactors to be built the construction of the first two units started in September 2008. It is expected that the two units enter into commercial operation in 2013. The second pair of units will follow closely. Yangjiang 1-6 and a further 14 units, along with six units at Lingao and Daya Bay, will be operated under regional DNMC management. [77]

In the other hand, the construction of the first unit of the Hongyanhe nuclear power plant in Donggang, 100 km North of Dalian, Liaoning, started in August 2007. The cost of the four

1 080 MWe CPR-1000 units is around US$ 6.6 billion. CNPEC is managing the project being the first nuclear plant in the Northeast of China. The entered into commercial operation of the first unit is planned for 2012-2014.

Haiyang, in Shandong province, had been mentioned as the likely site of further Generation III nuclear power reactors in China. This site will become the second site for a pair of Westinghouse AP1000 reactors. The cost of the construction and operation of these units will be around US$3.25 billion. The construction work is expected to start on September 2009. Construction of the 6-unit Ningde nuclear power plant started at the beginning of 2008. The first four CPR-1000 units will cost US$7,145 billion and will have over 70% of Chinese participation. The first unit is expected to enter into commercial operation in 2012, with the others three following to 2015. [77]

The Taishan nuclear power plant in Guangdong province will have six Areva's 1 650 MWe EPR units, starting with two of them stretched to 1 700 MWe. In November 2007, Areva finalized a contract with CGNPC for the construction of the first two units and for the supply of fuel and other materials and services for them until 2026. The two units are expected to enter into commercial operation in 2013 and 2014 respectively. The whole project, including fuel supply, totals €8 billion. The nuclear power reactors themselves are reported to be about €3.5 billion.

In February 2006, an agreement was signed between CNNC and China Huadian Corp for the construction of the first two units of the Fuqing nuclear power plant at Qianxe in the Sanshan area of Fuqing city, at a cost of US$2.8 billion. Commercial operation of both units is expected to be in 2013 and 2014.

In July 2006, a US$3.1 billion agreement was signed for the construction of the first two units of the Bailong nuclear power plant in Fangchenggang city of Guangxi autonomous region of south China, with CGNPC expecting construction to start by end of 2010, subject to NDRC approval. [77]

In October 2006, a preliminary agreement for the construction of two further 1 060 MWe AES-91 reactors at Tianwan in Lianyungang city of Jiangsu province was signed with Russia's Atomstroyexport. Construction will start when both units were commissioned. In November 2007, a further agreement for the construction of these units was signed. CGNPC's Lianyungang nuclear power project is planned to have four units of 1 000 MWe class to be constructed in phases. This is in Jiangsu Province close to CNNC's Tianwan nuclear power plant and involving the Jiangsu Nuclear Power Company. A proposal has been submitted to the NRDC and preparations for the project are proceeding as planned.

In November 2006, an agreement was signed by CNNC to proceed with the construction of the first two units of the Rushan nuclear power plant at Hongshiding near Weihai in Shandong Province. The cost of the two units is around US$3.2 billion. The first unit is expected to enter into commercial operation by 2015. Six units totaling 6 000-8 000 MWe are envisaged at the site.

The Wuhu nuclear power plant in the Bamaoshan area of Anhui Province is planned to have six 1 000 MWe units to be constructed in phases. CGNPC's proposal for the first two units of phase 1 has been submitted and some preparatory work is under way, but development is some years away.

Table 78. Other nuclear power reactors under construction

Name	Type	Status	Capacity (MWe)		Date Connected
			Net	Gross	
Fangjiashan 1	PWR	Under Construction	1 000	1087	
Fangjiashan 2	PWR	Under Construction	1 000	1087	
Fuqing 1	PWR	Under Construction	1 000	1087	
Fuqing 2	PWR	Under Construction	1 000	1087	
Haiyang 1	PWR	Under Construction	1 000	1115	
Hongyanhe 1	PWR	Under Construction	1 000	1080	
Hongyanhe 2	PWR	Under Construction	1 000	1080	
Hongyanhe 3	PWR	Under Construction	1 000	1080	
Hongyanhe 4	PWR	Under Construction	1 000	1080	
Lingao 3	PWR	Under Construction	1 000	1087	2010/08/31
Lingao 4	PWR	Under Construction	1 000	1086	
Ningde 1	PWR	Under Construction	1 000	1087	
Ningde 2	PWR	Under Construction	1 000	1080	
Qinshan 2-3	PWR	Under Construction	610	650	2010/12/28
Qinshan 2-4	PWR	Under Construction	610	650	2011/09/28
Sanmen 1	PWR	Under Construction	1 000	1115	
Taishan 1	PWR	Under Construction	1 700	1750	
Yangjiang 1	PWR	Under Construction	1 000	1087	
Yangjiang 2	PWR	Under Construction	1 000	1087	
Total 19			18 920		

Source: IAEA, 2008.

CGNPC expects to spend US$9.5 billion on its Lingao 2, Yangjiang and Taishan nuclear power plants by 2010 and to have 6 000 MWe on line by then, with 12 000 MWe under construction. Work is under way at all these sites and also at Ningde. It is also making efforts to start on the Lufeng nuclear power plant at Shanwei in Guangdong and Wuhu in Anhui province, but awaits NDRC approval.

Local authorities are expecting to have 34 000 MWe nuclear capacity on line by 2020, providing 20% of the province's power and 16 000 MWe under construction. From 2010 it expects to commission three units per year and from 2015 four units per year.

In November 2007, Huaneng Nuclear Power Development Company signed an agreement with CGNPC to build four CPR-1000 reactors at Shidaowan in Shandong Province to a cost of US$8 billion. A letter of intent regarding the construction of the first two units was signed in 2008. Construction is expected to start in 2012-2013.

CNNC said, in December 2006, that "it planned to build four 1 000 MWe units at Heyuan, inland in North East Guangdong, at a cost of US$6.4 billion, but no timing was mentioned for the initiation of the construction work".

In 2006, CNNC signed agreements in Liaoning, Hebei, Shandong and Hunan Provinces and six cities in Hunan, Anhui and Guangdong Provinces to develop nuclear power projects. CNNC has pointed out that there is room for 30 GWe of further capacity by 2020 in coastal areas and maybe more inland such as Hunan "where conditions permit". In October 2007, CNNC's list of projects included Chuanshan (Jiangsu Province), Jiyang (Anhui Province), Hebao Island (Guangdong Province), Shizu (Chongqing Province), Xudabao (Liaoning Province) and Qiaofushan (Hebei Province), among others.

It is important to single out that the investments in nuclear power in 2008 was 71% higher in comparison with previous year and this investment allow the reduction in 22% in the level of investments in coal. The Chinese energy plan foresee an investment of US$36.500 million in the construction of new nuclear power reactors and other nuclear facilities for the next 15 years.

To support this plan the amount of uranium needed are estimated in 4,058 tons for 2010 and 8,769 tons for 2020. The mining exploration programme is giving good results in many parts of China but not in an immediate manner. For this reason, China is considering the establishment of mix societies for the exploitation of uranium mines in Kazajstán, Niger and Namibia, besides the long term contract with Australian companies for the supply of uranium.

In the case of Taiwan, the State company Taiwan Power is studying the possible construction of six new nuclear power reactors in form of three groups of two units in four locations that harbor the current nuclear power reactors in operation and under construction. The objective is starting it operation in the next 10 or 12 years. The current government has changed the policy of the previous government to block the conclusion of the two ABWR units of Lungmen, now 88,6% built. The conclusion of these two nuclear power reactors has priority over the construction of new units.

THE NUCLEAR POWER PROGRAMME IN INDIA

India has a flourishing and largely indigenous nuclear power programme and is, without any doubt, the next biggest potential nuclear energy market behind China driven by the fast-expanding, fuel-deficient economies of both countries. India ranks sixth in the world in terms of energy production.

Coal has been and is the primary energy source in India as it accounts for 55% of India's energy production. This abundant fossil fuel, which within India accounts for 247.85 billion tons of reserves as of 2005, can last for some 80 years at the current level of consumption. If domestic coal production continues to grow at the current rate of 5% per year, however, India's total extractable coal reserves would run out in around 40 years. With only half a percent of global reserves within India, oil nonetheless constitutes over 35% of the primary energy consumption in India. India's present level of oil consumption is about 114 million metric tons of oil equivalent out of which India produces 25%, i.e., 29 million metric tons. India's per capita consumption of oil and gas is one-third the global average. The reserves of crude oil are merely 739 million metric tons, which can sustain the current level of production for 22 years.

India's production of natural gas in 2006, which was almost negligible at the time of independence in 1949, is at the level of around 87 million standard cubic meters per day. Natural gas constitutes about 9% of India's energy production, as compared to about 25% in the world. India already imports 20% of its natural gas and this is predicted to go up to about 75% by 2020. (Roul, 2007)

India is also promoting the use of renewable energies with the purpose to achieve a target of 20%-25% in the coming years. The share of renewables in the energy mix of the country is now around 5%. Another important component in the energy balance of the country is hydro with 25% share.

The use of nuclear energy for electricity production started in 1954 when India's First Prime Minister, Jawaharlal Nehru, said "It is perfectly clear that atomic energy can be used for peaceful purposes". In 1969, after years of effort, India's first atomic power plant went critical, in Tarapur, Maharashtra. In 1974, India tested an atomic bomb and it is now a nuclear weapon state.

Nuclear power for civil use is well established in India since the second half of the 20th century and the country is self-sufficient in nuclear power reactor design and construction. It is important to stress that energy independence is India's first and highest priority. While successive federal governments have been seeking energy security by 2012 for India, the current President Abdul Kalam goes further to prescribe energy independence by 2032. To address this critical challenge, the base of the country's energy supply system has steadily shifted from non-renewable to renewable sources, as well as towards development of nuclear energy sources. (Roul, 2007)

According with WANO and IAEA sources, electricity demand in India has been increasing rapidly in the last years. In 2002, India produced 534 billion kWh which was almost double the 1990 output, though still representing only 505 kWh per capita. In 2005, India produced 599 billion kWh and increase of 12.2% with respect to 2002. The per capita figure is expected to almost triple by 2020, considering a 6.3% annual growth. In 2006, nuclear power supplied 15.6 billion kWh representing 2.6% of total India's electricity production. It is expected that nuclear contribution be 25% by 2050. In 2006, almost US$9 billion was committed for power projects, including 9 354 MWe of new generating capacity, taking forward projects to 43.6 GWe with a value of US$51 billion.

India has, in 2009, a total of 17 nuclear power reactors in operation with a net capacity of 3 779 MWe and producing 13.2 billon of kW, which represent around 2.5% of the total generation of electricity in the country, almost the same that in 2006. In 2008, the nuclear share was 2.03%.

Its civil nuclear power strategy has been directed towards the complete independence of the country with respect the nuclear fuel cycle, due to its resistance to sign and ratify the NPT and due to the development of a military nuclear programme to acquire nuclear weapons capability. As a result, India's nuclear power programme proceeds largely without fuel or technological assistance from other countries. The indigenous nuclear power reactors used in India to produce electricity is one of the reason why. during the 1990s, the capacity factors associated to the operation of these units were between some of the lowest in the world. However, this situation changes in recent years. In 1995, the capacity factor was 60%, which was still a low capacity factor but much higher than in the past. In 2001-2002, the capacity factor increased up to 85%. In 2007, the load capacity factor was 81.6%.

India's nuclear energy self-sufficiency extends from uranium exploration and mining through fuel fabrication, heavy water production, reactor design and construction, to reprocessing and waste management. It has a small FBR and is building a much larger one. It is also developing technology to utilize its abundant resources of thorium as nuclear fuel. It is expected to have at least 20 000 MWe nuclear capacity on line by 2020, subject to an opening of international trade.

Table 79. Nuclear power in India 2008

	Number of units connected to the electric grid	Nuclear electricity generation MWe	Nuclear percentage of total electricity supply (%)
India	17	3 779	2.3%

Source: CEA, 8th Edition, 2008 and WANO data base, March 2008.

Table 80. Annual electrical power production for 2008

Total power production (including nuclear)	Nuclear power production
649 935.6 GWhe	13 168.5 GWhe

Source: IAEA PRIS, 2009.

A summary of the information regarding the use of nuclear energy for electricity generation in India is shown in Table 79-81.

The situation of the nuclear power reactors in operation in India is summarized in Table 81.

Finally, it is important to note that India has six nuclear power reactors under construction with a net capacity of 2 976 MWe, 10 units planned with a net capacity of 9 760 MWe and 15 units proposed with a net capacity of 11 200 MWe.

Table 81. India's operating nuclear power reactors

Reactor	Type	Net capacity MWe	Commercial operation
Tarapur 1	BWR	150	Oct. 1969
Tarapur 2	BWR	150	Oct. 1969
Kaiga 1	PHWR	202	Nov. 2000
Kaiga 2	PHWR	202	March 2000
Kaiga 3	PHWR	202	May 2007
Kakrapar 1	PHWR	202	May 1993
Kakrapar 2	PHWR	202	Sept. 1995
Madras 1	PHWR	202	January 1984
Madras 2	PHWR	202	March 1986
Narora 1	PHWR	202	January 1991
Narora 2	PHWR	202	July 1992
Rajasthan 1	PHWR	90	Dec. 1973
Rajasthan 2	PHWR	184	April 1981
Rajasthan 3	PHWR	202	June 2000
Rajasthan 4	PHWR	202	Dec 2000
Tarapur 3	PHWR	490	Aug. 2006
Tarapur 4	PHWR	490	Sept. 2005
Total 17		3 779	

Source: CEA, 8th Edition, 2008.

The Evolution of the Nuclear Power Sector in India

India is one of the developing countries with an important nuclear power programme for peaceful purposes. It has also an important military nuclear programme.

The Atomic Energy Establishment was set up at Trombay, near Mumbai, in 1957, and renamed as Bhaba Atomic Research Centre (BARC) ten years later. The first two nuclear power reactors BWR type built in India were Tarapur 1 and 2 with a net capacity of 150 MWe each. The construction work started in October 1964, were connected to the electric grid in February 1969 and entered into commercial operation in October the same year. Both reactors were supplied by General Electric and were built on a turnkey contract. They have been using imported enriched uranium from the former Soviet Union and later on from Russia. However, late in 2004, Russia declined to supply further fuel for India. They underwent six months refurbishment over 2005-2006 and, in March 2006, Russia agreed to resume fuel supply.

Plans for building the first PHWR reactor CANDU type were finalized in 1964. The nuclear power reactor Rajasthan 1 was built in collaboration between AEC Ltd and the Nuclear Power Corporation of India Ltd (NPCIL). The construction work started in August 1965. The Rajastahn 1 reactor was down-rated early in its life and has operated very little since 2002. Due to different problems the reactor, with a net capacity of 90 MWe, was connected to the electric grid in November 1972 and entered into commercial operation in December 1973. The unit was shut down in 2004 and it is waiting the final decision of the Indian's government about the future of the reactor.

The construction of the second Canadian CANDU nuclear power reactor at Rajasthan nuclear power plant (Unit 2) started up in April 1968The unit was connected to the electric grid in November 1980 and entered into commercial operations in April 1981. Units 3 and 4 at Rajasthan nuclear power plant, Units 1, 2 and 3 at Kaiga nuclear power plant, Units 1 and 2 at Kakrapar nuclear power plant, Unit 2 at Madras nuclear power plant and Units 1 and 2 at Narora nuclear power plants, all of them with a net capacity of 202 MWe, were indigenously designed and constructed by NPCIL, based on a Canadian nuclear power design.

It is important to single out some of the characteristics of the PHWR nuclear power reactors produced by NPCIL. First, the PHWRs produced and operated by Canada and India is substantially more complex than the PWR and BWR units more widely used around the world for energy power generation. Of course, the PHWRs use natural uranium as fuel, unlike the PWRs and BWRs which need enriched uranium. There are more systems and auxiliary equipment in PHWRs which have to work reliably in order to allow the adequate operation of the reactor. Following the Pokhran I test in 1974, relations with Canada ceased and all equipment and materials required for the normal operation of nuclear power reactors were manufactured in India.

In December 1984 and in April 1985, the two nuclear power reactors Kakrapar 1 and 2 started to be constructed. Both units were connected to the electric grid in November 1992 and in March 1995 and entered into commercial operations in May 1993 and in September 1995 respectively.

The construction of the nuclear power reactors Kaiga 1, 2 and 3 started in September and December 1989 and March 2002 respectively. The units were connected to the electric grid in September 2000, December 1999 and April 2007 and entered in commercial operations in November and March 2000 and in May 2007 respectively.

Tarapur 3 and 4 started its construction in May and March 2000, were connected to the electric grid in June 2006 and 2005 and started commercial operation in August 2006 and in September 2005 respectively.

Russia is supplying the country's first large nuclear power plant, comprising two WWER-1000 (V-392) reactors, under a Russian-financed US$3 billion contract. The AES-92 units are built by NPCIL and will be commissioned and operated by it. Russia will supply all the enriched fuel, though India will reprocess it and keep the plutonium. (Roul, 2007)

It is important to note that plans have been adopted for the India specific safeguards to be administered by the IAEA in relation with eight nuclear power reactors. These are, beyond Tarapur 1 and 2, Rajastahn 1 and 2, Rajasthan 3 and 4 (2010), Kakrapar 1 and 2 (2012) and Narora 1 and 2 (2014).

The Energy Policy in India

The energy policy of the government of India aims at "ensuring in a judicious manner adequate energy supplies at an optimum cost, achieving self sufficiency in energy supplies and protecting the environment from the adverse impact of utilizing energy resources". The main elements of the energy policy are: a) "accelerated exploitation of domestic conventional energy sources, viz. coal, hydro, oil, gas and nuclear power; b) energy conservation and management with a view to increasing energy productivity; c) optimizing the utilization of existing energy capacity in the country; d) development and exploitation of renewable sources of energy to meet the energy requirement of rural communities; e) intensification of research and development activities in the field of new and renewable energy sources; and f) organization of training for the personnel engaged at various levels in the energy sector".

The Department of Atomic Energy (DAE), under the direct control of the Prime Minister of India, has formulated an approach and perspective on the nuclear power in the country. The approach foresees three stage for the implementation of the nuclear power programme adopted by the government. The first stage calls for setting up of natural uranium fuelled PHWRs. In the second stage calls for the use of the FBRs utilizing a uranium-plutonium fuel cycle. In the third stage, the programme foresees the use of the FBRs utilizing thorium as a fuel.

It is important to note that India's natural uranium deficiency has resulted in a commitment to this ambitious, technically challenging three-stage programme designed to exploit the country's thorium reserves, which at an estimated 290 000 metric tons are the second largest in the world. (Roul, 2007)

Nuclear Safety Authority and the Licensing Process

The Atomic Energy Regulatory Board (AERB) was formed in November 1983 by the government of India, in exercise of the powers conferred by the Atomic Energy Act of 1962. The purpose is "to carry out regulatory and safety functions as envisaged in the Act". As per its constitution, AERB has the power "to enforce rules and regulations framed under the Atomic Energy Act for radiation safety in the country". AERB also has the authority "to administer the provisions of the Factories Act, for industrial safety of the units of DAE.

AERB has been delegated with powers "to enforce some of the provisions of the Environmental Protection Act, at DAE installations". Enforcement of safety related regulation at all nuclear facilities lies with AERB empowered by the government of India.

The AERB conducts in-depth reviews so that nuclear facilities do not pose any radiological risk to the public and plant personnel. The authorization process involves various major activities like site approval, construction, commissioning, operation and decommissioning of nuclear installations. The authorization process is an ongoing process starting with site selection and feasibility study, continuing through the construction and operation of the facility until the decommissioning of the nuclear power plant. The applicant is required to provide all relevant information, such as safety principles, analysis, criteria and standards proposed for each major stages and quality assurance demonstrating that the nuclear power plant will not pose any undue radiological risks to site personnel and the public.

AERB has advisory committees for site selection, design review and authorization and licenses for commissioning. The advisory committees are assisted by unit level safety committees, which undertake detailed safety assessments at the design and commissioning stages of nuclear facilities. AERB then issues its authorization based on the recommendations of the advisory committee. Safety assessments during plant operation are done by the Safety Committee for Operating Plants (SARCOP). Authorization is granted only for a limited period and further authorization is required beyond that period. Authorization also includes explicit conditions that the applicant must adhere to. AERB also ensures that all the nuclear facilities have put in place an emergency preparedness procedure and organization I case of a nuclear accident.

The main national law in force in India related with its nuclear power programmes is the "AtomicEnergy Act 1962". Additional rules, codes and regulations covering the entire nuclear fuel cycle are the following: a) Radiation Protection Rules, 1971; b) Atomic Energy (arbitration procedure) Rules, 1983; c) Atomic Energy (working of mines, minerals and handling of prescribed substances) Rules, 1984; d) Atomic Energy (safe disposal of radioactive waste) Rules, 1987; e) Atomic Energy (factories) Rules, 1996; f) Atomic Energy (control of irradiation of foods) Rules, 1996.

The AERB had revealed about 130 incidents where safety had been compromised in various nuclear power reactors warranting urgent corrective measures in the Bhabha Atomic Research Centre, Indira Gandhi Centre for Atomic Research, Uranium Corporation of India, Heavy Water Board, Indian Rare Earths Limited and several other nuclear facilities. The CIRUS reactor had an inherent problem of radiation leakage. CANDU reactors suffered from heavy leakage of water. Dhruva reactor experienced fuel leakage, attributed to imperfect design architecture. Radioactive waste from the Tarapur nuclear power plant endangered lives of about 3,000 villagers living nearby. Also, there is a tremendous pressure on nuclear reactors safety from outside like terrorists attacks, taking into account a series of terrorist attacks carried out in the country in the last years.

Nuclear Fuel Cycle and Waste Management

It s important to single out the following regarding nuclear fuel for Indian nuclear power reactors. India's long-standing civilian nuclear plans call for extensive reprocessing of spent nuclear fuel from current nuclear power reactors in operation to harvest plutonium. The plutonium would then be used in a new generation of reactors to breed uranium-233 from blankets of thorium that would surround the plutonium fuel. Many decades into the future, the dream is to have a thorium-based fuel cycle that would ensure India's energy independence. Scientists and technocrats in India have constantly pointed to thorium and its FBRs as a future solution to the foresee shortage of energy for the coming years. Thorium, a naturally occurring radioactive metal, as well as uranium, can be used as fuel in a nuclear power reactor. Given a start with some other fissile material (U-235 or Pu-239), a breeding cycle similar to but more efficient than that with U-238 and plutonium (in slow-neutron reactors) can be set up. Without doubt, thorium presents some advantages over uranium but big problems too. Thorium found in India, as well as in Australia, is about 3 times more abundant than uranium. The hitch with using thorium as a fuel is that breeding must occur before any power can be extracted from it, and that requires neutrons and the only practical scheme, at the moment, is to combine the thorium with conventional nuclear fuels (made up of either plutonium or enriched uranium, or both), the fissioning of which provides the neutrons to start things off.

Previous work on thorium elsewhere in the world did not lead to its adoption, largely because its performance in water reactors did not live up to expectations. Using thorium to prevent plutonium build up, requires the fuel to be configured differently than in most past experiments that were carried out in India and other countries. Those trials incorporated highly enriched uranium (now discouraged because of proliferation concerns) and presupposed the spent nuclear fuel would be reprocessed to extract its fissile contents. Neither practice is now envisaged.

India faces severe challenges regarding the operational safety of all kinds of nuclear installations, from uranium mines to nuclear power plants. Despite of the opinion of the government that the management and disposal of nuclear waste has been carried out fairly satisfactorily and according with international criteria, criticisms on the overall activities of nuclear energy still remain. Public protests against Uranium Corporation of India Ltd's (UCIL) have prevented it from opening up any new mine since 1985. In 2004, UCIL has tried thrice to set up new uranium mines in Andhra Pradesh, Meghalaya and Jharkhand but has not got permission anywhere. The Andhra Pradesh and Meghalaya governments have agreed to UCIL's proposal in principle, but have withheld permission because of public pressure and nuclear activist campaigns focusing on UCIL's poor safety record in Jaduguda in Jharkhand. Independent studies have demonstrated the irresponsible handling of uranium ore by UCIL that had put some 50,000 people in Jaduguda at risk and caused genetic deformities in the area.

At the same time, there are also serious problems to do with treating and disposing of the large volumes of highly radioactive waste generated not only by the operation of nuclear power reactors, but also by nuclear installations that extract plutonium or produce nuclear fuel. There is also the question of cost of decommissioning nuclear power reactors after their useful life. Safety of nuclear reactors has also become an issue of concern for the public opinion in India.

The Public Opinion

There is no strong public opposition to the use of nuclear energy for electricity generation. There are few individuals/groups which talk in public against nuclear power.

Development of nuclear power with indigenous resources is a matter of national pride. However, there is public concern about safety, environment, waste management and economics of nuclear power. In a survey carried out in March 2009 by Global Management Consultancy Accenture, the strongest support for nuclear energy as a means of reducing fossil fuel reliance, whether standalone or in combination with renewable, came from respondents in India (67%). When asked if their country should start using or increase the use of nuclear power, either outright or if their concerns were addressed, 96% respondents in India, say yes.

Looking Forward

It is important to single out that India is amongst the top 10 countries of the world in terms of production of electricity by hydro, coal, oil and gas, and it is nowhere near the top 10 with respect to nuclear power generation. (Roul, 2007)

The target, since 2004, has been for nuclear power to provide 20 000 MWe by 2020. However, in 2007, the prime Minister referred to this "as modest and capable of being doubled with the opening up of international cooperation". There should be no doubt that on the basis of indigenous fuel supply only, the 20 000 MWe target is not attainable, or at least not sustainable without uranium imports. At this point, even if a 20-fold increase takes place in India's nuclear power capacity by 2031-32, the contribution of nuclear energy to India's energy mix is, at best, expected to be 5-6%.

Another important element to be considered is how to expand the use of nuclear energy for electricity generation in the future, taking into account the particular political situation that the country is facing due to its military nuclear programme. On this issue, Indian's politicians are divided. Some are of the opinion that the expansion in the use of nuclear energy for electricity generation should be done using exclusively indigenous nuclear technology. Others are of the opinion that foreign advanced nuclear technology should also be used, but without putting in danger the energy independence of the country.

NPCIL is evaluating a site for up to 6 000 MWe nuclear capacity at Pati Sonapur in Orissa State. Major industrial developments are planned in that area and Orissa was the first Indian State to privatize electricity generation and transmission. State demand is expected to reach 20 billion kWh per year by 2010. NPCIL is also planning construction of a 1 600 MWe nuclear power plant in the Northern State of Haryana, one of the country's most industrialized region, by 2012. The state has a demand of 8 900 MWe, but currently generates less than 2 000 MWe and imports 4 000 MWe. The US$2.5 billion nuclear power plant would be sited at the village of Kumaharia, near Fatehabad and paid for by the state government.

Apparently, in anticipation of easing nuclear trade restriction with foreign countries, the National Thermal Power Corporation (NTPC) brought forward consideration of a 2 000 MWe nuclear power plant to be in operation by 2014. It would be the utility's first nuclear power plant and also the first nuclear power plant not built by NPCIL.

However, it is important to note that the contribution of nuclear power to India's overall power generation now is negligible, even less than what wind energy generates due to the relative small capacity of its nuclear power reactors in operations.

The approved nuclear power programme included the construction of 31 new units. Six units are now under construction. Long-term plans foresee to build another 25 units. Around 19 of them will be constructed between 2010 and 2020. These units are the following: Ten 700 MWe PHWRs units; three 500 MWe FBRs units and up to six 1 000 MWe WWERs units, adding about 20 000 MWe to the electricity generating capacity of the country; the other six units are proposed and will enter into commercial operation after 2020. According with India's plans, nuclear energy share should reach 20 GWe by 2020[152]. Now India gets less than 3% of its electricity from nuclear, but it is, along with China and Russia, one of the leaders in the construction of new nuclear power reactors, boasting six of the world's 54 units under construction (11%).

It s important to single out that, in 2005 and 2006, India connected the first two units using its new 540 MWe (gross, 490 MWe net) HWR design at Tarapur site (Unit 3 and 4). India is also designing an evolutionary 700 MWe HWR and is developing the Advanced Heavy Water Reactor (AHWR), a heavy water moderated boiling light water cooled, vertical pressure tube type reactor, which has passive safety systems and optimized to use thorium fuel.

Another important development in India is in the field of FBRs. In India, a Fast Breeder Test Reactor (FBTR) has been in operation since 1985 and the 500 MWe Prototype Fast Breeder Reactor (PFBR) is now under construction at Kalpakkam.

President of NPCIL declared recently that the new agreement with the USA will contribute to reach the goal of the country to construct up to 2020, a total of 25 new units by a total of 20 960 MWe of new nuclear capacity. On the other hand, a similar agreement has been adopted with France and with the Russian Federation. The agreement signed between India and Russia would allow Russia to build four nuclear power reactors in Southern India. The agreement follows the conclusion of a landmark nuclear deal between the USA and India in 2008, which opened the way for nuclear trade between India and other nations. The President of NPCIL indicated that "its company is in discussion with Areva, AtomStroyExport, General Electric and Westinghouse for the purpose to purchase, to each one of the companies, of two nuclear power reactors with a net capacity of 1 000 MWe each and with the possibility of extending the purchase to six or eight units in the future". The negotiations more outposts are the ones carried out with Areva and AtomStroyExport. However, the discussion with the USA companies has move forward and two sites have been identified for the future construction of two nuclear power reactors with US's companies. These sites are located in Gujarat and Andhra. No commercial agreements with General Electric-Hitachi or with Westinghouse have been discussed until now.

For 2009, NPCIL will acquire 2,000 tons of uranium, it will contract long term uranium provisions and it glides to invest up to US$1,000 million in actions of four foreign mining companies. In occasion of the signature with Russia of an important agreement for the

---

[152] Some experts estimate that this share will be only 17 GWe. India's future plans, however, are even more impressive: an eight-fold increase by 2022 in order to reach 10% of the electricity supply and an overall 70-fold increase to 2052 with the purpose to reach 26%. A 70-fold increase figure certainly sounds remarkable, but it works out to be an average growth rate of 9.5% per year, which is a bit less than the average global nuclear growth from 1970 through 2002.

construction of four new nuclear power reactors, the government of India opens the nuclear activity to the participation of the private industry. After the Nuclear Suppliers Group, in September of 2008, authorized the nuclear trade with India, several countries, among them the USA, France and Canada, are carrying out a strong campaign to actively take part in the future nuclear power development in India, that is considered to be of the order of several tens of thousands of million dollars in the next decades.

Finally, it is important to stress that the Head of the Atomic Energy Commission of India informed the IAEA General Conference, in September 2008, that India is developing an Advanced Heavy Water Reactor, the AHWR-LEU, adapted for the use of thorium and enriched uranium up to 19,75% instead of plutonium. The reactor has levels of security similar to the reactors of the next generation and produces half of the waste produced by current model of LWRs. The power of the AHWR-LEU reactor will be 300 MWe, what makes it capable to be used in small electrical grid typical of developing countries.

The Nuclear Power Programme in Pakistan

The current installed electricity generation capacity in Pakistan stands at 19 400 MWe and, in 2005, a total of 94 billion kWh was produced. In that year, the primary energy consumption was distributed in the following manner: gas 50.3%, oil 29.8%, hydro 11.0%, coal 7.6% and nuclear 1.2%. The installed generation capacity would be increased to 162 590 MWe by 2030, if the foresee energy consumption increase significantly in the coming years. In order to satisfy this foresee increase in the demand of electricity, the Pakistan's government has planned to make major shift to coal, nuclear and renewable resources to achieve the set target of electricity generation.

At present Pakistan is mainly depending on hydropower generation that has dropped to 2 000 MWe power due to water shortages in the country. Pakistan is now looking towards wind, coal, nuclear and solar resources to increase the power generation with the purpose to reduce the 1 400 MWe power shortfall foresee by 2010. At present, the country is facing a power shortfall between 1 500 MWe and 2 000 MWe. Electricity consumption per capita is 402 kWh, which represents less than one-sixth of the world average of 2 516 kWh.

Pakistan has a relative small nuclear power programme compose by two units in operation with a total net capacity of 425 MWe[153], which produced, in 2007, a total of 1.7 billion of kWh representing 2.34% of the total electricity produced by the country in that year. In 2008, the nuclear share was 1.9%. One nuclear power reactor is under construction with a net capacity of 300 MWe, two units planned with a net capacity of 600 MWe and two more units proposed with a net capacity of 2 000 MWe. However, the government has plans to increase the nuclear power programme substantially in the coming years, due to a foresee increasing in the electricity demand and limited indigenous energy resources. Pakistan has an ambitious plans to increase nuclear power capacity to 8 800 MWe by 2030. [102]

---

[153] According with Inside WANO, Vol. 14, No 3, 2006, page 14, the nuclear capacity installed in Pakistan is 462 MWe.

Evolution of the Nuclear Power Sector in Pakistan

The first nuclear power reactor constructed in Pakistan is a small 125 MWe Canadian PHWR. The construction of the Kanupp nuclear power reactor started in August 1966, was connected to the electric grid in October 1971 and entered into commercial operation in December 1972. It is important to note that this reactor has been facing with many equipment problems and remained out of service for long periods of time, reducing its participation in the electricity production in the country. The load net capacity of the Kanupp unit was 32.8% in 2007.

The second unit is the Chasnupp 1, a 300 MWe PWR supplied by China. The construction of this nuclear power reactor started in August 1993 in Punjab, was connected to the electric grid in June 2000 and entered in commercial operation in September of the same year. The Chasnupp nuclear power plant achieved an annual load net capacity factor of 74.2%. Construction of its twin, Chasnupp 2, with a net capacity of 300 MWe, started in December 2005. It is reported to cost US$860 million, with US$350 million financed by China. An electric grid connection is expected in 2011.

A summary of the information regarding the use of nuclear energy for electricity generation in Pakistan is shown in Table 82.

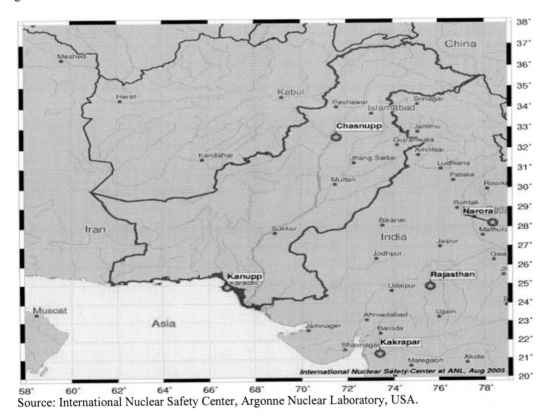

Source: International Nuclear Safety Center, Argonne Nuclear Laboratory, USA.

Figure 134. Location of nuclear power plants in Pakistan.

Table 82. Nuclear power reactors in operation in Pakistan

	Type	MWe	Construction start	Commercial operation
Karachi	PHWR	125	August 1966	December 1972
Chasma 1	PWR	300	August 1993	September 2000
Total (2)		425		

Source: CEA, 8th Edition, 2008.

The Chinese's government has assured Pakistan's authorities that the country will support to set up at least four new nuclear power reactors (1 300 MWe each) to meet its growing energy demands in the future. The Pakistani's government request for more Chinese nuclear power reactors to bridge the widening gap between the demand and supply of energy. Islamabad and Beijing have been engaged in nuclear energy cooperation for many years. The government has plans to built, with the cooperation of Chinese's government, of 0.9 GWe of new nuclear capacity by 2015 and a further 7.9 GWe by 2030.

The Energy Policy in Pakistan

In 2005, an Energy Security Plan was adopted by the government, calling for a huge increase in generating capacity to 162 590 MWe by 2030. It includes plans for lifting nuclear capacity to 8 800 MWe, 900 MWe of this by 2015 and a further 1 500 MWe by 2020. Plans include four Chinese nuclear power reactors of 300 MWe each and seven of 1 000 MWe, all PWRs. There were tentative plans for China to build two 1 000 MWe PWR units at Karachi as KANUPP 2 and 3 but, in 2007, China propose the supply of its CNP-1000 type, which is the only one able to be exported. Pakistan is now exploring the possibility of smaller units with higher local content.

It is important to note that the energy sector in Pakistan needs for an energy policy that consider all alternative and renewable energy sources available in the country. Numerous countries are pushing to develop nuclear power generation capacity, with the purpose to increase their electricity generation. In a carbon constrained world, with increasing global awareness of the risks of climate change, nuclear power is held up as a clean and efficient way to meet economic development objectives while limiting carbon emissions. Furthermore, nuclear power is often seen as a means of ensuring greater self-reliance and independence from petroleum imports from unstable neighbors or region. In the case of Pakistan, the promises of nuclear power generation are, in the opinion of some experts, largely exaggerated through 2030. While it remains true that Pakistan currently has an electricity generation capacity shortage and will need considerably more capacity by 2030, there is ample potential supply from numerous sources. Traditional sources such as natural gas and hydro will continue to be important for Pakistan, but increasingly, the potential of renewable energy will be harnessed. Pakistan is extremely well endowed not only with large-scale hydro, but also world-leading solar and wind resources. The government has recognized this, by establishing the Alternative Energy Development Board, and has increased the amount of investments in this sector. With a portfolio approach encompassing traditional and renewable energy sources, along with energy efficiency measures, the role of nuclear power in the mix of electricity generation by 2030 will not be relevant, even if the high estimates are achieved by Pakistan. The resulting contribution would represent only 3-6% of total electricity generation. Furthermore, Pakistan's overall contribution to global carbon emissions remains miniscule at 0.4%, so substitution through an aggressive nuclear energy programme does not suggest meaningful progress on the climate change agenda. (Stephenson and Tyan, 2007)

Nuclear Safety Authority and the Licensing Process

The nuclear regulatory infrastructure was established in 1965, when the first research reactor PARR I was commissioned. The nuclear regulatory regime further improved when the first nuclear power reactor was commissioned in 1971 at Karachi. A nuclear safety and licensing division was established in Pakistan Atomic Energy Commission (PAEC) which functioned as the de facto regulatory authority till it was upgraded to Directorate of Nuclear Safety and Radiation Protection (DNSRP) after the promulgation of Pakistan Nuclear Safety and Radiation Protection Ordinance of 1984.

As State party of the International Convention on Nuclear Safety, in 1994, the government of Pakistan established an independent nuclear regulatory authority entrusted with the implementation of the legislative and regulatory framework governing nuclear power and radiation use in the country, with the purpose to separate the regulatory functions from the promotional aspects of the nuclear programme, following recommendations of the IAEA. As a transitory measure Pakistan Nuclear Regulatory Board (PNRB) within PAEC was established to oversee the regulatory affairs. Complete separation of promotion and regulatory functions and responsibilities was achieved in 2001, when the President of Pakistan promulgated the Pakistan Nuclear Regulatory Authority Ordinance No. III of 2001. Consequently, Pakistan Nuclear Regulatory Authority (PNRA) was created, dissolving the Pakistan Nuclear Regulatory Board and Directorate of Nuclear Safety and Radiation Protection. The main mission of PNRA is "to ensure safe operation of nuclear facilities and to protect radiation workers, general public and the environment from the harmful effects of radiation by formulating and implementing effective regulations and building a relationship of trust with the licensees and maintain transparency in its actions and decisions".

PNRA is a competent and independent body for the regulation of nuclear safety, radiation protection, transport and waste safety in Pakistan, and also empowered it to determine the extent of civil liability for damage resulting from any nuclear incident. PNRA "devises, adopts, makes and enforces such rules, regulations, orders or codes of practice for nuclear safety and radiation protection as may, in its opinion, be necessary. It plans, develops and executes comprehensive policies and programmes for the protection of life, health and property against the risk of ionizing radiation, and regulates the radiation safety aspects of: a) exploitation of any radioactive ore; b) production, import, export, transport, possession, processing, reprocessing, use, sale, transfer, storage or disposal of nuclear substance, radioactive material or any other substance as the PNRA may, by notification in the official Gazette, specify; c) equipment used for production, use or application of nuclear energy for generation of electricity; d) or any other uses".

Nuclear Safety Directorate (NSD) is one of the Directorates of Executive Wing of PNRA. Its main responsibilities include the following: a) licensing of nuclear power plants, including modifications, periodic safety reviews and re-licensing; b) licensing and inspections of nuclear grade equipment manufacturing facilities; c) establishing and maintaining regulatory framework for nuclear safety; d) review and assessment; e) self assessment; d) coordinating with Regional Directorates in activities related to nuclear safety; f) maintaining and disseminating information on nuclear safety within PNRA; and g) preparation of regulations, working procedures and guidelines.

The Radiation Safety Directorate (RSD) "regulates and supervises matters related to radiation protection". Its mission is "to ensure that the harmful effects of radiation on human health and the environment arising from licensed activities are as low as reasonably achievable". Like the other technical Directorates of PNRA, RSD works independently and give recommendations after careful assessment and evaluation of radiation safety situation at all licensed nuclear facilities. These are submitted for consideration and approval of PNRA for implementation as regulatory decisions.

Radiation protection work of RSD covers a number of different areas. These are: a) preparation of regulations and supporting regulatory guidelines for different users of radiation; and b) evaluation of radiation protection programme at nuclear power plants and application of radiation in medicine, research and industry for control of occupational exposure. In addition, RSD "maintains round the clock National Radiation Emergency Coordination Center (NRECC), to deal with accidents involving radiation at nuclear and radiation facilities". In case a radiation related accident occurs, technical support group, comprising of experts from various directorates of PNRA is summoned at NRECC for supervisory and advisory role on management of control of spread of radiation. Furthermore, RSD takes part in joint international projects with the IAEA, the United Nations Scientific Committee on Effects of Atomic Radiation (UNSCEAR) and International System of Occupational Exposure (ISOE) with the purpose "to improve the national infrastructure for radiation protection". It also participates in IAEA programmes for updating Basic Safety Standards and radiation protection practices.

Nuclear Fuel Cycle and Waste Management

Pakistan, as a non-recognized nuclear weapon state, has the control of the whole nuclear fuel cycle. Enriched fuel for the PWRs is imported from China. The government has set a target of producing 350 tons of U-3O8 per year from 2015 to meet one third of anticipated requirements. Low grade ore is known in central Punjab at Bannu Basin and Suleman Range.

A small uranium centrifuge enrichment plant at Kahuta has been operated by the PAEC since 1984 and does not have any apparent civil use. It is not under safeguards and, for this reason, it is assumed that the facility is used mainly for military purpose.

In 2006, PAEC announced that it was preparing to set up separate and purely civil conversion, enrichment and fuel fabrication plants as a new US$1.2 billion Pakistan Nuclear Power Fuel Complex which would be under IAEA safeguards and managed separately from existing facilities. At least the enrichment plant would be built at Chak Jhumra, Faisalabad, in the Punjab about 2013, then be expanded to be able to supply one third of the enrichment requirements for a planned 8 800 MWe generating capacity by 2030. However, if Pakistan cannot obtain exemption for Nuclear Suppliers' Group trade sanctions in order to build more nuclear power capacity and obtain more uranium in the near future, there may be no point in proceeding with this civil fuel complex. [86]

It s important to note that PAEC has responsibility for radioactive waste management. A Waste Management Centers are proposed for Karachi and Chashma. A National Repository for low and intermediate-level wastes is due to be commissioned by 2015.

Spent nuclear fuel is currently stored in pools at each nuclear power reactor sites. Longer-term dry storage at each site has been proposed. The question of future reprocessing remains open.

Looking Forward

Pakistan nuclear power programme is a modest one and include the construction of 5 new units in the future. One unit is already under construction, two more are planned and two units have been proposed. The implementation of the nuclear power programme will depends of the position adopted by the Chinese's government in supporting this programme. For the time being, China begun designing two additional nuclear power reactors for Pakistan, which will come up at its Chashma Nuclear Power Complex. The decision came soon after Pakistani president Asif Ali Zardari's visit to China in the last week of February 2009. The state-run Shanghai Nuclear Engineering Research and Design Institute has announced that it began designing the third and forth generators for the complex on March 1. The new nuclear power reactors will have a capacity of 325 MW each. China has already built one reactor at Chashma and is in the process of putting up a second one, which will be installed in 2009-2010.

The Chinese company, Shanxi Diesel Engine Heavy Industry Co. Ltd, has produced the emergency diesel generation system for the nuclear power plants in Pakistan after official inspections. Another Chinese company, China Zhongyuan Engineering Corp. is the general contractor for the plants.

Reports in the Pakistani media suggested that Islamabad is expecting Beijing to fund 85% of the construction cost for the third and fourth units in the form of suppliers' credit but until now the position of the Chinese in the participation in the financing of these reactors has not been defined. The country's public sector development programme for 2008-2009 had set the cost of the third and fourth units at US$1.61 billion.

THE NUCLEAR POWER PROGRAMME IN ARMENIA

Armenia produced, in 2007, around 2.35 million of kWh using nuclear energy. This amount represents 43.5% of the total electricity produced by the country. Of the 5.9 billion kWh gross generation in 2006, nuclear supplied 44%, hydro 31% and gas 25%. Electricity consumption per capita is about 1 700 kWh per year. The former Soviet Union supported the construction of two V-230 reactors (each of 407.5 MWe gross and 376 MWe net) on solid basalt which were supplied power the first unit from 1976 and the second unit from 1980. Design life was 30 years. These two units were the first nuclear power reactors designed to be built in a region of high seismicity. For this reason the two units were modified accordingly and were designated as V-270. Armenia has in operation only Unit 2 of its two nuclear power reactors at Metsamor nuclear power plant. Plans for units 3 and 4 at the site were abandoned after the Chernobyl accident.

It is important to note that both units were stopped in 1988 after an earthquake affected Armenia. However, the earthquake did not damage the structure of the nuclear power plant. Unit 1 continues out of service. This unit, after 13 years operation, is now being decommissioned. In 1993, it was decided to restart the second unit due to the severe economic crisis and this was achieved in 1995, after 6.5 years shutdown. Since then, the IAEA has been participating in safety improvements at the plant, which is now scheduled to be closed in 2016.

At present, Armenia is considering building a new nuclear power reactor that substitutes Unit 2 of the Metsamor nuclear power plant. In 2007, Russia offered to build a new 1 000 MWe nuclear power plant in Armenia in return for minority ownership of it. The energy minister announced a feasibility study for a new unit at Metsamor, which would cost some US$2 billion. The investigation was carried out with Russia, the USA and the IAEA. The USA has expressed willingness to help build a new nuclear power plant to go on line about 2016, and the figure of $3 billion has been mentioned for a 1 000 MWe unit. In February 2009, the government announced a tender for a new 1 000 MWe unit at Metsamor, which it expected to cost about US$5 billion all up. Russia and Turkey were invited to tender, to help stability in the region. In May 2009, Worley Parsons was chosen to administer the project and a US$460 million management contract was signed in June. In the same month, the government passed legislation providing for construction of up to 1 200 MWe of new nuclear capacity at Metsamor from one or more reactors. Parliament approved a new unit of 1 200 MWe in October 2009 on the basis of projected cost of US$4 to US$5 billion. A Russian-Armenian joint stock company is being set up by the Ministry of Energy and Natural Resources with Atomstroyexport with the purpose to build a 1 060 MWe Russian VVER unit. Construction is due to begin in 2011-12, with commissioning by 2017. The new unit will has a 60-year service life.

THE POSSIBLE INTRODUCTION OF A NUCLEAR POWER PROGRAMME IN INDONESIA

Indonesia's population is served by power generation capacity of only 21.4 GWe, which produce 127 billion kWh per year[154]. This gives a per capita electricity consumption of 475 kWh per year. With an industrial production growth rate of 10.5%, electricity demand is estimated to reach 175 billion kWh in 2013 and 450 billion kWh in 2026. About 45% of Indonesia's electricity is generated by oil and gas, so as well as catering for growth in demand in its most populous region, the move to nuclear power will free up oil for export. Remaining electricity supply is from coal (36%), hydropower (12%) and geothermal energy (7%). [26] The oil and gas have been playing an important role in the Indonesian economy, together accounting for over 85% of commercial net energy and about 30% of the country's export revenues. However, unlike coal, other hydrocarbon reserves, especially oil, are not so abundant. At present a low reserve margin with poor power plant availability results in

---

[154] Production data of 2005.

frequent blackouts affecting most of the country. Indonesia has no nuclear power programme under implementation so far but has three research reactors in operation.

The Indonesian economy is growing as average about 7% in the last years. At the same time, the government has been fairly successful in controlling the inflation rate at less than 10% for the last ten years. How this good economy results has been achieved? This has been the result of the implementation of Indonesian Long Term Development Programme. This programme promotes the development of national economy through developing a more productive industry supported by a strong agriculture.

It is expected that this development create a strong basis for a self-sustaining growth in economy and other development sectors. The self-sustaining development has to be maintained by reliable sufficient supply of ever-increasing demand of energy. The Outline of State Policy gives emphasis on the importance for Indonesia's sustained economic and social development of meeting its rapidly growing energy needs efficiently, and of minimizing the adverse environmental and social impacts of energy use. Some key government policies are: a) intensification; b) diversification; c) conservation; d) energy pricing; e) environmental protection; f) active private sector participation.

In Indonesia, the energy sector is of great importance in the development of the country's economy. Indonesia's current per capita energy consumption is relatively low as compared to other ASEAN countries. Most energy resources are located outside the island of Jawa. However, the island with its large population and industry constitutes the major area of energy demand. For this reason, the government has adopted a policy of promoting development of the energy sector in a way which maximizes economic efficiency, provide regional development and employment opportunities.

The Energy Policy in Indonesia

According with the Indonesia National Energy Policy, the following are the four main objectives to be reached: 1) "securing the continuity of energy supply for domestic use at prices affordable to the public; 2) enhancing the life quality of the people; 3) stimulating economic growth; 4) reserving an adequate supply of oil and gas for export to provide source of foreign exchange to fund the national development programmes".

To meet these objectives, the Indonesian's government adopts the following three policy measures: 1) "intensification: to increase and expand exploration of energy sources available in the country; 2) diversification: to reduce dependence on only one type of fuel (i.e. oil), and later to replace it with other available fuels; and 3) conservation: to economize energy production and utilization".

Implementation of the energy policy covers several aspects such as issuance of regulations, standards, energy pricing incentives and disincentives and the application of appropriate technologies. The technologies that have to be considered are identified as follows: a) technologies to support a more sustainable energy supply, through the harnessing of the potential renewable sources; and b) clean and efficient energy technologies to support environmental programme and sustainable development.

It is important to note that a 2001 power generation strategy showed that introduction of a nuclear power plant on the Java-Bali electric grid would be possible in 2016 for 2 GWe rising to 6-7 GWe in 2025, using proven 1 000 MWe technology with 85% capacity factor and investment cost \$2 000/kWe. The Java-Bali interconnected system accounts for more than three quarters of Indonesia's electricity demand. [26]

Three research reactors are operated by the National Atomic Energy Agency (BATAN). One of these reactors is used to support the introduction of a nuclear power programme in the country in the future. It is a 30 MWe research reactor at the Serpong Nuclear Facility near Jakarta. The facility started up in 1987.

The Basic Law on the Use of Nuclear Energy for Peaceful Purposes was passed in 1964, with the adoption of the Decree No. 31. This law contains the main regulations for the introduction and use of atomic energy for peaceful purpose in the country. The law stipulates that "BATAN is the highest implementing and supervisory agency in the use of nuclear energy for peaceful purpose in Indonesia". It is further stipulated that "BATAN shall have the duty to conduct researches and implement the use of nuclear technology for safety concern, health and welfare of the people".

As a follow up of the Decree No. 33, government Regulation No. 33 was issued in 1965 for the formation of BATAN, an agency having the following duties and obligations: a) "to build facilities and develop nuclear technology in Indonesia; b) to regulate and supervise the use of nuclear technology in Indonesia; c) to prepare and increase the quality of experts in the nuclear field; and d) to give general and scientific information concerning nuclear energy to the people".

Nuclear Fuel Cycle and Waste Management

Indonesia has front-end capabilities in ore processing, conversion and fuel fabrication, all at a laboratory scale. There have been no experiments in reprocessing, but there is a radioactive waste programme for spent fuel from the research reactors.

The Public Opinion

It is important to note that the introduction of a nuclear power programme in Indonesia is not free of public resistance. Government officials have consistently brushed away complaints about the region's unstable tectonics and the project's high costs. Environmentalists warn that on top of frequent earthquakes and occasional tsunamis, Indonesia has more environmentally sound sources of alternative power to chose from, including geothermal and natural gas and, for this reason, they do not see the need to introduce a nuclear energy programme in the country.

For the people of Central Java, the prospect of living next to a nuclear power plant is regarded as a nightmare that could become a worrying reality. For this reason, thousands of people from Jepara, Pati and Kudus carried out a large demonstration to oppose the government plan.

It is true that the role played by activists from various non-governmental organizations has been very important in raising public awareness to the potential dangers of the use of nuclear energy for electricity generation. It is also true that the issue of safety has been at the core of public anxiety over the government plan for the introduction of a nuclear power programme in the country. No one denies that the lack of energy constitutes one of the key problems hampering economic development in Indonesia. Everyone in his or her right mind would also recognize the growing demand for energy if Indonesia is to sustain its economic growth. It is all understandable that the demand for electricity and the need to secure a long-term electricity supply, is more pressing in Java and that after 2016, Java and Bali alone will need an additional 1 500 to 2 000 MW capacity to satisfy the foresee increase in the energy demand.

However, dismissing the people's concerns - as voiced by some government officials - by accusing them of lacking understanding and information is indeed a display of arrogance that is rejected by the majority of the Indonesian's population. They do not fear the prospect of living next to a nuclear power plant simply because of the Chernobyl nuclear accident. The opposition displayed by the people of Central Java, and by others across the nation, is in fact based on very rational grounds. First and foremost, there are safety fears. There is strong doubt - even distrust - that whoever administers the nuclear power plant will have the ability and absolute commitment to ensure the safety of the plant. Second, the concern over safety is also based on the fact that Indonesia is sitting on the "Ring of Fire". As earthquakes have become more and more frequent, it is clear that any plan to build a nuclear power plant needs to take this concern very seriously. Third, there are also concerns over corruption that could undermine the safety of the plant. Fourth, does Indonesia really need nuclear energy as a source of electricity? The people often hear politicians proudly claim that Indonesia is a country rich in natural resources. For this reason, the people need to know why the government cannot think about the use of other energy alternatives beside nuclear energy.

If the government insists on building a nuclear power plant ignoring the people's concerns, then the people's resistance could increase and that is, of course, a recipe for new tension in society-state relations.

Looking Forward

Feasibility studies for the introduction of a nuclear power programme in Indonesia have been conducted since the end of 1970s. The first pre-feasibility study was conducted in 1978-1979 with the assistance of the government of Italy. In 1985, work began on updating the studies with the assistance of the IAEA and of the government of the USA, France and Italy. These updated reports and the analytical capabilities developed by the Indonesian partners during the process of these cooperation's, have become the foundation for the present planning activities. In 1989, the government initiated a study focused on the Muria Peninsula in central Java and carried out by BATAN. It led to a comprehensive feasibility study for the construction of a nuclear power plant, completed in 1996, with Ujung Lemahabang as the specific site for the construction of the plant, selected for its tectonic stability.

On December 30, 1993, two years after the starting date, the Indonesian's competent authorities submitted the feasibility study report and preliminary site data report to BATAN. The general conclusion was as follows:

1) there is no hindrance in integrating either 900 MWe or 600 MWe nuclear power reactor into Jawa-Bali electric grid system in early 2000s;

2) generation cost of 600 MWe nuclear power reactor is competitive to that of similar size of coal-fired power plant using de-SOx and de-NOx equipment;

3) the conventional financing scheme is capable of producing electricity at a cheaper cost than both build, operate and own and build, operate and transfer financing scheme;

4) practically, every type of reactor, PWR, BWR and PHWR, with proven technology can be appropriately constructed in Indonesia;

5) open fuel cycle, direct disposal of spent nuclear fuel, should be selected at least two decades after the operation of the first nuclear power reactor as the current Indonesian strategy;

6) Ujung Lemahabang is the best candidate site for the first nuclear power reactor, with Ujung Grenggengan and Ujung Watu as the first and the second alternative site respectively.

Recommendations resulting from reviews by an independent consultant had been followed up and integrated into the final report of the feasibility study. The followings are some general points derived from the reports: a) "the introduction of nuclear power programme to Jawa-Bali electric system represents an optimal solution. The role of the nuclear power plant to be constructed under the nuclear power programme is to stabilize the supply of electricity, conserves strategic oil and gas resources and protects the environment from deleterious pollutants; b) Ujung Lemahabang, an area of approximately 500 hectares at the tip of the Muria peninsula at the North coast of central Jawa, has appeared to be the best candidate sites for the construction of the first nuclear power reactor from both technical and economical point of view".

A 2001 power generation strategy showed that introduction of a nuclear power reactor on the Java-Bali electric grid would be possible in 2016 for 2 GWe rising to 6-7 GWe in 2025, using proven 1 000 MWe technology with 85% capacity factor and investment cost US$2,000/kWe. The Java-Bali interconnected system accounts for more than three quarters of Indonesia's electricity demand.

Late in 2003, BATAN was reported to have narrowed the choice of the type of nuclear power reactors to be built to two: a South Korean 1 000 MWe PWR or a Canadian 700 MWe PHWR - probably the KSNP+ (OPR-1000) and ACR-700 respectively. Subsequent reports point to the Korean OPR-1000 option as the best alternative for Indonesia and suggest an increasing sense of urgency due to power shortages.

Under the 2006 Law on Nuclear Reactors, the project may be given to an Independent Power Producer to build and operate, on one of three sites on the central North coast of Java. Plans are to call tenders for two 1 000 MWe units, Muria 1 and 2, leading to decision in 2010 with construction starting soon after and commercial operation from 2016 and 2017. Tenders for Muria Units 3 and 4 are expected to be called for in 2016, with the purpose to start up commercial operation by 2023. Other two units are expected to be constructed in this site in the future.

The government has US$8 billion earmarked for the construction of four nuclear power reactors with a total capacity of 6 GWe to be in operation by 2025. Under current plans, it aims to meet 2% of power demand from nuclear by 2017 and up to 4-5% in the future. It is estimate that nuclear generation cost would be about US$4 cents/kWh compared with US$7 c/kWh for oil and gas.

In July 2007, Korea Electric Power Corp and Korea Hydro and Nuclear Power Co. (KHNP) signed a memorandum of understanding with Indonesia's PT Medco Energi Internasional to progress a feasibility study on building two 1 000 MWe OPR-1000 units from KHNP at a cost of US$3 billion. This was part of wider energy collaboration. In addition, BATAN has undertaken a pre-feasibility study for a small Korean SMART reactor for power and desalination on Madura. However, this awaits the building of a reference plant in Korea. In addition, the Gorontalo Province on Sulawesi is reported to be considering a floating nuclear power plant from Russia.

The Japanese and Indonesian governments signed a cooperation agreement in November 2007 relating to assistance to be provided for the preparation, planning and promotion of Indonesia's nuclear power development and assistance programme for public relations activities. The IAEA is reviewing the safety aspects of both Muria and Madura proposals, in close cooperation with Indonesia's Nuclear Technology Supervisory Agency.

However, it is important to be aware that introducing nuclear power in Indonesia presented a number of obstacles. Some of these obstacles are: a) the lack of trained personnel; b) the lack of an appropriate entity to implement the nuclear power programme; and c) failure to address problems rose by an IAEA study of 1997-2002. Other elements that need to be considered are the large unexploited reserves of geothermal power, the lack of appropriate infrastructure and the highly inefficient use of power. (Gunn, 2008)

THE POSSIBLE INTRODUCTION OF A NUCLEAR POWER PROGRAMME IN THAILAND IN THE FUTURE

As the Thai economy continues its steady growth, policymakers are drafting energy strategies with a view toward keeping the lights on for the next 15 years. To do that Thailand must nearly double its electricity production to about 55 000 MWe each year. Energy officials in Bangkok have drawn up no less than nine 15-year plans to add between 27 000 MWe and 36 790 MWe to the electric grid by 2021. All of the plans call for the use of coal and nuclear energy, with coal set to generate a maximum of 21 000 MWe to limited the negative impact in the atmosphere of the emission of CO_2. Nuclear power would account for about 4 000 MWe in every plan. In 2005, the peak demand was about 20 GWe and some 132 billion kWh per year was supplied. Thailand consumes each year nearly 70% of electricity produced by gas and about 15% by coal. While the reliance on gas is already considered high, Thai authorities has warned that 90% of Thailand's power could come from gas in the future, if the country does not embrace either coal or nuclear energy[155].

Depending on one energy source for the generation of electricity is not a wise decision and make the country extremely vulnerable to price changes. To secure energy supply in a cost-effective way and, at the same time, to avoid that the country be hostage to gas imported, analysts say that "the most viable option is coal, a fossil fuel source responsible of global

---

[155]A power plant fueled by natural gas has the benefit of, on average, producing about half as much carbon dioxide, a third as much nitrogen oxides and 1% as much sulfur oxides as a coal-fired power plant.

warming and the use of nuclear energy for the production of electricity". Until 2009, Thailand has no nuclear power reactor in operation.

In future years, as domestic natural-gas production in the Gulf of Thailand peaks, Bangkok may have to import increasingly expensive liquefied natural gas from Middle Eastern countries like Iran. "Certainly it's a concern to be so reliant on natural gas", says Norkhun Sitthiphong, the Ministry of Energy's Deputy Permanent Secretary who is overseeing the electricity plan. He stresses that "natural gas is subject to volatile world market oil prices and we know exactly what can happen when they rise quickly".

Interest by Thailand in nuclear power was revived by a forecast growth in electricity demand of 7% per year for the next twenty years. Capacity requirement for the production electricity in 2016 is forecast at 48 GWe. As gas prices rise, the Atomic Energy Commission and its Office of Atoms for Peace (OAP) are assessing the feasibility of introducing a nuclear power programme in the country in the coming years. The Electricity Generating Authority of Thailand (EGAT) would probably be the organization responsible for the construction of a nuclear power plant in the country in the coming years. Thailand's National Energy Policy Council commissioned a feasibility study for the construction of nuclear power plant in the country in the future, and among the options considered was the construction of additional 5 000 MWe of nuclear generating capacity starting up in 2020-21. [26]

The National Energy Policy Council (NEPC) finally approved Thailand's Power Development Plan 2007-2021 (PDP 2007) and the Thai cabinet acknowledged and approved the PDP in June 2007. One major issue stipulated in the PDP 2007 is "the introduction of a nuclear power programme and to the allocation to this programme the responsibility to supply 2 000 MW of electricity in 2020 and another 2 000 MW in 2021".

To develop the preparatory work for the introduction of a nuclear power programme in the country, the NEPC appointed the Nuclear Power Infrastructure Preparation Committee (NPIPC). In June 2007, the Energy Minister announced "that EGAT will proceed with plans to build a 4 000 MWe nuclear power plant and has budgeted some US$53 million between 2008 and 2011 with the purpose to carry out the necessary preparatory work, half of it coming from oil revenues". [26]

It is expected that the construction work will commence in 2015, with the objective to start up the commercial operation of the nuclear power plant in 2020. The expected capital cost related with the construction of the first nuclear power plant in the country is US$6 billion and the electricity cost is expected to be about US$6 cents/kWh, slightly less than from coal.

The government plans to establish safety and regulatory infrastructure by 2014 in order to support the introduction of the nuclear power programme. The nuclear technologies that are under consideration are the following: a) Toshiba BWR technology; b) Mitsubishi PWR design; c) Areva PWR technology; d) GE BWR design; and e) Russian PWR technology.

According to an energy and power development plan that will run through to 2021, it will take an estimated 13 years to complete construction of the first nuclear power plant. The sites for the plants have yet to be confirmed, but Ranong, Chumphon and Surat Thani Provinces on the southern coast of Thailand have been cited as possible locations in which the nuclear power plant could be built.

THE POSSIBLE INTRODUCTION OF A NUCLEAR POWER PROGRAMME IN VIETNAM IN THE FUTURE

Vietnam produces electricity through the use of two fossil fuel: coal and oil. At the same time, and due to insufficient national production of electricity, the country needs to import electricity from other countries such as Laos, Cambodia and China. But bearing in mind the prices increase of coal and oil occurred across the globe in 2007 and 2008, the Vietnamese government has no other choice than to reduce imported electricity in the short term, and use of all renewable energy sources available in the country in order to diminish, as much as possible, the cost of the electricity supply.

However, the use of renewable energy sources like wind or solar energy for the supply of electricity is not an easy task. The reason is the following: The use of wind and solar source for electricity generation is limited because their usage depends on climactic and seasonal conditions, as well as of spaces large enough to collect solar energy. What is the alterative? The only available alternative is the use of nuclear energy for electricity generation in order to satisfy a demand estimated at 300 billion kWh by 2020.

Vietnam is rich of hydro and coal resources for electricity generation and recently, natural gas has been established as a resource for gas-turbine generation. The generation mix in 2000 and 2005 is shown in Figure 192, where the proportion of gas-turbine is increasing rapidly, while the relative contribution of hydro is decreasing. Coal fire generation is increased for the last five years due to investment from the private sector, which shared 21% of generation mix for 2005.

The country produced 56.5 billion kWh gross in 2006 from 11.4 GWe of capacity in different plants, giving a per capita consumption of 445 kWh per year. In 2006, around 36% of electricity came from hydro, 28% from gas, 13% from coal and the demand is growing rapidly. In mid 2008, capacity was about 12.5 GWe but electricity demand is significantly higher than this, resulting in rationing the supply of energy. According to the government, at the end of 2006, electricity demand is expected to grow 15% and, for this reason, plan to increase generating capacity to 25 GWe was adopted. The country has one research reactor at Da Lat, operated with Russian assistance.

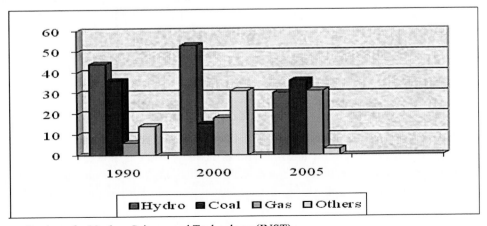

Source: Institute for Nuclear Science and Technology (INST).

Figure 135. Generation mix in Vietnam from 1990 to 2005.

Table 83. Participation of the different energy sources in the generation of electricity in Vietnam in 2006

Type	Number of plants	Capacity MW	Percent of the total
Hydro	13	4 460	36.5
Gas-oil	7	3 399	27.8
IPP	8	2 572	21.0
Coal	3	1 545	12.6
Diesel	-	245	2.1

Source: Vietnam Atomic Energy Commission.

Vietnam does not have yet any nuclear power reactor in operation but has plans to introduce a nuclear power programme in the coming years. Looking at the neighbors in the region there is a diverse consideration on the need to introduce a nuclear power programme to ensure energy development. North-East Asia, including China, is rapidly introducing a large scale nuclear power programme. Other countries within the ASEAN members have carried out a feasibility study for the introduction of a nuclear power programme in the coming years. Since 2002, Vietnam has carried out a pre-feasibility study for the construction of its first nuclear power plant.

The Nuclear Energy Plan in Vietnam

In the early 1980s, two preliminary nuclear power studies were undertaken, followed by another which reported in 1995 that "around the year 2015, when electricity demand reaches more than 100 billion kWh, nuclear power should be introduced for satisfying the continuous growth in the country's electricity demand in that time and beyond". More recently, a national energy plan approved by Vietnam's National Assembly includes at least 2 000 MWe of nuclear power capacity to be commenced sometime after 2012. This follows a feasibility study in 2002 and establishment of nuclear cooperation agreements with Russia, South Korea and the USA, the first related principally to Vietnam's 500 kW Da Lat research reactor. [26]

A law on nuclear energy was approved by the National Assembly on the June 3, 2008, paving the way for the construction of the first nuclear power reactor in the country. At the same time, the government of Vietnam signed an agreement with the government of Japan through which this last country will assist Vietnam in the construction of its first nuclear power reactor to be built in 2015.

On January 3, 2006, the Prime Minister approved the "Long-term Strategy for Peaceful Utilization of Atomic Energy up to 2020", emphasizing the important goals for building the first nuclear power reactor in Vietnam. To achieve this goal it was decided to create the necessary infrastructure to ensure the safe and effective operation and exploitation of the nuclear power reactor to be built in the country, and for the long-term national programme for nuclear power development. In this regards the following actions has been implemented:

a) the establishment of the Vietnam Atomic Energy Commission in 1976 under the Ministry of Science and Technology;

b) Vietnam Electricity has been designated as the company responsible for building and operating the nuclear power reactors to be built in the country;

c) the establishment of the National Steering Committee for the development of the nuclear power in Vietnam on 5 March 2002;

d) the Nuclear Energy Programme Implementing Organization (NEPIO) of Vietnam will be responsible for the conducting of all preparation works to ensure the successful introduction of the nuclear power programme in the country;

e) the establishment of an independent regulatory authority called the "Agency for Radiation Protection and Nuclear Safety Control (VARANSAC)" on 19 May 2003.

In February 2006, the government announced that a 2 000 MWe nuclear power plant would be on line by 2020. Another feasibility study was carried out in 2009 and formal approval will then be required to open a bidding process in 2012, with a view to start the construction of the first nuclear power plant by 2015. The plant is expected to start its commercial operation 4-5 years later. This general target was confirmed in a nuclear power development plan approved by the government in July 2007, with the target being raised to 8 000 MWe by 2025.

In October 2008, two nuclear power reactors with a total capacity of 2 000 MWe are planned for the Southern Ninh Thuan Province. The plant is expected to enter into commercial operation about 2018. The approved plan contemplate the construction of another two nuclear power reactors with a capacity of 2 000 MWe to be constructed at Vinh Hai. These plants would be followed by the construction of further units with a capacity of 6 000 MWe by 2030. Negotiations with Westinghouse are reported and, in February 2009, China Guangdong Nuclear Power Group signed a nuclear energy cooperation agreement with Vietnam in order to support the introduction of a nuclear power programme in the country.

Though Vietnam was facing a serious electricity shortage, the country could not speed up the project because it needed time to prepare sufficient facilities, human resources, technology and legal documents for the introduction of a nuclear energy programme.

According to EVN, by 2025, nuclear power should account for 10% of the country's total electricity output according with the nation's electricity development plan for the period 2006-2025, approved by the Vietnam government.

The Public Opinion

The Vietnam Nuclear Power Institute surveyed local residents about the construction of a nuclear power reactor in the selected site and most of them agreed with the construction. Nevertheless, it is important to note that after the Chernobyl nuclear accident, the reputation of nuclear energy was low in Vietnam, but this not means that there is any important antinuclear movement in the country. However, the possibility to introduce a nuclear power programme in the coming years often raises a discussion in the press of contradicting opinions expressing a serious public concern on reliability of operation of a nuclear power plant. Even many people consider that nuclear techniques should not the first choice, if there is another energy alternative that can be used in order to satisfy the foresee increase in demand of energy.

The Situation of the Nuclear Power Programme in Philippines

The Philippines produced 56.5 billion kWh of electricity in 2005 and 56.7 billion kWh (gross) of electricity in 2006. Around 27% came from coal, 29% from gas, 8% from oil, 17.5% from hydro and 18.5% from geothermal.

The Philippines is one of the few countries in the world that has attempted in the past to venture into using nuclear energy for electric generation but without success. The Philippines started to use nuclear energy for peaceful purposes in June 1958 with the approval of the Republic Act 2067 known as "The Philippine Science Act of 1958". This Act created the Philippine Atomic Energy Commission (PAEC) and empowered it "to conduct or cause the research and development of, among others, processes, materials and devices used in the production of atomic energy".

Under the US Atoms for Peace programme, the Philippines received a small research reactor, which went online in 1963. Supplied by the US company General Atomics, the facility was originally a 1MWt open general-purpose reactor until it was converted to a 3MWt TRIGA-type design in 1988. It had previously used 93% HEU, but its conversion to the TRIGA design enabled it to use LEU. Operated by the PAEC and located within the campus of the University of the Philippines Diliman in Quezon City, the reactor was used for radioisotope production, neutron spectrometry, neutron-activation analysis, reactor physics and training purposes. However, shortly after being restarted in 1988, the PRR 1 reactor pool suffered a serious leak which led to the reactor being shut down permanently. Efforts were undertaken in the 1990s in collaboration with the IAEA to repair it, but these fell victim to the financial constraints encountered by the PNRI and, in 2002, a decision was made to cease repairs and decommission the reactor. The decommissioning will take place under the auspices of the IAEA Research Reactor Decommissioning and Dismantling programme within which the PRR 1 has been chosen as the model reactor upon which to demonstrate the decommissioning process. The irradiated HEU materials test reactor aluminum plate fuel was shipped back to the US in 1999, although as of 2009 the un-irradiated and slightly irradiated TRIGA fuel rods remain on the PRR 1 reactor site. [104]

Throughout the 1960s, the Philippine government requested the IAEA for assistance in examining the prospects and feasibility of using nuclear energy for electricity generation in the country. After the studies conducted by the Philippine government in collaboration with the IAEA, the Philippine Congress passed on June 15, 1968 the Republic Act 5207 known as the "Atomic Energy Regulatory and Liability Act of 1968". This act gave PAEC the power "to issue license for the construction, possession or operation of any atomic energy facility in the Philippines, including nuclear power plants for electricity generation". On June 23, 1971, Administrative Order No. 293 was approved by which a Coordinating Committee for Nuclear Power Study (CCNPS) was established. This committee was mandated "to prepare the initial work for a new and updated feasibility study regarding the possible use of nuclear energy for electricity generation in the country".

On September 10, 1971, the Charter of the National Power Corporation (NPC) was revised with the enactment of Republic Act 6395. This act authorized the NPC "to construct, operate and maintain power plants for the production of electricity from nuclear, geothermal and other sources". Consequently, in 1972, the CCNPS undertook a second feasibility study with assistance from the United Nations Development Programme. (Raymond, 1999)

The Energy policy in Philippines

Declaration of Policy of the Department of Energy Act of 1992 states, in its section 2, the following: "It is hereby declared the policy of the state: 1) to ensure a continuous, adequate and economic supply of energy with the end view of ultimately achieving self-reliance in the country's energy requirements, through the integrated and intensive exploration, production, management and development of the country's indigenous energy resources, and through the judicious conservation, renewal and efficient utilization of energy to keep pace with the country's growth and economic development and taking into consideration the active participation of the private sector in the various areas of energy resource-development; and 2) to rationalize, integrate and coordinate the various programmes of the government towards self-sufficiency and enhance productivity in power and energy without sacrificing ecological concerns."

Based on this main two main elements, more specific policy thrusts have been formulated. These include the following: a) sustaining momentum in the exploration and development of oil and other indigenous resources; b) diversifying sources of energy imports while ensuring a balance between cost and supply stability; c) enhancing private sector participation in energy projects; d) moving towards deregulation of domestic energy pricing; e) promoting fuel substitution and diversification in power generation; f) formulating and strictly implementing comprehensive rehabilitation programmes; g) intensifying promotion of energy conservation and energy-efficient technologies; h) enhancing assessment of and planning for the energy needs of the countryside development; i) integrating environmental concerns in the planning and implementation of energy projects; j) promoting technological cooperation among energy research and development institutions and strengthening coordination of the national energy research and development systems; k) rationalizing operation of energy institutions and strengthening coordination of energy policy and programme execution.

It is important to note that the policy of achieving self-reliance through indigenous sources of energy can pave the way for the eventual use of nuclear energy which is a cheap and, to a certain extent, renewable source of energy as the plutonium from spent fuel rods can be reprocessed and reused as fuel for nuclear power plants. Thus, using nuclear energy can decrease the country's dependence on imported oil. The Philippine Energy Plan for 1998-2035 prepared by the Department of Energy envisions that nuclear power shall provide 600 MW of electricity in the 2021-2025 period.

The Baatan Nuclear Power Plant

In response to the 1973 oil crisis, the Philippines decided to build the Bataan nuclear power plant with two units. Construction of Bataan 1, a 620 MWe Westinghouse PWR, began in 1976. Unfortunately, construction work on Unit 1 was stopped in June 1979 after the Three Mile Island nuclear incident. A commission was formed to look into the plant's safety.

Consequently, after five months of inquiry, the Commission recommended that "additional safety upgrades be incorporated in the original design of the reactor. Construction work resumed in September 1980. The inclusion of additional safety upgrades in the reactor design lengthened the construction timetable by an additional 18 months and increased the cost of construction to around US$1.95 billion. In 1985, the nuclear power reactor was finally completed and the first batch of nuclear fuel was delivered". (Raymond, 1999)

However, in 1986, with the assumption to power of a new Administration, the nuclear power reactor was mothballed. Afterwards, the Philippine government set up a Presidential Committee on Philippine Nuclear Power Plant. The Committee subsequently filed a lawsuit against Westinghouse on charges of bribery of President Marcos in connection with the plant's contract.

Meanwhile, in 1987, the President of Philippines issued the Executive Order 128, which reorganized the government and reconstituted PAEC as the Philippine Nuclear Research Institute under the Department of Science and Technology.

On March 4, 1992, the Philippine government and Westinghouse entered into a compromise agreement. Under this agreement, Westinghouse would upgrade and refurbish the plant over a three-year period at a cost of US$400 million and operate the plant for up to 30 years. In return, Westinghouse will be paid a management fee of US$40 million a year and US$0.29 cents for every kWh generated by the plant. Westinghouse would also pay the Philippines US$10 million in cash and US$75 million in discounts on the upgrade and credits on non-nuclear-related equipment for the Philippines' power development programme. However, these efforts failed. In April 2007, the Philippine government made the final payment for the nuclear power plant and decided to convert it into a natural gas-fired power plant, but this was impractical, and the plant has simply been maintained. It is important to stress that for over 30 years, until meeting obligations in April 2007, Filipino taxpayers paid US$155,000 a day in interest on the plant that never produced a kilowatt of power. (Gunn, 2008)

Source: Photograph courtesy of Arkibong Bayan (www.arkibongbayan.org).

Figure 164. Bataan nuclear power plant.

Nuclear Safety Authority and the New Nuclear Power Programme

The PNRI is mandated "to regulate the safe utilization of nuclear energy in order to ensure public safety and environmental protection". It is important to single out that PNRI retains the dual mandate of promoting and regulating the peaceful application of nuclear energy, which is in clear contradiction to what the IAEA and the nuclear international community are recommended in the field of nuclear safety infrastructure. With respect to its regulatory role, the PNRI is empowered "to perform the following functions: a) establish and issue regulations and orders with respect to atomic energy, facilities and materials for the protection, health and safety of the workers and the public; b) make inspections to ensure compliance with the requirements for ensuring the protection, health and safety of both the workers and the public; c) issue licenses to qualified persons/users; d) modify, amend, suspend or revoke any license; e) establish and issue regulations and orders for the safe transport of atomic energy materials and facilities; f) regulate and license the acquisition, distribution and use of radioactive materials and for this issue and promulgate rules, regulations or orders; g) issue rules and regulations establishing standards and instructions to govern the shipments, possession and use of radioactive materials, to protect health, minimize danger to life and property, promote the national defense and security, and protect the public".

PNRI undertook the first site selection studies for the first nuclear power plant to be built in the country with the active participation of the IAEA. In the mid-70s, the PAEC assumed a different role vis-à-vis the nuclear power programme. It exercised regulatory control over the construction of the first Philippine nuclear power plant. To undertake this awesome task, the PAEC drew the needed expertise from its own scientists on the nuclear area, and the academe and private consultancy companies on the non-nuclear areas in order to evaluate the safety documents submitted by the operator. With the mothballing of the nuclear power programme, the PNRI concentrated its regulatory function on the regulation of radioactive materials and honed its expertise in this area including radiation protection, emergency preparedness and response, and security aspects. (De la Rosa, 2007)

Since the law exempts the facilities of the PNRI from regulatory control, the PNRI instituted an internal regulatory programme for its radiation facilities and radioactive materials in its possession. The promotional departments referred to as users are under the supervision of a Radiation Safety and Security Board which submits to the regulatory department of PNRI applications for a permit to use radioactive materials in its installations. The regulatory department evaluates such application and conducts a site survey, if necessary, and issues the permit when all requirements are met by the user. This programme hopes to enhance radiation safety in the use of radioactive materials within the PNRI, and strengthens safety culture among PNRI scientists.

However, it is important to note that a draft Nuclear Regulatory Act has been prepared for submission to the Philippine Legislature in order to reorganize the regulatory authority in the country. It provides for the creation of a separate nuclear regulatory authority to have regulatory control over radioactive materials and electronic devices emitting radiation. The present PNRI will remain as the national nuclear research institution to promote the peaceful applications of nuclear energy. For this reason, the PNRI will no longer be constrained in fully promoting the benefits of nuclear energy. At the same time, a clear delineation of the roles and functions of the two independent nuclear agencies will benefit in strengthening the national nuclear programme. The new regulatory authority will have the independence to implement its nuclear regulations in matters relating to nuclear and radiation safety. The creation of a separate and independent nuclear regulatory authority is an important requirement that the government must address before it embarks on a new nuclear power programme.

Finally, the recruitment of additional manpower and human resource development would be a priority programme for the new regulatory authority. Inasmuch as the government is seriously assessing the possibility of including the nuclear power option in its energy plan by 2018, the most important component of the nuclear power programme that should be considered by the government is the availability of human resources and the training of the staff the new regulatory authority, the prospective operator, among others.

The Public Opinion

According with reference 108, in the 1992 National Survey conducted by the Social Weather Stations, almost half (48%) of the respondents were aware of the Bataan nuclear power plant. In terms of the respondents' geographical location, awareness was greater in urban centers (63%) than in rural areas (32%). Furthermore, awareness was low in rural Visayas (29%) and rural Mindanao (23%). In terms of the people's perceptions as to the safety of operating the Bataan nuclear power plant, it is interesting to note that a slight majority (52%) viewed the plant as being unsafe. Male respondents (55%) were more skeptical as to the safety of operating the plant as compared to female respondents (48%). It is also interesting to note that the younger respondents were more apprehensive as regards the issue of safely operating the plant. Sixty-two percent of the youth (18-24 yrs. old) believed that the plant was unsafe while only 52% of the intermediate young (25-34 years old) and 44% of the older respondents (45 years and older) believed so. Respondents in Metro Manila were more apprehensive (60%) about the operation of the plant as compared to other locales.

Looking Forward

In 2007, the Philippines Department of Energy (DOE) set up a project to study the development of nuclear power, in the context of an overall energy plan for the country. Nuclear energy for the production of electricity was considered as one of the energy sources that could reduce the country's dependency on imported oil and coal. In its 2008 update of the national energy plan, 600 MWe of nuclear energy capacity was projected on line in 2025, with further 600 MWe increments in 2027, 2030 and 2034 to give 2 400 MWe.

In 2008, an IAEA mission visited the nuclear power plant responding to a request presented by the government and advised that Bataan 1 could be refurbished and economically and safely be operated for 30 years. Refurbishment is estimated to cost US$800 million. The IAEA also recommended "the adoption of a policy framework for nuclear power development in the country for the coming years. With regard to the commercial viability of the Bataan nuclear power plant, the IAEA mission indicated that "the operation of the plant will represents a significant savings over the lifetime of other plants (gas turbine, coal, combined cycle and oil)". It estimated that operating Bataan nuclear power plant could generate a savings from "a low of P 1.06 billion for geothermal to a high of P 3.6 billion for gas turbines annually". It also pointed that "nuclear energy is generally cheaper and more stable over longer periods of time compared to other fuels". (Raymond, 1999)

In December 2008, the National Power Corporation announced that it would commission Korea Electric Power Corp (KEPCO, parent company of KHNP, to conduct an 18-month feasibility study on commissioning the nuclear power plant located in Bataan.

At the same time, the Department of Energy is considering how to rebuild local skills in nuclear sciences and engineering due to lack of available skill professional in the field of nuclear power in the country[156]. In view of the lack of government policy on nuclear power for the past 20 years and with the retirement of trained staff, the nuclear institutions in the Philippines have not been able to sustain its knowledge base on nuclear power, and the universities also stopped its graduate programme in nuclear engineering. The situation is exacerbated by the fact that young engineers are discouraged to pursue a programme in nuclear engineering due to lack of professional perspectives. This situation, if not addressed soon, will have a significant impact on future nuclear power programme to be developed in the country.

THE POSSIBLE INTRODUCTION OF A NUCLEAR POWER PROGRAMME IN BANGLADESH IN THE FUTURE

Bangladesh is an agrarian country, with very little industrial development. Because of the fast population growth, the amount of per capita cultivable land is dwindling very fast. In order to survive as a nation and to be a prosper country in the 21st century, Bangladesh will have to shift from an agrarian economy to an industrial economy. How this shift can be done? This shift can be done only if the energy sector can be developed significantly in the coming years, with the purpose to increase electricity generation and energy supply. Electrification of the whole country should be taken as the top most priority. To meet the growing demands of electricity in the domestic and industrial sectors, Bangladesh will have to come up with a plan for massive production of power. The country is not self-reliant on energy production and needs to reduce its dependency on foreign oil gradually, while explore the feasibility of developing alternative sources of energy, such as nuclear energy. (Khale Quzman, 2007)

Bangladesh produced 22.6 billion kWh in 2005 from some 4 GWe installed power capacity, giving per capita consumption of 114 kWh per year. Around 89% of electricity comes from natural gas. Electricity demand is now rising rapidly. In 2006, Bangladesh produced 24.3 billion kWh (gross). It is important to single out that electricity demand is now rising rapidly and the government aims is to increase capacity to 7 GWe by 2014 in order to satisfy the foresee electricity demand.

One of the energy alternatives considered by Bangladesh to be included in the energy mix of the country in the coming years is the use of nuclear energy for electricity generation. Building a nuclear power plant in the West of the country was initially proposed in 1961. Since then a number of feasibility reports have affirmed the technical and economic

---

[156] The state-owned National Power Corporation originally had 710 nuclear engineers who were trained by Westinghouse and Ebasco Overseas Corp. in the 1980s, but this has declined to about one hundred, many of whom are due to retire in the next five to ten years.

feasibility of the use of nuclear energy for the production of electricity in the country. The Rooppur site in Pabna district was selected, in 1963, for the site of the first nuclear power plant to be built and for this reason land was acquired. The government gave formal approval for the construction of a 125 MWe nuclear power reactor in 1980 but the construction of the reactor never started. The country has a Triga 3 MW nuclear research reactor operational since 1986.

With growth in electricity demand and the need to increase in electric grid capacity in the following years, a much larger nuclear power plant looked feasible. For this reason, the government, in 1999, expressed its firm commitment to build a nuclear power plant in Rooppur. In 2001, it adopted a national Nuclear Power Action Plan and, in 2005, it signed a nuclear cooperation agreement with China. In 2007, the Bangladesh Atomic Energy Commission proposed the constructions of two 500 MWe nuclear power reactors for Rooppur by 2015, quoting likely costs of US$0.9-US$1.2 billion for a 600 MWe unit and US$1.5-US$2 billion for 1 000 MWe unit. [26]

In April 2008, the government reiterated its intention to work with China in building the Rooppur nuclear power plant and China offered funding for the project. The IAEA has approved a Technical Assistance Project for Rooppur nuclear power plant to be initiated between 2009 and 2011, and it now appears that a 600 MWe plant is envisaged to finally be built[157]. The main argument used by the government to built the nuclear power plant at Rooppur site is a massive electricity shortages that Bangladesh is facing which have hit its booming textile industry, with generation of 3 000 MWe at peak times still 2 000 MWe short of actual demand.

Russia, China and South Korea had earlier offered financial and technical help to introduce a nuclear power programme in the country. In May 2009, Bangladesh and Russia signed a deal, which could lead to construction of the first nuclear power reactor in the country. The two sides signed a memorandum of understanding on peaceful use of nuclear energy, which officials said was the first step toward construction of a nuclear power plant. "We are looking into safety and cost and then we will decide which country will build the first nuclear power plant in Bangladesh," power minister spokesman Afrazur Rahman said to news papers. "If it goes ahead, the plant would have a capacity of 600 to 1000 MWe", Rahman said. Power outages are frequent, particularly in the summer months from April to October, when supply is diverted to farms for irrigation. It is important to note that Bangladesh's gas reserve are also fast depleting, forcing the country to look for alternative sources of energy and the use of nuclear energy for electricity generation.

THE POSSIBILITY TO INTRODUCE A NUCLEAR POWER PROGRAMME IN AUSTRALIA IN THE FUTURE

Australia has the world biggest uranium reserve and is one of the main uranium exporter country supplying about a fifth of the uranium oxide market. Australia is already the biggest

---

[157] The World Bank estimated that Bangladesh needed US$10 billion in investment for its electricity supply in the next decade in order to satisfy the foresee increase in electricity demand.

uranium exporter to Japan, but Canberra is expanding its uranium export markets to fast-growing economies, including India and Russia.

Australia produced 255 billion kWh from 46 GWe of capacity in 2006, with 23 billion kW per year being embedded in aluminum exports. Final consumption was 187 billion kWh, hence per capita consumption is around 9 100 kWh per year. Coal-fired power plants accounts for 80% of Australia's electricity needs, making it the world's biggest emitter of greenhouse gas per capita, 12% from gas and 7% from hydro. [26] This situation gives it a high output of CO_2, which is the main reason used by the government to start the consideration of a possible nuclear electricity generation in the future. Australia has operated a research reactor since 1956 and has now commissioned its 20 MWt replacement.

It is important to single out that Australia interest in the use of nuclear energy for the electricity generation started in the 1970s, when the Australian government sought tenders for building a nuclear power reactor at Jervis Bay. Designs from UK, USA, Germany and Canada were short listed, but a change in leadership led to the project being cancelled in 1972. However, until 1983, there were various plans and proposals for building an enrichment plant. [26]

At the end of 2006, a report prepared by a taskforce indicated that "nuclear power would be 20%-50% more expensive than coal-fired power and with renewables it would only be competitive if low to moderate costs are imposed on carbon emissions (A$15-40 or US$12-30 per ton CO_2). Nuclear power is the least-cost low-emission technology that can provide base-load power and has low life cycle impacts environmentally". The first nuclear power plants could be running in 15 years, and looking beyond that, 25 reactors at coastal sites might be supplying one third of Australia's (doubled) electricity demand by 2050. Certainly the challenge to contain and reduce greenhouse gas emissions would be considerably eased by investment in nuclear power plants. Emission reductions from nuclear power could reach 18% of national emissions in 2050.

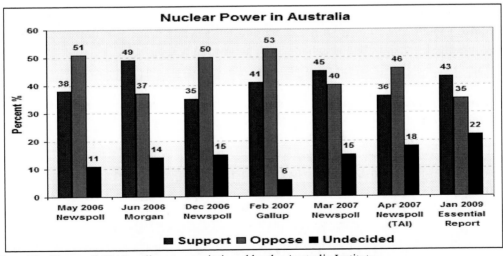

Note: The Newspoll (TAI) poll was commissioned by the Australia Institute.

Figure 136. The outcome of the polls carried out in Australia during the period 2006-2009.

Early in 2007, a private equity company, Australian Nuclear Energy, said "it was examining the prospects for a nuclear power plant". For the purpose to use nuclear energy for the electricity generation in the future, Australia signed recently with Japan a plan to strengthen bilateral cooperation on the peaceful use of nuclear power.

The Public Opinion

Several polls have been carried out related with the possible use of nuclear energy for the generation of electricity in the last few years. The outcome of these polls is shown in Figure 136.

From Figure 136 can be stated that the support for the introduction of a nuclear power programme in the country has not changed over the last three years. People support of the use of nuclear energy for the production electricity reached a peak in February 2007 (41%) while the peak for those that were oppose to the use of this type of energy was reached in the same year (53%). However, in January 2009, there were more people that support the introduction of a nuclear power programme in the country (43%) than those that were against (35%). It is important to note, however, that this situation was the same that the ones presented in June 2006 and in March 2007.

There are two trends which are probably occurring. The proportion of the Australian people that is undecided on whether to accept or reject the introduction of a nuclear power programme in the country in the future is slowly and consistently increasing through time. In May 2006, the percentage of the undecided people was only 11% while, in January 2009, the number increases to 22%. In net terms, can be said that it is likely that an increasing number of people are becoming more open to the possibility of the use of nuclear energy for the production of electricity in Australia and that a large chunk of this increase is coming from a reduction in the size of opposition to nuclear power. Overall opposition to nuclear power appears to be weakening, and while that seems to have led to an increase in the size of those undecided about nuclear power, it is too early to tell whether it has also flowed through to actual increased support, even though the latest poll certainly suggests it has.

However, the Prime Minister from Australia expressed, in October 2009, "that the government will not support the use of nuclear energy for the generation of electricity in spite of the results of a recent opinion poll that showed that 49% of Australians supports the use of nuclear energy as alternative source of energy, while 43% is openly against". The results of this survey show a change of the public opinion that, in 2006, supported the nuclear energy option: 38% was I favor, while 51% was against. The government has made clear his position: it considers that the country has other energy alternatives that can be used for the generation of electricity.

THE POSSIBILITY TO INTRODUCE A NUCLEAR POWER PROGRAMME IN NEW ZEALAND IN THE FUTURE

New Zealand produced, in 2003, some 41 billion kWh per year from 8.4 GWe installed capacity including 5.25 GWe from hydro, 1.38 GWe from gas-fired, 1.0 GWe from coal-fired and 0.42 GWe from geothermal, for an average per capita consumption of about 10 000 kWh per year. Around 58% of electricity comes from hydro, the mayor source of energy available in the country, 24,5% from gas, 7.6% from coal, 6.7% from geothermal, 2% from wind and 1.3% from biomass. New Zealand has depended primarily on hydro-electric power for its electricity generation for many years, but scope for expansion is limited and evens the reliability of present capacity depends on capricious rainfall, as in 2001 and 2003. [26] However, it is important to single out that hydro output has not increased over the last 15 years, and that growth in demand since 1990 has been mostly met by gas-fired power plant, at least until the 1 000 MWe state-owned Huntly plant shifted to using coal for 80% of its energy.

In 1968, for the first time the national power plan identified the likely need for the introduction of a nuclear power programme in the country in a decade or more ahead, since readily-developed hydro-electric sites had been utilized. Plans were made and a site at Oyster Point on the Kaipara harbor near Auckland was selected for the construction of the first nuclear power plant. Four 250 MWe nuclear power reactors were envisaged, to supply 80% of Auckland's needs by 1990. But then the Maui gas field was discovered, along with coal reserves near Huntly and the project was abandoned by 1972. In 1976, a Royal Commission was set up to enquire further into the question of the possible use of nuclear energy for the generation of electricity in the country in the future. In the Commission's report, presented in 1978, it said "that there was no immediate need for New Zealand to embark upon a nuclear power programme, but suggested that early in 21st century a significant nuclear programme should be economically possible".

THE POSSIBILITY TO INTRODUCE A NUCLEAR POWER PROGRAMME IN MALAYSIA IN THE FUTURE

Malaysia produced 87 billion kWh in 2005, 93% of this from fossil fuels - mostly natural gas, and the rest from hydro. A comprehensive energy policy study, including consideration of a possible introduction of nuclear power programme, is now underway and will be completed in 2010. The state-owned utility is tentatively in favor of the introduction of nuclear power programme in the country in the coming years. In August 2006, the Malaysian Nuclear Licensing Board stated that "plans for the use of nuclear energy for the production of electricity after 2020 should be brought forward and two nuclear power reactors built much sooner". This intention has since been reiterated from the Ministry of Science, Technology and Innovation.

The Malaysian Institute for Nuclear Technology Research (MINT) has operated a 1 MW Triga research reactor since 1982. In April 2007, MINT was renamed the Malaysian Nuclear Agency (or Nuclear Malaysia) to reflect its role in promoting the peaceful uses of atomic energy, including the use of this type of energy for the production of electricity.

Chapter 6

THE CURRENT SITUATION AND PERSPECTIVES IN THE USE OF NUCLEAR ENERGY FOR ELECTRICITY GENERATION IN THE LATIN AMERICAN REGION

Latin America has important oil and gas reserves and is using, in a great scale, hydropower for the generation of electricity. Latin America accounts for 11% of world reserves of oil and 5% of natural gas. Venezuela is the Latin American country with the largest quantity of oil (80 billion barrels, in 2007) and natural gas reserves (4.6 trillion m^3). The current oil reserves in Venezuela can covers 68,3 years at the current production rate. After Mexico, Venezuela is also the main oil producer (2.24 million barrels a day in 2009), while Argentina heads the list of natural gas producers (48.4 billion m^3 in 2009). Argentina and Bolivia are the main natural gas exporters in South America, but the highest gas reserves/production ratios are located in Bolivia, Ecuador, Peru and Venezuela. Brazil has great potential due to oil and gas reserves discovered off the Southeast Brazilian coast. Brazil is the second largest ethanol producer in the world, using sugarcane as feedstock, which is extremely productive and efficient. In Ecuador, Venezuela, Bolivia and Peru, it exceeds 130 years, but in Argentina it is only 10 years. Total production attributed to the Caribbean is summarized in the production by Trinidad and Tobago, which liquefies its natural gas and exports practically the entire production to the American market. [109]

Brazil is the country with the highest electricity installed capacity and the highest electricity demand in the whole Latin American region followed by Mexico, Argentina and Venezuela. The potential use of nuclear energy for electricity production is an option that cannot be excluded from any future energy balance study to be carried out by several Latin American countries. The purpose of these studies is to find out the most adequate energy mix balance in order to satisfy the future increase in electricity demand.

In Latin America, the current plans for nuclear expansion are ambitious but involve a very limited number of countries. Argentina and Brazil, countries with two nuclear power reactors in operation, may seek to double or triple existing nuclear capacity, according with the energy plans already approved by their respective governments. Mexico, which operates also two nuclear power reactors, may build as many as eight more reactors by 2025, according with preliminary plans under consideration. Chile, Venezuela and Uruguay are discussing plans for the introduction, for the first time, of a nuclear power programme in the coming years and, for this reason, are carrying out preliminaries studies on this subject.

With the exception of Mexico, interest in nuclear power around the hemisphere is driven by a desire to find alternatives to the unstable production of electricity using the current hydroelectric power capacity installed in several Latin American countries. Rising electricity demand and prices have also tightened natural gas supplies. In additional to this specific situation, in the specific case of Chile, the gas cutoff from Argentina due to deficit in the supply of gas from that country and the nationalization of Bolivia's natural gas production in 2006, increase the interest of Chile in the use of nuclear energy for the production of electricity as a supplement to risky tight gas supplies.

In Mexico, the electricity demand is projected to grow 6% annually and even if the country builds the eight new nuclear power reactors under consideration, nuclear power will only account for 12% of the total electricity generation. The rest of the demand of electricity should be satisfy by other energy source and will not relieve the country's overwhelming dependence on oil and natural gas for its total energy consumption. The reason is very simple: Nuclear power only produces electricity, whereas oil and natural gas are used for many other purposes. In addition of what have been said before, the construction of new nuclear power reactors could take years to be finished taking into account that Mexico's first reactor, in Laguna Verde, took twenty years from initial bids to operation.

Argentina also faces high electricity demand. With a supply shortfall anticipated in 2010, most observers expect that the gap will increase considerably thereafter. And like Mexico, Argentina relies heavily on conventional fossil-fuel power plants and hydroelectric power. According to an August 2006 announcement by Planning Minister Julio de Vido, "Buenos Aires will spend US$3.5 billion to refurbish its larger reactor at Embalse (in Córdoba) and complete construction of a third reactor, Atucha 2, by 2010". The country's ambitious plans to build five more nuclear power reactors by 2023 could double nuclear power's share of electricity generation. But nuclear power will not solve Argentina's energy needs. Hydroelectric capacity—of which only about 20% is used—will need to expand to meet the country's annual goal of 40 000 MWe in electrical generation by 2025.

Brazil is also considering building a handful of nuclear power reactors in the next two decades. Its national energy plan calls for four new units to be built by 2025. In the next 50 years, industry officials have reportedly suggested Brazil's nuclear capacity could reach 60 GWe of nuclear capacity. This would require building about 58 more nuclear power reactors—a tall order considering that construction on the next reactor, Angra 3, was delayed two years by environmental concerns. Brazil now relies on hydropower for 92% of its electricity production. The idea, according to Francisco Rondinelli, head of Brazil's nuclear association, would be "to diversify at least 30% of electricity generation equally into nuclear energy, gas and biomass".

Venezuela and Uruguay are also interested in the use of nuclear energy for the production of electricity but have a long way to go. Venezuelan President Hugo Chávez has sought nuclear cooperation from Brazil, France, Iran and Russia, but few suppliers appear to be biting, perhaps because Venezuela's plans are not well defined. Meanwhile, Uruguay, which gets virtually all its electricity from hydropower, has mentioned nuclear power as a future option, but national laws banning nuclear energy would need to be overturned. Moreover, its withdrawal of plans for a natural gas plant in 2005 because of the cost (US$200 million) and time of construction (26 months) suggests that a nuclear power reactor, which costs billion of dollars and takes at least four to five years to construct, may not be in the minds of many Uruguayan authorities and representatives of the private energy industry.

From Figure 164 can be stated that hydropower is and will continue to be the main energy source for electricity generation in the Latin American region in the next 10 years, followed by natural gas. Nuclear power will increase its share but will be the fifth energy source for electricity generation after 2013.

There are no other available energy alternatives that the use of nuclear energy for electricity generation to satisfy the future energy shortfall that some of the Latin American countries could be facing in the coming years. Even without a spurt in nuclear power generation, the expansion of transmission grids between countries could be one potential solution to unpredictable energy supplies. The proposed Central American Electrical Interconnection System (SIEPAC) will connect 1,100 miles (1,770 kilometers) of transmission lines between Central American countries and Mexico. Brazil and Argentina have successfully covered shortfalls by selling each other electricity in off seasons. Another alternative to be considered is the extension of natural gas pipelines.

It is important to single out that the Latin American region is not an exception to a global trend that sees nuclear energy as clean, green and homegrown. But like anywhere else, new nuclear power plants will require major political, financial and public support. The unknown variable now is the current economic crisis. Its anticipated dampening effect could provide breathing room to develop the intellectual support and physical infrastructure required for nuclear power. Or, it could extinguish the nuclear enthusiasm that has been growing for the past few years.

A secure supply of energy is of crucial importance for the development of vital economy sectors in all Latin American countries. Electricity supply and the possibilities for industrial development are particularly closely linked and are indispensables to ensure the economic development of the Latin American countries.

Fossil-fuelled power plants are sources of pollution and they have become an essential factor in considerations of environmental protection, not only at the local level but regionally and internationally as well. It is becoming increasingly unlikely that any country in the region can avoid considering environmental concerns when setting a national energy policy for electricity production. For some of the Latin American countries the only available technology for the production of electricity that, at the same time, takes into account relevant environmental considerations is nuclear energy.

Besides to have national energy policy for the development of the energy sector in most of the Latin American region, the adoption of a policy concerning relations with neighboring countries is increasing in importance, particularly when a country decide to introduce a nuclear energy programme for electricity generation, or are thinking to expand an ongoing nuclear power programme. There is no doubt that the adoption of a regional energy policy is an important complement of any national energy policy adopted by any State of the region. The importance of a regional energy policy in the Latin American region is growing, particularly in recent years, as shown by the number of regional associations and alliances being formed for various purposes, some of them to ensure energy supply in the future and at affordable price. This applies also in the case of nuclear power programmes, within which many topics related with regional cooperation could yield direct benefits. According with reference 1, the following are important components of any regional energy policy related with the introduction of a nuclear power programme in the Latin American region.

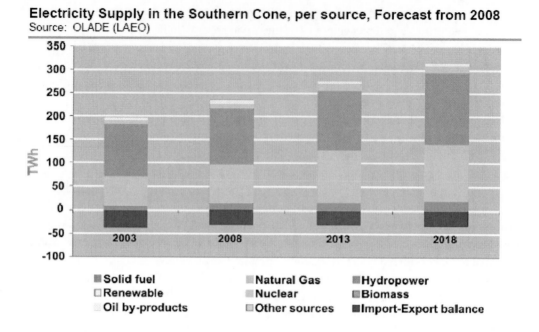

Figure 164. Electricity supply in the Southern Cone per source forecast from 2008 to 2018.

The Evolution of the Nuclear Power Sector in the Latin American Region

Higher world market prices for fossil fuels, the reduction of the current world fossil fuels reserves and the climate changes affecting all countries in the region, have put the use of nuclear energy for electricity generation again on the agenda of some Latin American countries without nuclear power plants currently in operation and have revived interest in countries with stagnating nuclear power programmes. Based on the abovementioned reality, at least three Latin American countries are considering seriously the expansion of their nuclear energy programmes for electricity generation, while at least three others are considering, for the first time, the use of this type of energy for the same purpose.

Because they are driven largely by demand, current high prices for fossil fuels are likely to be more permanent than were in the 1970s. High world market prices for fossil fuels have the greatest impact on countries that are highly dependent on energy imports, particularly developing countries with relatively scarce financial resources. Several Latin American countries are included in this group.

Several developing countries from different regions with sizable domestic fossil fuel resources have recently begun analyzing the introduction of nuclear power programmes for electricity generation in the period 2015-2020. For some of these countries, the immediate impact of increased oil prices is not the same as for others, particular for those that are net importer oil countries. For countries with sizable domestic fossil fuel resources, such as Venezuela, nuclear power programmes for electricity generation could be a vehicle to increase export revenues by substituting domestic demand for natural gas (and to a lesser extent oil) by nuclear power. The additional export earnings as a result of the substitution of domestic demands of fossil fuel by the use of nuclear energy for electricity generation may well finance the construction of a country's first nuclear power plants, or additional nuclear power reactors.

In the Latin America region, only 2.3% of its electricity comes from nuclear sources in 2009, according with the IAEA and WANO sources. However, if expansion plans in Argentina, Brazil and Mexico succeed, and the intention of constructing new nuclear power plants expressed by some other countries from the region, such as Chile, Uruguay and Venezuela are materialized in the near future, then that proportion could be more than double in a decade.

The experience in the use of nuclear energy for electricity generation in the Latin American region indicates that the technical and economic performance of nuclear power is not dependent on whether it is privately or stated owned. Some countries, including Argentina, Brazil and Mexico developed efficient electricity generation programmes within the public sector, whereas others countries, mainly outside the region such as the USA, have attained similar results under a majority private ownership or a mix of private and public ownership.

How was the evolution of the nuclear sector in the Latin American regions? Three Latin American countries introduced a nuclear power programme during the 20th century with some degree of success. These countries were the following: Argentina, Brazil, and Mexico[158]. Argentina was the first country to introduce a nuclear power programme in the region. The country started to build its first nuclear power plant, Atucha 1, in 1964. Afterwards, built the Embalse nuclear power plant, which entered into operations in 1984. Atucha 2, the second unit in the Atucha nuclear power plant, has yet to be completed.

Mexico was the second country in the region to introduce a nuclear power programme. It began to build its only nuclear power plant at Laguna Verde in 1969, but did not manage to use the plant commercially until three decades later.

In 1974, Brazil joined the group when it started the construction of Angra 1 in the vicinity of Rio de Janeiro. Operations got underway in 1984. Construction began on Angra 2 in 1976, in the same complex as Angra 1 and with support from the Federal Republic of Germany. In 1984, started the construction of Angra 3, which has not been finished.

Summing up can be stated the following: During the 20th century four Latin American countries developed nuclear energy programmes three of them, Argentina, Brazil and Mexico, with some degree of success. Cuba after an initial successfully nuclear power programme in the 1970s had to stop it definitively in the 1990s.

The three challenges of utmost priority for the region, if nuclear expansion is going to take place in the coming years, would be:

1) to reach a rational nuclear power development throughout the region. This requires a deep, long-term analysis of all energy alternatives available in the region for power generation, their feasibility and convenience, on a country-by-country but also on a regional basis. Decisions should be made taking into account the sustainability and optimization of natural energy resources based on true energy needs, rather than on a search for prestige or interest in following trends;

---

[158] Cuba tried to introduce an important nuclear power programme in the 1970s but did not manage to conclude it due to a lack of funds and of partnerships ready to participate in the conclusion of the construction of the first unit in the 1990s. The Cuban's nuclear power programme was definitively suspended in the 1990s.

2) to achieve an adequate integration of new-coming states, to keep the highest standards on nuclear safety and security. This implies shared responsibilities between new-comers and suppliers. The first undertaking should be necessarily seen as a long-term process, when the regulatory framework, the infrastructure to support the introduction of a nuclear power and the build-up of human capabilities should be progressively developed. This is a key to the nuclear success, given that human capital and nuclear safety culture cannot be measured in years, but in decades;

3) to ensure that nuclear power growth in the region is done within a strict, non-proliferation framework. This implies to keep working an effective scheme of verification and control to avoid that sensitive materials or technologies could end up in the hands of terrorist groups. It would be useful a full adherence to the IAEA Additional Protocol.

THE NUCLEAR POWER PROGRAMME IN ARGENTINA

Argentina is the major natural gas producer in the region, exporting to Chile, Uruguay and Brazil. However, there are some problems with the supply of gas to some countries in the region due to gas production problems in Argentina.

The largest oil reserves in the Southern Cone are to be found in Argentina. No oil is produced in Uruguay and Paraguay while oil production in Chile is small. Forecast show that with an increase in the demand of gas in the Southern Cone, Chile, Uruguay and Paraguay will have to import oil from another region, since the Argentinean reserves shall be used internally.

Argentina was a pioneer in the use of nuclear energy for peaceful purpose and for the production of electricity in Latin America. The original Argentinean plan for the use of nuclear energy for electricity production contemplated the construction of a total of six nuclear power reactors, but the plan was never fully implemented. During the 1980s, interest in the use of nuclear energy for electricity production decreased because of the country's abundant supply of natural gas. In the 1990s, the state decided not to proceed with an activity requiring such large investments.

The paralysis of the nuclear industry had an impact on energy output. The share of electricity generated by nuclear energy in Argentina fell from 15% in the 1980s to 6,2% in 2007 and to 6.18% in 2008. In 2006, gross electricity production was 115 billion kWh, 50% of this from gas and 33% from hydro. Nuclear power accounted for 7.6% of the country's generation mix in March 2009, a little bit higher than in 2007 and in 2008, according to a report prepared by power consultancy Fundelec. Installed capacity is about 35 GWe.

Argentina's electricity production is largely privatized and is regulated by the National Electricity Regulator Entity (ENRE). The country's atomic energy commission (CNEA, Comisión Nacional de Energía Atómica, National Atomic Energy Commission) was set up in 1950 and resulted in a spate of activity centered on nuclear research and development, the construction of several research reactors, including a 5 MW unit commissioned in 1968. In 2008, five research reactors are operated by CNEA and others institutions.

According with the IAEA, nuclear power development in Argentina has been aimed at achieving medium and long-term energy autonomy, improving domestic technology and diversifying energy supply. Due to the moderate amount of non-renewable energy sources available in the country, fossil fuels reserves are expected to become scarce in the medium-term even with conservative economic growth perspectives. Hydroelectric power is being developed, but they will be completely exhausted in the medium-term. Other renewable sources are far from reaching commercial operation and are not competitive yet. Therefore, nuclear power plants shall surely play an increasingly important role in future electricity generation in the country in the coming years.

The Energy Policy in Argentina

Due to its special characteristics, the activities related to the use of nuclear energy for peaceful purposes in Argentina needs to be subject to national (or federal) jurisdiction and regulated as an organic and indivisible system. For this reason, the National Congress is empowered to establish the laws concerning the subject, through Section 75 paragraphs 18 and 32 of the Constitution. Within this context, Act No 24804, 1997 or "National Law of the Nuclear Activity", is the legal framework for the peaceful uses of nuclear energy in the country. Article 1 of the Act No 24804, 1997, sets that "concerning nuclear matters the state will establish the policy and perform the functions of research and development and of regulation and control, through the National Atomic Energy Commission and the Nuclear Regulatory Authority".

Moreover, the mentioned law sets that "any nuclear activity either productive or concerning research and development, that could be commercially organized, can be carried out both by the state and the private sector.

Safety assessments carried out by the Nuclear Regulatory Authority from the beginning of Atucha I and Embalse nuclear power reactors operation indicate "that no objection is found for continuing their commercial operation in compliance with the regulatory standards in force in the country, and usual international standards for the nuclear industry. The company in charge of the operation of the nuclear power reactors foresees that the construction of Atucha 2 will be completed and its commercial operation initiated within a time term no longer than six years".

The Evolution of the Nuclear Power Sector in Argentina

Argentina started to show interest in the use of nuclear energy for electricity generation in 1964. The feasibility study for the construction of the first nuclear power reactor was carried out in 1965. The construction of Atucha 1 (see Figure 137) started in 1968 and was carried by Siemens and Kraftwerk Union (KWU) from Germany. The unit was connected to the electric grid in March 1974, and entered into commercial operation in June of the same year. It has a pressure vessel, unlike any other existing heavy water reactor, and it now uses slightly enriched uranium (0.85%) as fuel, which has doubled the burn-up and consequently reduced operating costs by 40%. Atucha 1 operated well until a major shutdown to make internal reactor repairs in 1989. The fuel channels were then completely replaced, but there have been more shutdowns since then. The 335 MW Atucha 1 unit has for several years seen unplanned shutdowns for repairs. Atucha 1 was a turnkey facility.

Source: Photograph courtesy of Wikipedia (Watsonpatricio).

Figure 137. Atucha 1 and Atucha 2 nuclear power reactors.

In 1967, a second feasibility study was undertaken for a larger nuclear power plant in the Cordoba region, 500 km inland, called Embalse (see Figure 138). The Embalse nuclear power reactor with a net capacity of 600 MWe started to be built in April 1974, was connected to the electric grid in April 1983 and entered into commercial operation in January 1984. In this case, a CANDU-6 type of reactor from AECL was selected, partly due to the accompanying technology transfer agreement, and was constructed with the Italian company Italimpianti running on natural uranium fuel. [73] The Embalse nuclear power reactor was constructed with a high participation of the local industry.

In 2007, both nuclear power reactors provided 8.6% of Argentinean's electricity output and had an estimated useful life of 30 to 40 years.

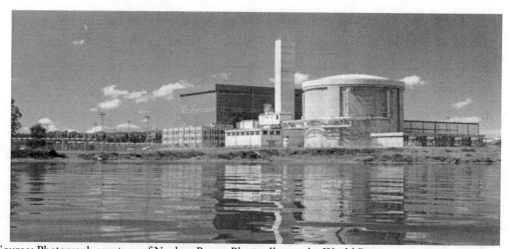

Source: Photograph courtesy of Nuclear Power Plants all over the World Page.

Figure 138. Embalse nuclear power plant.

In 1979, a third nuclear power reactor, Atucha 2, was ordered following a government decision to have four more units coming into operation during the period 1987-1997. It was a Siemens design, a larger version of Unit 1, and construction started in 1981 by a joint venture of CNEA and KWU. However, work proceeded slowly due to lack of funds and was suspended in 1994 with the plant 81% complete. [73] In 2003, plans for completing the 692 MWe Atucha 2 were presented to the government. The Siemens design of the Atucha 2 is unique to Argentina. For this reason, the Argentinean competent authorities were seeking expertise from Germany, Spain and Brazil to complete the unit for some US$ 400 million.

The original cost of Atucha 2 was estimate to be around US$1.5 billion and was initially scheduled to be finished in 1987. The pressure vessel was eventually installed at the end of 1999. The IAEA reported several main items to complete, including the electro-mechanical, instrumentation and control, heavy-water inventory and first core fuel deliveries. In 1995, the Argentine government announced that the construction of Atucha 2 would restart at a cost of about US$700m. The government had discussed updated designs for Atucha 2 with AECL in order to increase the level of operational safety of the reactor. Atucha 2 is expected to entry into operation in 2011. The entry into operation of Atucha 2 will increase in 3% the energy to be provided to the national electrical grid and will increase the nuclear share from the current 6.2% up to 9%. The operating nuclear power reactors in Argentina are shown in Table 84.

The sites of the nuclear power plants constructed in Argentina are shown in the following map.

Summing up can be stated that the expansion of the nuclear power programme in Argentina include one nuclear power reactor under construction (Atucha 2), one planned and one proposed each one with a net capacity of 740 MWe.

For many years, the country has achieved a sustained nuclear development in spite of considerable political, financial and technical difficulties. This development includes a complete control of fuel cycle activities for natural uranium using heavy water. After the initial problems affecting the normal operation of Atucha 1, in the last years the unit operated with high load factors, a high degree of self-sufficiency and strict safety standards. The production of electricity in Argentina by all sources in 2005 was the following (see Figure 140).

From Figure 140 can be easily see that fossil fuels and hydro are the two main energy sources for the production of electricity in the country following by nuclear energy. In this last case, the total production of electricity using nuclear energy in 2008 is shown in Table 85.

Table 84. Operating nuclear power reactors in Argentina

Reactors	Location	Model	Net MWe	First power
Atucha 1	Buenos Aries	PHWR - Siemens	335	1974
Embalse	Cordoba	PHWR - Candu 6	600	1983
Total (2)			935	

Source: CEA 8th Edition, 2008.

Source: International Nuclear Safety Center at Argonne National Laboratory, USA.

Figure 139. Sites of the Atucha and Embalse nuclear power plants.

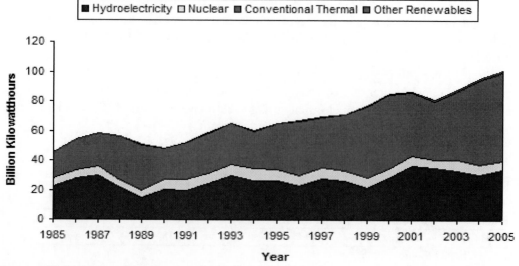

Source: EIA International Energy Annual.

Figure 140. Argentina's electricity generation by sources.

Table 85. Annual electrical power production for 2008

Total power production (including nuclear)	Nuclear power production
110 538.6 GWhe	6 835.1 GWhe

Source: IAEA, PRIS, 2009.

Nuclear Safety Authority and the Licensing Process

In 1994, the Nuclear Regulatory Authority was formed and took over all regulatory functions from the national Nuclear Regulatory Board and CNEA. As well as radiation protection, it is responsible for safety, licensing and safeguards activities in Argentina.

The regulatory system in place considers that the organization known as "Responsible Organization", is fully responsible for the radiological and nuclear safety of the installation that is under their responsibility, as well as for it the physical protection and for the safeguard activities carried out by the IAEA in the country. The mere compliance with the regulatory standards does not exempt the Responsible Organization from the mentioned responsibility. For this reason, the regulatory standards are not prescriptive but, on the contrary, they are "performance-based" standards, that is to say, "they establish the fulfillment of safety objectives; the way of reaching these objectives is based on engineering experience, on the qualification of designers, constructors and operators and on suitable decisions taken by the Responsible Organization itself". Therefore the Responsible Organization must demonstrate and convince the Nuclear Regulatory Authority that the installation is safe.

A basic aspect of the regulatory system is the approach adopted, in which the Responsible Organization deals with the design, construction, commissioning, operation and decommissioning stages of the nuclear power plant, being completely responsible for radiological and nuclear safety of the installation as well as for physical protection and safeguards.

The regulatory standards establish that "the construction, commissioning, operation or decommissioning of a nuclear power plant shall not be initiated without the corresponding license, previously required by the Responsible Organization and issued by the Nuclear Regulatory Authority". The validity of such licenses is subordinated to the compliance with the stipulated conditions included in the corresponding license, and to the standards and requirements issued by the Nuclear Regulatory Authority. The non-compliance with any of these standards, conditions or requirements may be enough reason for the Nuclear Regulatory Authority to suspend or cancel the corresponding license validity, according to the sanction regime in force.

The nuclear power plant staff must be adequately trained and qualified according to their functions and responsibilities in the installation. The Nuclear Regulatory Authority requires also that the nuclear power staff assigned to safety related tasks be licensed.

The regulatory system considers licenses for construction, commissioning, operation and decommissioning, which establish the conditions the Responsible Organization must fulfill at each stage. The construction license is issued when standards and requirements of sitting, basic design and expected safety operation conditions have been complied prior with the beginning of this stage. It is a document through which the Nuclear Regulatory Authority authorises the nuclear power plant construction under the established conditions, and shall be fulfilled by the Responsible Organization.

The applicable standards, consistent with international recommendations on the subject, establish "the safety criteria to be observed in the design of the installation and define the timetable and type of mandatory documentation that shall be presented together with the application for a construction license" (Standard AR 3.7.1). In particular, the nuclear power plants design "must comply with the radiological criteria related to accidents" (Standard AR 3.1.3).

Once the construction license is requested by the Responsible Organization, a continuous interaction between the constructor or operator of the future installation and the Nuclear Regulatory Authority is initiated. It is an iterative process, as complex as the demands involved. It should be emphasized that the Responsible Organization's capability to carry out its responsibilities is evaluated since the construction stage.

Before Act No 24804, 1997 was passed, the Nuclear Regulatory Authority authorized the commissioning by issuing a specific authorization once the corresponding requirements were fulfilled. The above mentioned law establishes "that the Nuclear Regulatory Authority shall issue a commissioning license". The commissioning license establishes the conditions for fuel and moderator loading, operation with increasing power up to its nominal value, as well as verifications and tests of the components, equipment and systems to determine whether they comply with the original design basis. To do so the Responsible Organization must appoint an Ad Hoc Commissioning Committee constituted by acknowledged specialists, "who will continuously evaluate the execution of the commissioning programme and recommend its prosecution, if applicable" (Standards AR 3.7.1, AR 3.8.1 and AR 3.8.2).

The operating license is issued when the Nuclear Regulatory Authority verifies that conditions, standards and specific requirements applicable to a particular installation are fulfilled. Such conclusion will be the result of analysing the submitted documentation and detailed studies, inspection reports carried out during construction and commissioning and the Ad Hoc Commissioning Committee recommendations.

The operating license is a document by which the Nuclear Regulatory Authority "authorizes the commercial operation of a nuclear installation under stipulated conditions, which shall be fulfilled by the Responsible Organization" (Standard AR 3.9.1). The non fulfilment of any of the imposed requirements without the corresponding Nuclear Regulatory Authority authorization would imply the application of sanctions that could lead to the operating license suspension or cancellation.

At the end of its lifetime and under the Responsible Organization's request, the Nuclear Regulatory Authority authorizes the ending of the nuclear power plant commercial operation and issues a decommissioning license. In this document, "conditions for the nuclear power plant safe dismantling are established, being the Responsible Organization in charge of planning and providing the necessary means for their fulfillment" (Standard AR 3.17.1).

The evaluations performed prior to issuing a nuclear installation license include mainly aspects of quality assurance, construction procedures, operation procedures, previsions for in-service inspections, etc. Besides, emergency plans shall be prepared in coordination with corresponding National, Provincial and Municipal authorities.

Nuclear Fuel Cycle and Waste Management

According with the IAEA, the Argentina uranium resources is about 15 000 tons of uranium, though the CNEA estimates that there is some 55,000 tons of uranium as exploration targets. In Argentina, uranium exploration and a little mining started in the mid of the 1950s. They are seven uranium mines in the country but, in 1997, the CNEA Sierra Pintada mine in Mendoza closed for economic reasons. Cumulative national production until then from open pit and heap leaching at seven mines was 2,509 tons. Reserves there and at Cerro Solo in the South total less than 8 000 tons of uranium. A resumption of uranium mining is part of the 2006 plan to expand the use of nuclear energy for the electricity production, in order to make the country self-sufficient. In 2007, CNEA reached agreement with the Salta provincial government in the North of the country to reopen the Don Otto uranium mine, which operated intermittently from 1963 to 1981. A 150 tons per year mill complex and refinery producing uranium dioxide operated by Dioxitek, a CNEA subsidiary, is located in Cordoba.

CNEA has a small conversion plant at Pilcaniyeu, near Bariloche, Rio Negro, with 60 tons per year capacity. All enrichment services are currently imported from the USA. [73]

Production of fuel cladding is undertaken by CNEA subsidiaries. Fuel assemblies are supplied by CONAUR SA, also a CNEA subsidiary, located at the Ezeiza Centre near Buenos Aires. The fuel fabrication plant has a capacity of 150 tons per year for Atucha type fuel and CANDU fuel bundles.

During the period 1983-1989, INVAP, a state-owned company, operated a small diffusion enrichment plant for CNEA at Pilcaniyeu. This was unreliable and produced very little low-enriched uranium. In August 2006, CNEA expressed its intention to restart, in 2007, on a pilot scale the operation of the enrichment plant, using its own Sigma advanced diffusion enrichment technology. The main reason given by the Argentinean authorities to restart the operation of the enrichment plant was to keep Argentina within the limited circle of countries recognized as having the right to operate enrichment plants, and thereby support INVAP's commercial prospects internationally.

Heavy water is produced by Empresa Neuquina de Servicios de Ingeneria (ENSI SE) which is jointly owned by CNEA and the Province of Neuquen where the 200 tons per year plant is located (at Arroyito). This was scaled to produce enough for Atucha 2 and the following reactors and so now has capacity for export. It will, nevertheless, need US$200 million expansion to supply 600 tons of heavy water to Atucha 2 by end of 2009. [73]

There are no plans for reprocessing used fuel, though an experimental facility was built in the early 1970s at Ezeiza nuclear center.

The Federal Radioactive Waste Management Act assigns responsibility to CNEA for radioactive waste management, and creates a special fund for the purpose. Operating plants pay a fee into this fund. Low and intermediate-level wastes including spent nuclear fuel from research reactors are handled at CNEA's Ezeiza facility. Spent nuclear fuel is stored at each nuclear power plant site. There is some dry storage at Embalse. CNEA is also responsible for plant decommissioning, which must be funded progressively by each operator of nuclear power reactors.

The Public Opinion

During the 1970s, 1980s and 1990s, the public opinion of Argentina was in favor to the use of nuclear energy for electricity production, as well as the use of nuclear energy for other peaceful purposes. However, this situation changed in the last decade due to certain problems associated with the mining of uranium and with the application of other processes associated with the use of Au and Pb that are very controversial in this moment. They had been used for a long time but today it is not possible anymore. For this reason, the former authorities of the Mendoza.

Province prohibited by law the transport and use in the entire province, sulfuric acid, which is an important product in the mining of uranium. This industry uses the mentioned product by tons. Regrettably, in that province the country have the biggest uranium open mine and this law prohibiting the adequate mine exploitation.

In addition to what have been said before, in ex-producer provinces of uranium arose the necessity to clean the tails of previous work, as well as to deal with ecologists and politicians that are against the use of nuclear energy for electricity generation. Other cities and provinces have passed laws and regulations prohibiting the use of nuclear energy for peaceful purposes, including for the generation of electricity, creating a very difficult situation to the central government for the expansion of the Argentinean nuclear power programme.

Looking Forward

Argentina is suffering an increasing energy demand and a shortcut of energy supply that cannot be overcome only with the use of conventional and renewable energy sources.

In August 2006, the Argentinean's government announced a US$3.5 billion strategic plan for the development and consolidation of the country's nuclear power sector. This involves completing Atucha 2 by 2012 and extending the life of Atucha 1 and Embalse nuclear power reactors. Extending the life of the Embalse CANDU-6 type reactor by 25 years, in partnership with AECL, is expected to cost US$400 million. Completing Atucha 2 by 2010 is expected to cost US$600 million, including 600 tons of heavy water with a cost of US$400 million. On August 2006, when the President of Argentina re-launches the nuclear power programme for the country, the Arroyito heavy water plant was re-opening. Until 2009, the plant has produced 340 tons of heavy water of the 600 tons needed to ensure the future operation of Atucha 2.

The goal to be achieved by this strategic plan is to allow nuclear power to be part of an expansion in generating capacity to meet foresee rising demand in the electricity consumption in the country in the future. Meanwhile, a feasibility study for the construction of a new nuclear power reactor will be undertaken with the aim to start construction after 2010, and US$2 billion has been projected for this purpose. In July 2007, Argentina Nuclearelectrica (NASA) signed an agreement with AECL for the construction of two 750 MWe CANDU-6 units and a US$60 million contract with the German company Siemens for the upgrade of the instrumentation and control systems of Atucha 2, which is now under construction. Existing systems have been acquired prior to the stoppage of the works for over 20 years and responded to an old technology and safety requirements. Once renewed with the introduction of new instrumentation and control technologies, Atucha 2 will comply with the latest security standards currently in use in the world.

One of the main objectives to be achieved by the nuclear power programme in the country is the development of the CAREM reactor[159]. In 1984, the CAREM concept, which proposed to build reactor modules with a capacity of 15 MWe, is characterized by a design of safe operation, a simple engineering construction, which allows its operation in countries without an advanced industrial developments. Its power is compatible with existing networks in these types of countries, making it suitable for remote and not interconnected areas, and maximizes manufacturing a number of components by reducing costs and construction times. The CAREM reactor could be used for electricity generation or as a research reactor or for water desalination.

Since the assumption of the new authorities in the CNEA in January 2008, and with the support of relevant national authorities, it was decided to support the construction of a prototype CAREM 25 with a power of 25 MWe. The purpose of the construction of this prototype is to serve as project learning for design and construction of a 300 MW nuclear power reactor based on the same concept. Until now Argentinean authorities has been able to launch the preparation of the project, the integration of its cadres and technicians, as well as developing the necessary procedures to start the construction in the last quarter of 2009.

In accordance with information release in September 2009, the Argentinean government confirmed the future construction of a new nuclear power plant integrated by two CANDU type reactors of 750 MWe capacity each. The main contract would be signed before the end of 2009 with AECL. The plant will be located in the Atucha nuclear power plant site. At the same time, it was informed that the works in Atucha 2 follow the approved programme and it is expected that the connection to the electric grid will be done in 2012. Coincident with the entrance into commercial operation of Atucha 2, the Embalse nuclear power reactor will be stopped for 18 months, to allow the extension of the useful life of the reactor for 30 additional years. Also in the area of Lima, next to the Atucha nuclear power plant site, the construction of the prototype of the CAREM reactor will be carried out (see Figure 141). The Minister of Federal Planning, Julio De Vido, confirmed "that in cooperation with AECL the necessary preliminary works related with the future construction of the new nuclear plant will be carried out in the coming years. The construction of the new nuclear power plant using Canadian technology could cost US$3,200 million. Its generating capacity will be the double of the capacity of Atucha 2 or Embalse. The Senate of the Argentinean Congress approved the proposal presented by the government regarding the expansion of the nuclear power programme by 47 votes in favor and 6 against.

Finally, it is important to note that NASA, CNEA and AECL have signed a new agreement that supplements the one signed in 2006 enlarging its scope. The new agreement foresees the joint development of activities regarding the new advanced CANDU type reactor (ACR1000). The three entities will collaborate in several programmes, among them the extension of the useful life of Embalse nuclear power reactor and the feasibility study for the construction of a new nuclear power reactor in Argentina with CANDU 6 reactor type. The three entities will also collaborate in some initiatives referred to heavy water pressure

---

[159] In Spanish CAREM means Central Argentina de Elementos Modulares. In English means Argentinean Central Modular Elements.

reactors, the production of components for fuel channels and the fuel cycle associated with CANDU type reactors.

THE NUCLEAR POWER PROGRAMME IN BRAZIL

Electricity consumption in Brazil has grown strongly since 1990, reaching a per capita consumption of 2 235 kWh per year. For this reason, Brazil has become the world's tenth largest energy consumer and the third consumer in the Western hemisphere, behind USA and Canada, according to the US Department of Energy. At the same time, Brazil contains the world's sixth largest uranium reserves. Since March 2006, Brazil has been the ninth country in the world to control the full nuclear fuel cycle. Coal has a small share in the Brazilian energy mix. The country has a large reserve of steam coal, but imports metallurgical coal.

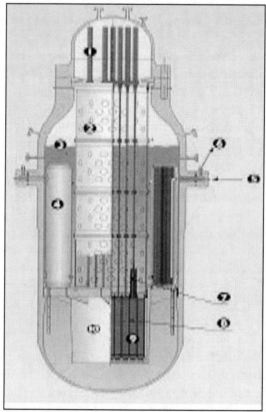

Source: Savoi generators.
1 Control drive 6 SG steam Outlet
2 Control drive 7 Absorbing element rod structure 8 Fuel element
3 Water level 9 Core
4 Steam generator 10 Core support
5 SG water inlet structure

Figure 141. The Carem reactor.

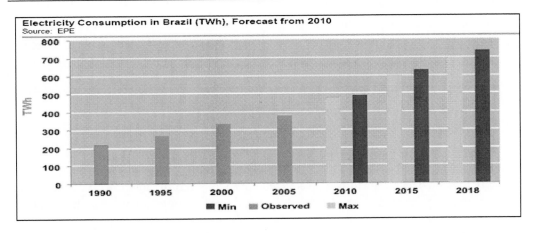

Figure 142. Electricity consumption in Brazil for the period 2010-2018.

Consumption of gas has grown by 100% in the past five years. The considerable availability of natural gas in neighboring countries, the reasonable domestic gas availability, high energy efficiency and relatively low price resulted in this fuel being used widely in Brazilian industry. Although Brazilian proven gas reserves remain modest, probable reserves are promising, principally those on the continental shelf. North Brazil will continue to import natural gas, mainly from Bolivia and possibly Peru. Hydrocarbon exploration activities in South Brazil provide optimistic expectations for new reserves of natural gas off the coast from Espirito Santo to Santa Catarina. [109]

Oil consumption in Brazil has been increasing at a rate of 2.5%. The country has become self-sufficient and, with new exploration possibilities, there is an expectation of large reserves, especially in the pre-salt layer in Santos Basin. [109] In the future, Brazil might become an oil exporter.

It is important to note that over the last ten years growth in electricity consumption in Brazil has been restricted due to the power crisis at the start of the 21th century. In 2005, Brazil had 93.2 MW installed capacity, including the Brazilian share of Itaipu, and has produced 402.9 TWh, 84% of which is hydropower. Growing demand and installed capacity of electricity is shown in Figure 142.

In 2006, gross production of electricity was 419 TWh, an increase of 4%, with net imports of 41 TWh being required. The participation of the different energy sources in the production of electricity in 2006 was the following: 83% of power is produced from hydro, 4% from gas and 3% from nuclear. In 2007, the participation of nuclear energy in the electricity production in the country was 2.8% generating 11.7 billion of kWh. In 2008, the nuclear share 3.12%. The total production of electricity by nuclear energy in 2007 was the following:

Table 86. Annual electrical power production for 2008

Total power production (including nuclear)	Nuclear power production
448 363.5 GWhe	14 003.8 GWhe

Source: IAEA PRIS, 2009.

Summing up can be stated that Brazil will continue to build hydropower plants, complemented by thermal power plants using natural gas, coal and those burning sugarcane bagasse. It will also have higher nuclear capacity and more alternative energy sources. [109]

The Evolution of the Nuclear Power Sector in Brazil

The history of Brazil's nuclear power programmes can be traced back to the early 1930s, when initial research activities in nuclear fission were carried out at the University of São Paulo. By the mid-1930s, Brazil had discovered vast deposits of uranium. In 1940, President Getúlio Vargas signed an agreement with the USA for cooperative mining, including mining for uranium and monazite. During the 1940s, Brazil signed three additional agreements with the USA. In exchange for monazite, the USA transferred nuclear technology. In the early 1950s, President Vargas encouraged the development of an independent national nuclear capabilities in the country in order to use nuclear energy for peaceful purpose. He offered to sell uranium or thorium to the USA in exchange for nuclear technology. Under President Vargas mandate, Brazil sought to purchase three ultracentrifuge systems for uranium enrichment from West Germany. After Vargas's death, Acting President João Café Filho reversed the nationalistic nuclear policy and allowed the USA to control uranium research and extraction for two years. President Juscelino Kubitschek sought to develop indigenous nuclear capabilities by appointing a Congressional Investigating Committee (Comissão Parlamentar de Inquérito, CPI) to examine USA nuclear ties with Brazil. The CPI urged Brazil to adopt an independent nuclear posture. As a result, President Kubitschek, in 1956, created the IPEN (Institute for Energy and Nuclear Research). Kubitschek's successor, President Jânio Quadros, continued the independent nuclear policy, which was based on natural uranium, as did his successor, João Goulart.

In 1970, the Brazil's government decided to seek bids for the construction of its first nuclear power reactor. The turn-key contract for Angra 1 was awarded to Westinghouse, and construction started in 1971 at a coastal site between Rio de Janeiro and Sao Paulo (see Figure 144). In March 1982, the reactor reached it first criticality and in April of this year was connected to the electric grid. The unit entered into commercial operation in January 1985.

Source: Photograph courtesy of Wikipedia (Sturm).

Figure 143. Angra nuclear power plant.

It is important to single out that Angra 1 suffered continuing problems with its steam supply system and was shut down for some time during its first few years of operation. Its lifetime load factor over the first 15 years was only 25%, but since 1999 it has been much better. [75] In 2007, the net load factor of Angra 1 was 68.2%. Angra 2 started to be built in January 1976, was connected to the electric grid in July 2000 an entered into commercial operation in February 2001. Angra 2 has performed well since it entered in commercial operation reaching a net load factor in 2007 of 85.9%. In 2007, the total net load factor of both units was 79.2%.

In 1975, the government adopted a energy policy to become fully self-sufficient in nuclear technology. For the implementation of this policy an agreement was signed with the Federal Republic of Germany for the supply of eight 1 300 MWe units over a period of 15 years. The first two units were to be built immediately, with equipment from Siemens-KWU. The rest were to have 90% Brazilian content under the technology transfer agreement. However, Brazil's economic problems impede the construction of the first two Brazilian-German nuclear power reactors. The construction of the reactors was interrupted and the whole programme was reorganized at the end of the 1980s. [75]

The construction of Angra 2 began in 1976 in the same complex as Angra 1 and with the support from the Federal Republic of Germany but was stopped until 1995. Construction of Angra 2 was resumed in 1995, with US$1.3 billion of new investment provided by German banks, Furnas and Electrobras. The two nuclear power reactors had an estimated useful life of 30 to 40 years.

Source: International Nuclear Safety Center at Argonne National Laboratory, USA.

Figure 144. Site of the Angra nuclear power plant.

Table 87. Operating Brazilian nuclear power reactors

Reactors	Model	Net MWe	First power	Commercial operation
Angra 1	PWR	520	1982	1/1985
Angra 2	PWR	1 275	2000	2/2000
Total (2)		1 896		

Source: CEA, 8th Edition, 2008.

In 1984, the construction of the third nuclear power reactor, Angra 3, was imitated but has not been finished yet[160]. However, according to the Ten-Year Plan for the Expansion of Electrical Energy 2006-2015, Angra-3 is now scheduled to come on stream in December 2012. [68]

The Brazilian Institute of Environment (Ibama) announced the concession of the environmental license to reinitiate construction of Angra 3 within the Angra nuclear power plant site. Angra 3 will have a thermal power capacity of 3 765 MW and an electric power of 1 350 MW. Angra 1, 2 and 3 together will have the capacity to supply the energy that will satisfy all the needs of the Rio de Janeiro city.

The Brazilian Strategic National Plan for the Nuclear Power Sector

The Brazilian nuclear power programme is now transformed into a strategic national plan, instead of just a government project. The expected action plans to be implemented under the strategic national plan are being finalized and its aim is to: a) increase the nuclear energy share of Brazil's electric-generating capacity in the coming years. A study undertaken by the Energy Research Enterprise, part of the Ministry of Mines and Energy, proposed that aside from Angra 3, four to six new 1 000 MW nuclear power reactors should be built by 2030, increasing nuclear generation from 2.8%, in 2007, to 5%-6% of the country's generating capacity; b) ensure self-sufficiency in the production of nuclear fuel, providing 100% of the country's demand. This requires searching for new uranium deposits, expanding both mining activities and the enrichment plant at Resende in São Paulo and enlarging the fuel assembly complex; c) strengthen regulations and create a new regulatory agency separate from CNEN; and d) develop a broad and consistent educational and training programme, joining the universities and the research institutes of CNEN and the Ministry of Science and Technology, to address all areas of the Brazilian nuclear programme, followed by a strategy of hiring highly skilled personnel to face any challenges.

The Brazilian government has plans to add 105 650 MW of electric energy capacity between 2015 and 2030 from diverse sources, including nuclear energy, in order to satisfy the foresee increase in electricity demand in that period.

---

[160] Most of the equipment for the third nuclear power reactor, Angra 3, has been kept in storage to a cost for the Brazilian government of US$20 million annually.

Nuclear Safety Authority and the Licensing Process

The main nuclear legislation is the National Policy on Nuclear Energy of 1962. Amending legislation passed in 1989 and 1999. The Brazilian nuclear regulatory authority is the Directorate of Radiation Protection and Safety (DRS) of CNEN. It is responsible for licensing and supervision of all nuclear facilities. The Brazilian Institute for the Environment is also involved with licensing facilities. The DRS also has a Licensing and Control Authority (SLC), with its major laboratory in Pocos de Caldas, Minas Gerais State. Other regional units are located in Angra dos Reis, Rio de Janeiro State; Caetite, Bahia State; Fortaleza, Ceara State; Goiania, Goias State; and Planalto Central District, Brasilia, Federal District.

According with the Law on Ionizing Radiation Protection and Safety, "a legal entity may start to work only after obtaining a license from the DRS and after being recorded into the unique Register of Legal Entities which carry out activity with sources of ionizing radiation, in accordance with the provisions of this Law (hereinafter: Register). A license referred in paragraph 1 above shall be granted, upon request, for a fixed term period of 1 to 5 years, depending on the type and strength of the radiation source, as well as the requirements of its utilization. Any modification in the conditions for carrying out activity, as established by this law and provisions adopted pursuant to this law, may be made only on the basis of a permission granted by the DRS and after their recording into the Register. The license may be issued only, if: 1) the installations and facilities which are used for activities involving ionizing radiation and in which temporary storage, depositing or disposal is being performed, as well as equipment, instruments and devices generating ionizing radiation shall meet the technical, safety and other conditions; 2) the professionals working with sources of ionizing radiation shall be provided with personal protection equipment, as well as with the required meters for measurement of the radiation intensity; 3) the professionals working with sources of ionizing radiation shall have adequate technical qualifications and shall meet the required health conditions for the work they perform; 4) there is a programme on provision of protection and safety, plan for procedures in case of incidents, as well as programmes for monitoring and evaluation of the exposure of the population, individuals who work on jobs connected with the sources of ionizing radiation and the patients".

According with Articles 9 and 10 of the law, "the Directorate shall not issue a license to a legal entity which applies for a license to carry out activity which does not justify the usage of sources of ionizing radiation from a social, economic, medical or other aspect established by this law. The Legal Entity which works with sources of ionizing radiation, shall cover the costs for the license issued by the Directorate, for depositing of the radioactive waste, as well as for the measurements carried out during the utilization of the sources, in accordance with this law".

A Nuclear Programme Coordination and Protection Commission was established and include representatives from every organization concerned with nuclear issues. The Commission is open to local government and others with relevant interests in nuclear matters.

Nuclear Fuel Cycle and Waste Management

Brazil started uranium mining active exploration in 1970s and 1980s. Brazil has known resources of 231,000 tons of uranium, which represents around 6% of world total reserves.

The three uranium mine sites are Pocos de Caldas, in Minas Gerais State, closed in 1997; Lagoa Real or Caetite, in Bahia State operating since 1999 with 340 tons of uranium per year capacity; and Itataia, in Ceara State operating since 2007.

Brazil has announced its intention to increase uranium production up to 1 360 tons per year by 2012, apparently by expanding Lagoa Real/Caetite mines to 670 tons per year and bringing Itataia mine into production at 680 tons per year. All mined uranium is used domestically, after conversion and enrichment abroad.

It is important to note that an uranium centrifuge enrichment programme started by the Brazilian Navy in the early 1980s, with the purpose to develop a nuclear propulsion programme to be applied for the construction of submarines. A demonstration plant was built at Ipero, and an industrial plant at Resende which will cater for much of the needs of the Angra reactors. The first cascade of the centrifuge enrichment programme commenced operation in 2006 and the second in 2007-2008. Stage 1 with four modules totaling 115,000 SWU per year and costing US$170 million was officially opened in 2006. Each module consists of four or five cascades of 5,000-6,000 SWU per year. The full stage 1 plant is expected to produce 60% of the fuel needs for Angra 1and 2 by 2012. Stage 2 will increase the installed capacity to 200,000 SWU. The centrifuges are domestically-developed and very similar to Urenco technology. Brazil has pledged to enrich uranium to only 3.5% U-235, the concentration required by its two power reactors. This would be too weak to fuel a bomb, which typically requires a concentration of 90% or above. (Liz and Milhollin, 2004)

A fuel fabrication plant designed by Siemens was built at Resende, with a capacity of 160 tons per year pellet production and 280 ton per year fuel assembly.

With respect to waste management the situation in Brazil is the following. CNEN is responsible for management and disposal of radioactive wastes. Legislation, in 2001, provides for repository site selection, construction and operation. Spent nuclear fuel produced by the Angra nuclear power plant is storage in the plant site. Brazil has no policy on reprocessing of the spent nuclear fuel.

Finally, it is important to single out that radical changes of the Brazilian nuclear policy, in the beginning of 1990s, determined the interruption of most research and development fuel cycle activities and the facilities in which these activities were carried out were shutdown.

The Public Opinion

After the Chernobyl disaster, public opposition to Brazil's nuclear power programme increased significantly. Important mass demonstrations against the use of nuclear energy for electricity generation took place in the country. At the same time, the scientific community expressed their opinions that the nuclear power programme adopted by the military government was unnecessary unless the objective was the production of nuclear weapons. For this reason, scientists demanded public debate of Brazil's nuclear policy and insisted that the nuclear industry should not be self-regulatory.

Because of public opposition, financial difficulties, foreign debt and the end of the military regime, of the eight new nuclear power reactors that were initially planned only two were constructed and another one is waiting to be finished. However, the Brazilian government is pushing an expansion of its nuclear power programme by concluding the construction of Angra 3 and the construction of a handful of new nuclear power reactors in the coming years.

Looking Forward

The government of Brazil decided, in 2008, to retake the construction of Angra 3, that were paralyzed during almost 20 years. With the environmental license, the State company Thermonuclear Electrobrás (Eletronuclear) will be able now to plan the resumption of the construction works. It is expected that Angra 3 will entry into commercial operation between 2012 and 2013.

At the same time, Brazil is changing two of the steam generators of Angra 1 at a cost of 560 millions reales[161] with the purpose to increase the efficient of the nuclear power plant. The government is considering building a handful of nuclear power reactors by 2030 in order to increase up to 8GWe of new nuclear power capacity. By 2060, the government hopes to have 60 GWe of nuclear capacity installed. The idea, according to Francisco Rondinelli, Head of Brazil's Nuclear Association, would be "to diversify at least 30% of electricity generation equally into nuclear energy, gas and biomass".

In November 2006, the government announced plans not only to complete Angra 3 but also build four further 1 000 MWe nuclear power reactors from 2015 at a single site. Angra 3 construction approval was confirmed by Brazil's National Energy Policy Council in June 2007 and received Presidential approval in July the same year. Environmental approval was granted in March 2009. The construction time is expected to take 66 months and its first power late 2014. In December 2008, Electronuclear signed an industrial cooperation agreement with Areva, confirming that this French company will complete Angra 3 and be considered for supplying further nuclear power reactors. Areva also signed a services contract for Angra 1.

In July 2008, the Brazilian's government said it expected to license construction of two further nuclear power reactors in the next twelve months, in the Northeast and Centre-South of the country. Electronuclear is looking at the Westinghouse AP1000 design, which is reported to be favored, the Areva-Mitsubishi Atmea-1 and WWER-1000 from Atomstroyexport.

From the economic point of view electricity produced from existing nuclear power plants is about 1.5 times more expensive than that the electricity from established hydropower plants. Electricity energy to be produced by Angra 3 is expected to be slightly over twice as expensive as the electricity produced by old hydropower plants, this is about the same as that from coal and cheaper than that from gas.

In the other hand, president of Electronuclear, Otho Luiz Pinheiro, had announced that "Brazil wants to construct six nuclear power reactors until 2030, the two first in the Northeastern region of the country"[162]. The construction of the first of these six reactors will start in 2017, according with the plan submitted for approval.

Finally, it is important to note that the French group GDF Suez announced the signature, in September 2009, of a nuclear cooperation agreement with the Brazilian companies

---

[161] Reales is the Brazilian national currency.
[162] Two nuclear power reactors were already included in a plan and four more were proposed.

Electrobrás and Electronuclear. The cooperation works will be centered in the technology and the exploitation of nuclear power plants, the selection of sites and the development of human resources, among others. GDF Suez has a long experience in nuclear matters, it possesses and operate seven nuclear power reactors in Belgium and it has a participation in two nuclear power plants in France. The operation of new nuclear power plants are among their objectives, maintaining their presence in Europe.

THE NUCLEAR POWER PROGRAMME IN MEXICO

Mexico is rich in hydrocarbon resources and is a net energy exporter. The country's interest in nuclear energy is rooted in the need to reduce its reliance on these sources of energy, which are the main responsible for climate change. For this reason, Mexico is adopting specific measures to ensure that in the coming years the country will increasingly rely on natural gas, in the use of nuclear energy for the production of electricity and in the use of renewables.

Energy growth was very rapid until late 1990s, but then leveled off for a few years. In 2007, electricity demand is expected to grow again at an average rate of almost 6% a year; around 257 billion kWh was generated in that year. The electricity supply is quite diverse, with gas supplying 126 TWh (49%), oil 52 TWh (20%), coal 32 TWh (12.5%), hydroelectric dams 27 TWh (10.5%) and nuclear energy 9.95 TWh (4.6%) in 2007. In 2008, the nuclear share was 4.04%. Per capita power use is about 1 800 kWh per year. Of a total of 54 GWe capacity installed in the country in 2006, the nuclear capacity was 1.36 GWe (gross), hydro 10.7 GWe, geothermal 960 MWe and the balance fossil fuels.

However, it is important to note the following: Mexico's close relationship to the USA economy means that it too is likely to see a negative impact from the current downturn. About 80% of Mexico's exports are sent to the USA, and in combination with depressed world oil prices and the global credit crunch, its dependence on the US economy has slowed the growth of the Mexican economy. A return to high world oil prices and recovery of the USA economy after 2010 are expected to support a return to Mexico's trend growth, with GDP increasing by an average of 3.4% per year from 2006 to 2030. (IEO, 2009)

Mexico has two nuclear power reactors in operation at Laguna Verde nuclear power plant, with a net capacity of 1 310 MWe. In 2008, the production of electricity using nuclear energy was 9 358.8 GWhe.

The Evolution of the Nuclear Power Sector in Mexico

The first Mexico's interest in the use of nuclear energy for peaceful purpose, including the production of electricity was in 1956, with the establishment of the National Commission for Nuclear Energy (CNEN, Comisión Nacional de Energía Nuclear)[163]. This organization

---

[163] CNEN was later transformed into the National Institute on Nuclear Energy (INEN), which in turn was split, in 1979, into the National Institute of Nuclear Research (ININ), Mexican uranium (URAMEX) and the National Commission on Nuclear Safety and Safeguards (CNSNS). URAMEX's functions were taken over by the Ministry of Energy in 1985.

took general responsibility for all nuclear activities in the country, except the use of radioisotopes and the generation of electricity. The Federal Commission Electricity (CFE, Comisión Federal de Electricidad), one of the two state-owned electricity companies, was assigned the role of introducing a nuclear power programme in the country. Preliminary studies to identify potential sites for the construction of nuclear power plants begun in 1966 and were carried out by CNEN and CFE. In 1969, CFE invited bids for proven nuclear power reactors designs with a capacity of around 600 MWe. In 1972, a decision to build the first nuclear power plant was adopted and, in 1976, construction began at Laguna Verde on two 654 MWe General Electric BWRs.

Laguna Verde Unit 1 (see Figure 145) started its construction in October 1976, was connected to the electric grid in April 1989 and went into commercial operation in July 1990. Its performance has been good since the beginning of the operation of the unit. In 2007, the net load factor of Unit 1 was 84.4%. Laguna Verde Unit 2 started to be constructed in June 1977, was connected to the electric grid in November 1994 and entered into commercial operation in April 1995. In 2007, the net load factor of Unit 2 was 82.6%. The initial cost of the Laguna Verde nuclear power reactor was estimate to be around US$550 million each. However, due to cost overruns, a decade of delays and heavy interest payments on foreign loans have pushed the cost of the project above US$3.5 billion and have contributed to the shelving of plans for the construction of other nuclear power reactors. Because of rapidly increasing costs, the opponents to the construction of the nuclear power plant indicated that each kW of electricity generated by Laguna Verde will cost double that produced in power plants using oil as fuel, information that was not dispute by the government.

It is important to single out that the Mexican industry is not in a position to supply major components for the Laguna Verde nuclear power plant. For this reason, the main components were acquired abroad. Initially, the main architect engineer for Unit 1 was the Electric Bond and Share Company (EBASCO). However, for Unit 2, CFE acted as architect engineer with the advice from EBASCO and General Electric (GE). Mexican companies undertook the civil engineering work and Mexican staff maintains the reactors and is operating the CFE's simulator.

Table 88. Annual electrical power production for 2008

Total power production (including nuclear)	Nuclear power production
231 396.4 GWhe	9 358.8 GWhe

Source: IAEA, PRIS data base.

Table 89. Operating Mexican nuclear power reactors

Reactors	Model	Net MWe	First power
Laguna Verde 1	BWR	655	1989
Laguna Verde 2	BWR	655	1994
Total (2)		1 310	

Source: CEA, 8th Edition, 2008.

Source: Photograph courtesy of Wikipedia (Theanphibian).

Figure 145. Laguna Verde nuclear power plant.

The nuclear power reactors in operation in Mexico are a second generation technology; had an estimated useful life of 15 to 20 years; generate around 4.6 % of the country's electricity and represents 2.74% of the install capacity of the CFE, Mexico's largest energy provider. In 2008, the total production of electricity by nuclear energy is shown in Table 88.

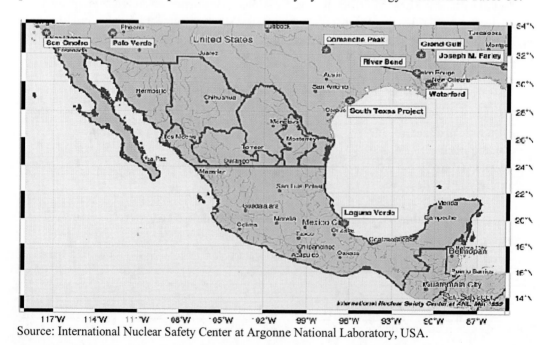

Source: International Nuclear Safety Center at Argonne National Laboratory, USA.

Figure 146. Site of the Laguna Verde nuclear power plant.

The Energy Policy in Mexico

The use of nuclear energy for the production of electricity is a proven alternative in Mexico, as demonstrated by the high availability, reliability and safety indicators at Laguna Verde nuclear power plant. Nuclear power is also a realistic option that complies with environmental requirements that are anticipated to be stricter in the future. However, there are currently no future firm plans to develop new nuclear power reactors in the short term due to the high initial investments required, which at the current time is not competitive with natural gas based power plants. [63] However, a proposal for the construction of two nuclear power reactors in the future was prepared by the national competent authorities for government approval.

Nuclear Safety Authority and the Licensing Process

The licensing process in force in Mexico consists of two steps. The first step starts with the application to build a nuclear power reactor by the utility presenting to the national regulatory authority (National Commission on Nuclear Safety and Safeguards) the application itself and the preliminary studies of sitting, environmental impact and quality assurance programme and concludes with the granting of the construction permit. The second step consist with the license for commercial operation.

If the above mentioned documents satisfy the CNSNS requirements, then the utility is required to present technical information on the nuclear power reactor to be built and the technical information regarding the type of reactor to be used. This information includes the construction procedures and fundamental safety systems to cope with operational transitions and postulated accidents. This information is evaluated by the CNSNS's technical personnel and a set of questions is transmitted to the utility, before the pouring of any concrete at the site. During the Laguna Verde experience, three provisional construction permits were granted to the CFE before the so-called "Definitive Construction Permit" was issued. [63] However, this limited work authorization has been eliminated in the procedure for the construction of future units.

During the construction of a nuclear power plant, the regulatory authority can inspects the construction works and has the legal authority to stop any construction activity if the agreed standards are not being fulfilled. After the evaluation of the documentation, the regulatory authority can issue the "Technical Basis" to grant the construction permit, addressed by the Ministry of Energy, the authority responsible to grant a construction permit under the current nuclear law.

The essential legal texts regulating nuclear power in the country are: a) Act on Nuclear Activities of 1984, which entered into force on 5 February 1985; b) Act on Third Party Liability for Nuclear Damage, published in the official gazette on 31 December 1974; c) General Radiological Safety Regulations, published in the official gazette on 8 November 1988; d) General Act on Ecological Equilibrium and Environmental Protection, published in the official gazette on 28 January 1988; and e) Mexican Official Guidelines NOM-012-STPS-1993 on Health and Safety at Work in Premises where Ionizing Sources are Handled, Stored or Carried, published in the official gazette on 15 June 1994.

Nuclear Fuel Cycle and Waste Management

The Ministry of Energy is responsible for nuclear fuel cycle policy and operations and can, by law, delegate some of these responsibilities to public entities such as CFE and the ININ. CFE has been authorized by the Ministry of Energy to negotiate uranium stock purchases, uranium enrichment and fuel fabrication contracts.

An interim waste repository managed by ININ collects all low-and intermediate-level radioactive waste produced by medical, industrial and other radioisotope applications. This repository will be replaced by a permanent one in the future. Another interim low and intermediate-level radioactive waste repository is operated by the Laguna Verde nuclear power plant to handle wastes coming from the plant operation. Spent nuclear fuel from the Laguna Verde nuclear power plant is being stored in the reactor's pools, which have been re-racked to increase the original capacity in order to accommodate all the spent nuclear fuel that the reactors will produce during their expected lifetime. [63] This solution gives CFE the time needed to study all possibilities before adopting a definitive solution, depending on future developments regarding the final disposal of high-level radioactive waste.

Mexico does not produce uranium due to the low cost of uranium currently available on the world market. For the next several years, the required uranium for reloads of Laguna Verde reactor will be obtained from the world market, as there are no plans for producing uranium locally. Some 2 000 tons of uranium reserves have been identified in Mexico, but they are too expensive to exploit at current world market prices.

Enrichment is provided by the US Department of Energy through a long-term contract. Fuel fabrication is currently done in the USA by GE. Four assemblies supplied by Siemens are being tested in the fourth cycle of Unit 1 of Laguna Verde nuclear power plant and there are plans to test four assemblies supplied by Asea-Brown Boveri (ABB-ATOM) in the near future also.

A fuel fabrication pilot plant is almost ready to start operation at the National Nuclear Research Institute using technology provided by GE. This pilot plant could produce up to 20 fuel assemblies per year for the Laguna Verde nuclear power reactors. However, it is important to stress that after some experience is gained with the operation of the plant and the fuel produced, the plant will probably be shut down since it is not economical to fabricate nuclear fuel on this scale.

A repository exists in Mexico for all the low and intermediate-level waste produced in both medical and industrial facilities. This repository will be closed in the near future to avoid social problems due to population growth in the vicinity of the facility. For the Laguna Verde nuclear power plant, the high-level waste is being stored at the plant site. As for the low and intermediate-level waste produced by the nuclear power plant, detailed site studies are now under way at the plant site in order to determine the engineering design basis for a "triple barrier" repository using the French approach. The repository is planned to have the capacity for the waste generated during the operating life of at least four nuclear reactor units, and could also include the waste generated by the medical and industrial facilities in the country. [63]

The Public Opinion

The construction of the Laguna Verde nuclear power plant and the use of this type of energy for electricity generation in Mexico met with severe political and public opposition, as well as financial and technical difficulties, since the construction works of the first nuclear power reactor started in 1976. It is a facility that many Mexicans, especially those in the State of Veracruz, do not want. Effort was made to make their displeasure known, even before the construction of the Laguna Verde nuclear power plant begins. People begun to march in the streets, organized voluntary power blackouts, scribble protests on the backs of electrical bills - anything they think will persuade the Mexican government to abandon the Laguna Verde project. Political opposition and the mass demonstrations against the use of nuclear energy for electricity generation that took place in 1986, force the government of Mexico to reduce to two nuclear power reactors, the initial plan for the construction of 20 units.

The safety operation of the Laguna Verde nuclear power plant was a matter of concern for different groups, which were against the construction of this plant. One of the opinions expressed by these groups was based on the resumed outdated technology whose safety had already been questioned and a volcano five miles away in an active volcanic zone raised additional safety issues. However, an engineer who has been director of the project for the last four years, Eng. Rafael Fernández de la Garza, dismisses such arguments as "gigantic myths". He said that "the Laguna Verde site was chosen precisely because it is one of the most seismically stable locations in Mexico and that we are too close to the sea for lava from any volcano to come here".

Opponents and supporters of the use of nuclear energy for electricity generation agree that the debate in Mexico City, 175 miles downwind from the nuclear power plant, has greatly stepped up after the Chernobyl nuclear disaster, which was the detonator for the emergence of a mass anti-nuclear movement in Mexico. "The majority of people are very frightened", Mr. Fernandez said. "No matter how we try to explain to them how Laguna Verde differs from Chernobyl nuclear power reactors and an accident of this type could not occur here, people remain terrified". Before Chernobyl nuclear accident there was really no organized opposition to the construction of the nuclear power plant in Laguna Verde, but Mexican authorities had a very difficult time to convince people that despite Chernobyl nuclear accident, nuclear energy remains one of the most efficient, secure and clean sources of electrical energy.

In October-November 1999, WANO inspected the nuclear power plant and evaluated activities in 72 different areas. All but nine received critical commentaries from the association. In January 2000, a series of numerous minor accidents at the nuclear power plant were reported to the CFE, indicating that the facility's accident simulator does not work properly, the staff lack adequate training and some of the equipment is obsolete.

According with the outcome of the survey carried out in 2005 by GlobeScan Incorporated for the IAEA, 32% of the Mexican participants support the construction of new power reactors, 28% are of the opinion that the country should use the current nuclear capacity for the generation of electricity but there are not supporting the construction of new units, while 23% are in favor to close all nuclear power reactors operating in the country.

Looking Forward

According with the Mexican competent authorities, there are until now no definite plans to expand the use of nuclear energy for the electricity production in Mexico, but for upgrading of the existing units in order to increase their output by around 10%. Several officials, including the Energy Secretary and the president of the national utility, had publicly called for an open discussion to re-initiate the nuclear power programme.

In March 2005, CFE was reported to be considering the construction of some 10 000 MWe of new nuclear capacity. The possible expansion of nuclear power capacity could include additional units at Laguna Verde nuclear power plant site as well as installations at two new sites. A year later, the CFE Director-General announced that the commission aimed to construct a new nuclear power plant by 2020 at the latest, to help meet rising electricity demand. He also stated that CFE was spending US$150 million on raising the capacity of the two Laguna Verde nuclear power reactors. A major retrofit project for Laguna Verde was announced in March 2007 when completed in 2010, the capacity of each unit will have been increased by 20% to about 785 MWe.

In February 2007, CFE signed contracts with Spain's Iberdrola and also Alstom with the aim to fit new turbines and generators to the Laguna Verde nuclear power plant at a cost of US$605 million. With the introduction of these turbines and generators the plant will produce 20% more power - about 280 MWe. With the approval from the CNSNS, the reactors could be uprated progressively from 2008 to 2010. Meanwhile, 11.6 MWe uprates to both units were achieved in 2007 through better flow control.

There should be no doubt that the Mexican's government support an expansion of its nuclear power programme, primarily to reduce dependence on natural gas. The most recent proposal is for the construction of two nuclear power reactors to come on line by 2015 with seven more to follow it by 2025. This later proposal has not been approved yet by the Mexican authorities. The purpose of the Mexican's government is to bring nuclear share of electricity up to 12% in the coming years. Cost studies show that the production of electricity using nuclear energy is competitive with gas, if the production cost is about US$4 cents/kWh in all scenarios considered.

In the longer term, Mexico may consider the construction of small nuclear power reactors such as IRIS to provide power and desalinate sea water for agricultural use. ININ have previously presented plans for the construction of a nuclear power plant consisting of three IRIS nuclear power reactors sharing a stream of sea water for cooling and desalination. With seven desalination units, 140,000m^3 of potable water could be produced each day, as well 840 MWe.

THE HISTORY OF THE NUCLEAR POWER PROGRAMME IN CUBA

Cuba depends heavily upon imported oil in order to satisfy around 50% of its total oil needs. The importation of oil has drained Cuba of its sparse hard currency. It is important to note that the country's production of electricity has been fraught with difficulties forcing power plants to work under their full capacity, leading to frequent blackouts. This figure has fallen further in the 2000s due to the relative decline in the Cuban economy since 1990s.

Table 89. Public service plants in Cuba and their generation in MW

| | | | Thermal power plants | Gas power plants[b] | Generators | | Hydro-electric | Other Thermal generation c) |
					Diesel Plants	New Technology		
Year	Total	Total [a]						
2000	4 286,5	3 436,6	3 064,5	240,0	74,7	-	57,4	849,9
2001	4 410,9	3 506,5	3 161,4	213,0	74,7	-	57,4	904,4
2002	3 959,6	3 387,2	3 040,0	213,0	76,8	-	57,4	571,9
2003	3 965,0	3 302,2	2 880,0	288,0	76,8	-	57,4	662,3
2004	3.763,5	3 303,2	2 880,0	288,0	76,7	-	58,5	459,8
2005	4.275,1	3 597,5	2 940,0	315,0	65,2	229,1	48,2	677,1
2006	5.176,0	4 730,0	2 940,0	405,0	68,6	1 268,2	48,2	445,5
2007	5.429,4	4 882,9	2 901,4	426,7	70,2	1 443,7	40,9	546,5
2008	5.388,9	4.849,9	2 298,0	455,0	88,5	1 940,8	60,1	546,5

Source: National statistic office, 2008.

Table 90. Comparative data on energy in Cuba

	1958	2007
Population (million of inhabitants)	5.6	11.4
Power installed (MW)	397	5, 861
Access to electricity (%)	56	96
Gross electricity generation (GWh)	2 550	16 694
Specific fuel consumption (g/kWh)	399	274
Total of consumers	772 000	3 200 00
Electricity consumption (kWh)	377	1 486

Source: Arrastía Avila (2008).

The development of the energy sector after 1959 can be divided into three distinct periods. The first one cover three decades between 1959 and 1989. This period is characterized by a rapid growth in Cuba's energy sector, facilitated by subsidized Soviet oil imports and other forms of financial support. The period included the country's largest buildup in energy generation infrastructure and highest rates of growth in energy consumption, based on oil and products imported from the former Soviet Union at highly subsidized prices. The second one is the so called "special period" during the 1990s. This period is characterized for an accelerated domestic oil production with the purpose to increase domestic production to compensate the drastic reduction of oil supply from Russia. Cuba began to use fuel oil produced by the country in the seven large generation power plants due to the lack of appropriate oil supply. The third period cover 1998 until today. This period is characterized by the supply of around 90,000- 100,000 barrels of oil per day from Venezuelan in change of Cuban support of different social programmes promoted by the Venezuelan government, the blackouts of 2004-2005, the Energetic Revolution of 2005-2006, and the Independent Power Production (IPP) arrangement with Sherritt, based on combined cycle gas turbines (CCGT). While the sector is significantly more stable than during the period of blackouts, after 2004 frequent interruptions in old oil-fired power stations, which only guaranteed an average availability of 60%, worsened with the impact of hurricanes on the high-voltage transmission lines and, for this reason, severe restrictions in the consumption of energy was adopted by the Cuban government in 2009 in order to avoid short-cuts of electricity to the population. The energy restrictions adopted affect significantly several economic sectors of the country.

At the same time, it is important to note that most of the new power installations in the country are emergency generators and motors that burn fossil fuels, both diesel and fuel oil. These technologies have had a positive impact on the environment because they have lower specific consumption rates (234 g/kWh) compared to large oil-fired power plants (284 g/kWh on average). Cuba has a generating capacity of 2 497 MW based on distributed generation – 1 280 MW corresponds to diesel generators and the rest are fuel oil motors (540 MW), CHP (529 MW) and renewable technologies (69 MW). The country also has a reserve of more than 6,000 small diesel generators installed in key centers of the economy and services to the population such as bakeries, shopping centers, hospitals, clinics, food production sites and schools, among others. The combined power of all these generators reaches the figure of 690 MW and the purpose is to interconnect them to the National Electrical Grid. (Arrastía Avila, 2008)

Cuba began to show an interest in nuclear energy before the 1959 Revolution. Its first nuclear accord was struck in June 1956. The agreement envisaged close cooperation with the USA in civilian uses of nuclear energy. The initiative for the agreement came from the USA, which was making an aggressive attempt to capture the world nuclear market under the "Atoms for Peace" policy. Under the terms of this agreement, a small nuclear power plant of 40 MW was to be built jointly by American Machine Foundry and Mitchell Engineering (Great Britain) in Pinar del Rio Province. However, the plan was canceled in the middle of 1960S after the break in diplomatic relation between Cuba and the USA. (Benjamin Alvarado, 1994)

The new Cuban's government inherited this interest in nuclear energy, particularly for the production of electricity, and decided to seek support from the former Soviet Union to introduce a nuclear power programme in the country. The first sign of links between the Cuban and the former Soviet governments in the nuclear field appeared in January 1967, when a Soviet-sponsored photo exposition, "Atomic Energy for Peaceful Purposes", was held at the Academy of Sciences in Havana. Nine months later a public announced was made almost simultaneously that the countries had reached an agreement on cooperation in the peaceful uses of atomic energy. Under the accord, the Soviet's government will provide Cuba with a research reactor for experimental and teaching purposes, as well as hardware to set up physics and radiochemical laboratories. However, the next major agreement came in April 1976 when Cuba and the former Soviet Union signed an agreement to construct two 440 MW nuclear power reactors in the South central Province of Cienfuegos and planned to build two more units in this site to complete four nuclear power reactors. Juragua's nuclear power reactors are the newest model known as the WWER-440 anti-seismic V/318 of Soviet design. These are the first Soviet designed reactors to be built in the Western Hemisphere in a tropical environment. Upon conclusion, the first reactor, Juragua 1, would generate approximately 15% of Cuba's energy demands.

In addition to the Juragua nuclear power plant, the initial plan for the use of nuclear energy for electricity generation in Cuba, contemplated the construction of two other nuclear power plants, one in the western part of the island in the Pinar del Rio Province and the second in the Eastern part in the Holguin Province. In total, 12 nuclear power reactors with around 4 800 MWe nuclear capacity were planned to be built in these three sites. The nuclear capacity planned to be built represented, at that time, the double of the energy capacity of Cuba.

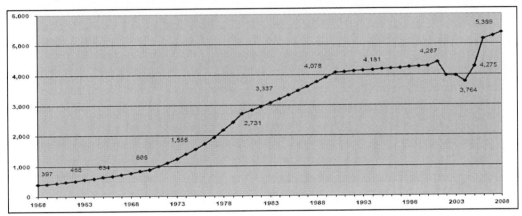

Source: Belt (2009).

Figure 147. Installed generation capacity MW during the period 1958 -2008.

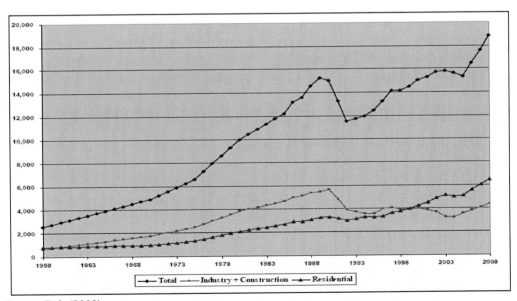

Source: Belt (2009).

Figure 148. Electricity consumption per key sector during the period 1958-2008 GWh.

It is important to single out that Cuban's officials expected that when the Juragua's nuclear power plant entry into commercial operation it will safe around 600,000 ton of oil annually per unit. When all units entered into commercial operation the Juragua nuclear power plant will save 2.4 million of ton of oil annually. However, the collapse of the former Soviet Union in the 1990s halted not only the construction at Juragua site, but also discourages the construction of the other two initially planned nuclear power plants.

The construction of the first nuclear power reactor of the Juragua's nuclear power plant began in 1983. The former Soviet Union supplied a majority of the nuclear reactor components, dispatched technicians to supervise the construction of the plant and trained Cuban engineers designated to operate the nuclear power reactor. Tentatively, the first unit of the Juragua's nuclear power plant was scheduled to be operational in late 1995 or early 1996. However, after the collapse of the former Soviet Union early in the 1990s, the construction work at the Juaragua's nuclear power plant site stop. The first attempt to resume the construction of the Juragua's nuclear power plant was made in April of 1992, when Russia agreed to complete construction at Juragua site, if the Cubans would begin repayment of Soviet loans for the construction of the plant and assume responsibility for all additional hard currency spending for the Juragua project. (Benjamin Alvarado, 1994)

In September 5, 1992, the President of Cuba announced that "Cuba was temporarily suspending construction of the nuclear power reactor at Juragua. In announcing his bitter and painful decision, the Cuban president blamed the Russians for demanding US$200 million for continued work on the project. Yet, on November 4, 1992, Russian and Cuban officials announced that they had agreed to resume the long-stalled construction of Cuba's first nuclear power plant with French assistance. Although this agreement called for resuming construction at the Juragua site, no concrete plans have ever been detailed. At the end of 1992, the civil construction of the first unit ranged from 90% to 97% complete with only 37% of the equipment installed. About 20% to 30% of the civil construction on the second unit was completed.

A second attempts to resume construction at Juragua site took place in October 1995. A high-level Russian delegation, with full backing of the government, arrives in Havana to conclude an agreement with the Cuban authorities to complete the construction of at least the first two nuclear power reactors. To raise the estimated US$800 million necessary to complete the reactors, Russia and Cuba decided to form a joint venture, open to the participation of third parties. Companies in Britain, Brazil, Italy, Germany and Russia expressed interest in an economic association. In order to keep the nuclear option alive and despite the difficulties to find a third partner to finance the conclusion of part of the Juragua's project, in September 1993, Moscow advanced US$30 million to Cuba to mothball the Juragua project in the Cienfuegos Province. (Benjamin Alvarado,1994) Russia provided Cuba this loan for maintenance and other support work at Juragua's nuclear power plant site.

It is important to note that although the decisions by the Cuban's government to suspend and then resume construction were unexpected, a closer look into the matter reveals that these were necessary and rational decisions to adopt. At the same time, it is obvious that Russia at that time has a vested long-term economic interest in the successful and safe completion of the Juaragua's nuclear power plant, because it is going to be the first Russia nuclear power reactor to be built in the Western Hemisphere and its success will likely increase the marketability of the Russian reactors. Alexander Nechayev, of the Russian engineering firm

Zarubezhatomenergostroy, stated that "under the present conditions, where the first reactor is almost 90% complete, it would be extremely unreasonable to end the construction".

Many efforts were made by the Cuban's government to convince other countries, as well as private or State companies, to provide the necessary resources to allow the Russian and Cuban counterparts the conclusion of at least the first two units. However, this association failed in its efforts forcing the Cuban's government to abandon definitively work at Juragua site in 1997, as well as the use of nuclear energy for electricity generation in the future.

It is important to stress that the IAEA provided, during the period 1991 to 1996, about US$680,000 to Cuba to develop the ability to conduct a safety assessment of Juragua's nuclear power reactors and in preserving or mothballing the reactors while construction is suspended. The IAEAs assistance stopped in 1997 when the government decided to abandon definitively its intention to use nuclear energy for electricity generation.

The construction of a nuclear power plant in Cuba has drawn staunch opposition from the USA for political and also due to possible safety problems. A 1992 GAO report addressing the current status of Juragua's nuclear power plant concluded that, "if the reactors were completed, the possibility of an accident was likely". The debate over safety at Juragua's nuclear power plant raises the possibility of a "Cuban Chernobyl" laying a mere 90 miles off the coast of Florida. Critics contend that if a serious or major accident were to take place, large amounts of nuclear fuel could spew into the atmosphere and surrounding waters. The radioactive fallout would create a "dead zone" with an 18-mile radius where nothing could survive; a 200-mile radius where there would be serious health risks and food production would be impossible; pockets of high contamination that could drift as far as 300-miles away, and a radioactive cloud creating serious ecological damage as far north as Tampa, Florida, with secondary fallout that the welding pipes for the cooling system were weakened by air pockets, bad soldering and heat damage. Russian officials responsible for construction and quality control defensively and flatly deny that Juragua's safety is a legitimate concern. They point to Finland's Loviisa Soviet-designed WWER nuclear power reactor as evidence of a safely-operating Soviet designed nuclear power facility. (Benjamin Alvarado, 1994) The Cuban's government also rejected the possibility of a nuclear accident like the Chernobyl one due to the different type of nuclear power reactor to be used in the Juragua nuclear power plant.

The US NRC Director of Operations, James Taylor, in testimony before the Senate Energy and Environment Subcommittee on Nuclear Regulation, told the panel "that senior NRC staffer Harold Denton had visited the site in 1989. He said that construction work observed then included the containment liner, turbine generator building and other components. No primary system components had been installed at the time of the visit". Taylor further stated that "NRC experts had concluded that the Cubans were knowledgeable in their respective areas of expertise and that the technical issues raised by critics could potentially be safety concerns". But he added that "since analysts have no way to evaluate the validity of the claims made by defectors and lack specific inspection knowledge, it is somewhat difficult to assert that they are safety concerns

As a result, the USA adopted a policy that opposes the completion of both nuclear power reactors, and discourages other countries from providing assistance except for safety purposes to Cuba's nuclear programme. This opposition stopped all efforts carried out by Cuba and Russia to find a third party ready to support the conclusion of the work at Juragua site. At the same time, the US Energy Department has refused to include Cuba in its US$180 million nuclear safety programme established with Russia and former Soviet Bloc States in Eastern Europe, totaling 59 nuclear power reactors, with the objectives to improve the safety of Soviet design nuclear power reactors operating in these countries.

THE POSSIBLE INTRODUCTION OF A NUCLEAR POWER PROGRAMME IN CHILE IN THE FUTURE

Chile, one of the strongest economies in Latin America and one of the healthiest among emerging countries worldwide, is strongly dependent on its neighbors for its energy needs. Copper mining, its main economic activity, is extremely energy-sensitive. Chile currently depends on Argentina for natural gas imports, but its Eastern neighbor is struggling to supply its domestic market and this situation is limiting the export of gas to Chile. Gas-rich Bolivia, which plans to quadruple its gas output to Argentina in the next two-and-a-half years, refuses to sell gas to Chile because of a dispute over Bolivia's sea access and due to its limited export capacity. Chile currently produces 7 500 MW of electricity a year, but studies show that the country growing economy will need an additional 5 000 MW within ten years to cover the needs of both households and the national mining sector.

Taking into account the above situation, the announcement made by Argentina that its natural gas supplies to Chile might be affected in the future[164] and the foresee increase in the electricity demand in the coming years, the government of Chile has decided to carry out preliminary studies related with the diversification of energy sources and to a possible introduction of a nuclear power programme in the future.

The introduction of a nuclear power programme will be controversial. Some parties coalition has strongly backed the use of nuclear energy for electricity production, while others has been more hesitant, promoting the construction of large damns in the Southern region as an alternative. However, while Chilean politicians fight over what kind of energy to back, leading Chilean business conglomerations, expressed interest in making long-term investments in the use nuclear energy for the production of electricity.

Chile's President said recently that "her government was still studying its nuclear energy development options, but it will be for the next government to make a decision as studies could take up to eight years. I will do all the work; the next government will have all the studies needed". Based on the results of these studies, the government should adopt a decision whether to use nuclear energy for electricity generation in the future or use other type of energy with the same purpose.

Why Chile is now interesting in the introduction of nuclear energy for electricity generation? Chile currently imports 72% of its energy needs in the form of oil, gas and carbon

---

[164] In 2004, Argentina's energy ministry promulgated Resolution No. 659/2004, which gave it the authority to grant natural gas supply privileges to domestic consumption rather than to exports. This was the first in a series of restrictions that affected natural gas shipments to Chile. One of the most severe such restrictions came on August 5, 2005, when the Argentine government imposed rationing that was the equivalent of 59% of Argentina's total shipments to Chile. On May 17, 2007, those restrictions reached a high point of 64% of total shipments. That means the supply of natural gas shrank by 14.1 million cubic meters at a time when Chile's daily imports were 22 million meters, according to the CNE. For this reason, Chile's electrical power plants are obliged to operate with diesel power in order to counteract the effects of a reduced supply of gas.

all of them high contaminant of the atmosphere that could increase the level of contamination of the city of Santiago de Chile, the capital of the country. This dependency puts the country in a vulnerable position given the volatility of international prices and supply interruptions, according to the National Energy Commission.

Although the use of nuclear power plants are, in the specific case of Chile, more expensive than the use of conventional energy power plants for the production of electricity, the nuclear option offers greater security in energy supply. Nuclear energy is not subject to market swings, as are coal and oil, and is not vulnerable to climactic changes, unlike hydroelectric sources. When considering the nuclear option, the Chilean's authorities should think about this situation, as well as the location of the country in an active earthquake zone of the region, before a final decision is adopted regarding the use or not nuclear energy for electricity generation.

To support its position, Chile asked the IAEA to provide all necessary advice and information on the possible use of nuclear energy for electricity production. According with the advice provided by the IAEA, it takes sometimes eight years to make a good decision, specifically in a country that has as many earthquakes as Chile. If electricity system planning includes the nuclear power option in a realistic manner, then the country should have an adequate total installed power capacity, in an integrated electric grid of over 7–10 times the unit size of the nuclear power reactor to be installed. Electricity grid size, grid mix, the extent of interconnection and the need for strengthening the electric grid stability, are important considerations in the assessment of whether to use or not nuclear energy for electricity generation. It is expected that a decision on this important matter for the economic development of Chile will not be taken before 2013. All arguments in favor or against the use of nuclear energy for electricity production are being discussed. Viability studies are also carried out on the use of solar, biomass and wind-power energy for electricity generation.

It is important to single out that, in addition to the studies to introduce a nuclear power programme in Chile, the Russian Atomic Agency Rosatom presented a proposal to the Chilean's government to transform the nuclear research reactor located in the Aguirre Nuclear Research Centre into a nuclear power plant for the production of electricity with a capacity of 10 MW. Rosatom has asked the University of Santiago de Chile to make the initial analysis of the research reactor. "They have requested us to make a diagnosis on the present conditions of the research reactor to know exactly in what conditions this reactor are, and we will analyze the outcome of this diagnosis with the present Executive Director of the Chilean Nuclear Energy Commission, Fernando Lopez, with whom we have been analyzing the possibility of transforming that reactor in order to present a proposal to the government". Based on this diagnosis the government of Chile will take a decision whether to support this proposal or not.

The establishment of a national consultative process that includes all interested parties, is one of the eight recommendations of a study on the regulatory framework and international experience in the use of nuclear energy for electricity generation carried out by the Radiological and Nuclear Security Agency of Finland (STUK), entity that identify the basic steps for the beginning of a successful introduction of a nuclear power programme in Chile, including the necessity of a development of all laws and regulations related with the use of this type of energy and the creation of an independent regulatory agency.

The study was proposed, in December 2008, to STUK by the National Energy Commission of Chile (CNE). The outcome of the study was included in a report that was presented in March of 2009 and revealed by the CNE in the second half of 2009. The report suggested the establishment of a national consultative process that includes parliamentarian, politicians, energy experts and the general public, in order to have the necessary transparency and the public trust, because the public acceptance on the nuclear option is an indispensable requirement to the future use of nuclear energy for the production of electricity. The report suggested the following recommendations to the considered by the government before a final decision is adopted related with the introduction of a nuclear power programme in the country in the future.

1) taking into account the world development of nuclear activities since the 1970s and the adoption of laws and regulations to regulate the use of nuclear energy in different countries, the work of the Chilean Nuclear Energy Commission, the Ministry of Health, the CNE and the Ministry of Energy, and the normative already in force in Chile to evaluate the environmental impact of the use of nuclear energy for the production of electricity, the STUK report concludes that "this base guarantees the preparation of the country for the introduction and development of the nuclear option in the future";

2) the beginning of a commercial nuclear power programme requires the development of the necessary specific laws and regulations to support the licensing process for the design, construction and operation of nuclear power plants;

3) the current regulatory system in place in the country is not enough to regulate the commercial operation of a nuclear power plant. It requires the establishment of an independent regulatory authority based on the Convention on Nuclear Safety and the adoption of a licensing process to grant the necessary authorization for the commercial operation of a nuclear power plant;

4) Chile needs to prepare the staff that will be in charge of all regulatory activities and clearly define its competence. It is important that the Chilean's authorities provide; to the national competent authorities; the corresponding computer programmes for carrying out the indispensable security analysis, with the purpose of supporting the work of the regulatory authorities; to establish an appropriate education system and training of the staff that will be involved in the introduction and operation of a nuclear power plant; and to begin a research programme relate with the management of the nuclear waste at national level;

5) the Commission of Environment should carry out the evaluation of the environmental impact of the introduction of a nuclear power programme in the country. This evaluation is necessary for the selection of the site and to provide the detailed knowledge on environmental conditions in the country before the nuclear power reactor entry into commercial operation;

6) it is necessary to carry out a feasibility study to define the nuclear power reactor design to be used as well as other conditions for the commercial operation of a nuclear power plant. For a successful nuclear power project without big surprises or delays and with the purpose of predicting the costs, it is necessary to select a proven nuclear power reactor design and a reliable supplier of the reactor;

7) the establishment of a national consultative process, including all interesting parties and the public;

8) to complete the evaluation of the Chilean nuclear infrastructure, in accordance with the IAEA recommendations, and to prepare an independent revision of this evaluation.

Without any doubt, the decision to introduce a nuclear power programme in the country in the future will depends of the outcome of the current studies that are underway in Chile, the position of politicians and businessmen and the public opinion on the use of nuclear energy for electricity generation.

THE POSSIBLE INTRODUCTION OF A NUCLEAR POWER PROGRAMME IN VENEZUELA IN THE FUTURE

Venezuela is a country sitting on top of some of the world's largest oil and gas reserves and with a large hydropower system. Venezuela is the Latin American country with the largest quantity of oil (77 billion barrels) and natural gas reserves (4.6 trillion m^3). After Mexico, Venezuela is also the main oil producer (2.24 million barrels a day in 2009)[165]. At the current rate of production the confirmed Venezuelan oil reserves will satisfy its energy needs for more than 68 years. In the case of gas, the reserve/ production ratio will satisfy the current country energy needs for more than 150 years. In 2007, Venezuela had 22.22 GWe of installed capacity. From this capacity installed around 61% correspond to hydropower and approximately 39% thermal.

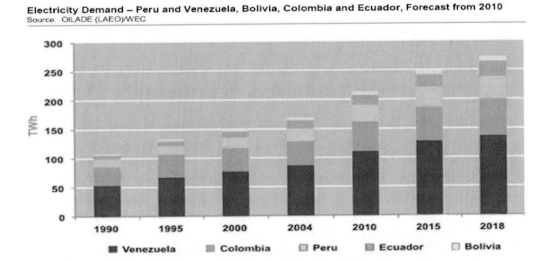

Figure 149. Venezuela electricity demand and the forecast for the period 2010-2016.

---

[165] In 2005, the oil production was 3.1.million of barrel per day.

The Venezuela electricity demand and the forecast for the period 2010-2016 is shown in Figure 149. However, and despite of the Venezuela's fossil fuels reserves, the government announced on May 21, 2007, that it would start to work on the research and construction of nuclear power reactors for electricity production in the future, in order to diversify energy sources and a possible solution to global warming.

During Russian President Dmitry Medvedev's visit to Caracas in late November 2008, Venezuelan Minister of Energy and Petroleum and the President of Rosatom signed a bilateral agreement on nuclear cooperation between these two countries. According to a Rosatom press release, "the agreement established a framework for the development of joint research in the field of controlled nuclear fusion, for the design, development, manufacture and use of nuclear research reactors for the production of radioisotopes for use in industry, medicine and agriculture and for the construction of nuclear power plants, for help Venezuela in the development of the infrastructure and legislative framework for the development of the peaceful use of nuclear energy in different economics sectors and for possible exploration and development of Venezuela's uranium and thorium deposits". The agreement specifies "any nuclear equipment and know-how supplied by Russia will not be used by Venezuela to produce nuclear weapons or other nuclear explosive devices, nor to achieve any military objectives and will be under the IAEA safeguards system"[166]. According to the Rosatom statement, "the agreement will not involve the transfer of any know-how or systems for chemical reprocessing of irradiated fuel, isotope enrichment of uranium or production of heavy water, its main components or any objects produced from them, nor uranium-enriched to 20% or above".

It is important to note that Venezuela has no nuclear infrastructure and few specialists in nuclear matters outside of the Venezuelan Institute of Scientific Research (IVIC). The country has one known research reactor, the RV-1 (3 MW), which was purchased from the US General Electric Company in 1956 and, subsequently, shut down in 1994. Past technical cooperation in the nuclear field between Venezuela and the IAEA can be considered as minimal. In 2007, the IAEA approved a project in Venezuela for human resource development and for general assistance in the uses of nuclear science and technology.

The introduction of nuclear energy for electricity generation in Venezuela is not an easy option and is not exempt of difficulties. These difficulties are, among others, the following:

1) the country has little experience in the use of nuclear energy for peaceful purpose and has no experience in the use of this type of energy for electricity production;

2) there is an insufficient number of professionals and technicians trained and with experience in the nuclear field and with the necessary knowledge to ensure the successful introduction of a nuclear energy programme in the near future;

3) the country has a vast oil and gas reserves and a great experience in the use of these fossil fuels for electricity generation and this situation should have to be taken into account to define which could be the scope of the nuclear power programme within the energy balance of the country;

---

[166] It is important to note that Venezuela is a state party of the NPT and has signed an IAEA safeguard agreement, but until now has not signed the IAEA Additional Protocol, which would give the Agency broader inspection powers in the country.

4) the participation of hydro energy in the energy balance of the country is very high, which make more expense the use of nuclear energy for electricity generation;

5) the possible public opposition to the use of nuclear energy for electricity generation, taking into account some of the points mentioned above.

THE POSSIBLE INTRODUCTION OF A NUCLEAR POWER PROGRAMME IN URUGUAY IN THE FUTURE

Uruguay has no oil and gas reserves and the difference between energy installed capacity and the maximum energy demand is very small. Taking into account this particular energy situation the President of Uruguay meet with the leaders of the political parties with the purpose to communicate the government plan for the study of the introduction of a nuclear power programme in the country in the future. The purpose of this study is to satisfy the foresee increase in the demand of electricity in the best economic manner and using all available energy sources. In the meeting, the President explained that the government has the intention to carry out in the first stage a preliminary technical studies in order to allow the government to determine if Uruguay is ready to use nuclear energy for electricity generation in the coming years. If the outcome of the study is positive, then the government will consolidate the nuclear authority already established. In the third stage the government will promote the necessary hiring to decide if the introduction of a nuclear power programme is possible and necessary. If the decision is to support the introduction of a nuclear power programme in the country, then all necessary actions and measures should be taken in order to ensure the successful introduction of this programme and the beginning of the construction of the first nuclear power reactor in an appropriate time.

The Uruguayan President also informed all political parties on the establishment of a commission, integrated by representatives of several ministries, the Armed Forces and three of the political opposition parties, with the purpose to present a report within 90 days on the possibilities of the use of nuclear energy for electricity generation in the country in the future.

It is important to note that the introduction of nuclear energy for electricity generation in Uruguay is not an easy option and is not exempt of difficulties. The main difficulties are, among others, the following: 1) the capacity of the electric grid is relative small and this is a situation that need to be in mind during the selection of the type and capacity of the nuclear power reactor to be installed; 2) the country has little experience in the use of nuclear energy for peaceful purpose and has no experience at all in the use of this type of energy for electricity production; 3) there is an insufficient number of capable professionals and technicians in the nuclear field and with the necessary experience to ensure the successful introduction of a nuclear power programme in the near future; 4) the possible public opposition to the use of nuclear energy for electricity generation; 5) the relative small size of the country.

It is difficult to reach, at this stage, an objective conclusion of whether the use of nuclear energy for electricity production is an alternative option for Uruguay future energy needs, but with the available information on this issue it is not difficult to said that the use of nuclear energy for the production of electricity in Uruguay is not a simple and cheap option.

Chapter 7

THE PERSPECTIVES OF THE INTRODUCTION OF A NUCLEAR POWER PROGRAMME IN THE MIDDLE EAST AND IN THE NORTH OF AFRICA

With growing urbanization, the growth of the industry sector, including the oil industry sector, and the expansion of the tourism sector in many countries in the Middle East, the consumption of power has been rising steadily in the last years. The table below provides data on electricity consumption per capita per year during the period 1980 and 2003. With the exception of Qatar, which was doing well already in 1980 and Iraq, which has gone through a period of sanctions and wars, all other countries in the table below show a triple-digit growth in energy consumption per capita. Many countries in the Middle East, particularly members of the Gulf Cooperation Council[167], have gone through a period of breathtaking growth and it is safe to surmise that their consumption of electricity has grown correspondingly.

Table 91. Electricity consumption per capita (kilowatt/hours) in selected countries

Country	1980	2003	Increase (%)
Qatar	10,616	17,489	64.7
UAE	6,204	14,215	129.1
Bahrain	4,784	10,830	126.4
Saudi Arabia	1,969	6,620	236.2
Jordan	366	1,585	333.1
Syria	433	1,570	252.6
Egypt	433	1,287	197.2
Morocco	254	560	120.5
Iraq	878	1,542	75.6

Source: United Nations Development Programme, Arab Human Development Report 2005.

---

[167] The following countries are member of the Gulf Cooperation Council: Bahrain, Kuwait, Oman, Qatar, Saudi Arabia and the United Arab Emirates.

In the Middle East, liquids fuel consumption is projected to increase by 3.6 million barrels per day from 2005 to 2030. Three major factors contribute to the growth in oil consumption in this region: first, although the population in the Middle East is relatively small, the nations of the region have recorded relatively high birth rates over the past several decades, so that at present a substantial portion of the population is young and reaching driving age, increasing the demand for personal motorization; second, energy use is heavily subsidized in many of the resource-rich nations of the region; and third, many of the world's major oil-exporting nations are in the Middle East, and as world oil prices have continued to rise, so too have their per-capita incomes. As standards of living have improved, demand for personal motorization has increased, and many nations of the region have seen double digit growth in automobile sales in recent years. (IEO, 2009)

According with IEO 2009 report, electric power generation in the Middle East region is projected to grow by 2.6% per year, from 0.6 trillion kWh in 2005 to 1.1 trillion kWh in 2030. The region's young and fast-growing population and a strong rise in projected national income are expected to result in a rapid increase in demand for electric power. Despite short-term supply issues in some Middle Eastern countries, natural gas is expected to remain the region's largest source of energy for electricity generation. In 2005, natural- gas-fired generation plants accounted for 56% of the Middle East region's total power supply. In 2030, the natural gas share is projected to be 65%, as the petroleum share of generation falls over the projection period. Petroleum is a valuable export commodity for many nations of the Middle East, and there is increasing interest in the use of domestic natural gas for electricity generation in order to make more oil assets available for export.

It is important to single out that the Middle East is the only region in the world where petroleum liquids are expected to continue accounting for a sizable portion of the fuel mix for electricity generation in the future. In 2005, the region as a whole relied on oil-fired capacity to meet 36% of its total generation needs, and it is expected that petroleum liquids will continue to provide 29% of the total generation needs in 2030. Oil-fired generation in the Middle East is projected to increase by an average of 1.6% per year from 2005 to 2030. Other energy sources make only minor contributions to the Middle East region's electricity supply. Israel is the only country in the region that uses significant amounts of coal to generate electric power and Iran is the only one projected to add nuclear capacity, with the completion of its Bushehr nuclear power plant expected to enter into commercial operation in 2015. Finally, because there is little incentive for countries in the Middle East to increase their use of renewable energy sources, renewables are projected to account for a modest 3% of the region's total electricity generation throughout the 2005 to 2030 period. (IEO, 2009)

In recent years, more than a dozen states in the Middle East have expressed their interest in the introduction of a nuclear power programme. These states have offered a number of official reasons to justify their interest in the use of nuclear energy for the production of electricity, including powering water desalination plants, diversifying their energy industry in the face of increasing energy demands, and furthering economic and scientific development, among others. One of the reasons why a group of states in the region have expressed their interest in the use of nuclear energy for electricity generation and for other peaceful purposes in different economic sectors, is the possible use of this type of energy for water desalination. States in the region have shows interest in developing nuclear energy, in particular for water desalination purposes, for more than a decade. Egypt proposed the construction of nuclear power plants for desalination purpose, first in 1964, then again in 1974 and 1983, but all of these plans were cancelled due to safety concerns following the 1986 Chernobyl nuclear accident.

Beyond the renewed interest in nuclear energy expressed by a group of states with prior nuclear energy pursuits, the following group of countries have indicated their interest to consider the introduction, for the first time, of a nuclear power programme in the future: Bahrain, Jordan, Kuwait, Oman, Qatar, Saudi Arabia, the United Arab Emirates (UAE) and Yemen. In addition to the interest of a group of Arab States in the introduction of a nuclear power programme in the coming years. there is also a renewed interest on this subject as part of broader regional calls to develop nuclear power. Arab League Secretary-General Amr Mousa stated, during the March 2006 league summit in Khartoum, Sudan, the following: "The Arab world's quick and decisive entry into the field of peaceful use of nuclear power is necessary". Months later, in December 2006, the Gulf Cooperation Council ordered a wide study for the development of a joint programme in the field of nuclear technology for peaceful purposes. The first stage of this study was completed in late 2007.

There should be no doubt that the interest of a group of Middle East states in the introduction of a nuclear power programme is not only related to the generation of electricity but is due also to the need to use oil reserves in the most productive and effective manner from the finance point of view in the future. Record oil revenues have driven an economic boom that is straining the region's power electric grids. To keep the export cash coming in, some of the largest oil and gas producers are looking at nuclear energy to minimize burning fuel for power at home. The only energy alternative that these countries could consider to achieve this goal is nuclear energy. "Nuclear energy for electricity production is the logical thing for countries in the Gulf to do", said Giacomo Luciani, director of the Swiss-based Gulf Research Center Foundation. "When oil was cheap and abundant, it was right to burn it for power. Right now it's irrational".

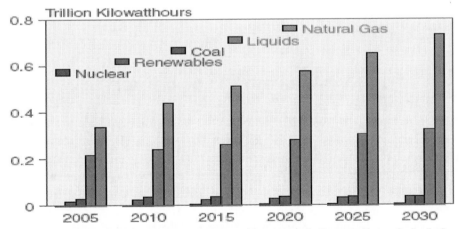

Sources: 2005: Derived from Energy Information Adminis-tration (EIA), *International Energy Annual 2005* (June-October 2007), web site www.eia.doe.gov/iea. Projections: EIA, System for the Analysis of Global Energy Markets/Global Electricity Module (2008).

Figure 150. Net electricity generation in the Middle East by source 2005-2030.

In total, thirteen countries throughout the Middle East have announced new or revived plans to explore the use of nuclear energy for electricity production. They spoke of the need for energy diversification to meet growing electricity demand and the economic and environmental benefits of nuclear power to achieve that goal. This surge of interest in the use of nuclear energy for the production of electricity is consistent with a worldwide trend likened to a nuclear renaissance.

Constraints to Nuclear Power

The following are the main constraints that exist in the region related with the introduction of a nuclear power programme:

1) all countries in the region have little experience in the use of nuclear energy for peaceful purpose and has no experience at all in the use of this type of energy for electricity production;
2) there is an insufficient number of capable professionals and technicians in the nuclear field and with the necessary experience to ensure the successful introduction of a nuclear power programme in the near future;
3) the political instability that exist in the Middle East;
4) the terrorist activities that exist in some part of the Middle East;
5) the possible public opposition to the use of nuclear energy for electricity generation;
6) the experience of several states in the region on the use of other energy sources for electricity production and the vast oil and gas reserves that exist in several countries in the Middle East.
7) the relative small size of some countries in the region that could make very difficult the final disposal of nuclear waste.

THE NUCLEAR POWER PROGRAMME IN IRAN

In 2006, Iran produced some 201 billion kWh (gross) from 31 GWe installed capacity, giving per capita consumption of 1 943 kWh per year. In Iran, electricity demand has been increasing by about 7% annually in recent years and the demand for energy to fuel the increase in electric power generation has pressured the country's supply infrastructure. At the beginning of 2008, unusually cold winter weather increased the demand for natural gas, both for power generation and for residential and commercial uses. The sharp increase in natural gas demand has, since 2006, resulted in large natural gas shortages at Iran's power plants during the winter, and many have switched to burning fuel oil and diesel to meet the power demand. The participation of the different energy sources in the production of electricity in the country in 2006 is the following: 74% of electricity comes from gas, 17% from oil and 9% from hydro.

Iran interest in the use of nuclear energy for the production of electricity dates back more than 30 years ago. In 1955, the first discussions on developing a nuclear power programme for Iran took place between the USA and Iran. However, the first concrete step was taken in

1957 when the USA signed an agreement with Iran for the development of a peaceful nuclear cooperation programme. This was promoted as part of the US Atoms for Peace programme that was supposed to provide technical assistance to the a group of States that agree not to develop a nuclear weapon programme in the future and to put all of their nuclear activities under international control. In the same year, the Central Treaty Organization (CENTO) that consisted of Iran, Pakistan, Turkey, Iraq, Britain and the USA moved its Institute of Nuclear Science from Baghdad to Tehran after Iraq withdrew from CENTO. In 1959, the government of Iran ordered establishment of a nuclear research center at Tehran University, Tehran Nuclear Research Center (TNRC) and began negotiating with the USA to purchase a 5 MW nuclear research reactor for the Center. To this date, the Center remains one of Iran's main nuclear research organizations. (Sahimi, 2004)

It is important to single out that over the next decade the USA provided nuclear fuel and equipment that Iran used to start up its research at the TNRC. The USA stops the supply of nuclear fuel in 1979. By the 1970s, France and Germany joined the USA in providing assistance to the Iranian nuclear power programme. By the mid-1970s, Iran had signed contracts with Western firms—including France's Framatome and Germany's Kraftwerk Union—for the construction of nuclear power plants and supply of nuclear fuel. (Bruno, 2008)

The first announcement on Iran's intention for constructing a nuclear power plant was made in December 18, 1972, when Iran's Ministry of Water and Power began a feasibility study for constructing this type of power plant in Southern of the country. The intention of Iran was to produce, by 1990, a total of 10 000 MW of electricity using nuclear power plants. However, a 1974 study by the Stanford Research Institute concluded that 2Iran would need, by 1994, to produce 20 000 MW of electricity by nuclear power plants". Thus, in March 1974, the Shah announced plans for generating 23 000 MW of electricity, as soon as possible, by the operation of 23 nuclear power reactors, to be constructed before 1994. To achieve his goal, the Shah established the Atomic Energy Organization of Iran (AEOI), appointed Dr. Akbar Etemad, a Swiss-trained physicist as its first chief, and announced "that the AEOI, like everything else, would be run directly under his command". (Sahimi, 2004)

The aim of the Iranian government was to have a full-fledged nuclear power industry with the capacity to produce 23 000 MWe of electricity with the purpose to convert Iran into a powerful modern state. At the same time, the use of nuclear energy for the production of electricity will allow Iran to export electrical power as well as gas and oil improving Iran's trade and current accounts, shoring up Iran's finances and benefiting its population. In Iranian eyes, pursuing civil nuclear energy is a key element of Iran's development strategy and its response to the looming energy crunch. Iran's nuclear power programme also has a research dimension (medical and university research, agriculture, etc.) that is essential for any country seeking to establish a scientific reputation. (El-Hokayen, 2006)

It is important to note that the nuclear power programme's size and some of its objectives raised questions in the USA and other countries about a nuclear proliferation risk. In addition, despite the Shah's claims that he did not want to build a nuclear weapon, his regime's interest in reprocessing plutonium and its insistence on a full right to reprocess nuclear spent fuel raised apprehensions, not only in the USA but in other countries in the Middle East and in other regions of the world as well. To achieve the above mentioned goal, the government of Iran wanted to buy some nuclear power reactors from the USA as well as from some Western Europe countries, but US presidents would not approve the sales without conditions limiting his freedom of action to use US supplied resources. As such, the USA pressed Tehran to accept a multinational reprocessing center to avoid an Iranian domestic capability. Secretary of State Henry Kissinger also became interested in a "buyback" option so that the USA could acquire spent nuclear fuel rods from Iranian reactors. With buyback, Kissinger believed, "Washington could ensure that any reprocessing occurred in the USA, not Iran".

During the second half of 1977, the Iran's government and the Carter administration reopened the nuclear negotiations and by mid-1978 the two sides had initialed an agreement. However, the agreement was ever signed by both governments. As in the 1976 draft, the agreement retained a US veto on reprocessing, but in light of Carter's tougher approach on that issue, it didn't include options for buyback or for the establishment of a multinational plant for reprocessing the spent nuclear fuel. Iran's spent nuclear fuel could be reprocessed in Western Europe, but only if the material could not be stored in Iran, the USA or Europe. All options would be subject to US law which includes determination of no significant increase in the risk of proliferation associated with approvals for reprocessing. It would be possible to return recovered plutonium in the form of fabricated fuel to Iran, but only "under arrangements which are deemed to be more proliferation resistant than those which currently exist". On March 3, 1975, Iran and the USA signed an agreement worth about US$15 billion, according to which the USA was ready, among other things, to build eight nuclear power reactors in Iran with a total capacity of about 8 000 MW. The agreement was signed by the US Secretary of State, Henry A. Kissinger, and Iran's Finance Minister, Houshang Ansari. The fuel for the nuclear power reactors was to be supplied by the USA. The agreement was never implemented.

Despite of the failure to implement the USA-Iran agreement in the mid 1970s, France and the Federal Republic of Germany signed an agreement for the construction of four nuclear power reactors in two selected sites. In May 1975, the construction of two 1 200 MWe PWR units, supplied by Siemens KWU, started. However, Western governmental support for Iran's nuclear power programme began to erode ahead of the Islamic Revolution of 1979. Pressure on France, which in 1973 signed a deal to build two units at Darkhovin and Germany, whose Kraftwerk Union began building a pair of reactors at Bushehr in 1975, led to the cancellation of both projects. (Bruno, 2008)

Source: Photograph courtesy of eideard.wordpress.com.

Figure 151. Busherh nuclear power plant.

Iran revived its nuclear power programme in 1991 with a bilateral agreement with China for the supply of two 300 MWe PWR units of Chinese design, similar to the Qinshan nuclear power plant. However, for different reasons, the plan was never implemented.

In 1994, Russia's Minatom and AEOI agreed to complete Unit 1 of the Bushehr nuclear power plant with a WWER-1000 model, using mostly of the infrastructure already in place. This long-awaited 915 MWe nuclear power reactor, being constructed by Atomstroyexport, is nearing completion and is expected to entry into commercial operation before 2011. A second unit is planned at the same site.

After two years delay due to Iran's reluctance to return spent nuclear fuel to Russia without being paid for it, two agreements were signed early in 2005 covering both supply of fresh fuel for the new Bushehr nuclear power reactor and its return to Russia after used it. Supply of the fuel was originally contingent upon Iran's signing the IAEA Additional Protocol to its safeguards agreement with the IAEA. Iran already signed the Additional Protocol on the 18 of December, 2003 but has not ratified it yet. The Russian agreement ensure the fuel supply to Iran for the foreseeable future, removing any justification for enrichment locally. It also means that the anticipated 6-7 TWh per year from the operation of the reactor will free up about 1.6 million tons of oil or 1,800 million cubic meters of gas per year, which can be exported for hard currency.

Russia's Atomstroyexport delivered, in December 2007, the first of 163 fuel assemblies for the initial core of the nuclear power reactor at Bushehr. The fuel is enriched to 3.62% or less and is under full IAEA safeguards.

At the same time, AEOI has announced that a new indigenous 360 MWe nuclear power reactor is going to be built at Darkhovin in Khuzestan Province in the Southwest of the country, at the head of the Gulf, where two Framatome 900 MWe units were about to be constructed in 1970s. It has also invited bids for two additional units of up to 1 600 MWe to be built near Bushehr, with the purpose to entry into commercial operation in 2016.

Iranian President Mahmoud Ahmadinejad inaugurated its first nuclear fuel production plant in 2009 with the aim to produce fuel for Iran's Arak heavy water reactor.

Since 2002, the IAEA has issued several reports about Iran's nuclear power programme, alternating praise and criticism of Iran's cooperation, and culminating with the February 2006 referral of the Iran issue to the UN Security Council requesting that Iran stop its enrichment programme. Several sanctions have been adopted by the Security Council due to the resistance of Iran to stop the enrichment programme. The following proposals could be used as a basis for a possible solution of the Iran case in the future: 1) the multi-nationalization of Iran's uranium enrichment activities, in other words, putting a UN-IAEA flag on the Natanz facility; 2) the establishment, under IAEA auspices, of international fuel centers that would guarantee fuel supply to all nations and deal with spent uranium; 3) the Russian proposal to enrich and take back spent uranium on Russian soil; and 4) the ratification by Iran of the IAEA Additional Protocol and adhering to the Comprehensive Nuclear Test Ban Treaty (CTBT). These steps would send positive signals to the international community and create positive momentum to reach a fair and sustainable solution of the Iran's nuclear power programme and the suspicious of the international community of the use by Iran of this programme for military purposes.

It is important to note that Iran has rejected the first three of the above proposals and has not yet ratified the IAEA Additional Protocol. The Iran's history with its pre-Islamic Revolution nuclear suppliers is summarized in the following sentences. Prior to 1979, the

USA, France and Germany, today the driving forces behind the international community's effort to obtain a permanent suspension of Iran's uranium enrichment activities, were Iran's major partners in developing its nuclear power sector. For Iranians, the memory of the bitter disputes over these suspended contracts between Iran and its Western partners in the aftermath of the Islamic Revolution acts as a powerful reminder of the need to assert complete control over Iran's nuclear power programme. Even Iran's current partners, Russia and China, are considered unreliable when it comes to nuclear issues. Iran's officials recalled several instances when these two countries, under USA pressure, reneged on their commitments at great financial and political cost for Iran. Iranian blamed the West in general, and the USA in particular, for what they perceive to be double standards in dealing with Iran's nuclear power programme. Israel, which did not adhere to the NPT, is suspected of having a large arsenal of nuclear weapons and maintains a policy of "nuclear ambiguity". The fact that the IAEA does not have access to Israel's nuclear programme amounts to intolerable uncertainty for Tehran. Moreover, it stands in the way of any effort to create a weapon mass destruction-free zone in the Middle East, Iran's stated preference. Similarly, the USA-India deal suggested too many that the USA prioritizes its own expedient interests over international norms and obligations. Under this deal, the USA proposes to support India's own civil nuclear power programme in exchange for some guarantees and concessions deemed insufficient by many non-proliferation experts and ignore that India is a non state party of the NPT and has developed nuclear weapons against the will of the international community. (El-Hokayen, 2006)

THE PERSPECTIVES OF THE INTRODUCTION
OF A NUCLEAR POWER PROGRAMME IN EGYPT

Egypt produced 115 billion kWh (gross) in 2006 from 18 GWe installed capacity, giving a per capita consumption of 1 350 kWh per year. For Egypt, gas was, in 2008, the dominant fuel used for the generation of electricity, accounting for 49.5%, followed by oil at 43.9% and hydro with a 5.3%. Egypt's 2008 market share of 10.10% is set to ease to 9.99% by 2013. Egypt's estimated 17.4TWh of hydro generation in 2008 is forecast to reach 21.2TWh by 2013, with its share in the regional hydro market falling from 58.5% to 37.6% over the period.

The first attempt to introduce a nuclear power programme in Egypt was made in 1964, In that year, a 150 MWe nuclear power plant with 20,000 m^3/day desalination was proposed to be built in the country and, in 1974, a 600 MWe nuclear power plant for electricity generation was proposed. The government's nuclear power plants authority was established in 1976, in order to implement the nuclear power programme and, in 1983, the El Dabaa site on the Mediterranean coast was selected for the construction of the first nuclear power reactor to be built in the country. The implementation of this plan was cancelled following the Chernobyl nuclear accident. More recently, the government carried out a feasibility study for a cogeneration plant for electricity and water desalination, updating it in 2003.

A new agreement for the peaceful uses of atomic energy was signed with Russia at the end of 2004 and a further one in March 2008, reviving Egypt's plans for the introduction of a nuclear power programme, including a water desalination plant in the country. The plan is supported by Rosatom. In 2006 a nuclear cooperation agreement was reached also with China. Egypt already operate a 2 MW Russian nuclear research reactor serviced by Russia, and a 22 MW Argentinean research reactor partly supported by Russia, which started up in 1997.

On the basis of the feasibility study for a cogeneration plant for electricity and potable water at El-Dabaa, in October 2006, the Minister for Energy announced that a 1 000 MWe nuclear power reactor would be built there by 2015. The US$1.5 to US$2 billion project would be open to foreign participation. In December 2008, the Energy and Electricity Ministry announced that "following an international tender, it had decided to award a US$180 million contract to Bechtel to choose the reactor technology, choose the site for the plant, train operating personnel and provide technical services over some ten years". However, Egypt's Electricity Ministry says, on May 3, 2009, that "the cooperation with US construction giant Bechtel as a consultant for the country's first nuclear power reactor was cancelled and the government has chosen Australia's Worley Parsons (WOR.AX) for the 10-year consultancy". The ministry confirmed that "Egypt aims to begin generating nuclear electricity in 2017 at one of five possible sites".

THE PERSPECTIVES OF THE INTRODUCTION OF A NUCLEAR POWER PROGRAMME IN ALGERIA

The power generating capacity in Algeria has reached 9 273 MW. It should exceed 11 000 MW by 2010 and 20 000 MW by 2015. Around 95% of the current power capacity is provided by gas-fired power plants. Domestic power demand should rise to 9 300 MW by 2010, compared to about 5,447 MW in 1996. Algeria produced 35 billion kWh (gross) of electricity in 2006, almost all from natural gas. The country is a major gas exporter to the EU and other countries.

In January 2007, Russia signed an agreement to study the introduction of a nuclear power programme in the country. During the period 2007-2008, other nuclear cooperation agreements were signed with Argentina, China, France and the USA. In February 2009, the government announced its plan to build its first nuclear power reactor in the coming years with the purpose to start its commercial operation about 2020. The plan contemplates the possibility to construct further units every five years thereafter.

The country has two research reactors in operation since 1995, at Draria and Ain Ouessara sites. One was built by INVAP of Argentina and the other by China. Algeria has big uranium-deposits but no uranium enrichment capacity.

THE PERSPECTIVES OF THE INTRODUCTION OF A NUCLEAR POWER PROGRAMME IN JORDAN

Jordan currently generates most of its energy needs from fossil fuels, 95% of which it imports from neighboring Arab countries at a cost of 20% of its GDP. It generated 11.6 billion kWh and imported 0.5 billion kWh of electricity in 2006 in order to cover the demand of electricity. Jordan has 2 030 MWe of generating capacity installed and expects to need an additional 1 200 MWe by 2015 due to a foresee increase in the electricity demand.

The energy minister has said that "the country expects to have a nuclear power plant operating by 2015 for electricity and desalination". Jordan's Committee for Nuclear Strategy has set out a plan for the introduction of a nuclear power programme in the country with the purpose to provide around 30% of electricity by 2030 or 2040, and to have an excedent in the production of electricity for exports to other countries in the region. With this objective in mind, in the middle of 2008, an agreement was reached between the Jordan Atomic Energy Commission (JAEC) and AECL with SNC-Lavalin to conduct a three year feasibility study on building a nuclear power reactor using natural uranium as fuel for power and water desalination. In August 2008, it was reported that the government intended to sign up for an Areva nuclear power reactor to be built in the country and discussions on this subject was held in November that year pointed to an 1 100 MWe unit, presumably from Atmea, the Areva-Mitsubishi joint venture which is developing such a unit for countries embarking upon for the first time in a nuclear power programme. Site selection is planned to be carried out in 2009, though options appear to be limited to 30 kilometers of Red Sea coast near Aqaba, and JAEC has confirmed that the site will be in this area. JAEC has said that "a tender is likely to be presented in the middle of 2010 with the purpose to start construction work in 2012". Further nuclear projects are likely to focus on water desalination.

In December 2008, JAEC signed a memorandum of understanding with Korea Electric Power Corp (KEPCO, parent company of KHNP) to carry out site selection and feasibility study on nuclear power and water desalination projects. This is related to Doosan Heavy Industries, Korea's main nuclear equipment maker, carrying out water desalination-related work in Jordan under a separate recent agreement, and KEPCO having won a tender to build a 400 MWe gas-fired power plant on a build-own-operate basis. KEPCO and Doosan are reported to be ready to offer Jordan their OPR-1000 nuclear power reactor for its first nuclear power plant.

Jordan has signed nuclear cooperation agreements with the USA, Canada, France and UK, in respect to both power and water desalination, and is seeking help from the IAEA on this subject. It has signed also a nuclear cooperation agreement with China, covering uranium mining in Jordan and nuclear power, and another with the Republic of Korea related to infrastructure, including nuclear power and water desalination. It has signed also a preliminary cooperation agreement with Russia. The agreements signed with China will allow Jordan to get a Chinese non-critical training system, according to the President of JNEC. The facility would be designed and constructed in coordination between the Chinese Atomic Energy Corporation and the Jordan University for Sciences and Technology, where the unit would be built. The second deal involved a blueprint for implementing an agreement initialed in August to build a research nuclear reactor in Jordan and to help the country extract uranium from at least two local areas.

In March 2009, the JAEC was evaluating proposals from four reactor vendors: KEPCO, Areva, Atomstroyexport and AECL with the purpose to select the type of nuclear power reactor to be built in the country in the coming years.

Uranium resources in Jordan are estimated in 80,000 tons plus another 100,000 tons in phosphate deposits, according with the information provided by the government.

Finally, it is important to single out that Jordan has hired, for US$12 millions, the Belgian company Tractebel Engineering for the environmental study of a place 25 kilometers to the

South of the port of Aqaba on the Red Sea. The place has been pre-selected for the installation of the first nuclear power plant in the country. At the same time, Jordan has signed, in 2009, with Argentina a cooperation agreement for the peaceful uses of the nuclear energy. The agreement covers the production and use of radioisotopes, the exploration of minerals, the design and the construction of research and power reactors and management of nuclear waste.

THE PERSPECTIVE OF THE INTRODUCTION OF A NUCLEAR POWER PROGRAMME IN LIBYA

In 2006, Libya produced 24 billion kWh (gross) of electricity, 59% of this from gas and 41% from oil. Libya currently has electric power production capacity of about 4.7 GW. It is important to single out that the existing power stations in Libya are being converted from oil to natural gas, and all new power plants built run on natural gas, in large part to maximize the volume of oil available for export purposes. Libya is also looking at potential wind and solar projects, particularly in remote regions where it is impractical to extend the electric grid.

The interest in Libya in the use of nuclear energy for the production of electricity started, in 1976, when negotiations were held between France and Libya for the purchase of a 600 MW nuclear power reactor. A preliminary agreement was reached between these two countries, but strong objection by the international community led France to cancel the project. In the 1980s, Libya discussed the construction of a nuclear power plant with the former Soviet Union and with the Belgian firm Belgonucleaire to provide engineering support and equipment for this proposed project but, in 1984, USA pressure led the firm to refuse the contract. Discussions with the former Soviet Union about the supply of nuclear power reactor projects continued until the collapse of this country early 1990s, but never produced a final agreement[168].

Early in 2007, it was reported that Libya was seeking an agreement for USA assistance in building a nuclear power plant for electricity and water desalination. In 2006, an agreement with France was signed for peaceful uses of atomic energy and, in mid 2007, a memorandum of understanding related to building a mid-sized nuclear power plant for seawater desalination was also signed. Areva TA would supply the nuclear power reactor.

Libya has a Russian 10 MW nuclear research reactor which is under IAEA safeguards. Moscow and Libya said, in November of 2008, that they were negotiating a deal for Russia to build a nuclear research reactor for the North African State and for the supply of fuel. Libyan's officials said that a document on peaceful nuclear cooperation was under discussion between the two countries. Under the deal, Russia would help Libya design, develop and operate a nuclear research reactor and will provide fuel for its operation.

Recently, Libya and France have signed a cooperation agreement on the peaceful use of nuclear energy which involves potentially big business for French companies. The French company Veolia signed a water desalination contract with Libya worth €2.1 billion over a period of 25 years.

---

[168] The nuclear power programme that Libya had declared to the IAEA consisted of a 10 MW IRT nuclear research reactor, a pool-type research reactor using 80% enriched uranium in operation and a critical assembly (100 W), both located at the Tajura Nuclear Research Center (TNRC).

Finally, on 8 March 2004, Russia, the USA and the IAEA removed 16 kilograms of highly enriched uranium fuel from Libya's Tajura Nuclear Research Center; the HEU fuel was airlifted by a Russian company to Dimitrovgrad, where it will be down-blended into low-enriched uranium fuel. The Tajura Soviet-supplied IRT-1 research reactor will soon be converted to low-enriched uranium; costs of fuel removal and conversion will be paid by the USA and the IAEA. [55]

Libya made several attempts to develop a clandestine military nuclear power programme but failed in this effort and the programme was definitively stopped in 2000s.

THE PERSPECTIVE OF THE INTRODUCTION OF A NUCLEAR POWER PROGRAMME IN QATAR

Qatar has invested heavily in its energy sector which depends mostly on natural gas. Domestic energy consumption, in 2009, is expected to average 17 million tons per year of oil equivalent, up from 15.3millon tons in 2007 and 13 million of tons in 1993. All the six Arab Gulf Co-operation Council states, except Qatar, face power shortages and budget deficits in 2009 due to a deficit in the production of enough gas to meet the demand for electricity of the five of these states. Qatar has a surplus of electricity production that it exports to fellow GCC states. Qatar's energy base is small. Oil consumption in 2009 is to average about 62,000 barrel per day, up from 50,000 in 2007 and 10,500 in 1985. Natural gas accounts for more than 85% of its energy and industrial consumption, excluding gas being exported by pipeline and in LNG and NGL forms, up from zero in the 1960s.

Initial Qatari interest in the introduction of a nuclear power programme was due to the fall in international oil and gas prices. If Qatar decided to go ahead with building a nuclear power plant, feasibility studies showed it would be unlikely to bring the plant into commercial into operation before 2018. Late in 2008, the Qatar's government announced that there was not yet a strong case for proceeding, especially in the absence of modern 300 to 600 MWe nuclear power reactors being available in the market. Qatar expects to need 7 900 MWe of installed capacity for the generation of electricity by 2010, in addition with a desalination capacity of 1.3 million cubic meters per day.

French power giant EdF signed a memorandum with Qatar, in early 2008, for the development of a bilateral cooperation in the peaceful use of nuclear energy in the country power programme.

THE PERSPECTIVE OF THE INTRODUCTION OF A NUCLEAR POWER PROGRAMME IN THE UNITED ARAB EMIRATES (UAE)

In 2006 the UAE produced 66.8 billion kWh (gross), 98% of it from gas. It has about 18 GWe capacity. Electricity demand is growing by 9% per year and is expected to require 40 GWe of capacity by 2020. In April 2008, the UAE adopted a comprehensive policy on the introduction of a nuclear power programme in the country for the coming years. The reason for the preparation of this programme is the expected escalating electricity demand from 15.5 GWe, in 2008, to over 40 GWe, in 2020, with natural gas supplies sufficient for only half of this. Imported coal was dismissed as an option due to environmental and energy security implications.

According with this policy, nuclear power emerged as a proven, environmentally promising and commercially competitive option, which could make a significant base-load contribution to the UAE's economy and future energy security. For this reason, around 20 GWe of nuclear capacity is envisaged from about 14 nuclear power reactors, with nearly one quarter of this operating by 2020. Two nuclear power reactors are envisaged for a site between Abu Dhabi and Ruwais and a third possibly at Al Fujayrah on the Indian Ocean coast.

In January 2008, three French companies Areva, Suez and Total signed a partnership agreement to propose to UAE the construction of two EPR units in the country in the coming years. Two of the companies, Suez and Total, would each invest up to 25% of the project with Abu Dhabi entities providing at least 50%. Suez would be the operator of the nuclear power plant and Areva would supply the nuclear power reactor and manage the fuel. It is important to sing out that Total and Suez are well established in the region and together operate a power and water desalination plant for Abu Dhabi, 100 km West of Dubai. The consortium's first EPR would not be operating before 2017. At the same time, France signed a nuclear cooperation agreement with UAE. A Memorandum of Understanding on nuclear energy cooperation between the USA and UAE was signed in April 2008. [26]

Accordingly, and as recommended by the IAEA, the UAE established a Nuclear Energy Programme Implementation Organization which has set up the Emirates Nuclear Energy Corporation (ENEC) as a public entity, initially funded with US$100 million, with the purpose to evaluate and implement a programme for the construction of nuclear power plants within UAE. In the last IAEA General Conference held in Vienna, Austria, in September 2009, the UAE representative informed to the conference that the government of the UAE has allocated US$40 000 million for the development of a peaceful nuclear sector and for the establishment of an regulatory national authority. The purpose of the authority is to supervise the implementation of the nuclear power programme.

EAU has taken this position keeping in mind the foresee increase of the electricity consumption in the country in the coming years as result to an expected economic and demographic growth. To use hydrocarbons to satisfy this demand would mean, for the EAU, the reduction of the oil exports that provide 80% of the revenues of the state. Abu Dhabi plans to build three or four nuclear power reactors in the first stage, with intervals of 18 months among them in way of being able to evaluate with more precision in each case the real energy necessities. EAU has committed formally to acquire abroad its nuclear fuels and has no intention to produce enriched uranium locally.

Secondly, the UAE will also draft a comprehensive national nuclear law which establishes a fully independent Federal Authority for Nuclear Regulation (FANR), to be headed by a senior US regulator. Thirdly, it will offer joint-venture arrangements to foreign investors for the construction and operation of future nuclear power reactors similar to existing Independent Water and Power Producer structures, which 60% have owned by the government and 40% by other partner.

The UAE is setting up a model of managing its nuclear power programme based on contractor services rather than more slowly establishing indigenous expertise. It is important to note that instead of developing domestic enrichment and reprocessing facilities, the UAE would seek to conclude long-term arrangements for the secure supply of nuclear fuel, as well as the safe and secure transportation and, if available, the disposal of spent nuclear fuel via fuel leasing or other emerging fuel supply arrangements.

The UAE is reported to have invited nine companies for construction of its first nuclear power plant. By 2020, it hopes to have three 1 500 MWe nuclear power reactors running and producing electricity at a quarter the cost of that from gas. Three consortia from USA and Japan, France and Korea presented, in July 2009, offers to build nuclear power reactors in EAU using ABWR, EPR and APR-1400 reactors designs respectively. In December 2009, a consortium of the Republic of Korea was selected by UAE government for the construction and operation of four nuclear power reactors by an estimated value of US$40,000 million to meet their growing demands for electricity. UAE government "determined that the team of (KEPCO by their initials in English) is better equipped to meet the requirements of the government in this ambitious programme, said the director of the Nuclear Energy Corporation of the UAE (ENEC), Khaldoon al - Mubarak in a statement. The first contract for the construction of four nuclear power reactors will worth US$20,400 million and is expected to facilitates additional contracts of other US$20,000 million to operate and maintain the reactors during the next 20 years, said the South Korean Ministry of Economy. The South Korean Presidential Office described the agreement as "the largest mega project in the Korean history".

The UAE has signed also deals with the USA and the UK for cooperating in peaceful use of nuclear energy. The London daily, al-Hayat, reported "an unprecedented expansion" in economic cooperation between the UAE and France to respond to the former's needs in the use of nuclear technology both for the production of power and the desalination of water.

THE PERSPECTIVES OF THE INTRODUCTION OF A NUCLEAR POWER PROGRAMME IN MOROCCO IN THE FUTURE

Morocco produced 22.6 billion kWh in 2005 mostly by coal and some by oil. It is foresee an increase in the electricity demand in the future as well as a need for water desalination. For this reason, the government has been preparing plans for building an initial nuclear power plant in 2016-2017 at Sidi Boulbra and Atomstroyexport is assisting with feasibility studies for this purpose. For water desalination it has completed a pre-project study with China, at Tan-Tan on the Atlantic coast, using a 10 MWt heating nuclear power reactor, which produces 8 000 m^3/day of potable water by distillation. Morocco has a 2 MW Triga research reactor.

Chapter 8

THE CURRENT SITUATION AND PERSPECTIVES IN THE USE OF NUCLEAR ENERGY FOR ELECTRICITY GENERATION IN THE AFRICAN REGION (SUB-SAHARAN REGION)[169]

In Africa, every aspect of human development (health, agricultural, educational or industrial) depends upon reliable access to modern energy sources. Elsewhere in Africa, but mainly in African countries below the Sahara desert, electricity production is unacceptably low. There is a great contrast in energy production between sub-Saharan Africa and other regions of the world, including the North part of the continent. Of the 53 countries in Africa, a limited number have large energy potential. Hydropower potential is the most evenly spread type of energy sources in the region, but concentrated mainly on the Congo River. Oil and gas are mostly concentrated in Algeria, Nigeria and Libya, two of them in North Africa; coal is mostly found in Southern Africa and geothermal potential exists in Eastern Africa[170]. There are only two nuclear power reactors in operation in Africa (South Africa) and there are, until December 2009, no nuclear power reactors under construction in the region.

As a result, in the absence of adequate trans-African resource-sharing arrangements and infrastructure, many African countries suffer from scarce energy resources and must pay high prices to import energy. For example, in Tanzania the energy crisis came to a point where urbanites had to endure 18-hour power cuts and some industrialists were considering closing their facilities down after water levels in the Mtera Dam fell below the permitted power-generation level of 690 cubic meters.

Africa's ongoing energy crisis, probably the biggest and most imminent threat to economic growth, comes at a time when the impact in the environment, due to the greenhouse gases produced by the operation of fossil fuel power plants, is putting again the use of nuclear energy for the production of electricity under serious consideration in many countries in all regions, including Africa.

The expansion of the uranium mining sectors to support the expansion nuclear power programme in several countries in other regions are also under serious consideration with the

---

[169] This Chapter will refer only to countries located below the Sahara's desert.

[170]

biggest uranium reserves, including some African countries. The African countries joining the booming global uranium trade are Tanzania (Mkuju River concession), Mozambique (Mavuzi concession), Zimbabwe (Kanyemba concession), Guinea, Zambia (Kariba Valley concession), Madagascar, Nigeria, Central African Republic, and Uganda. Tanzania is among the newcomers to the global uranium exploration and mining drive, with significant energy rich uranium deposits along the Mkuju River being explored by Australian uranium company Mantra Resources. Zimbabwe has been discussing plans to invest in nuclear energy since the 1990s and there were plans to acquire a nuclear reactor from Argentina. Uranium was discovered in the Kanyemba area close to the country border with Zambia and Mozambique. Zimbabwe currently requires a daily supply of 2 100 MWe, but generates only 400 to 450 MWe from its hydroelectricity and thermal power plants and was looking at other forms of renewable energy to close the current gap between demand and supply of energy.

It is important to stress the following: Every country's energy mix is compose with the type of energy available in the country and involves a range of national preferences and priorities that are reflected in national energy policies. These policies represents a compromise between energy demand and supply to avoid shortages of energy, as much as possible, environmental quality, energy security, cost, public attitudes, safety and security, available skills, and production and service capabilities, among others. Relevant national energy authorities and industry businessmen must take all of these into account when formulating an energy policies and strategies.

One of the available types of energy used in many other regions for the generation of electricity is nuclear energy. Therefore, it could be very interesting to investigate whether nuclear power can safely alleviate energy shortages and optimize an energy mix consistent with the national interests of some African countries, taking into account their technological development. Namibia and Nigeria are two African countries seriously considering the introduction of a nuclear power programme in the future to satisfy the foresee increases in the demand of electricity. In the specific case of Nigeria can be stated, without any doubt, that the country has a great experience in the energy sector and is one of the main oil producer and exporter countries in the African region.

THE EXPLOITATION OF URANIUM

The attitudes of African decision makers, experts and public about the use of nuclear energy for electricity generation range from negative or cautious to positive or enthusiastic. Supporters perceive nuclear power as a technology that would allow the continent to demonstrate technical progress and technological development.

The search for cleaner energy sources, including the use of nuclear energy for electricity generation, is motivated by widespread concern that Africa is more vulnerable than other regions to climate change, such as desertification, food shortages, epidemics, insufficient water supply, coastal erosion and increased refugees. Abundant uranium resources in the region are an element that was considered essential for the development of a successful nuclear power programme in the region. According to a 2005 IAEA's report, "Africa maintains 18% of the world's known recoverable uranium resources". However, it is important to single out that the existence of uranium reserves is not a guarantee for the successful introduction of a nuclear power programme in Africa.

Australian Company Omega Corp is reported to have obtained licenses to mine the Kariba project in Zambia, the Kanyemba project in Zimbabwe and the Mavuzi concession in Mozambique. Omega Corp operates in the region through its subsidiary, Mantra Resources. In the Kariba Valley, a company called Africa Energy Resources, a subsidiary of Australian firm Australian Energy Ventures, started drilling in May 2006, and there are other Australian companies conducting further exploration. So far, Paladin Resources, one of the world's biggest uranium mining companies, has clinched a 10-year deal to mine uranium at Kayelekera in Malawi, and at least four other mining deals are awaiting approval in Namibia, ranked as Africa's second largest producer of uranium. Malawi approved the Keyelekera project in the country's Northern region district of Karonga after months of resistance from environmentalists and human rights organizations. Paladin Resources of Australia will extract 10 000 tons of uranium from the site for the next 10 years. Experts project that the Keyelekera plant will produce an estimated 200 tons of sulphuric acid a day for use in the leaching process which converts uranium ore to the required 10 000 tons of yellow cake.

Uranium is increasingly perceived in the region as an important part of a country's national wealth equal to oil and gas. For this reason, African countries with uranium deposits do not simply want to ship ore and yellowcake to industrialized nations. There is a growing sentiment that energy-starved Africa is destined to take advantage of its uranium resources and to use this important mineral to help rapidly to establish a modern and contemporary economy. For example, South African officials increasingly talk about the need to ensure that its abundant uranium resources are exploited for the benefit of the country rather than for that of foreign companies.

CONSTRAINTS TO NUCLEAR POWER

A nuclear power infrastructure includes manufacturing facilities, legal and regulatory frameworks, and expanded institutional measures to ensure safety and security and appropriate human and financial resources, among others relevant components. To build an adequate infrastructure to support the introduction of a nuclear power programme require careful planning, preparation and investment over at least a 10 to 15 year period. The construction of a nuclear power plant also requires a large upfront investment, usually US$2 to $3.5 billion per reactor. These are daunting tasks and requirements for any country, but particularly for the African countries seeking the introduction of a nuclear power programme in their energy balance in the future.

In addition, it is important to note that prevalence of interstate conflicts, ethnic strife, insurgencies, corruption and crime create a hostile environment in some African countries for the safe and secure introduction of a nuclear power programme. Examples of such risks are abundant. In April, armed men attacked a uranium prospecting camp maintained by the French company Areva in Northern Niger, killing a security guard and wounding three other people. In the Congo, only a thin barbwire fence protects Africa's first nuclear research reactor, which possesses a totally outdated control room and an unguarded radioactive waste storage building. Through decades of war, dictatorship and political upheaval, the Congo has repeatedly been accused of illegally selling its natural uranium or for not preventing smuggling schemes involved this sensitive material. As recently as November 2007, armed gunmen gained access to a major nuclear research facility in South Africa and reached the emergency control room before guards chased them away.

Subsequently, instilling law and order and the creation of the necessary infrastructure must be a prerequisite for the introduction and development of a nuclear power programme in any African country. The IAEA is working with some African States to enhance nuclear security by improving controls and detection equipment, upgrading physical protection and providing emergency assistance and training staff. For their part, African States must get more closely integrated into the international regimes. For example, 23 African States still have not fulfilled their obligations to conclude comprehensive safeguard agreements under the NPT, and only 25 African States acceded to the 1987 Convention on the Physical Protection of Nuclear Material. Worse yet, just two countries have ratified the Convention's 2005 amendment, which contains 12 fundamental principles of nuclear security.

The lack of sufficient capable and experience professionals and technicians in almost all African States in the field of nuclear power is one the most important constraints that exist in the region for the introduction of a nuclear power programme in the near future. The preparation of the staff that is going to support the introduction of a nuclear power programme in any country of the region is a long-term goal that needs to be in the mind of all politicians and businessman when the use of nuclear energy for the generation of electricity is discussed.

THE FUTURE ROLE OF NUCLEAR POWER IN AFRICA

Africa produced, in 2004, a total of 105 GWe of electricity. The electricity produced by nuclear energy was 1.8 GWe representing only 1.7 % of the total electricity production. This was produced in only one country of the region[171]. It is foresee that, in 2010, the total electricity produced by the African countries will be between 115-135 GWe and the total electricity produced by nuclear energy will be almost the same than in 2004 (1.8 GWe). The nuclear share will be between 1.3% and 1.6%. The expected electricity production, in 2020, will be between 143 to 207 GWe and the total electricity to be produced by nuclear energy will be between 2.1 to 4.1 GWe, representing between 1.5% and 2% of the total electricity to be produced by the region. In 2030, it is expected that the total production of electricity will be between 181 and 316 GWe and the total electricity produced by nuclear energy will be between 2.1 and 9.1 GWe, representing between 1.2% and 3% of the total. From the mentioned table can be concluded that only a very small part of the projected increases in nuclear power capacity in the world will be located in Africa.

Since Africa has not achieved economies of scale, the most rational approach to nuclear power would start with the construction of small nuclear power reactors whose output would better match existing electric grid capacity. One such reactor is already on the market: A Russian floating nuclear power plant, operated by a Russian team and, by cable, connected to a country's electric grid[172]. More largely, Russia is offering medium-sized nuclear power

---

[171] This country is South Africa.

[172] The offer of this type of nuclear power plant is under consideration by Namibia.

reactors to Namibia and, in exchange, Russia is hoping to participate in Namibia's uranium mining projects. Vnestorgbank and Tenex recently got licenses to prospect and produce uranium in Namibia.

Another small type of nuclear power reactor on the market is the PBMR. South Africa wants to export its design to its neighbors, while China is rapidly finalizing a somewhat different PBMR design that it will attempt to export to developing countries. Hungry for natural resources, China has launched a massive campaign to exploit Africa's riches uranium reserves, focusing on Niger as a strategic source of uranium for its nuclear power programme. As a fledging giant with a self-sustained nuclear record, India is also highly visible in uranium–rich African countries such as Zambia, offering to help develop nuclear strategies and policies and hawking its competitively priced PHWR. The Republic of Korea has decided to construct, by 2008, a one-fifth-scale demonstration plant of its 330 MWe SMART PWR, which will also include a demonstration desalination facility. The size of this type of reactor could be adequate to the electric grid of many African countries.

Meanwhile, African countries are demonstrating an increased interest in regional cooperation as a way to establish the required economies of scale for nuclear power generation. This cooperation may involve interconnected electric grids, collective facilities, cooperative education and training programmes, shared expertise in safety and security and common management practices and skilled labor pools, among others. In this sense, regionalization is looked upon as a useful mechanism for money, efficiency and reliability. It is important to stress that for several neighboring countries, each may find that nuclear power in not part of its least cost energy strategy when the analyses are done separately. But when the least-cost strategy is analyzed for the group of countries taken together, it might turn out to include nuclear power. However, it is important to note that regionalization can yield expected results provided a climate of trust between participating countries something that not always is present in Africa.

African countries are poised to benefit from nuclear power in many ways. But nuclear power is not a quick way to fix the continent's energy problems. In order to succeed, African governments must go beyond the traditional framework of a technical programme and apply considerable effort toward ameliorating the problems that plague their industrial infrastructure, public governance, educational system and other institutions in the public and private sectors. If well organized, the pursuit of nuclear power could become a rewarding endeavor for the continent and serve the future energy needs of its people at least for a group of countries. But the international community must be aware that a lack of expertise, oversight, political conflicts and safety and security measures, among others, could increase the likelihood of nuclear proliferation or terrorism both within and outside the continent.

THE NUCLEAR POWER PROGRAMME IN SOUTH AFRICA

Electricity consumption in South Africa has been growing rapidly since 1980. Total generating capacity in South Africa is 41.3 GWe and 94% of the total electricity produced by the country comes from coal-fired power plants largely under the control of the state utility Eskom. In 2005, the total electricity generation was 245 billion kWh, with 2 billion kWh net exports and 198 billion kWh of final consumption. In 2006, around 10 billion kWh, or 4.4%

of the total production of electricity, was produced using nuclear energy. In 2007, around 12.7 billon of kWh was produced using nuclear energy, representing 5.5% of the total electricity produced in the country. Early in 2008, electricity demand in South Africa is uncomfortably close to its generating capacity. In, 2008, the nuclear share was 5.25%.

In South Africa operates the only two nuclear power reactors in the Sub-Saharan region, both located at Koeberg with a net capacity of 1 842 MWe. It is important to note that South Africa accounts for 60% of all of Africa's energy production[173], despite of the fact that due to increase electricity consumption and the extension of its power grid to rural constituencies, the country has been struggling with serious power shortages in the last several years. To overcome these shortages in the future, South Africa is considering an ambitious expansion of its current nuclear power programme, which involves the construction of seven nuclear power reactors of the type Pebble Bed Modular Reactor (PBMR) for 2025[174]. The total capacity to be installed is around 7 565 MWe.

The Evolution of the Nuclear Power Sector in South Africa

South Africa has two French (Framatome) nuclear power reactors. The construction of these two units started in 1976 and they are both located at the Koeberg site, East of Cape Town. The nuclear power reactors supplied, in 2008, around 5.25% of the country's total electricity. The first nuclear power reactor started to be constructed in July 1976, was connected to the electric grid in April 1984 and entered into commercial operation in July of the same year. The second unit started to be constructed in July 1976, was connected to the electric grid in July 1985 and entered into commercial operation in November of the same year.

Eskom invested US$160 million to improve the safety operation of the PWR units up to international safety standard level. The two French-built Koeberg PWR required the investments to bring them in line with the updated safety reference for EdF 900 MW class PWR. Eskom has decided to follow EdF's practice of bringing all the units in its PWR series up to the safety level of the series youngest unit, and if possible, to the level of more recent series. Eskom also plans to spend US$2 million over the next five years to rehabilitate buildings and other structures that suffer from saltwater and climate. The conditions in Koeberg are some of the most severe for seaside nuclear power plants.

The South African State owned utility Eskom is heavily involved in the development of the PBMR. According with current plan, the government has the intention to construct three nuclear power reactors of this type in the coming years. The construction of the first nuclear power reactor will start in 2009 and it in is expected to enter into commercial operation in 2014. In November 2004, a contract was awarded to Mitsubishi Heavy Industries (MHI) of

---

[173] Africa as a whole generates only 3.1% of the world's electricity.

[174] PMBR is the South African design of nuclear power reactors which, according with South African experts, is much safer and proliferation-resistant than other nuclear power reactor designs. The Pebble Bed Modular Reactor (PBMR) is a new type of high temperature helium gas-cooled nuclear reactor, which builds and advances on world-wide nuclear operators' experience of older reactor designs. The most remarkable feature of these reactors is that they use attributes inherent in and natural to the processes of nuclear energy generation to enhance safety features. According with WANO and IAEA sources, three nuclear power reactors are already planned and four others are propose in South Africa

Japan for the basic design, research and development of the PBMR helium driven turbo generator system, as well as the core barrel assembly. The British company BNFL had invested US$15 million to obtain a 20% equity stake in the enterprise. The now Japanese owned Westinghouse took over 15% of the equity stake from BNFL. Peco Energy – later Exelon Corp - of the USA had acquired a 12.5% stake. In December 2001, Exelon said that they were considering building a PBMR reactor in the USA in parallel to those proposed in South Africa. However, following the change in management at Exelon, the company withdrew from the PBMR project in April 2002. The other partners in the development of the PBMR are the South African Industrial Development Corporation, which is owned by the South African government and Eskom. Negotiations with the French reactor builder Areva for shared research and development into the modular high temperature reactor failed. The reason of the failure of the negotiations was the following: According with French industry representatives, "a reactor design between 125 to165 MW may increase the unit cost of electricity and make it uneconomic".

Eskom supplies about 95% of South Africa's electricity and more than 60% of whole Africa electricity production. Early in 2008, regional electricity demand is almost the same than the country supplies capacity, so that South Africa power exports have been curtailed, while domestic demand is being managed by major cutbacks in industrial use, expected to lead to a significant decline in economic growth. Eskom is spending some US$39 billion on building new coal and gas turbine power plants by 2012, belatedly to address the current energy crisis and, by 2025, it expects to double its generating capacity to 80 GWe. About half of the increment in electricity capacity in the country, in 2025, is expected to be nuclear.

Table 92. Operating South African nuclear power reactors

Reactors	Type	Net MWe	First power
Koeberg 1	PWR	921	1984
Koeberg 2	PWR	921	1985
Total (2)		1 842	

Source: CEA, 8th, Edition, 2008.

Source: Photograph courtesy of Wikipedia (Pipodesign Philipp P Egli).

Figure 152. Koeberg nuclear power plant.

Table 93. Annual electrical power production for 2008

Total power production (including nuclear)	Nuclear power production
242 222.8 GWhe	12 713.4 GWhe

Source: IAEA PRIS, 2009

The Energy Policy in South Africa

In August 2007, a draft nuclear energy policy was prepared indicating that nuclear energy is the type of energy source that should be used to satisfy the significant increase in the demand of electricity for South Africa in the coming years and, at the same time, reduce its dependency of coal-fired power plants. In January 2008, the Department of Minerals and Energy and Eskom released a new policy document, "National Response to South Africa's Electricity Shortage". The plan includes work on the country's electricity distribution structure and the fast-tracking of electricity projects by independent power producers. At the same time, the new plan outlines the importance of reducing demand by pricing electricity correctly as well as promoting energy efficiency and deterring and, if necessary, outlawing, energy inefficiency. Eskom aims to reduce demand by about 3 000 MWe by 2012 and a further 5 000 MWe by 2025, through an aggressive campaign which will include promoting the use of solar-powered geysers as well as liquid petroleum gas for cooking. The government is also set to introduce a rationing scheme that will reward and penalize customers based on their energy usage.

The White Paper on Energy Policy adopted by the South Africa government at the end of June 2008, "calls for the achievement of energy security through the diversification of primary energy sources". Further, South Africa's electricity generation capacity has to be increased significantly in the next few decades to facilitate economic growth and social progress, while remaining sensitive to climate change. This presents an opportunity to promote diversity in primary energy sources, considering that the use of nuclear energy is increasingly being recognized worldwide as one of the strategies to mitigate greenhouse gas emissions and global warming, since it is an important carbon-free source of power. South Africa also possesses sizeable uranium reserves and has an extensive uranium mining industry, making the country one of the important producers of uranium in the world. The presence of this primary energy source in South Africa is a key element of security of energy supply nationally.

According with the Nuclear Energy Policy and Strategy for the Republic of South Africa of 2007, "the only economically viable alternative to coal as base load generation on a large scale is, therefore, nuclear energy. Nuclear energy is attractive for a number of reasons amongst which are the following: a) South Africa has sizeable uranium reserves and a vibrant mining industry; b) the extraction of uranium ore does not present any major challenges; c) value addition in the form of beneficiation of uranium ore and the implementation of a strong nuclear energy programme would lead to job creation and the development of skilled workforce; d) a solid regulatory framework, which would facilitate a structured development of the nuclear sector, already exists in South Africa; and e) South Africa has good non-proliferation policy credentials and as such pursuit of a peaceful nuclear energy programme can be done within national and international nuclear non proliferation obligations.

The Nuclear Energy Policy objectives to be achieved are the following:

1) "promotion of nuclear energy as an important electricity supply option through the establishment of a national industrial capability for the design, manufacture and construction of nuclear energy systems;

2) establishment of the necessary governance structures for an extended nuclear energy programme;

3) creation of a framework for safe and secure utilization of nuclear energy with minimal environmental impact;

4) contribution to the country's national programme of social and economic transformation, growth and development;

5) to guide in the actions to develop, promote, support, enhance, sustain and monitor the nuclear energy sector in South Africa;

6) attainment of global leadership and self-sufficiency in the nuclear energy sector in the long term;

7) exercise control over unprocessed uranium ore for export purposes for the benefit of the South African economy;

8) establishing of mechanisms to ensure the availability of land (nuclear sites) for future nuclear power generation;

9) allow for the participation of public entities in the uranium value chain;

10) promoting energy security for South Africa;

11) improvement of the quality of human life and to support the advancement of science and technology;

12) reduction of greenhouse gas emissions;

13) skills development related to nuclear energy".

The Energy Policy and Strategy for the Republic of South Africa identified a group of principles that the government should use to guide its activities in the nuclear sector. These principles are the following:

1) "nuclear energy shall be used as part of South Africa's diversification of primary energy sources to ensure security of energy supply;

2) nuclear energy programme shall contribute to economic growth and technology development in South Africa through investment in infrastructure, creation of jobs and development of skilled workers;

3) nuclear energy shall form part of South Africa's strategy to mitigate climate change and global warming;

4) all activities undertaken in pursuit of nuclear energy shall be in a manner that mitigates their impact on the environment;

5) all nuclear energy sector activities shall take place within a legal regulatory framework consistent with international best practice;

6) nuclear energy shall be used only for peaceful purposes and in conformity with national and international legal obligations;

7) in pursuing a national nuclear energy programme there shall be full commitment to ensure that nuclear and radiation safety receives the highest priority;

8) South Africa shall endeavor to use uranium resources in a sustainable manner. To the extent possible technologies chosen for nuclear power plant shall be those that allow for maximum utilization of uranium resources including the use of recycled uranium;

9) government shall encourage the development of appropriate institutional arrangements to ensure the development of human resources competent to discharge the responsibility of managing a nuclear infrastructure;

10) South Africa shall strive to acquire technology know-how and skills to enable design development, construction and marketing of its own nuclear power reactor and fuel cycle systems. To this end an industrial support base for the nuclear sector shall be developed as appropriate, taking into account the scale of the national programmes. Technology transfer shall be optimized in any procurement of nuclear and related equipment;

11) all facets of the nuclear energy sector shall always be subjected to appropriate safeguards and security measures;

12) government shall support research, development and innovation in the use of nuclear technology. Government shall also support participation in global nuclear energy technology innovation programmes;

13) government shall put in place effective mechanisms to protect and safeguard the South African nuclear energy industry Intellectual Property rights and innovative technology designs;

14) government shall create programmes to stimulate public awareness and inform the public about the nuclear energy programme;

15) government will ensure that adequate funding will be made available to support the technology development initiatives that are essential to the implementation of this policy. In addition, where appropriate, price support mechanisms can be implemented to enable the ongoing operations of key technologies".

The Nuclear Reactor Technology Development

The nuclear reactor technology to be used in the expansion of the nuclear power in South Africa will be PWR, while the PBMR is developed for both electricity and heat. By 2016, the local manufacturing of nuclear components and equipment should be under way and the PBMR commercialized, all with a view to exports this nuclear technology as well as for local use in the generation of electricity. Eskom has put US$1 billion to support the development of the PBMR. Conversely, export of unprocessed uranium will be restricted and a strategic stockpile will be maintained in order to support the expansion of the nuclear power programme.

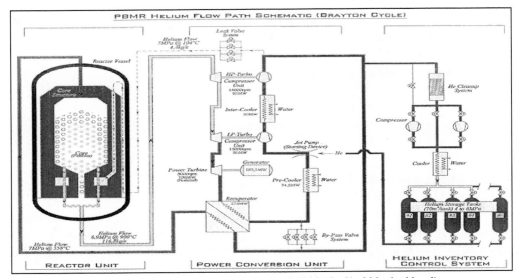

Source: Eskom(http://www.eskom.co.za/nuclear_energy/pebble_bed/pebble_bed.html)

Figure 153. The PBMR nuclear power reactor.

It is important to note that the PBMR draws on well-proven German expertise and aims for a step change in safety, economics and proliferation resistance. The capacity of the units will be around 165 MWe and will have a direct-cycle gas turbine generator and thermal efficiency of about 41%.

The construction cost of the PBMR reactors is expected to be modest and generating cost competitive. In 2003, environmental approval was given for construction of the demonstration PBMR unit at Koeberg and the fuel plant at Pelindaba near Pretoria. After the demonstration pilot plant is in operation, the South African government has said that it wants to order seven units totaling at least 7 565 MWe. It is expected that one quarter of South Africa's electricity will be provide by the operations of the seven PBMRs units.

Nuclear Safety Authority and the Licensing Process

In 1948, the Atomic Energy Act was adopted. Based on this act, the Atomic Energy Corporation (AEC) was created with the purpose to deal with all elements of the South African nuclear power programme. In 1963, the Nuclear Installations Act provided all necessary elements for licensing all types of nuclear installations to be constructed in the country in the future. In 1982, the Nuclear Energy Act made the AEC responsible for all nuclear matters associated with the South African nuclear power programme, including the enrichment of uranium. An amendment to the act created the autonomous Council for Nuclear Safety, later on called "National Nuclear Regulator (NNR)", with the purpose to deal with all safety issues related with the operation of all nuclear installations covering the full fuel cycle from mining to waste disposal. It is focused on health and safety.

The Nuclear Energy Act of 1999 gives responsibility to the Minister of Minerals and Energy for nuclear power generation, management of radioactive wastes and the country's international commitments in the nuclear field. The Nuclear Energy Corporation of South Africa (NECSA) is a state corporation established under the act.

The Department of Minerals and Energy (DME) is the state agency with overall responsibility for nuclear energy in the country and administers the above acts.

The Department of Environmental Affairs (DEA) is the state agency responsible for environmental assessment of nuclear energy projects. DEA has a cooperative agreement with the NNR in order to evaluate all nuclear projects and its impact in the environment.

Nuclear Fuel Cycle and Waste Management

Building upon 23 years of experience with nuclear power, an extensive programme to develop all aspects of the nuclear fuel cycle was prepared. With uranium mining already well established, the next step is to ensure the technology for conversion, enrichment, fuel fabrication and also reprocessing of used fuel. These components of the fuel cycle are considered as strategic priorities related to energy security and the most important elements of a military nuclear programme for the production of a nuclear weapon. South Africa had stopped enriching uranium in 1997 following the dismantling of its nuclear weapons programme, but it reportedly has some residual capabilities that can be used with this purpose.

The Public Opinion

In November 2008, Global Management Consultancy Accenture carried out a survey of more than 10,000 persons in 20 countries on the use of nuclear energy for electricity production. According with the result of this survey, 55% of persons surveyed in South Africa supported the use of nuclear energy for the production of electricity as a means of reducing fossil fuel reliance. When asked if their country should start using or increase the use of nuclear power, either outright or if their concerns were addressed, 88% of respondents in South Africa said yes.

Looking Forward

Eskom plans to double its total generating capacity to 80 000MW over the next two decades, with nuclear power making up about half of the new capacity. The state supplier is considering bids from France's Areva and the US's Westinghouse Electric to build a new nuclear power plant that could start generating electricity from 2016, and has said that "it could build more nuclear power plants by 2025". South Africa is also going ahead with the R17-billion PBMR project, one of the most technologically advanced capital investment projects undertaken in the country since 1994. The PBMR project entails the building of a demonstration reactor at Koeberg near Cape Town and a pilot fuel plant at Pelindaba near Pretoria. Construction is due to start in 2009, with the first fuel to be loaded four years later. If successful, another ten units could be built. However, until now three additional nuclear power reactors have been planned and four more have been proposed. The PBMR project is supported by the government, Eskom, the Industrial Development Corporation and the US companies Westinghouse and Exelon.

THE POSSIBLE USE OF NUCLEAR ENERGY FOR ELECTRICITY GENERATION IN NIGERIA IN THE FUTURE

Nigeria's nuclear ambitions began shortly after it gained independence in 1960, and many steps were taken to put Nigeria on the road to nuclear development. However, concerns over what parts of the country would get nuclear research facilities and nuclear power reactors delayed progress until today. The new Director General of the IAEA held talks, in December 2009, with Nigerian Vice-President Goodluck Jonathan and Alhassan Bako Zuma, Minister for Science and Technology, on the future cooperation between the IAEA and Nigeria, which plans to build its first nuclear power reactor with Russian assistance in 2017. "We are getting people from Moscow to build it for us the first nuclear power reactor. I just came back two days ago from Moscow and it is the Russians who are going to build the nuclear facilities for us, working with Nigerians", Science and Technology Minister Alhassan Bako Zuma said at a press conference with the Director General of the IAEA.

Russian and Nigeria representatives have moved ahead in their co-operation on building a nuclear power plant in Nigeria in the coming months. Representatives of Gazprom and the Nigerian government signed a nuclear cooperation agreement on June 25, 2009 in the Nigerian city of Abuja. Russia wants to build a nuclear power plant in Nigeria and also be granted authority to find uranium in the country. Sergei Kiriyenko, the head of Rosatom, said that "Russian will have prepared an initial proposal to construct a nuclear power plant in Nigeria by the end of 2010". Construction is expected to begin in 2011 with power production to begin in 2017. The nuclear power plant is expected to provide up to 4 000 MW of energy by 2025. Nigeria's former Minister of Science and Technology insisted in November 2008 "that Nigeria's nuclear power programme will not use foreigners, but would depend primarily on local labor, skills and expertise.

In March 2009, Russia signed a nuclear energy cooperation agreement with Nigeria, that provided for domestic uranium exploration and mining. An additional agreement, signed in June 2009, gave Russia access to Nigeria's gas reserves in exchange for the construction of a Russian nuclear power plant and a new research reactor. One nuclear power plant can save the country more than 50,000 barrels of oil per day.

There are already two nuclear research centers at Ahmadu Bello University, in Zaria, and another in the capital, Abuja.

Chapter 9

CONCLUSIONS

It is an undisputed reality that energy constitutes the motive force of the civilization and it determines, in a high degree, the level and rate of the economy development of any country. For this reason, the use of all type of energy for the electricity generation, including nuclear energy, is a crucial element to ensure the economic development of any country.

However, the generation of electricity using fossil fuels is a major and growing contributor to the emission of carbon dioxide to the atmosphere, which is seriously affecting the climate all over the world. Considering the different energy alternatives that the world can use now to satisfy the foresee increase in energy demand, particularly electricity demand in the coming years, there are only a few realistic options for the reduction of CO_2 emissions. These options are, among others, the following: a) increase efficiency in electricity generation and use; b) expand use of all available renewable energy sources such as wind, solar, biomass and geothermal, among others; c) massive introduction of new advanced technology like the capture carbon dioxide emissions technology at fossil-fueled (especially coal) electric generating plants, with the purpose to permanently sequester the carbon produced by these plants; d) increase use of new types of nuclear power reactors for electricity generation that are inherent safe and proliferation risk-free; and e) energy saving.

The amount of total energy produced and of energy used per capita in the world is increasing. The world total energy requirements increased by a factor of 2.5 between 1970 and 2006, from 6 181 GW to 15 311 GW per year. However, it is expected that the demand of electricity probably be triple from now until 2050 and the main reasons for this increase are a boost in the world population, in the percentage of these population living in big cities and an improving in their quality of live. One of the main problems that the world is now facing is how to satisfy this foresee increase in the electricity demand, using all available energy sources in the most efficient manner, but without increasing the emission of CO_2 to the atmosphere with the purpose to reduce to the minimum the impact in the climate all over the world.

It is important to stress that every country's energy mix involves a range of national preferences and priorities that are reflected in national energy policies and strategies. These policies and strategies represent a compromise between expected energy shortages, environmental quality, energy security, cost, public attitudes, safety and security, available skills and production and service capabilities. Relevant national competent authorities must take all of these elements into account when formulating an energy policy and strategy for the country.

One of the alternatives that several countries has in their hands to satisfy the foresee increase in the electricity demand in the future is the use of nuclear energy for the generation of electricity. However, the use of nuclear energy for this purpose can be affected by the following reasons:

a) nuclear power generate stronger political passions more than any other energy alternatives such as natural gas, coal, hydropower and oil;

b) because of the front-loaded cost structure, high interest rates, or uncertainty about interest rates, weaken the use of nuclear energy for electricity generation more than for energy alternatives. Nuclear power's front-loaded cost structure also means that the cost of regulatory delays during construction is higher for nuclear power than for other energy alternatives;

c) the strength, breadth and durability of commitments to reducing CO_2 emissions will also influence nuclear power's growth;

d) the nuclear industry is a global industry with good international cooperation and hence the implications of an accident anywhere will be felt in the industry worldwide;

e) nuclear terrorism may have a more far reaching impact than comparable terrorism directed at other fuels power plants;

f) while a nuclear power plant in itself is not a principal contributor to proliferation risk, proliferation worries can affect public and political acceptance of nuclear power;

g) among energy sources, high level radioactive waste is unique to nuclear power and the management of these wastes a great concern for the public opinion in almost all countries.

The nuclear industry has proclaimed that nuclear energy for electricity generation will again play an important role in the energy balance in several countries, particularly in North America and Europe, because in Asia nuclear energy played, is still playing and will continue to play an important role in the production of electricity, at least in a group of countries of the region. However, it is important to stress that nuclear power will only become more relevant if concerns about plant safety, nuclear waste disposal and the risk of proliferation can be solved to the satisfaction of the public.

From the 31 countries that operate now nuclear power plants, six of them, USA, France, Japan, Germany, Russia and the Republic of Korea, produce three quarters of the nuclear electricity in the world and, with the exception of Germany, all of them have plans to expand their nuclear power programmes in the coming years. However, the role of nuclear power in the overall energy sector remains very limited even in these countries. In France, the most nuclear country in the world that generates 76.18% of its electricity with nuclear power plants, in 2008, the use of nuclear energy for electricity production only provides 17.5% of its final energy needs. Like most of the other countries, France remains highly dependent on fossil fuels to cover over 70% of its final energy consumption. None of the other five largest nuclear countries cover more than 7% of their final energy by nuclear power, the US and Russia less than 4%.

The current contribution of nuclear energy to the total electricity generation varies according with the region. In Europe, nuclear generated electricity accounts for 26.03% of the total electricity produced. In North America it is approximately 19%, whereas in Africa and Latin America it is 1.84% and 2.61% respectively. In Asia, nuclear energy accounts for 9.46% of electricity generation and it is expanding in several countries, particularly China, India, Republic of Korea and Japan.

During 2009, there were 436 nuclear power reactors in operation in 31 countries from all regions with a net capacity of 367 310 MWe and 54 nuclear power reactors under construction in 14 countries with a net capacity of 46 288 MWe. During the period 1950-2007, 119 nuclear power reactors were shut down in 18 countries totalizing 35 150 MWe (net). The countries with the highest number of nuclear power reactors shut down are USA with 28 units, following by the UK with 26 units, Germany with 19 units and France with 11 units. Three nuclear power reactors were connected to the grid in India, China and Romania adding 1 852 MWe to the electric grid in these countries; started the construction of seven nuclear power reactors totalizing 5 195 MWe in China, the Republic of Korea, Russia and France, and nine new orders for the construction of nuclear power reactors were approved in China, the Republic of Korea, Japan and Russia totalizing 11 660 MWe.

According with the IAEA sources, during 2008, no new nuclear power reactor was connected to the electric grid, while in the same period three old units were closed (Hamaoka Units 1 and 2 in Japan and Bohunice Unit 3 in Slovakia). At the same time, the construction of 10 new nuclear power reactors begun: six in China, two in the Republic of Korea and two in Russia, increasing the number of nuclear power reactors under construction from 33 units in 2007 to 54 units in 2009.

In 2009, North America has 122 nuclear power reactors in operation and one nuclear power reactors under construction. Western Europe has 130 operating nuclear power reactors and only two under construction one in Finland and another in France. Russia has 31 nuclear power reactors in operation and nine under construction and Eastern Europe has 35 operating nuclear power reactors and six under construction. Asia has 103 nuclear power reactors in operation and 34 nuclear power reactors under construction. Latin America has six nuclear power reactors in operation and one nuclear power reactor under construction. Africa has two nuclear power reactors in operation no nuclear power reactor under construction.

It is important to stress that the IAEA has foresee that the participation of nuclear energy in the world's energy balance will drop from 16% to 8-10% in 2030, if no decision is adopted by the EU, Canada and the USA to build more nuclear power reactors for electricity generation in the coming years.

Summing up can be stated the following:

1. Europe remained the largest nuclear energy user for electricity generation among all regions since 1980. However, this situation could change in the future;

2. North America was the second largest user of nuclear energy for electricity production since 1980;

3. with the largest regional population and fast growing economies, Asia became the third largest nuclear energy user for electricity production since 1980;

4. Latin America registered almost the same growth in its nuclear energy use for electricity generation since 1980 and it is fourth larger user of nuclear energy in the world;

5. Africa registered the lowest nuclear share and this situation will not change in the future.

BIBLIOGRAPHY

Albright, D., *Phased International Cooperation with North Korea's Civil Nuclear Programmes*, Institute for Science and International Security (ISIS), March 19, 2007.

An Energy Policy for Europe, Communication from the Commission to the European Council and the European Parliament, Brussels, Belgium, 10.1.2007 COM(2007) 1 final, SEC(2007).

An European Strategy for Sustainable, Competitive and Secure Energy, European Commission, Brussels, COM (2006) 105 final, March 2006.

Arrastía Avila; M.; *Distributed generation in Cuba. Part of a transition towards a new energy paradigm*, Cogeneration and On-Site Power Production, November–December, 2008.

Asia's Nuclear Energy Growth, Promoting the Peaceful Worldwide Use of Nuclear Power as a Sustainable Energy Resource, WNA, February 2007.

A Technology Roadmap for Generation IV Nuclear Energy Systems, GIF-002-00, U.S. DOE Nuclear Energy Research Advisory Committee and the Generation IV International Forum, December 2002.

Basic Infrastructure for a Nuclear Power Project, IAEA-TECDOC-1513, IAEA, Vienna, Austria June 2006.

Belgium, IAEA Country File, IAEA, Vienna, Austria, December 2002.

Belt, J.A.B., *The Electric Power Sector in Cuba: Potential Ways to Increase Efficiency and Sustainability*, USAID Energy, April 2009.

Benjamin Alvarado, J., and Belkin, A., *Cuba's Nuclear Power Programme and Post-War Pressures*, The Nonproliferation Review, Winter 1994.

Blohm-Hieber, U., *Europe Strategic Vision*, IAEA Bulletin 49-2, Vienna, Austria, March 2008.

Bollen, J.C. and Eerens, H.C., *The Effect of a Nuclear Energy Expansion Strategy in Europe on Health Damages from Air Pollution*, Netherlands Environmental Assessment Agency (MNP), Bilthoven, the Netherlands; 2007.

Bruno, G., *Iran's Nuclear Programme*, Council on Foreign Relations, USA, September 2008.

Bulgaria, IAEA Country File, IAEA, Vienna, Austria, December 2004.

Canada, IAEA Country File, IAEA, Vienna, Austria, December 2003.

Canada, Nuclear Energy Agency, Nuclear Energy Data, France, March 2006.

Canada's Nuclear Renaissance: Implications for Public Policy, Summary Report, Public Policy Forum, Ottawa, Canada, July 2008.

Canada Uranium Production and Nuclear Power, Promoting the peaceful worldwide use of nuclear power as a sustainable energy resource, World Nuclear Association, USA, June 2008.

Choosing the Nuclear Power Option: Factors to be Considered, STI/PUB/1050, IAEA, Vienna, Austria, January 1998.

Considerations to Launch a Nuclear Power Programme, GOV/INF/2007/2, IAEA, Vienna, Austria, 2007.

Current Issues in Nuclear Energy: Nuclear Power and the Environment, International Nuclear Societies Council (INSC), American Nuclear Society (ANS), Illinois, USA, January 2002.

Czech Republic, IAEA Country File, IAEA, Vienna, Austria, December 2003.

De la Rosa A.M., *The Evolution of the Role of the Philippine Nuclear Research Institute in the National Nuclear and Radiation Safety Regime*, Philippine Nuclear Research Institute, Quezon City, Philippines, International Conference on the Challenges Faced by Technical and Scientific Support Organizations in Enhancing Nuclear Safety, Aix-en-Provence, France 23-27 April 2007 IAEA-CN-142/49, 2007.

D'haeseleer, W., *Belgium's Energy Challenges Towards 2030*, Final report, Commission Energy 2030, June 19, 2007.

El-Hokayen, E., *Iran's Nuclear Energy Programme: Policies & Prospects*, Joint Conference of the Pugwash Conferences and the Center for Strategic Research, Tehran, Iran, 25 April 2006.

Emerging Nuclear Energy Countries, Promoting the peaceful worldwide use of nuclear power as a sustainable energy resource, World Nuclear Association, April 2008.

Energy, Electricity and Nuclear Power: Developments and Projections —25 Years Past and Future, IAEA, Vienna, Austria, December 2007.

Europe's Vulnerability to Energy Crises, World Energy Council, London, United Kingdom, 2008.

Falter J.W., *Public Opinion on Nuclear Energy in Germany*, L' énergie nucléaire et les opinions publiques européennes, *www.ifri.org/files/**Energie**/Nucleaire2.pdf.*

Fernandez Vazquez, E. and Pardo Guerra, J.P.; *Latin America Rethinks Nuclear Energy*, IRC Americas, September 13, 2005.

Finland, IAEA Country File, IAEA, Vienna, Austria, December 2004.

France, IAEA Country File, IAEA, Vienna, Austria, December 2004.

French Nuclear Power Programme, Promoting the peaceful worldwide use of nuclear power as a sustainable energy resource, World Nuclear Association, April 2007.

Fundamental Safety Principles, Safety Fundamentals No. SF-1, IAEA, Vienna, Austria, 2006.

Germany, IAEA Country File, IAEA, Vienna, Austria, December 2004.

Global Public Opinion on Nuclear Issues and the IAEA, Final Report from 18 Countries prepared for the IAEA by GlobeScan Incorporated, October 2005.

Grimstom, M.C., *Nuclear Energy: Public Perceptions and Decision-Making,* World Nuclear Association Annual Symposium, London, United Kingdom, 4-6 September 2002.

Gunn, G., *Southeast Asia's Looming Nuclear Power Industry*, Japan Focus, Nagasaki University, February 2008.

Hesketh, K., Worral, A. and Weaver, D., *Future Challenges for Nuclear Energy in Europe*, Europhysics News, Vol. 35, No. 6, 2004.

Holt, M. and Behrens C.E., *Nuclear Energy in the United States*, Congressional Research Service, July 23, 2003.

Homamnn, A. and Streit, M., *Nuclear Energy in Switzerland: It's Coming Back!*, New Challenges for the Swiss Nuclear Society, Proceedings of the International Youth Nuclear Congress 2006, Stockholm, Sweden, Olkiluoto, Finland, 18 – 23 June, 2006.

Huber, P. W. and Mills, M. P., *Why the US Needs More Nuclear Power*, The Manhattan Institute, USA, City Journal, Winter 2005.

Hungary, IAEA Country File, IAEA, Vienna, Austria, December 2003.

International Energy Outlook 2007, Energy Information Administration, DOE/EIA-0484 (2007), USA, May 2007.

International Energy Outlook 2009, Energy Information Administration, DOE/EIA-0484(2009), USA, May 2009.

International Status and Prospects of Nuclear Power, Report by the Director General, GOV/INF/2008/10-GC (52)/INF/6, August 12, 2008.

Italy, IAEA Country File, IAEA, Vienna, Austria, December 2003.

Kaneko, K., Suzuki A., Choi, J-S., and Fei E., *Energy and Security in Northeast Asia: Proposals for Nuclear Cooperation,* Institute on Global Conflict and Cooperation, IGCC Policy Papers 37 (University of California, Multi-Campus Research Unit), 1998.

Keay, M., *Nuclear Power in the UK: Is it necessary? Is it viable?*, Oxford Institute for Energy Studies, Oxford Energy Comment, October 2007.

Khale Quzman, Md., *The Energy Challenge for 21st Century Bangladesh*, Georgia Southwestern State University, USA, 2007.

Khripunov. I; Africa *pursuit of nuclear power,* Bulletin of the Atomic Scientific, 28 November 2007.

Kotler, M.L. and Hillman, I.T., *Japanese Nuclear Energy Policy and Public Opinion*, Japanese Energy Security and Changing Global Energy Markets: An Analysis of Northeast Asian Energy Cooperation and Japan's Evolving Leadership Role in the Region, The Center for International Political Economy and The James A. Baker III Institute for Public Policy Rice University, May 2000.

Kovan, D., *Nuclear power in Europe*, Nuclear News, December 2005.

Kovryzhkin, Y.L., *Nuclear Power in Ukraine: Past, Present, Future*, World Nuclear Association, 2004.

Libya sends Tajura HEU to Russia, Prepares to Convert Reactor to LEU, Nuclear Fuel, March 15, 2004.

Lithuania, IAEA Country File, IAEA, Vienna, Austria, December 2004.

Liz, P. and Milhollin. G., *Brazil's Nuclear Puzzle*, Wisconsin Project on Nuclear Arms Control, USA, October 2004.

Long-Term Program for Research, Development and Utilization of Nuclear Energy, Atomic Energy Commission, Japan, November 2000.

Managing the First Nuclear Power Plant Project, IAEA TECDOC 1555, IAEA, Vienna, Austria, May 2007.

Marques de Souza, J.A., *The Past and Future Role of Nuclear Power in Reducing Greenhouse Gas Emissions in Brazil*, International Joint Meeting "The Role of Nuclear Power to Mitigate Climate Change" Sociedad Nuclear Mexicana, Latin American Section of ANS, Acapulco, Mexico. July 18-21, 1999.

McDonald, A.; *Nuclear Power Global Status*, IAEA Bulletin 49-2, Vienna, March 2008.

Medlock III, K.B., Hartley, P., *The Role of Nuclear Power in Enhancing Japan's Energy Security,* The James A. Baker III Institute for Public Policy of Rice University and Tokyo Electric Power Company Inc., September 2004.

Mexico Nuclear Energy Program, OECD Nuclear Energy Agency, France, June 2007.

Milestones in the Development of a National Infrastructure for Nuclear Power, IAEA Nuclear Energy Series No. NG-G-3.1, Vienna, Austria, September 2007.

Mochovce 3 and 4 Basic Facts, New Clear Power, Slovenské elektrárne, a.s. Slovakia, 2008.

Morales Pedraza, J., *The Current Situation and the Perspectives of the Energy Sector in the European Region*, Energy in Europe: Economics, Policy and Strategy, Nova Science Publishers, Inc, 2008:

National Energy Strategy, Lithuanian Energy Institute with the cooperation of the Danish Energy Agency, Lithuania, 2003.

Nuclear Country Notes, 2007 Survey of Energy Resources, World Energy Council, 2007.

Nuclear Energy in Finland, Energy Department, Ministry of Trade and Industry, Helsinki, Finland, October 2002.

Nuclear Energy in Italy, Promoting the peaceful worldwide use of nuclear power as a sustainable energy resource, World Nuclear Association, January 2008.

Nuclear Energy in Ukraine, Soviet Plant Source Book, July 1997.

Nuclear Overview on the Libya nuclear programme, James Martin Center for Nonproliferation Studies at the Monterey Institute of International Studies, October 2007.

Nuclear Power in Argentina, Promoting the peaceful worldwide use of nuclear power as a sustainable energy resource, World Nuclear Association, August 2007.

Nuclear Power in Belgium, Promoting the peaceful worldwide use of nuclear power as a sustainable energy resource', World Nuclear Association, March 2008.

Nuclear Power in Brazil, Promoting the peaceful worldwide use of nuclear power as a sustainable energy resource, World Nuclear Association, November 2007.

Nuclear Power in Bulgaria, Promoting the peaceful worldwide use of nuclear power as a sustainable energy resource, World Nuclear Association, April 2008.

Nuclear Power in China, Promoting the peaceful worldwide use of nuclear power as a sustainable energy resource, World Nuclear Association, April 2008.

Nuclear Power in China, Promoting the peaceful worldwide use of nuclear power as a sustainable energy resource, World Nuclear Association, February 2008.

Nuclear Power in Czech Republic, Promoting the peaceful worldwide use of nuclear power as a sustainable energy resource, World Nuclear Association, February 2007.

Nuclear Power in Germany, Promoting the peaceful worldwide use of nuclear power as a sustainable energy resource, World Nuclear Association, January 2008.

Nuclear Power in Hungary, Promoting the peaceful worldwide use of nuclear power as a sustainable energy resource, World Nuclear Association, July 2007.

Nuclear Power in Japan, Promoting the peaceful worldwide use of nuclear power as a sustainable energy resource, World Nuclear Association, April 2008.

Nuclear Power in Korea, OECD Nuclear Energy Agency, France, June 2007.

Nuclear Power in Korea, Promoting the peaceful worldwide use of nuclear power as a sustainable energy resource, World Nuclear Association, January 2008.

Nuclear Power in the Netherlands, Promoting the peaceful worldwide use of nuclear power as a sustainable energy resource, World Nuclear Association, March 2008

Nuclear Power in Pakistan, Promoting the peaceful worldwide use of nuclear power as a sustainable energy resource, World Nuclear Association, January 2008.

Nuclear Power Programme Planning: An Integrated Approach, IAEA-TECDOC-1259, IAEA, Vienna; Austria, December 2001.

Nuclear Power in Romania, Promoting the peaceful worldwide use of nuclear power as a sustainable energy resource, World Nuclear Association, March 2008.

Nuclear Power in Russia, Promoting the peaceful worldwide use of nuclear power as a sustainable energy resource, World Nuclear Association, April 2008.

Nuclear Power in Slovakia, Promoting the peaceful worldwide use of nuclear power as a sustainable energy resource, World Nuclear Association, February 2008.

Nuclear Power in Slovenia, Promoting the peaceful worldwide use of nuclear power as a sustainable energy resource, World Nuclear Association, May 2007.

Nuclear Power in South Africa, Promoting the peaceful worldwide use of nuclear power as a sustainable energy resource, World Nuclear Association, May 2008.

Nuclear Power in Spain, Promoting the peaceful worldwide use of nuclear power as a sustainable energy resource, World Nuclear Association, May 2007.

Nuclear Power in Sweden, Promoting the peaceful worldwide use of nuclear power as a sustainable energy resource, World Nuclear Association, February 2008.

Nuclear Power in Switzerland, Promoting the peaceful worldwide use of nuclear power as a sustainable energy resource, World Nuclear Association, March 2008.

Nuclear Power Situation in Switzerland, OECD Nuclear Energy Agency, France, March 2006.

Nuclear Power in Ukraine, Promoting the peaceful worldwide use of nuclear power as a sustainable energy resource', World Nuclear Association, January 2008.

Nuclear Power in the United Kingdom, Promoting the peaceful worldwide use of nuclear power as a sustainable energy resource, World Nuclear Association, April 2008

Nuclear Power in the USA, Promoting the peaceful worldwide use of nuclear power as a sustainable energy resource, World Nuclear Association, June 2008.

Nuclear Safety Review for the Year 2006, GC(51) /INF/2, IAEA, Vienna, Austria, July 2007.

Nuclear Technology Review 2007, Report by the Director General, IAEA, GC (51)/INF/3, Vienna, Austria, July 2007.

Pakistan Country Profile. Looking into the future, Inside WANO, The Magazine of the World Association of Nuclear Operators, Vol. 14, No 3, 2006.

Parker, L. and Holt, M., *Nuclear Power: Outlook for New U.S. Reactors,* Energy Policy Resources, Science and Industry Division, CRS Report for Congress, March 9, 2007.

Philippines, Preventing Nuclear Dangers in Southeast Asia and Australia, An IISS Strategic Dossier, Chapter 8, 2008.

Poland, IAEA Country File, IAEA, Vienna, Austria, December 2005.

Ponce de León Baridó, P., *The Potential of Nuclear Power in Portugal: Determining if and when to Include Nuclear Power in Portugal's Energy Supply Portfolio,* Engineering Systems Analysis for Design, December 15, 2006.

Promotion and Financing of Nuclear Power Programmes in Developing Countries, Report to the IAEA by a Senior Expert Group, IAEA, Vienna, Austria, February 1987.

Raymond, J.Q., *Using nuclear energy: A Philippine experience,* Department of Political Science, University of the Philippines (Diliman), Philippines, May 1999.

Regional Energy Integration in Latin America and the Caribbean, Work Programme of the World Energy Council, London, United Kingdom, 2008.

Remes, M., *Finland Debates Energy Future. Modern Nuclear Power under Scrutiny,* Department for Communication and Culture, Ministry for Foreign Affairs of Finland, April 2008.

Romania, IAEA Country File, IAEA, Vienna, Austria, December 2002.

Roul A., *India's Nuclear Power. Assisting Energy Independence or a Dangerous Experiment?* Eco World Nature and Technology in Harmony, 2007.

Russian Federation, IAEA Country File, IAEA, Vienna, Austria, December 2004.

Sacchetti, D., *Generation Next,* IAEA Bulletin 49-2, Vienna, Austria, March 2008.

Sahimi, M., *Iran's Nuclear Energy Program. Part V: From the United States Offering Iran Uranium Enrichment Technology to Suggestions for Creating Catastrophic Industrial Failure,* Payvand, December 7, 2004.

Safety of Nuclear Power Reactors, Promoting the peaceful worldwide use of nuclear power as a sustainable energy resource, World Nuclear Association, September 2007.

Saprykin, V., *Authorities and Society: Cooperation for Safe Nuclear Power Engineering, Nuclear Energy in Ukraine: Problems of Its Development, Security, and Public Support in the Context of the European Integration,* the Razumokov Centre, National Security and Defense magazine, June 2005.

Schneider, M., *Nuclear Power in France. Beyond the Myth,* The Greens, European Free Alliance in the European Parliament, December 2008.

Schneider, M. and Froggatt, A., *The World Nuclear Industry Status Report 2007*, Greens-EFA Group in the European Parliament, Brussels, November 2007.

Selecting Strategies for the Decommissioning of Nuclear Facilities: A Status Report, NEA, OECD, Paris, France, 2006.

Shkodrova, A., *Nuclear Argument Threatens Bulgaria's EU Plans*, Sofia, Bulgaria, ENS, June 15, 2004.

Slovakia, IAEA Country File, IAEA, Vienna, Austria, November 2005.

Slovenia, IAEA Country File, IAEA, Vienna, Austria, December 2003.

Spain, IAEA Country File, IAEA, Vienna, Austria, December 2003.

Spain, Nuclear Energy Agency, European Commission Directorate-General for Energy and Transport, Brussels, Belgium, June 2007.

Soviet-Designed Nuclear Power Plants in Russia, Ukraine, Lithuania, Armenia, the Czech Republic, the Slovak Republic, Hungary and Bulgaria, Nuclear Energy Institute, Washington, DC, USA, Fifth Edition, 1997.

Survey of Public Opinion for Green Electricity Production in Slovenia, University of Ljubljana, Faculty of Mechanical Engineering, Ljubljana, Slovenia, April 2005.

Stephenson, J. and Tyan, P., *Is Nuclear Power Pakistan's Best Energy Investment? Assessing Pakistan's Electricity Situation;* a paper for the Nonproliferation Policy Education Center, 2007.

Sweden, IAEA Country File, IAEA, Vienna, Austria, December 2003.

Switzerland, IAEA Country File, IAEA, Vienna, Austria, December 2003.

The Decommissioning and Dismantling of Nuclear Facilities Status, Approaches and Challenges, NEA, OECD, Paris, France, 2002.

The New Economics of Nuclear Power, WNA report, 2005.

The Netherlands, IAEA Country File, IAEA, Vienna, Austria, December 2002.

The Role of Nuclear Power in Europe, World Energy Council, London, UK, January 2007.

The Sustainable Nuclear Energy Technology Platform. A Vision Report, The European Commission, Directorate- General for Research Uratom, Belgium, 2007.

Thomas, S., *The Economics of Nuclear Power: Analysis of Recent Studies*, Public Services International Research Unit (PSIRU), University of Greenwich, United Kingdom, July 2005.

Thorium-Based Nuclear Power: An Alternative?, Laka Foundation, Amsterdam, The Netherlands, Nuclear Monitor, 21 February, 2008.

Ukraine, Promoting the peaceful worldwide use of nuclear power as a sustainable energy resource', World Nuclear Association, February 2002.

United Kingdom, IAEA Country File, IAEA, Vienna, Austria, December 2002.

United States of America, Nuclear Energy Agency, Nuclear Energy Data, USA Country File, France, June 2007.

Uranium is Energy. The Nuclear Power Plants of EnBW, EnBW Energie Baden-Württemberg AG, Karlsruhe, Germany, June 2007.

US Public Opinion About Nuclear Energy, National Survey by Bisconti Research, Inc. in cooperation with NOP World for the Nuclear Energy Institute, May 5-9, 2005.

Van der Zwaan, B.C.C., *Prospects for Nuclear Energy in Europe*, Energy Research Centre of the Netherlands (ECN), Amsterdam, The Netherlands, Nuclear Energy in Europe, 2006.

Vovk, E., *Nuclear Power in Russia*, Public Opinion Foundation, Russia, February 2006.

Westra, M., *Is Fusion the Future?*, IAEA Bulletin 48/2, IAEA, Vienna, Austria, March 2007.

Wikdahl, C.E., *Nuclear Power in Sweden, Finland and Europe*, Deane Conference on the Future of Nuclear Power, Chicago, USA, 27-28 March, 2008.

INDEX

C

D

E

F

M

N

O

P

S